# Analog Integrated Circuits
# for Communication
## Principles, Simulation and Design

**Second Edition**

Donald O. Pederson · Kartikeya Mayaram

# Analog Integrated Circuits for Communication

## Principles, Simulation and Design

### Second Edition

 Springer

Donald O. Pederson
University of California
Berkeley, CA
USA

Kartikeya Mayaram
Oregon State University
Corvallis, OR
USA

ISBN  978-1-4419-4324-8          e-ISBN  978-0-387-68030-9

Printed on acid-free paper.

9 8 7 6 5 4 3 2 1

springer.com

# Note to Instructors

The input files for Spice, which are used in various examples within this book, are available from http://eecs.oregonstate.edu/ karti/book/spicefiles as well as springer.com. Instructors can also request the solutions manual from springer.com.

# Preface

## Second Edition

Since the first edition of this book, there has been a significant growth in the communication integrated circuits (ICs) market. Radio frequency (RF) ICs with operating frequencies of several GHz or higher are commonplace for wireless communication applications. Several excellent text books on RFIC design are also available now. The focus of these books is on design using first-order analysis techniques with little or no simulation to back up the analysis. This lack of emphasis on simulation is due to the fact that Spice cannot be efficiently used to simulate RF circuits operating in the GHz frequency regime. Furthermore, simulation programs that can be used to simulate these high frequency circuits are not available in the public domain. In several cases, it is not even clear as to which RFIC simulator is the best for a particular application and how a circuit can be accurately simulated. This can be frustrating, particularly so, for a student new to the analysis and design of RF circuits.

The strength of the first edition of this book was in the tight coupling between the first-order circuit analysis and Spice simulation results. This is also the focus of this edition and, therefore, the book continues to address lower frequency circuits incorporating classical amplitude and frequency modulation schemes. The concepts governing these circuits are the same as those for the higher frequency circuits. For this reason the content does not deviate significantly from the first edition. Also, some of the circuit examples that were more representative of discrete implementations have been retained since the focus is on basic concepts.

Although noise and distortion are both important issues in communication circuits, the primary focus of this book is on distortion. The topic of noise in RF circuits is adequately covered in several books on analog and RFIC design and has been omitted here. This was a difficult decision for us and we recognize that this omission may not be appealing to some readers.

The changes to the second edition include a new Chapter 1 that covers the basics and provides a foundation for the rest of the book. As a result, some of the earlier chapters from the first edition have been revised. Also, the text and explanations in other chapters have been modified where necessary to reflect the new organization of the book. The text and equations for other chapters have been updated as well to remove errors and improve the presentation. Chapter 15 from the first edition has been eliminated. A discussion on compression/intercept points and matching networks has been added. In addition, several problems have been included at the end of every chapter. These problems emphasize the concepts presented in the chapter and give students an opportunity to put in practice these concepts.

Professor Pederson and I had started a discussion on the revisions to this book several years back. However, due to other commitments at my end the second edition could not be completed earlier. While working on this edition, I have fond memories of the wonderful collaboration that we had writing the first edition. Although Professor Pederson is no longer with us, the new edition is in keeping with the modifications he wanted to see. He was an exceptional teacher, researcher, mentor, colleague, friend, and visionary who is sorely missed.

I thank Professors Robert Dutton, Terri Fiez, and Yannis Tsividis for their suggestions and encouragement. I have tried to incorporate as many of their suggestions as possible. As with any revision, the intent has been to eliminate errors from the previous edition. However, new errors may have crept in during the revision process. I would appreciate the readers making me aware of errors they find in this edition.

Corvallis, OR                                              *Kartikeya Mayaram*
July 2007

# Preface

## First Edition

This book deals with the analysis and design of analog integrated circuits that form the basis of present-day communication systems. The material is intended to be a textbook for class use but should also be a valuable source of information for a practicing engineer. Both bipolar and MOS transistor circuits are analyzed and many numerical examples are used to illustrate the analysis and design techniques developed in this book. A set of problems is presented at the end of the book which covers the subject matter of the whole book.

The book has originated out of a senior-level course on nonlinear, analog integrated circuits at the University of California at Berkeley. The material contained in this book has been taught by the first author for several years and the book has been class tested for six semesters. This along with feedback from the students is reflected in the organization and writing of the text. We expect that the students have had an introductory course in analog circuits so that they are familiar with some of the basic analysis techniques and also with the operating principles of the various semiconductor devices. Several important, basic circuits and concepts are reviewed as the subject matter is developed.

The approach taken is as follows: first-order analysis techniques are developed first using basic principles and simple device models. Then circuit simulation is used to corroborate the analysis techniques. This procedure provides insight into the operation of circuits and a systematic way of getting an initial design of a circuit. The circuit simulation program SPICE has been extensively used to verify the results of first-order analyses, and for detailed simulations with complex device models. In this manner the student can appreciate the shortcomings of the hand analysis and can resort to simulations when necessary. Simulation results can only be interpreted once one has an understanding of how a circuit operates and this is reflected by the manner in which the material is presented. SPICE input files are given for all the circuits that have been analyzed so that the students can quickly duplicate the input file and verify the results.

The material contained in this book is covered in a 15-week semester course at the University of California at Berkeley. A chapter-by-chapter summary of the topics covered is given below.

Chapter 1 considers the large-signal performance of emitter-coupled pair and source-coupled pair circuits. The concept of harmonic distortion is introduced and series expansions are used as a method for obtaining the distortion components and the results are verified with SPICE simulations.

Another technique is introduced in Chapter 2 for computing the distortion components. The transfer characteristic of an amplifier is described by a power series and the harmonic distortion factors are derived. The concept of intermodulation distortion is also developed.

In Chapter 3 the distortion modification/generation due to source resistance and nonlinear beta in bipolar transistor circuits is considered.

Chapter 4 describes how distortion is modified in feedback amplifiers. The concept of feedback is reviewed and applied to several example circuits with particular attention to establishing correctly the appropriate loop-gain value.

Distortion and power transfer calculations in the basic IC output stages are the concern in Chapter 5. Class A, Class B, and Class AB output stages are described. An alternate method for distortion calculation is also presented.

Chapter 6 deals with transformers which are essential components in many baseband output stages and which also form the basis for tuned circuits and bandpass amplifiers. The basic low-frequency transformer is developed together with the circuit models and parameters which describe its electrical performance. In this chapter the analysis of transformer-coupled amplifiers and output stages is provided.

Tuned circuits are reviewed in Chapter 7. Elementary circuits and evaluation techniques are introduced. Major emphasis is given to circuits employing inductive transformers.

The design of simple bandpass amplifiers is considered in Chapter 8. Synchronous tuning, cascading, and bandwidth shrinkage are described and applied to the design of a multistage bandpass amplifier. The concept of cross modulation is also introduced.

Chapter 9 describes basic electronic oscillators. Simple and special circuits are used as a basis to provide insight into the design of oscillator circuits. The development is based both on the negative-resistance approach to oscillator analysis as well as the feedback approach.

The concept of bias-shift limiting is introduced in Chapter 10. Steady-state operation of the oscillator corresponds to Class C operation. Single device bipolar and MOS circuits are described.

Chapter 11 deals with relaxation and voltage-controlled oscillators. A graphical analysis of the basic oscillator is used to develop ideas that are helpful in understanding the operation and the design of these oscillators.

Relaxation oscillator examples in both BJT and MOS technologies are presented and techniques for controlling the period of oscillation are also described leading to voltage (current) controlled oscillators.

Analog multipliers, mixers, and modulators are considered in Chapter 12. The emitter-coupled pair is first introduced as a simple analog multiplier and extended to explain the development and operation of a four-quadrant multiplier. The concepts of mixing and modulation are developed and various examples are given.

Chapter 13 deals with demodulators and detectors which form an integral part of any communication circuit. Various techniques for AM and FM demodulation are described.

Phase-locked loops which are extensively used as frequency synthesizers and demodulators are examined in Chapter 14. The basic operation of a phase-locked loop is described using macromodels suitable for SPICE simulations.

Chapter 15 describes rectifier, regulator, and voltage-reference circuits. Simple rectifier circuits are introduced. The concept of regulation is developed and series and switching regulators are described. The regulator circuits use voltage references and some basic reference circuits are presented.

We are pleased to acknowledge the many comments and suggestions provided by our colleagues and students, especially Professor R. G. Meyer. We also appreciate the contributions of Ms. Elizabeth Rhine, Ms. Susie Reynolds, and Ms. Gwyn Horn for their excellent formatting and compositing of the manuscript.

Donald O. Pederson
Kartikeya Mayaram

# Contents

# 1

# Review of Communication Systems, Transistor Models, and Distortion Generation

## 1.1 Introduction

The purpose of any communication system is the transmission of information (speech, video, data) from a source to a receiver through a medium or channel. The channel for transmission can be wires as in wired communications (telephone lines, cable television) or free space (air) as in wireless communications (commercial radio and television, cell phones, pagers). The focus of this book is on wireless communication systems. The transmitter broadcasts the source information through free space using radiated power from an antenna. This power is picked up at the receiving end by an antenna and is processed by the receiver block.

Since the information signals occupy similar frequency bands (the base band), one cannot transmit these signals over a single communication channel without resorting to multiplexing the signals either in the time domain or frequency domain. Amplitude modulation (AM) and (frequency modulation FM) are two analog techniques for multiplexing signals in the frequency domain and are used extensively in commercial radio broadcast systems, AM and FM radios, respectively. The base band signals are translated to different locations (channels) at higher frequencies in the frequency spectrum using these modulation techniques. A secondary signal called the *carrier* is used for the frequency translation. Another benefit of the modulation schemes is that the low frequency base band signals are up converted to a higher frequency with the aid of the carrier and this makes their transmission through free space easier.

## 1.2 Amplitude and Frequency Modulations

The operation of modifying some property or characteristic of a carrier signal with an information signal is called *modulation* and the modified carrier is called a *modulated signal*. The process of recovering the original

D.O. Pederson and K. Mayaram, *Analog Integrated Circuits for Communication*, DOI 10.1007/978-0-387-68030-9_1,
© 2008 Springer Science+Business Media, LLC

information signal is referred to as *demodulation*. Details on various modulation/demodulation techniques can be found in [1], [2]. Consider the carrier to be a sinusoidal voltage signal of amplitude $V_c$ and radian frequency $\omega_c$. The carrier signal is then given by Equation (1.1).

$$v(t) = V_c \cos \omega_c t \qquad (1.1)$$

The carrier can be modulated by varying its amplitude resulting in amplitude modulation or frequency which corresponds to frequency modulation.[1] Consider AM, then

$$v(t) = V_c \left(1 + m \cos \omega_m t\right) \cos \omega_c t \qquad (1.2)$$

where the carrier has been modulated with a single-tone sinusoid of frequency $\omega_m$. In the above equation, $m$ is referred to as the *modulation index* and is the ratio of the amplitude of the modulating signal to the carrier amplitude.[2] The time-domain waveform of an amplitude modulated signal is shown in Figure 1.1a.

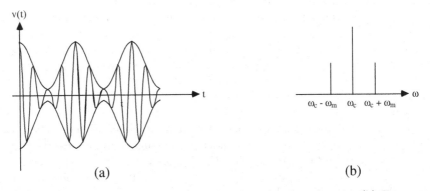

**Fig. 1.1.** (a) An amplitude modulated signal in the time domain. (b) Frequency spectrum of the signal.

Applying trigonometric identities, Equation (1.2) can be rewritten as

$$v_i = V_c \cos \omega_c t + \frac{V_c m}{2} \cos \left(\omega_c - \omega_m\right) t + \frac{V_c m}{2} \cos \left(\omega_c + \omega_m\right) t \qquad (1.3)$$

From the above expression, it is clear that two AM sidebands appear in the frequency spectrum as shown in Figure 1.1b. The sideband at $\omega_c - \omega_m$ is called the *lower sideband* and the sideband at $\omega_c + \omega_m$ is called the *upper sideband*.

---

[1] Another property of the carrier that can be varied is the phase which results in phase modulation (PM). Both FM and PM belong to the general class of angle modulation. Our focus here is only on frequency modulation.

[2] In order to properly demodulate the original signal, the modulation index must be less than one.

To achieve amplitude modulation as in Equation (1.2), a multiplication is needed and circuits for performing this operation are described in Chapter 13.

In FM, the frequency of the carrier is modulated and the modulated signal is given by

$$v(t) = V_c \cos \left( \omega_c t + \frac{\Delta \omega}{\omega_m} \cos \omega_m t \right) \tag{1.4}$$

where $\Delta \omega / \omega_m$ is defined as the modulation index (MDI) and $\Delta \omega$ is the maximum frequency deviation. A FM signal is shown in the time and frequency domains in Figures 1.2a and 1.2b, respectively. From the frequency spectrum in Figure 1.2b it is clear that the bandwidth of the signal increases as the modulation index increases.

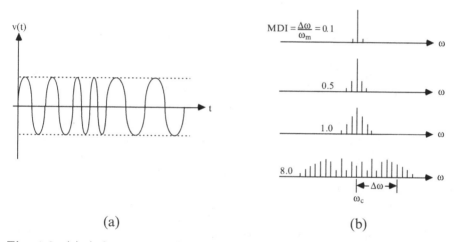

(a)                                         (b)

**Fig. 1.2.** (a) A frequency modulated signal in the time domain. (b) Frequency spectrum of the signal.

For FM broadcast in the United States, the FCC has allocated a 20 MHz band from 88 MHz to 108 MHz divided into 100 channels. Each channel has a bandwidth of 200 kHz. The maximum frequency deviation $\Delta f$ is 75 kHz and the maximum modulating frequency $f_m$ is 15 kHz resulting in a modulation index of 5. The bandwidth for an angle-modulated signal can be approximated by *Carson's rule* and is $2(\Delta f + f_m)$. For FM broadcast the maximum bandwidth is 180 kHz which is within the FCC specified limits.

## 1.3 The Super-Heterodyne Receiver

There are several radio receiver architectures which are described in [4], [5]. The most common form is the superheterodyne configuration shown in

Figure 1.3. The signal input, with a frequency $\omega_s$, is usually first amplified in a tunable, bandpass amplifier, called the radio-frequency (RF) amplifier, and is then fed into a frequency-translation circuit called the mixer along with an oscillator signal,[3] which is 'local' to the receiver, having a frequency $\omega_{lo}$. The local oscillator (LO) is also tunable and is 'ganged' with the input bandpass amplifier such that the difference between the input signal frequency and that of the local oscillator is (approximately) constant.

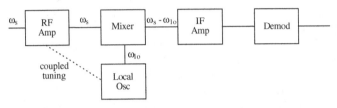

**Fig. 1.3.** The superheterodyne receiver configuration.

In operation, the mixer circuit produces sum and difference frequency components $(\omega_s \pm \omega_{lo})$ at the output, cf., Chapter 13. Usually, the sum frequency is rejected by sharply tuned circuits and the difference frequency component is subsequently amplified in a fixed-tuned bandpass amplifier. The difference frequency is called the intermediate frequency (IF) and the fixed-tuned amplifier is called the IF amplifier. The advantage of this superheterodyne configuration is that most amplification and outband rejection occurs with fixed-tuned circuits which can be optimized for gain level and rejection. Another advantage is that the fixed-tuned amplifier can provide a voltage-controlled gain to achieve automatic gain control (AGC) with input signal level.

Commercial AM broadcast in the United States occupies the 535 kHz-1605 kHz band (with an AM expanded band from 1605 kHz to 1705 kHz). The channel bandwidth is 10 kHz and the IF frequency is 455 kHz. For an AM radio station at 770 kHz the LO frequency for a superheterodyne receiver will be $770 + 455 = 1225$ kHz assuming that the LO is at a higher frequency than the input signal (referred to as high-side injection).

## 1.4 Transistor Models

Circuit simulators, such as Spice [3], are routinely used for analyzing integrated circuits. These simulators rely on accurate models for semiconductor devices to predict the performance of a circuit. In this book, simplified device models are used for a first-order analysis of circuits. Advanced and more

---

[3] The process of frequency translation with the aid of another signal is called *mixing* or *heterodyning*.

accurate models can be used in the circuit simulator for a detailed final analysis. An excellent discussion of the different semiconductor devices and models frequently used in integrated circuit design can be found in [6], [7]. A brief description of the simplified large-signal and small-signal models for the bipolar junction transistor (BJT), and metal oxide semiconductor field-effect transistor (MOSFET) is provided in this section.

### 1.4.1 Bipolar Junction Transistor (BJT) Model

A circuit symbol for a NPN transistor including the terminal currents and voltages is shown in Figure 1.4a. The collector current $I_C$ in a BJT is a function of the base-emitter voltage $V_{BE}$

$$I_C = I_S \exp\left(\frac{V_{BE}}{V_t}\right) \tag{1.5}$$

where $I_S$ is the saturation current and $V_t$ is the thermal voltage and is given by

$$V_t = \frac{kT}{q} \tag{1.6}$$

At room temperature, $T = 300°K$, $V_t \approx 25.85$ mV. From Equation (1.5), $V_{BE}$ can be expressed as

$$V_{BE} = V_t \ln\left(\frac{I_C}{I_S}\right) \tag{1.7}$$

It is useful to remember that for a factor of 10 change in collector current, $V_{BE}$ changes by 25.85 mV$\times$2.303 $\approx$ 60 mV.

The relationships between the base, collector and emitter currents are given by the following equations.

$$I_B = \frac{1}{\beta}I_C \tag{1.8}$$

$$I_C = \frac{\beta}{\beta+1}|I_E| \approx |I_E|$$

where $I_E$ is a negative quantity for the npn transistor operated in the normal active region and $\beta$ is the dc beta.[4]

In Figure 1.4b, the simplest form of the Ebers-Moll model for the bipolar transistor is shown [3], [8]. Note that the ohmic base resistor, $r_b$ (Spice parameter RB), the ohmic collector resistor, $r_c$ (Spice parameter RC), and the effects of base-width modulation ($V_A, r_o,$ and $r_\mu$) have been neglected. In this

---

[4] In this book, the ratio of the dc collector current and the dc base current is defined as dc beta, $\beta = I_C/I_B$. The ratio of $i_c$, the change of $I_C$, and $i_b$, the change of $I_B$, is defined as ac beta, $\beta_{ac} = i_c/i_b = \Delta I_C/\Delta I_B$.

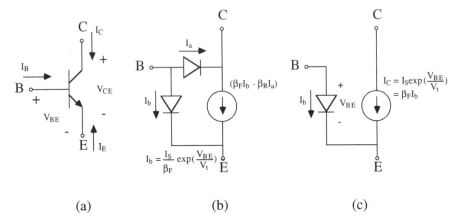

(a)                    (b)                    (c)

**Fig. 1.4.** (a) Circuit symbol for a NPN bipolar junction transistor. (b) Ebers-Moll large-signal model for the bipolar transistor. (c) Simplified large-signal model.

figure $\beta_F$ and $\beta_R$ are the forward and reverse current gains, respectively. Since the base-collector junction is reverse biased in the forward active region, $I_a$ is small and the simpler large-signal model for the BJT as shown in Figure 1.4c can be used.

The small-signal model is useful to establish a variational response of the transistor at a given operating point. The small-signal circuit model is derived from the large-signal characteristics and the complete intrinsic small-signal model (ignoring the parasitic elements) for the BJT is shown in Figure 1.5. The details of the model can be found in [6] and equations for the various circuit elements are provided in (1.9).

$$g_m = \frac{I_C}{V_t} \tag{1.9}$$

$$r_\pi = \frac{\beta_{ac}}{g_m}$$

$$r_o = \frac{V_A}{I_C}$$

$$C_\pi = g_m \tau_F + C_{je}$$

where $I_C$ is the collector current at the dc operating point, $V_A$ is the Early voltage (Spice parameter VA), $\tau_F$ is the forward transit time (Spice parameter TF). $C_{je}$ and $C_{jc}$ are the depletion region capacitances associated with the base-emitter and base-collector junctions, respectively.

### 1.4.2 MOSFET Model

A circuit symbol for a N-channel MOSFET including the terminal currents and voltages is shown in Figure 1.6a. In this book, a simplified model is used

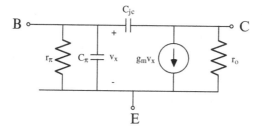

**Fig. 1.5.** Small-signal circuit model for the bipolar transistor.

for the MOS transistor and more advanced MOSFET models are described in [9], [10], [11]. The MOSFET has three distinct regions of operation - the off region (cut off), the resistance region (linear or triode), and the saturation region. The idealized drain current versus voltage equations for these three regions of operation are:

For $V_{GS} < V_T$, the off region, $\quad$ (1.10)
$$I_D = 0$$
For $V_{DS} < V_{GS} - V_T$, the resistance region, $\quad$ (1.11)
$$I_D = k'\frac{W}{L}\left[(V_{GS} - V_T)V_{DS} - \frac{1}{2}V_{DS}^2\right]$$
For $V_{DS} > V_{GS} - V_T$, the saturation region, $\quad$ (1.12)
$$I_D = \frac{k'}{2}\frac{W}{L}(V_{GS} - V_T)^2$$

where $V_T$ is the threshold voltage of the MOS device, $k'$ is called the gain factor of the device (KP in Spice), and $\frac{W}{L}$ is the width-to-length ratio of the specific device. The $V_{BS}$ dependence of the drain current is included through the body effect, and the threshold voltage for a nonzero body bias ($V_{BS} \neq 0$) is given as

$$V_T = V_{T0} + \gamma\left(\sqrt{V_{SB} + 2\phi_f} - \sqrt{2\phi_f}\right) \quad (1.13)$$

where $V_{T0}$ is the threshold voltage for a zero body bias (VTO in Spice), $\gamma$ is the body effect parameter (GAMMA in Spice), and $2\phi_f$ is the surface potential at strong inversion with $V_{SB} = 0$ (Spice parameter PHI). The drain current equations are also modified to include the effect of channel-length modulation, $\lambda$, (LAMBDA in Spice) [6]. In addition more advanced MOSFET models that adequately model small geometry transistors, such as BSIM3 or BSIM4 [10], [11] can be used in Spice. For our purposes, as we did for the BJT devices, these additional effects are initially neglected and are brought in later, often using circuit simulation. Note that gate current for a MOS transistor is zero under dc operation. A simplified large-signal equivalent circuit model in shown in Figure 1.6b.

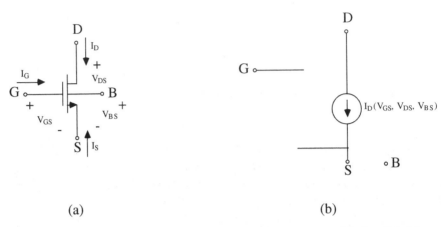

(a)                                        (b)

**Fig. 1.6.** (a) Circuit symbol for a N-channel MOS transistor. (b) Simplified large-signal model for the transistor.

To establish the variational response of the MOSFET at a given operating point it is helpful to assume that operation is restricted to the saturation region. The small-signal circuit model for a MOS transistor (only the intrinsic transistor) is shown in Figure 1.7. Once again, the details of the model can be found in [6], whereas equations for the various circuit elements are provided in (1.14).

$$g_m = k'\frac{W}{L}(V_{GS} - V_T) = \sqrt{2I_D k'\frac{W}{L}} = \frac{2I_D}{V_{GS} - V_T} \qquad (1.14)$$

$$g_{mb} = \frac{\gamma}{2\sqrt{2\phi_f + V_{SB}}}g_m$$

$$r_o = \frac{1}{\lambda I_D}$$

where $I_D$, $V_{GS}$, $V_{SB}$ are the drain current, gate-source voltage, and source-bulk voltage at the dc operating point, respectively. $C_{gs}$ and $C_{gd}$ are the gate-source and gate-drain capacitances, respectively.

## 1.5 Distortion Generation and Characterization

The above current-voltage relationships for the semiconductor devices show that these characteristics are not linear. The inherent nonlinearity of transistors results in an output which is a "distorted" version of the input. The

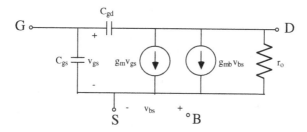

**Fig. 1.7.** Small-signal circuit model for the MOS transistor.

distortion due to a nonlinear device is illustrated in Figure 1.8. For an input voltage $v$ the output current is $i = F(v)$ where $F$ denotes the nonlinear transfer characteristics of the device; the dc operating point is given by $V_0$. Sinusoidal input signals of two different amplitudes are applied and the output responses corresponding to these inputs are also shown.

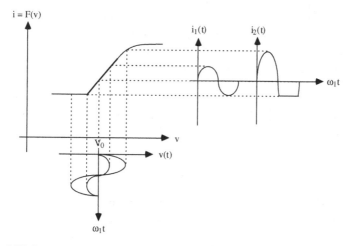

**Fig. 1.8.** I-V characteristics of a nonlinear device with input and output waveforms. The output is distorted for a large amplitude input signal.

For an input signal of small amplitude the output faithfully follows the input, i.e., the output signal waveform is an amplified version of the input waveform. However, for large amplitude signals the output is distorted; a flattening occurs at the negative peak value. The distortion in amplitude results in the output having frequency components that are integer multiples of the input frequency, *harmonics*, as shown in Figure 1.9. These harmonics are generated by the device or circuit nonlinearity and this type of distortion is referred to as harmonic distortion.

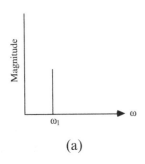
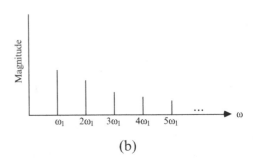

(a)                                                    (b)

**Fig. 1.9.** Frequency spectrum of (a) the input, and (b) the output of a nonlinear device or circuit.

A quantitative measure of the distortion 'generation' can be obtained with a Fourier analysis of the input and output waveforms. A direct comparison of the magnitudes (amplitudes) and phases of the Fourier components illustrates the distortion produced by a circuit block. An established procedure to accomplish this specifies that the input be a single sinusoid, i.e., a pure 'tone.' In actual measurements of an amplifier stage, the fundamental output is measured and compared to the magnitudes of the harmonics which are generated. Harmonic analyzer instruments are available to accomplish the measurement and consist of sensitive electronic voltmeters and tunable, very selective filters.

Harmonic distortion parameters, or factors, of a waveform are defined in terms of the ratio of the amplitude of a harmonic and the amplitude of the fundamental. The second-harmonic distortion factor, $HD_2$, is

$$HD_2 = \frac{|b_2|}{|b_1|} \tag{1.15}$$

where $|b_2|$ is the magnitude of the amplitude of the second harmonic and $b_1$ is the amplitude of the fundamental. The third-harmonic distortion factor is

$$HD_3 = \frac{|b_3|}{|b_1|} \tag{1.16}$$

where $b_3$ is the amplitude of the third harmonic.

The total harmonic distortion (THD) of a waveform is defined as the ratio of the root-mean-square (rms) amplitude of the sum of the upper harmonics to the amplitude of the fundamental.

$$THD = \frac{\sqrt{b_2^2 + b_3^2 + \cdots}}{|b_1|} \tag{1.17}$$

The root mean square of the harmonics must be taken, since the effective value of a sum of components, each with a different frequency, is needed. $THD$ can also be expressed in terms of the individual harmonic distortion factors.

$$THD = \sqrt{HD_2^2 + HD_3^2 + \dots} \qquad (1.18)$$

It must be emphasized that the distortion and distortion components are dependent upon the level of the input signal. If a larger-input sinusoid is used in the example above, the distortion is greater since a greater range of the nonlinear characteristic is traversed.

One way to investigate distortion is to use one of the Spice simulators.[5] An input waveform can be specified, usually of only certain pre-programmed types, and a desired output waveform as well as the input waveform can be plotted. A direct comparison illustrates the distortion. Often, it is helpful to choose appropriate scaling of the waveforms in making the comparison.

Distortion calculations in Spice are commonly performed with a Fourier analysis using the .FOUR command. A transient analysis[6] using the .TRAN statement is first performed for a sufficiently long time interval to ensure that all the startup transients die off. The circuit waveforms are then periodic and the circuit is said to be in a periodic steady state. The determination of the steady state is done by the user and could be subject to errors. A good check is to make sure that the time-domain waveform of interest is periodic. Then a discrete Fourier transform (DFT) is applied to the last time period of the simulated time-domain waveform data (.FOUR analysis). A fundamental frequency is specified and Spice reports the magnitude and phase of the harmonics of this fundamental frequency.[7]

As is brought out in the following chapters, great care must be taken in using the Spice programs to determine small distortion components. Distortion calculations using transient analysis can result in several errors. An excellent discussion of these can be found in [12]. It is important that the circuit settles to a periodic steady state for some fundamental frequency before Fourier analysis is used. This may require a long simulation time interval for circuits that exhibit widely separated time constants, or for which the specified fundamental frequency is very small. The time-domain simulation must be performed with tight tolerances when low power harmonics are of interest. The default value of the accuracy parameter, RELTOL, in Spice is 0.001 or 0.1%. Thus, simulated distortion components below 1% must be accepted with caution. Since the DFT algorithm introduces interpolation and aliasing errors a sufficient number of time points must be used within one time period.

---

[5] Several versions of Spice are used in this book: Spice2, Spice3. The choice depends on the features to be emphasized.

[6] A sinusoidal source at a given frequency is used as an input for an amplifier circuit.

[7] In Spice2, the dc component and the first nine harmonics are printed. This is also the default number in Spice3. A different number of harmonics (N) can be requested in Spice3 using NFREQS = N in the interactive mode.

When sufficient time points per period are used[8] together with a small value of RELTOL, say $1 \times 10^{-6}$, accurate values for small harmonic components can be obtained with Spice.

Furthermore, simulation of several communication circuits (mixers, oscillators) in the time-domain may not be possible when there is a wide spread in the signal frequencies and the circuit time constants. In such cases, frequency-domain simulations based on the *harmonic-balance method* are preferred [13], [14], [15].

In addition to the harmonic distortion factors, there are other measures of distortion that are described in later chapters.

## 1.6 Noise

Noise from semiconductor devices plays a very important role in the design of communication circuits. The dynamic range of a receiver is limited by noise at the low end of a signal and by the circuit nonlinearities (distortion) for large signals.

Circuit noise calculations are typically performed by linearizing the circuit at a dc operating point and calculating the noise contributions from each noise source based on a small-signal analysis [6]. The *linear* noise analysis is applicable only to amplifiers or circuits operating linearly where small-signal analysis is appropriate. However, circuits such as mixers and oscillators operate under large-signal conditions as discussed in later chapters. The circuit nonlinearities have a direct influence on the noise performance of the circuit. In a mixer, nonlinearities result in an up/down conversion of noise [16] which affects the overall noise performance of the circuit. For oscillators the noise manifests itself as phase noise or jitter [17]. Phase noise is an indicator of the spectral purity and timing accuracy of a signal. Although techniques have been

---

[8]    The parameters of the .TRAN statement in the Spice input file must be carefully selected. This statement is given by

.TRAN TSTEP TSTOP TSTART TMAX

The last two parameters in the above command (TSTART and TMAX) are optional. The analysis time interval is 0 to TSTOP and the results can be saved or viewed from a user defined time TSTART which has a default value of 0. TSTEP is the time increment used for printing or plotting the output results and TMAX is the maximum allowed time step for the simulator. Spice uses an internal time step for solving the circuit equations based on accuracy requirements. When TMAX is not specified, the step size cannot exceed the smaller of TSTEP or (TSTOP-TSTART)/50. If TMAX is defined, then TSTART must also be set as the order of these parameters in the .TRAN statement is important. It is useful to specify TMAX when time steps smaller than TSTEP are needed for accuracy. In this book, typically 20 to 100 time points per period have been used.

developed for analyzing noise in mixers [16] and oscillators [18], one typically uses a commercial circuit simulator, such as Spectre, for these analyses. The interested reader is referred to the other textbooks [17], [19] for details on both linear and nonlinear noise analyses.

## Problems

**1.1.** Read the paper by Razavi [4]. Based on this paper answer the following questions
(a) Explain the problem of the image frequency in a superheterodyne receiver architecture.
(b) If the input signal is at a frequency of 800 kHz and the intermediate frequency is 455 kHz then what are the LO and image frequencies?
(c) List two receiver architectures that do not suffer from the image problem.

**1.2.** Draw the block diagram for a homodyne (direct conversion) receiver architecture. List the advantages and disadvantages of both superheterodyne and homodyne receivers. Refer to the paper by Razavi [4].

**1.3.** For a mixer in a superheterodyne receiver there are two possible LO frequency choices $f_{RF} + f_{IF}$ (high-side injection) or $f_{RF} - f_{IF}$ (low-side injection). For an AM radio (530 kHz to 1610 kHz with an IF of 455 kHz), calculate the tunable range of the LO for both low- and high-side injections.

**1.4.** For the GSM (European cellular system) standard the receive band is 925-960 MHz. What should the tunable range of the LO be for
(a) one IF at 71 MHz,
(b) two IF, the first one at 250 MHz and the second at 10.7 MHz.

**1.5.** The output spectrum (first five harmonics) of a nonlinear circuit with a sinusoidal input of 1 V at 100 kHz is listed in Table 1.1. Using this data answer the following questions.
(a) What is the percentage HD2?
(b) What is the percentage HD3?
(c) What is the percentage THD?

| Frequency (kHz) | Magnitude |
|---|---|
| 100 | 0.800 |
| 200 | 0.040 |
| 300 | 0.100 |
| 400 | 0.020 |
| 500 | 0.050 |

**Table 1.1.** Frequency spectrum of output signal for Problem 1.5.

**1.6.** A waveform has only second and third harmonics. If $HD_2 = 3\%$ and $HD_3 = 4\%$ what is THD?

**1.7.** Study the usage of the Spice .TRAN and .FOUR statements from [3] or the Spice User's Guide. Apply these commands to the analysis of a simple transistor circuit.

# 2

# Large-Signal Performance of the Basic Gain Stages in Analog ICs

## 2.1 The Emitter-Coupled Pair

The basic stages of analog integrated circuits are the emitter-coupled pair (ECP) and the emitter follower (EF) for bipolar circuits and the source-coupled pair (SCP) and the source follower (SF) for MOS circuits. In the next several chapters, these basic stages are examined from the standpoint of their large-signal performance. Of particular interest is the distortion produced in the output waveforms relative to the inputs. In a later chapter, available power output and efficiency of dc power conversion are considered for output stages.

A primitive EC pair (ECP) is shown in Figure 2.1a and consists of two (hopefully) matched bipolar transistors, a current source, two load resistors and a voltage source, $V_{CC}$. These are the essential elements of this circuit. (The current source $I_{EE}$ is returned to a voltage source $-V_{EE}$.) The input can be applied at either or both of the two base leads of the transistors. The output voltage can be taken from either of the two collectors, separately, or the difference voltage, $V_o = V_{o1} - V_{o2}$, can be obtained. At the input, a differential input voltage, $V_D$, can be used together with a common-mode input voltage, $V_C$, These arrangements are also illustrated in the figure.

In Figure 2.2, an actual ECP is shown such as might be used in an integrated circuit. Note that the lower npn transistor-resistor circuitry supplies the $I_{EE}$ current-source function, while the upper pnp transistors supply the load and load biasing elements. For the purposes of this chapter, as well as most of this book, it is not necessary to treat the entire practical stage, at least not initially while the basic large-signal performance of the circuit is under examination. In fact, our usual procedure when investigating an actual circuit will be to examine carefully the function of all elements, replacing them with idealized elements if possible, and to retain and concentrate on the essential large-signal aspects of the circuit.

D.O. Pederson and K. Mayaram, *Analog Integrated Circuits for Communication*, DOI 10.1007/978-0-387-68030-9_2,
© 2008 Springer Science+Business Media, LLC

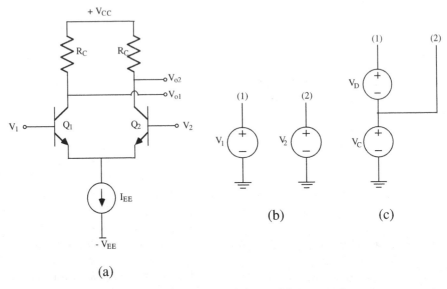

(a)

**Fig. 2.1.** (a) A primitive EC pair circuit. (b) and (c) Input voltage arrangements.

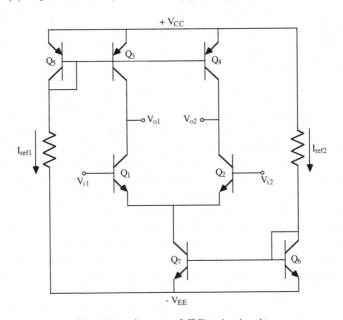

**Fig. 2.2.** An actual EC pair circuit.

## 2.2 The Large-Signal DC Transfer Characteristic of the ECP

In this section, the basic electrical performance characteristics of the EC pair are developed. First, the bias state and the small-signal properties of the ECP about a typical operating point are reviewed. The large-signal properties then are developed. In both cases, only the low-frequency behavior is examined. The effects of charge storage on the time and frequency responses can be examined later using circuit simulation programs.

Consider the ECP of Figure 2.3. Values for the two essential transistor parameters, $\beta$ (dc beta) and $I_S$, are given in the figure. The signal inputs, $(V_1, V_2)$, are two ideal voltage sources, and we let both have static (dc bias) values of 0 V. If the transistors are identical, the 1 mA common-emitter current source draws from each emitter a current of 0.5 mA. If we assume that the beta of the transistors is large and, therefore, that $I_B$ is negligible, the collector currents are also 0.5 mA. For the values of $R_{C1} = R_{C2} = 10$ k$\Omega$ and $V_{CC} = 10$ V, the dc values of the collector voltages are $V_{CC} - I_C R_C = 10 - 0.5$ mA $\times$ 10 k$\Omega = 5$ V. The voltage at the common emitters will be $-V_{BE}$. From Figure 2.3, $I_S = 1 \times 10^{-16}$ A and

$$V_{BE} = V_t \ln \left( \frac{I_C}{I_S} \right) = 25.85 \text{ mV} \times \ln \left( \frac{0.5 \times 10^{-3}}{1 \times 10^{-16}} \right) \tag{2.1}$$
$$= 0.756 \text{ V}$$

The $I_B$ approximation above must be stressed. We use this approximation constantly and often forget the fact.

Next, incremental (small-signal) components, $v_1$ and $v_2$, are added to $V_1$ and $V_2$, respectively. The input voltages can then be expressed as $V_{i1} = V_1 + v_1$ and $V_{i2} = V_2 + v_2$. For the situation at hand, $V_1$ and $V_2$ remain equal to zero, and $v_2$ is also assumed to be zero. To establish the variational response of the stage at the given operating point, it is helpful to introduce the simple small-signal circuit model for the transistors as shown in Figure 2.4. Note that the independent voltage sources are short circuits for this case, while the independent current source is an open circuit. Consistent with the assumption of an idealized ECP, we neglect for now the presence in each transistor of the ohmic base resistor, the ohmic collector resistor, and the effects of base-width modulation. The input resistance $R_{in}$ is calculated simply to be

$$R_{in} = \frac{v_1}{i_{in}} = 2r_\pi = 2\beta_{ac} \frac{1}{g_m} = 10.3 \text{ k}\Omega \tag{2.2}$$

where $1/g_m = 25.85$ mV/$I_C = 51.7$ $\Omega$. For this example, the value of $\beta_{ac}$, the ac beta, is assumed a constant. Therefore, $\beta_{ac} = \beta = 100$ (from Figure 2.3). The incremental voltage gain from $v_1$ to $v_{o2}$ is

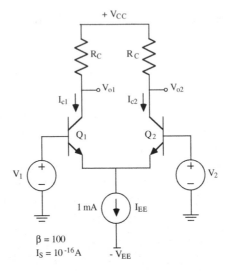

**Fig. 2.3.** An EC pair circuit.

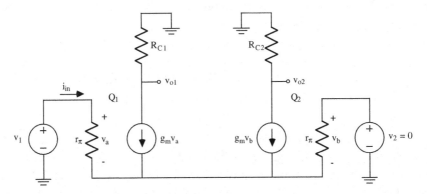

**Fig. 2.4.** Small-signal circuit of the EC pair.

$$a_v = \frac{v_{o2}}{v_1} \tag{2.3}$$

$$= \frac{1}{2} g_m R_{C2}$$

$$= 96.7$$

(Often, a notation for the transconductance of the ECP itself, $G_m$, is used. For the present example, $G_m = \frac{g_m}{2}$).

The output resistance looking back into the stage including $R_{C2}$ is simply $R_{C2}$, since $r_o$ and $r_\mu$ are assumed absent (infinite). The effects due to a finite value of $r_o$ are considered in Chapter 5.

The output resistance with respect to the output $v_{o1}$, looking back at $R_{C1}$, is again $R_{C1}$. The small-signal voltage gain from $v_1$ to $v_{o1}$ is $-96.7$.

Return now to the dc input voltage $V_1$ and change its value. The output voltage variation with $V_1$ could be determined by actual measurements or, for our purposes, can easily be determined using a circuit simulator such as Spice. It is to be recognized that even for the idealized circuit of Figure 2.3 and even with assumed simple device models, we are faced with the solution of a set of nonlinear algebraic equations which describe the circuit. In Figure 2.5, the circuit is shown with the simplest form of the Ebers-Moll model (Figure 1.4a). Although the setting up of the equilibrium circuit equations is straightforward, the solution of the nonlinear equations to obtain values, say, of $V_{o2}$ for a set of values of $V_1$, is tedious to say the least. Fortunately, for the restricted case where no saturation of the BJTs is permitted, a closed-form solution of the circuit equations can be found. This aspect is taken up in Section 2.4. In general, however, computer programs or circuit simulators must be used to solve (approximately) these equations.

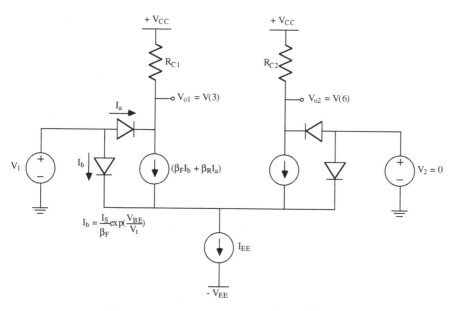

**Fig. 2.5.** Large-signal circuit of the EC pair.

The Spice input file for the ECP of Figure 2.3 is shown in Figure 2.6a. (A second ECP is included to permit easy comparison of common-output variables for two different situations.) Shown in Figure 2.6b are the dc node voltages in the circuit for input voltages of 0 V.

For the top circuit of Figure 2.6a, $V_1$ could initially be chosen to range from $-V_{EE} = -10$ V to $V_{CC} = 10$ V with a calculation interval of 0.1 V. However, the output plots for $V_{o1}$ and $V_{o2}$ are virtually piece-wise-linear step functions. The region near the breaks is illustrated in Figure 2.7. The input

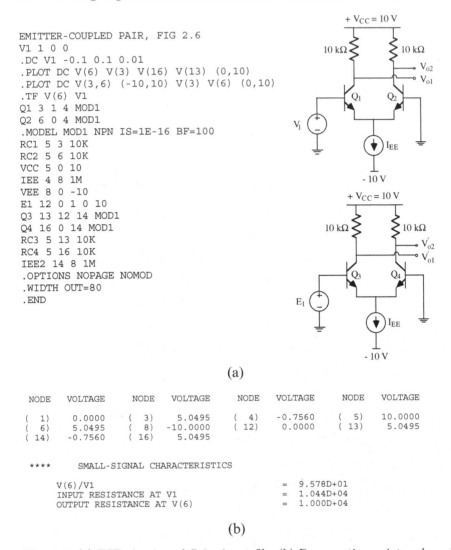

```
EMITTER-COUPLED PAIR, FIG 2.6
V1 1 0 0
.DC V1 -0.1 0.1 0.01
.PLOT DC V(6) V(3) V(16) V(13) (0,10)
.PLOT DC V(3,6) (-10,10) V(3) V(6) (0,10)
.TF V(6) V1
Q1 3 1 4 MOD1
Q2 6 0 4 MOD1
.MODEL MOD1 NPN IS=1E-16 BF=100
RC1 5 3 10K
RC2 5 6 10K
VCC 5 0 10
IEE 4 8 1M
VEE 8 0 -10
E1 12 0 1 0 10
Q3 13 12 14 MOD1
Q4 16 0 14 MOD1
RC3 5 13 10K
RC4 5 16 10K
IEE2 14 8 1M
.OPTIONS NOPAGE NOMOD
.WIDTH OUT=80
.END
```

(a)

| NODE | VOLTAGE | NODE | VOLTAGE | NODE | VOLTAGE | NODE | VOLTAGE |
|------|---------|------|---------|------|---------|------|---------|
| ( 1) | 0.0000 | ( 3) | 5.0495 | ( 4) | -0.7560 | ( 5) | 10.0000 |
| ( 6) | 5.0495 | ( 8) | -10.0000 | ( 12) | 0.0000 | ( 13) | 5.0495 |
| ( 14) | -0.7560 | ( 16) | 5.0495 | | | | |

```
****      SMALL-SIGNAL CHARACTERISTICS

        V(6)/V1                              =  9.578D+01
        INPUT RESISTANCE AT V1               =  1.044D+04
        OUTPUT RESISTANCE AT V(6)            =  1.000D+04
```

(b)

**Fig. 2.6.** (a) ECP circuit and Spice input file. (b) Dc operating point and small-signal characteristics of EC pair.

voltage source ranges from -0.1 V to 0.1 V with an increment of 10 mV and provides the output curves, A and B. For the other ECP ($Q_3, Q_4$), the voltage-controlled voltage source multiplies this range by 10, -1 V to +1 V with an increment of 0.1 V and provides output plots C and D. (After $V_{o1}$ drops near $V_1 = 0$, it ultimately starts to increase as $V_1$ increases due to saturation of transistor, $Q_1$.) The output plots A and B for the ±100 mV input range are clearly better to study the details of the nonlinear output response. Notice that the outputs "clamp" for input voltages of about ±0.1 V. This is to be expected. From the observation that $V_{BE}$ changes by 60 mV for each decade change of $I_C$, it is reasonable to expect that $V_{02}$ will range from near $V_{CC}$ to near zero for only a few decades of current change. For only two decades the corresponding change in $V_1$ should be only 120 mV. This aspect is treated in more detail in a following section. From the voltage-transfer characteristic, A of Figure 2.7, which is repeated with a printout of values in Figure 2.8, the voltage gain, $a_v$, of the stage at the operating point of $V_1 = 0$ V is the slope of the plot at that point. From the figure, $a_v = 94.6$, which compares well with the small-signal value calculated earlier. This result, as well as the previous small-signal calculations, can be checked with values obtained from a .TF Spice simulation as given in Figure 2.6b: $a_v = 95.8$, $R_{in} = 10.4$ k$\Omega$, and $R_{out} = 10$ k$\Omega$.

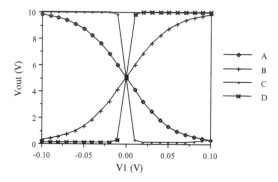

**Fig. 2.7.** Output voltages of EC pair.

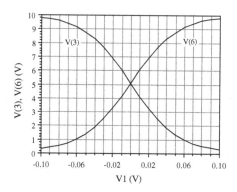

**Fig. 2.8.** V(3) and V(6) shown on an expanded scale.

As mentioned in Chapter 1, Spice can also be used to predict the distortion component generation. The input file for the circuit is specified to be a pure tone as listed in Figure 2.9. Again, two ECPs are simulated, with the lower listing having an input voltage which is 10 times that of the upper ECP listing due to the scaling parameter of E1. For the upper ECP listing, $V_1$ is assumed to be a sinusoid with a frequency of 100 kHz and a zero-to-peak amplitude of 10 mV. The frequency specification is arbitrary since no energy storage elements are included in the circuit or the devices. A new control line in the circuit input file, the .FOUR line, is also introduced. As shown, the line contains the value of the input frequency and the name of the outputs for which the Fourier analysis is requested. The output waveforms $V(6)$ and $V(16)$ from the Spice2 output are shown in Figure 2.10. $V(6)$ is also multiplied by 5 to achieve a better comparison. $V(6)$ is seen to be most sinusoidal in form; however, $V(16)$ for the large drive is 'squashed' at the extremes. In Figure 2.11, from Spice2, are the amplitudes and (relative) phase of the first nine harmonics of the output waveform of $V(6)$ as well as the values relative to the fundamental. It is clear from the values of Figure 2.11 that the harmonic content of the output waveform of the ECP is very small. Accurate values for the small harmonic contents were obtained by using sufficient time points per period together with a small value of RELTOL of $1 \times 10^{-6}$ in Figure 2.9.[1]

From Figure 2.11, the value of $HD_2$ is nearly zero. This is to be expected. The shape of the dc transfer characteristics of Figures 2.7 and 2.8 are antisymmetrical about the quiescent operating point ($V_1 = 0$ V, $V_{o2} = 5$ V). Thus, the even harmonics should not be generated. This aspect is developed more fully below. From Spice2 for a sinusoidal input of 10 mV, $HD_3 = 0.295\%$.

Distortion in a sinusoidal waveform below a few percent is difficult to observe, as seen in this example from the waveform $V(6)$ in Figure 2.10. It must be emphasized that the distortion and distortion components are dependent upon the level of the input signal. If a larger-input sinusoid is used in the example above, the distortion is greater since a greater range of the nonlinear characteristic is traversed. In Figure 2.10, the output voltage waveform $V(16)$ from a Spice2 run is shown where the input is a sinewave with an amplitude of 0.1 V. (See the Spice input file of Figure 2.9.) The output voltage is severely distorted with respect to a sine wave, and a very distinct flattening of the peaks of the output excursion is produced. This phenomenon occurs very often in nonlinear circuits. The Fourier components of the output waveform are given in Figure 2.12. Again notice that only the odd harmonics are appreciable due to the antisymmetrical nature of the transfer characteristic and as expected from the equal top and bottom flattening of the output voltage waveform in Figure 2.10. From Figure 2.12, $HD_3 = 15.6\%$ and $HD_5 = 3.1\%$.

---

[1] For this simulation, TSTEP in the .TRAN statement is set to 0.1 $\mu$s resulting in 100 time points per period.

```
EMITTER-COUPLED PAIR, FIG 2.9
V1 1 0 0 SIN 0 10M 100K
.TRAN 0.1U 20U
.PLOT TRAN V(6) V(16)
.FOUR 100K V(6) V(16)
Q1 3 1 4 MOD1
Q2 6 0 4 MOD1
.MODEL MOD1 NPN IS=1E-16 BF=100
RC1 5 3 10K
RC2 5 6 10K
VCC 5 0 10
IEE 4 8 1M
VEE 8 0 -10
E1 12 0 1 0 10
Q3 13 12 14 MOD1
Q4 16 0 14 MOD1
RC3 5 13 10K
RC4 5 16 10K
IEE2 14 8 1M
.OPTIONS RELTOL=1E-6
.OPTIONS NOPAGE NOMOD
.WIDTH OUT=80
.END
```

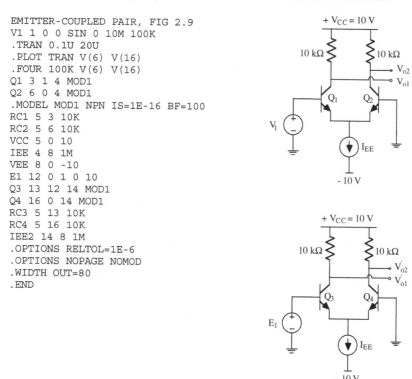

**Fig. 2.9.** Circuit and Spice input file for transient simulation of EC pair.

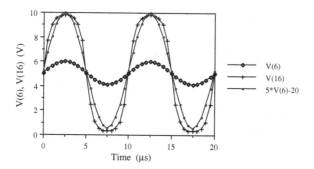

**Fig. 2.10.** Transient response of EC pair.

```
FOURIER COMPONENTS OF TRANSIENT RESPONSE V(6)        V1A = 10 mV, Tstep = 0.1 μs
DC COMPONENT =    5.049D+00
HARMONIC   FREQUENCY    FOURIER     NORMALIZED      PHASE      NORMALIZED
  NO         (HZ)      COMPONENT    COMPONENT       (DEG)      PHASE (DEG)

   1       1.000D+05   9.432D-01    1.000000        0.000        0.000
   2       2.000D+05   2.016D-05    0.000021       81.879       81.879
   3       3.000D+05   2.783D-03    0.002951       -0.092       -0.092
   4       4.000D+05   8.276D-06    0.000009      -84.203      -84.203
   5       5.000D+05   1.453D-05    0.000015      -24.081      -24.080
   6       6.000D+05   4.856D-06    0.000005      -54.029      -54.029
   7       7.000D+05   4.817D-06    0.000005      -66.910      -66.909
   8       8.000D+05   5.779D-06    0.000006      -63.846      -63.846
   9       9.000D+05   5.961D-06    0.000006      -52.771      -52.771


      TOTAL HARMONIC DISTORTION =       0.295120   PERCENT
```

**Fig. 2.11.** Fourier components of V(6).

```
FOURIER COMPONENTS OF TRANSIENT RESPONSE V(16)       V1A = 100 mV, Tstep = 0.1 μs
DC COMPONENT =    5.049D+00
HARMONIC   FREQUENCY    FOURIER     NORMALIZED      PHASE      NORMALIZED
  NO         (HZ)      COMPONENT    COMPONENT       (DEG)      PHASE (DEG)

   1       1.000D+05   5.453D+00    1.000000       -0.005        0.000
   2       2.000D+05   4.578D-04    0.000084      -79.883      -79.878
   3       3.000D+05   8.519D-01    0.156230       -0.028       -0.023
   4       4.000D+05   3.488D-04    0.000064      -67.455      -67.450
   5       5.000D+05   1.711D-01    0.031378       -0.131       -0.126
   6       6.000D+05   3.626D-04    0.000066      -60.112      -60.107
   7       7.000D+05   3.396D-02    0.006228       -0.609       -0.604
   8       8.000D+05   3.754D-04    0.000069      -53.168      -53.164
   9       9.000D+05   6.618D-03    0.001214       -2.802       -2.797


      TOTAL HARMONIC DISTORTION =      15.947613   PERCENT
```

**Fig. 2.12.** Fourier components of V(16).

## 2.3 Large-Signal Circuit Analysis of the ECP

If the ECP is designed properly and if the input voltage is small, the transistors do not enter saturation during an input cycle. For the idealized stage, the equivalent circuit reduces to that shown in Figure 2.13. In turn, the equilibrium equations of the circuit can be solved to provide a closed-form solution for the voltage transfer characteristic.

The necessary condition to avoid saturation of the transistors is, for $V_2 = 0$,

$$V_1 - (V_{CC} - I_C R_C) < V_{BCon} \approx 0.8 \text{ V} \qquad (2.4)$$

The value 0.8 V is an appropriate zero-order model for currents of the order of mA. If $V_1$ is constrained to be less than, say, 1 V, the required relation is approximately

$$V_{CC} - I_C R_C \approx 0 \qquad (2.5)$$

For the values of $I_{EE}, R_{C1}$, and $R_{C2}$ of Figure 2.9, no saturation will occur. The following equations describe the circuit.

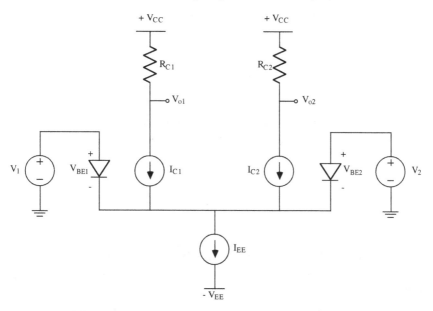

**Fig. 2.13.** Large-signal equivalent circuit of EC pair.

$$V_1 = V_{BE1} - V_{BE2} \tag{2.6}$$

The relation between the collector current and its $V_{BE}$ is next used

$$I_{C1} = I_S \exp\left(\frac{V_{BE1}}{V_t}\right) \tag{2.7}$$

$$I_{C2} = I_S \exp\left(\frac{V_{BE2}}{V_t}\right) \tag{2.8}$$

$$V_1 = V_t \ln\left(\frac{I_{C1}}{I_{C2}}\right) \tag{2.9}$$

$$I_{C1} = I_{C2} \exp\left(\frac{V_1}{V_t}\right) \tag{2.10}$$

If the base currents are negligible, the expressions for the two collector current outputs as a function of the differential input voltage are

$$I_{C1} + I_{C2} \approx I_{EE} \tag{2.11}$$

$$I_{C1} = I_{EE} - I_{C2} = \frac{I_{EE}}{1 + \exp\left(-\frac{V_1}{V_t}\right)} \tag{2.12}$$

$$= \frac{I_{EE}}{1 + \exp(-d)}$$

$$= \frac{I_{EE}}{2}\left[1 + \tanh\left(\frac{d}{2}\right)\right]$$

$$I_{C2} = I_{EE} - I_{C1} = \frac{I_{EE}}{1 + \exp\left(\frac{V_1}{V_t}\right)} \tag{2.13}$$

$$= \frac{I_{EE}}{1 + \exp(d)}$$

$$= \frac{I_{EE}}{2}\left[1 - \tanh\left(\frac{d}{2}\right)\right]$$

where $d = V_1/V_t$. The third expressions of (2.12) and (2.13) are obtained by adding and subtracting $\frac{1}{2}$ and by multiplying and dividing by $\exp(\frac{V_1}{2V_t}) = \exp(\frac{d}{2})$ or $\exp(\frac{-V_1}{2V_t}) = \exp(-\frac{d}{2})$. Plots of these expressions are given in Figure 2.14. The output voltages can be obtained from $V_o = V_{CC} - I_C R_C$. Note in the figure that the input voltage is scaled with respect to $V_t$.

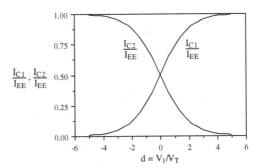

**Fig. 2.14.** $I_{C1}$ and $I_{C2}$ as a function of $d$.

As expected, the nature of the plots is the same as that produced by the Spice simulation of the last section. Spice has accomplished the same thing using numerical methods. The Ebers-Moll model in Spice is the same as those of Figure 2.13 when no transistor enters saturation. In Figure 2.14, the slopes of the curves at $V_1 = 0$ V provide the small-signal gains (transconductances) of the circuit.

Another output of interest is the differential output voltage. For $R_{C1} = R_{C2} = R_C$,

$$V_{o1} - V_{o2} = (V_{CC} - I_{C1}R_C) - (V_{CC} - I_{C2}R_C) \tag{2.14}$$

$$= -R_C(I_{C1} - I_{C2})$$

$$= -I_{EE}R_C \left[ \frac{1}{1 + \exp(-d)} - \frac{1}{1 + \exp(d)} \right]$$

where again $d = V_1/V_t$. Using alternate forms of (2.12) and (2.13), we obtain

$$V_{o1} - V_{o2} = -I_{EE}R_C \tanh\left(\frac{d}{2}\right) \tag{2.15}$$

The plot of this function obtained from Spice2 is shown in Figure 2.15. Also shown are plots of the two single-ended output voltages with a different scaling. The shapes of the differential and the single-ended outputs are the same, as expected.

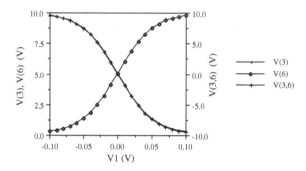

**Fig. 2.15.** V(3), V(6) and V(3, 6) as a function of V1.

## 2.4 Series Expansions to Obtain Distortion Components

For small values of $V_1$ with respect to $V_t$, the exponential functions of (2.12) and (2.13), or the tanh functions of (2.12), (2.13) and (2.15), can be expanded in a power series, and only the first few terms will be significant. For $I_{C2}$,

$$\frac{I_{C2}}{I_{EE}} = \frac{1}{1 + \exp(d)} = \frac{1}{2} - \frac{1}{4}d + \frac{1}{48}d^3 - \frac{1}{480}d^5 + \ldots \tag{2.16}$$

We now introduce a sinusoidal input and include it in the power series. For a dc value of $V_1 = 0$,

$$v_1 = V_{1A} \cos \omega_1 t \tag{2.17}$$

where $V_{1A}$ is the zero-to-peak amplitude of the input sinusoidal tone, $\omega_1$. The powers of the sinusoidal terms using (2.17) in (2.16) can be converted to harmonic terms using trigonometric identities, e.g., $\cos^3 x = \frac{1}{4}\cos 3x + \frac{3}{4}\cos x$ and $\cos^5 x = \frac{1}{16}\cos 5x + \frac{5}{16}\cos 3x + \frac{5}{8}\cos x$. The result is

$$\frac{I_{C2}}{I_{EE}} = \frac{1}{2} - \frac{1}{4}\frac{V_{1A}}{V_t}\cos\omega_1 t + \frac{1}{48}\left(\frac{V_{1A}}{V_t}\cos\omega_1 t\right)^3 + \ldots \tag{2.18}$$

$$= \frac{1}{2} - \frac{1}{4}\left[\frac{V_{1A}}{V_t} - \frac{3}{48}\left(\frac{V_{1A}}{V_t}\right)^3\right]\cos\omega_1 t \tag{2.19}$$

$$+ \frac{1}{4\times 48}\left(\frac{V_{1A}}{V_t}\right)^3\cos 3\omega_1 t + \ldots$$

Equation (2.19) can be expressed as

$$\frac{I_{C2}}{I_{EE}} = b_0' + b_1\cos\omega_1 t + b_2\cos 2\omega_1 t + b_3\cos 3\omega_1 t + \ldots \tag{2.20}$$

where the amplitudes of the Fourier coefficients are

$$b_0' = \frac{1}{2}$$

$$b_1 = -\frac{1}{4}\left[\frac{V_{1A}}{V_t} - \frac{1}{16}\left(\frac{V_{1A}}{V_t}\right)^3\right]$$

$$b_2 = 0$$

$$b_3 = \frac{1}{192}\left(\frac{V_{1A}}{V_t}\right)^3$$

The use of the notation $b_0'$, rather than $b_0$, is explained in the next chapter. Notice that a compression term appears in the fundamental due to the cubic term. The distortion factors are

$$HD_2 = \frac{|b_2|}{|b_1|} = 0 \tag{2.21}$$

$$HD_3 = \frac{|b_3|}{|b_1|} \approx \frac{1}{48}\left(\frac{V_{1A}}{V_t}\right)^2 \tag{2.22}$$

$$THD = \frac{\sqrt{b_2^2 + b_3^2 + \ldots}}{|b_1|} \tag{2.23}$$

The compression terms in $b_1$, are neglected for simplicity. Again, more attention is given to this aspect in later chapters.

From the previous example where $V_{1A} = 10$ mV, the value of $HD_2$ is zero and $HD_3 = 0.315\%$ from (2.22). This is close to the value determined from

Spice2, 0.295%, using RELTOL $= 1 \times 10^{-6}$. Notice that $HD_3$ goes up as the square of $V_{1A}$, the input amplitude, although the compression term in $b_1$ ultimately comes into the scene. From the developments above, all even-order harmonics should be zero. In a Spice output, these even-order harmonics are small, but not zero. The nonzero values are due to 'numerical noise' in the Fourier calculations.

The same results for $HD_3$ are obtained for the other two output possibilities, $V_{o1}$ and $V_o = V_{o1} - V_{o2}$. The only change occurs in the sign (phase) of the components of the Fourier expansion of the outputs.

For a large sinusoidal input, Spice simulations must be used to obtain values of the harmonics, or a special computer program can be written to produce the results. The results from the latter investigation are plotted as a family of curves as shown in Figure 2.16. Again an input voltage normalized with respect to $V_t$ is used, and the plots of the Fourier coefficients, $b_i$, are made in terms of a normalized collector current, $I_C/I_{EE}$. It is to be emphasized that these curves apply only to the cases where the transistors never saturate and where the effects of internal ohmic resistances and basewidth modulation in the transistors are negligible.

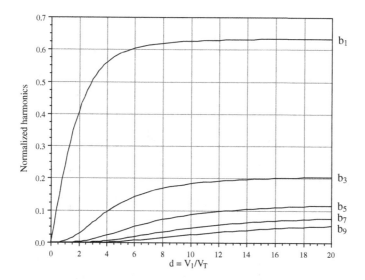

**Fig. 2.16.** Normalized harmonics of the collector current in ECP.

As an example in the use of the curves, assume an input of $d = V_1/V_t = 4.0$ ($V_{1A} \approx 100$ mV). A truncated series expression of (2.16) is not appropriate for this large drive. From the curves of Figure 2.16, using $d = 4$, the normalized value of the fundamental component of collector current is $b_1' = 0.56$. The normalized value of the third harmonic is $b_3' = 0.09$. The value of $HD_3$ is thus 16.1%, which compares well with the value of 15.6% from the Spice2 data of

Figure 2.12. A closed-form expression for the total harmonic distortion factor THD is not available.

In Section 11.1, the small- and large-signal aspects of the ECP with an input offset voltage, $V_1 = 0$ and $V_2 > 0$ are introduced in conjunction with the evaluation and design of a near-sinusoidal oscillator.

## 2.5 The Source-Coupled Pair

The source-coupled pair (SCP) is the basic gain stage for MOS analog ICs. A typical stage including representative biasing elements is shown in Figure 2.17. As with the ECP, it is helpful in establishing the small- and large-signal performance of the SCP to simplify the stage and retain only the essential components. Such an idealized stage is shown in Figure 2.18. As expected, the same bias and load elements are present as for the bipolar case. The commonality of bipolar and MOS circuit configurations is achieved when the different devices both have the same polarity and are enhancement-type devices.

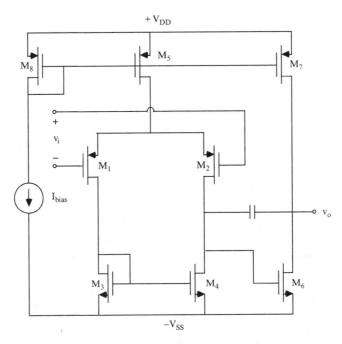

**Fig. 2.17.** A typical MOS source-coupled pair stage.

In this section, the basic electrical performance characteristics of the SC pair are developed. The same procedure is followed that is used in studying the

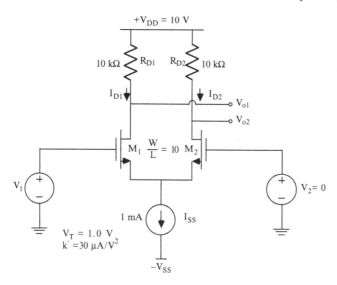

**Fig. 2.18.** An idealized SC pair circuit.

ECP in the earlier sections of this chapter. First, the bias state and the small-signal properties of the SCP about a typical operating point are reviewed. The large-signal properties are then developed. In both cases only the low-frequency behavior is examined. The effects of charge storage on the time and frequency responses can be examined later using circuit simulation programs.

For the SCP of Figure 2.18, values for two essential MOS device parameters $(V_T, k')$ are given in the figure and are described below. The width-to-length ratios $(W/L)$ of the devices are also given. The signal input is taken to be a single voltage source on the left. For now it is assumed to have a static (dc bias) value of 0 V. Consequently, $V_1$ can be considered to be a differential input signal with a zero common-mode component. If the transistors are identical, the common-source current source, $I_{SS} = 1$ mA, draws from each MOS source a current of 0.5 mA. Since the gate currents are zero, the drain currents are also 0.5 mA. For the values of $R_{D1} = R_{D2} = 10$ k$\Omega$ and $V_{DD} = 10$ V, the dc values of the drain voltages are $V_{DD} - IR_D = 10 - (0.5 \text{ mA})(10 \text{ k}\Omega) = 5$ V. The voltage at the common sources is $-V_{GS}$. For the quiescent, no input signal case, the gate-to-source bias voltage is denoted $V_{GG}$. Therefore, the quiescent value of the common-source voltage is $-V_{GG}$.

Next an incremental (small-signal) value of $v_1$ is added to $V_1$. To establish the variational response of the stage at the given operating point it is helpful to assume that operation is restricted to the saturated region of operation. The small-signal circuit model for a MOS transistor, as used in Figure 2.19a, consists only of a dependent-current generator $g_m v_a$, where $g_m$ is given by

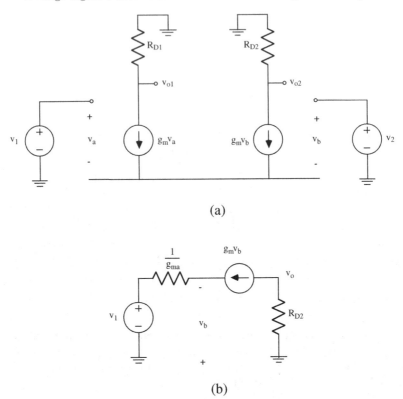

(a)

(b)

**Fig. 2.19.** (a) Small-signal equivalent circuit of SC pair. (b) Modified small-signal circuit for determining the incremental voltage gain.

$$g_m = \sqrt{2I_D k' \frac{W}{L}} \tag{2.24}$$

The calculated value of $g_m$ is 0.55 mA/V for the device and circuit values of Figure 2.18, for $V_1 = V_2 = 0$ and $\frac{W}{L} = 10$.

As usual, in Figure 2.19a the independent voltage sources are short circuits for the incremental case while the current source is an open circuit.

The input resistance $R_{in}$ of the SCP is infinite. The output resistance with respect to the output node $v_{o2}$, looking back into the stage across $R_{D2}$, is simply $R_{D2}$ since the device output resistance, $r_o$, is infinite because channel-length modulation has been neglected. (The effects of $r_o$ are taken up in Chapter 5.)

The incremental voltage gain from $v_1$ to $v_{o2}$ is easily determined from a modification of the small-signal circuit, as shown in Figure 2.19b. On the left, a Thevenin equivalent for the common-drain transistor is used. (In effect, the left-hand transistor is a source follower driving the right-hand transistor).

$$a_v = \frac{v_{o2}}{v_1} = 0.5 g_m R_{D2} \tag{2.25}$$

(For the SCP itself, as for the ECP, the transconductance is $G_m = \frac{g_m}{2}$.)

For the values of $g_m$ and $R_D$ of Figure 2.18, $a_v = +2.7$. This is a modest value relative to the earlier ECP example. The small value is due to the much smaller value of $g_m$ for the MOS device relative to a BJT for the same current level.

The output resistance with respect to the output node $v_{o1}$ looking back at $R_{D1}$ is again $R_{D1}$. The small-signal voltage gain from $v_1$ to $v_{o1}$ is $-2.7$.

Return now to the dc input voltage $V_1$ and change its value. The output voltage variation with $V_1$ can be determined by actual measurements, using circuit simulation, or for a special restriction by circuit analysis. Although setting up the equilibrium circuit equations is straightforward, the solution of piece-wise, nonlinear equations to obtain values, say, of $V_{o2}$ for a set of values of $V_1$ is tedious to say the least. Fortunately, for the special case where the MOS devices are restricted to operation in the saturation region, a closed-form solution of the circuit equations can be found. This development is taken up shortly.

For the general case, computer programs must be used to solve (numerically) the circuit and device equations, i.e., circuit simulation is used. The Spice input file for the SCP of Figure 2.18 is shown in Figure 2.20. Several input and output possibilities are made available with comment lines. The node voltages of the dc (quiescent) operating state are also shown in Figure 2.20. The output voltage plot of $V_{o2}$ versus $V_1$, which is a voltage-transfer characteristic, with an input voltage range of $\pm 5$ V is shown in Figure 2.21. Notice that the level of the input voltage to achieve clamping of the output voltage is $\pm 2$ V, much greater than the $\pm 0.1$ V for the ECP. The small-signal voltage gain of the stage is the slope of the transfer characteristic at the quiescent operating point, $a_v = 2.71$, approximately the same as the calculated value for the small-signal analysis above. The small-signal characteristics of this SCP, obtained from the .TF command in Spice, are also given in Figure 2.20.

For bipolar devices, a single equation can be used for both the off and normal active regions. This is not case for MOS devices. None-the-less, a closed-form solution for the transfer characteristic of the SCP can be developed if operation is restricted to the saturation region of operation. As is common for nonlinear situations, the ease of the solution of nonlinear simultaneous equations depends upon the sequence of operations. The following equations, together with the drain current equation for the saturated region, describe the circuit.

$$V_1 = V_{GS1} - V_{GS2} \tag{2.26}$$

$$I_{D1} + I_{D2} = I_{SS} \tag{2.27}$$

```
SOURCE-COUPLED PAIR, FIG 2.20
V1 1 0 0 SIN 0 0.2 100K
.DC V1 -5 5 0.5
.PLOT DC V(6) V(16)
.TF V(6) V1
*.TRAN 0.1U 20U
*.PLOT TRAN V(6) V(16)
*.FOUR 100K V(6) V(16)
M1 3 1 4 8 MOD1 W=100U L=10U
M2 6 0 4 8 MOD1 W=100U L=10U
.MODEL MOD1 NMOS VTO=1.0 KP=30U
RD1 5 3 10K
RD2 5 6 10K
VDD 5 0 10
ISS 4 8 1M
VSS 8 0 -10
E1 12 0 1 0 10
M3 13 12 14 8 MOD1 W=100U L=10U
M4 16 0 14 8 MOD1 W=100U L=10U
RD3 5 13 10K
RD4 5 16 10K
ISS2 14 8 1M
.OPTIONS RELTOL=1E-6
.OPTIONS NOPAGE NOMOD
.WIDTH OUT=80
.END
```

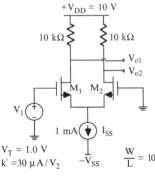

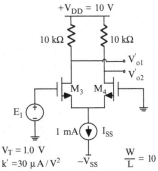

| NODE | VOLTAGE | NODE | VOLTAGE | NODE | VOLTAGE | NODE | VOLTAGE |
|------|---------|------|---------|------|---------|------|---------|
| ( 1) | 0.0000 | ( 3) | 5.0000 | ( 4) | -2.8257 | ( 5) | 10.0000 |
| ( 6) | 5.0000 | ( 8) | -10.0000 | ( 12) | 0.0000 | ( 13) | 5.0000 |
| ( 14) | -2.8257 | ( 16) | 5.0000 | | | | |

```
****     SMALL-SIGNAL CHARACTERISTICS

0       V(6)/V1                              =   2.739D+00
0       INPUT RESISTANCE AT V1               =   1.000D+20
0       OUTPUT RESISTANCE AT V(6)            =   1.000D+04
```

**Fig. 2.20.** Circuit and Spice input file for SC pair simulations. The dc operating point and small-signal transfer characteristics are also shown

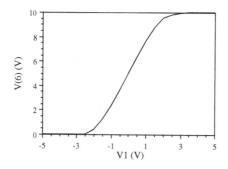

**Fig. 2.21.** Dc characteristics of SC pair.

Equation (1.12) is first solved for the gate-to-source voltages.

$$V_{GS1} = V_T + \sqrt{\frac{I_{D1}}{\frac{k'}{2}\frac{W}{L}}} \tag{2.28}$$

$$V_{GS2} = V_T + \sqrt{\frac{I_{D2}}{\frac{k'}{2}\frac{W}{L}}}$$

The (differential) input voltage is from (2.26):

$$V_1 = V_{GS1} - V_{GS2} \tag{2.29}$$

$$= \sqrt{\frac{I_{D1}}{K}} - \sqrt{\frac{I_{D2}}{K}}$$

where a constant, $K = \frac{k'}{2}\frac{W}{L}$, is introduced for convenience. Equation (2.27) is introduced by noting that each drain current can be considered having a dc component and an incremental component.

$$I_{D1} = I_{DA} + i_{d1} = I_{DA} + i_d \tag{2.30}$$

$$I_{D2} = I_{DA} + i_{d2} = I_{DA} - i_d$$

As shown the incremental components of the two drain currents, $i_{d1}$ and $i_{d2}$, are equal and opposite because of (2.27).

$$I_{SS} = I_{D1} + I_{D2} = I_{DA} + i_d + I_{DA} - i_d \tag{2.31}$$

$$= 2I_{DA}$$

$$I_{DA} = \frac{1}{2}I_{SS}$$

Using these relations in (2.29), we obtain after a bit of manipulation,

$$V_1^2 = \frac{I_{SS}}{K} \left\{ 1 - \sqrt{\left(1 + \frac{i_d}{I_{DA}}\right)\left(1 - \frac{i_d}{I_{DA}}\right)} \right\} \tag{2.32}$$

$$i_d = \sqrt{\frac{KI_{SS}}{2}} V_1 \sqrt{1 - \frac{KV_1^2}{2I_{SS}}} \tag{2.33}$$

$$\frac{i_d}{I_{SS}} = \frac{d}{2}\sqrt{1 - \left(\frac{d}{2}\right)^2} \tag{2.34}$$

where $d$ is the normalized input voltage.

$$d = \frac{V_1}{V_{GG} - V_T} \tag{2.35}$$

$$= \sqrt{\frac{2}{\frac{I_{SS}}{K}}} V_1$$

As noted earlier, $V_{GG}$ is the quiescent value of the gate-to-source voltage and is defined by the following:

$$I_{DA} = K\left(V_{GG} - V_T\right)^2 \tag{2.36}$$

For the example of Figures 2.18 and 2.20, $V_{GG} = V(4) = +2.83$.

The value of $(V_{GG} - V_T)$ can be considered to be the net drive above the threshold voltage. Expression (2.34) is plotted in Figure 2.22 and has the expected shape with respect to the voltage transfer characteristics obtained from circuit simulation. The total drain currents are obtained by adding or subtracting $i_d$ from $I_{DA} = \frac{I_{SS}}{2}$. The output voltages can be obtained from $V_o = V_{DD} - I_D R_D$.

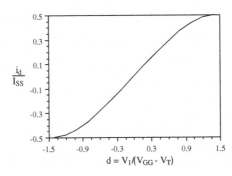

**Fig. 2.22.** $i_d$ as a function of $d$ (Equation (2.34)).

In Figures 2.23 and 2.24, output voltage waveforms and harmonic outputs from Spice simulations are given for two input amplitudes, 0.2 V and 2 V, both with a quiescent input (no incremental input signal) of 0 V. As with the ECP, the transfer characteristic of the SCP is antisymmetric about the operating point; therefore, even harmonics are not present in the output even for heavy overdrive (clipping of the output waveform). For $V_{1A} = 2$ V, the negative-going excursion of $V(6)$ reaches the resistance region of operation, $V_{DS} = 0.44 < V_{GS} - V_T = V_{GG} - V_T = 2.83 - 1.0 = 1.83$. For the sinusoidal input level of 0.2 V, $HD_3 = 0.035\%$ and for the 2 V input, $HD_3 = 4.8\%$.

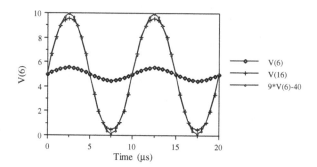

**Fig. 2.23.** Transient response of SC pair.

## 2.6 A Series Expansion to Obtain Distortion Components for the SCP

For small values of $V_1$ with respect to $V_{GG} - V_T$, the square-root function of (2.34) can be expanded in a power series and only the first few terms are significant.

$$f(x) = \sqrt{1-x} \tag{2.37}$$
$$= 1 - \frac{1}{2}x - \frac{1}{8}x^2 - \frac{1}{16}x^3 - \cdots$$

Using this expansion with (2.34), we obtain

$$\frac{i_d}{I_{SS}} = \frac{d}{2}\left[1 - \frac{1}{2}\left(\frac{d}{2}\right)^2 - \cdots\right] \tag{2.38}$$

We next introduce a sinusoidal input voltage and include it in the power series. For a dc value of $V_1 = 0$,

```
FOURIER COMPONENTS OF TRANSIENT RESPONSE V(6)      V₁A = 0.2 V, Tstep = 0.1 μs
DC COMPONENT =    5.000D+00
HARMONIC   FREQUENCY     FOURIER      NORMALIZED     PHASE      NORMALIZED
   NO        (HZ)       COMPONENT     COMPONENT      (DEG)      PHASE (DEG)

    1      1.000D+05    5.442D-01     1.000000       0.000       0.000
    2      2.000D+05    1.216D-05     0.000022      82.459      82.460
    3      3.000D+05    1.925D-04     0.000354      -0.702      -0.702
    4      4.000D+05    4.882D-06     0.000009     -85.351     -85.351
    5      5.000D+05    3.879D-06     0.000007     -56.312     -56.312
    6      6.000D+05    2.566D-06     0.000005     -52.444     -52.444
    7      7.000D+05    2.541D-06     0.000005     -68.088     -68.087
    8      8.000D+05    3.129D-06     0.000006     -64.488     -64.488
    9      9.000D+05    3.236D-06     0.000006     -52.681     -52.680

      TOTAL HARMONIC DISTORTION =       0.035480    PERCENT


FOURIER COMPONENTS OF TRANSIENT RESPONSE V(16)     V₁A = 2.0 V, Tstep = 0.1 μs
DC COMPONENT =    4.994D+00
HARMONIC   FREQUENCY     FOURIER      NORMALIZED     PHASE      NORMALIZED
   NO        (HZ)       COMPONENT     COMPONENT      (DEG)      PHASE (DEG)

    1      1.000D+05    4.779D+00     1.000000       0.000       0.000
    2      2.000D+05    1.145D-02     0.002397      89.602      89.602
    3      3.000D+05    2.279D-01     0.047692      -0.031      -0.031
    4      4.000D+05    8.565D-03     0.001792     -90.372     -90.371
    5      5.000D+05    1.128D-02     0.002361    -180.000    -180.000
    6      6.000D+05    4.908D-03     0.001027      89.405      89.405
    7      7.000D+05    3.492D-03     0.000731      -0.659      -0.658
    8      8.000D+05    1.905D-03     0.000399     -87.883     -87.882
    9      9.000D+05    7.303D-04     0.000153    -171.094    -171.093

      TOTAL HARMONIC DISTORTION =       4.786271    PERCENT
```

**Fig. 2.24.** Fourier components of the output voltage for two input amplitudes. V(6) for an input of 0.2 V and V(16) for an input of 2 V.

$$v_1 = V_{1A} \cos \omega_1 t \tag{2.39}$$

where $V_{1A}$ is the zero-to-peak value of the input sinusoidal tone. The powers of the sinusoidal terms using (2.39) in (2.38) can be converted to harmonic terms using trigonometric identities. The result, adding in the dc component of $I_{D2}$, is:

$$
\begin{aligned}
I_{D2} &= I_{DA} - i_d \\
&= I_{SS} \left[ \frac{1}{2} + \frac{1}{2} \frac{V_{1A}}{V_{GG} - V_T} \cos \omega_1 t \right. \\
&\qquad \left. - \frac{1}{64} \left( \frac{V_{1A}}{V_{GG} - V_T} \right)^3 \cos 3\omega_1 t - \ldots \right]
\end{aligned}
\tag{2.40}
$$

where compression and expansion terms from higher-order terms are neglected.

The above expression is of the form

$$I_{D2} = b'_o + b_1 \cos \omega_1 t + b_2 \cos 2\omega_1 t + b_3 \cos 3\omega_1 t + \ldots \tag{2.41}$$

The $b_i$ are the coefficients of the harmonic terms. As expected, the even harmonics in (2.40) are not present. The distortion factors are:

$$HD_2 = \frac{|b_2|}{|b_1|} = 0 \tag{2.42}$$

$$HD_3 = \frac{|b_3|}{|b_1|} = \frac{1}{32}\left[\frac{V_{1A}}{(V_{GG} - V_T)}\right]^2 \tag{2.43}$$

$$= \frac{1}{16}\frac{K}{I_{SS}}V_{1A}^2$$

In the previous Spice example where $V_{1A} = 0.2$ V, $V_T = 1$ V, $K = 0.15$ mA/V$^2$ and $I_{SS} = 1$ mA, the value of $HD_2$ is zero and $HD_3 = 0.0375\%$ which compares well with the Spice result. For $V_{1A} = 2$ V, the Spice result is $HD_3 = 4.8\%$ while the predicted value from (2.43) is 3.75%. As noted earlier, this level of input drives the transistors into the resistance region of operation and the analysis above is suitable only for a rough approximation.

## Problems

**2.1.** Two emitter-coupled pairs are shown in Figure 2.25.

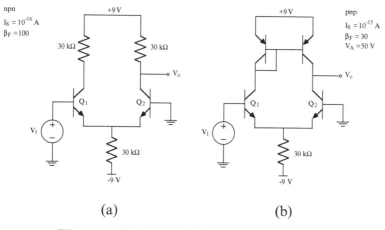

**Fig. 2.25.** Emitter-coupled pairs for Problem 2.1.

(a) For the resistive load circuit of Figure 2.25a, determine the bias state of the circuit for $V_1 = 0$ V.
(b) Repeat (a) for the circuit with an active load as in Figure 2.25b. Note

that $V_A = 50$ V for the pnp units.

(c) Determine the small-signal performance of the circuit of (a) about the quiescent bias point.

(d) Estimate the small-signal gain of the active-load circuit.

(e) Verify your results with appropriate Spice runs.

(f) Estimate the harmonic distortion of the circuit of Figure 2.25a with an input sinusoidal amplitude of 50 mV.

(g) Check your distortion estimate with a Spice run.

(h) Use Spice to determine the distortion for the circuit of Figure 2.25b with a 1.5 mV sinusoidal input.

(i) Use Spice to investigate the distortion generation of the active-load circuit if the parameter $V_A = 100$ V is added for the npn transistors. For the input, use an offset voltage of 5 mV and a sinusoidal amplitude of 1.5 mV.

**2.2.** An emitter-coupled pair is shown in Figure 2.26.

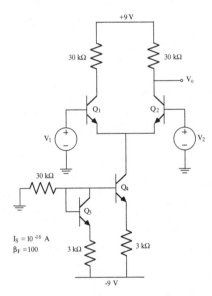

**Fig. 2.26.** Emitter-coupled pair for Problem 2.2.

(a) For $V_1 = V_2 = 0$ V, determine the quiescent bias state of the circuit.

(b) If the signal input is $V_1$ and the output voltage is $V_o$ as shown, what are the input and output resistances and the small-signal voltage gain of the circuit?

(c) Verify your estimates of (a) and (b) with Spice.

(d) Plot the (large-signal) voltage-transfer characteristic of the circuit. Check the value of gain estimated in (b) both from the above plot and from a .TF output of Spice.

(e) Estimate the harmonic distortion of the circuit with a sinusoidal input amplitude of 60 mV. Verify your estimate with Spice.
(f) Repeat Parts (a) through (e) with $V_2 = 0.03$ V.

**2.3.** An emitter-coupled pair is shown in Figure 2.27.

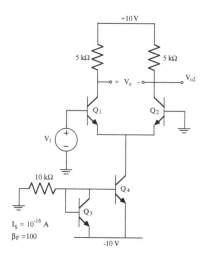

**Fig. 2.27.** Emitter-coupled pair for Problem 2.3.

(a) Determine the bias state of the circuit for $V_1 = 0$ V.
(b) Calculate the small-signal gain from the input $V_1$ to the output $V_{o2}$.
(c) For $V_1 = 20$ mV $\cos(2\pi 10^4 t)$, estimate the large-signal output response for both $V_{o2}(t)$ and $V_o(t)$.
(d) Estimate the harmonic distortion of $V_{o2}(t)$ and $V_o(t)$ using the series expansion method and graphical data (Figure 2.16) for an input amplitude of 20 mV.
(e) Verify Parts (a) through (d) with Spice.

**2.4.** A MOS source-coupled pair is shown in Figure 2.28.
(a) Determine the voltage transfer characteristic of the stage.
(b) Estimate the small-signal voltage gain and the level of HD3 for an input amplitude of 1 V.
(c) Verify the results of (b) with Spice.

**2.5.** For the MOS source-coupled pair shown in Figure 2.29,
(a) Determine the bias state of the circuit for $V_i = 0$ V.
(b) Estimate the small-signal voltage gain and the level of HD3 for an input amplitude of 1 V.
(c) Verify the results of (a) and (b) with Spice.

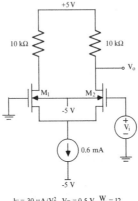

$k' = 30\ \mu A/V^2,\ V_T = 0.5\ V,\ \frac{W}{L} = 12$

**Fig. 2.28.** MOS source-coupled pair for Problem 2.4.

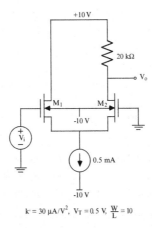

$k' = 30\ \mu A/V^2,\ V_T = 0.5\ V,\ \frac{W}{L} = 10$

**Fig. 2.29.** MOS source-coupled pair for Problem 2.5.

# 3

# Amplifier Power Series and Distortion

## 3.1 General Power Series Description

In the previous chapter, a power-series expansion of the transfer characteristics of a stage is obtained from the circuit analysis of the stage using simplified circuit models for the transistors. In order to use a truncated set of first few terms of the power series, a restriction must be made to relatively small input-signal amplitudes, e.g., for the ECP stage, $V_1/V_t \ll 1$.

In general, the transfer characteristic of an amplifier can be described by a power series. The coefficients of the series can be obtained from an analysis such as those of the earlier examples, from actual measurements of the transfer characteristics of an amplifier, followed by a polynomial approximation (see Section 3.4) or from the results of a circuit simulation, again with a polynomial characterization. In the following, we assume that the general amplifier is as illustrated in Figure 3.1, where we use for convenience a voltage-source input, $V_i$, to a lowpass amplifier, and the output variable is taken to be $V_o$. The presence and effects of a source resistance and/or a load resistance are assumed to be incorporated into the amplifier itself. Energy-storage effects within the amplifier are neglected. Therefore, the developments and results are valid only for low frequencies in the case of a dc-coupled amplifier or for the "midband" frequency region for an ac-coupled amplifier.

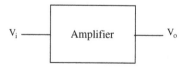

**Fig. 3.1.** A general amplifier configuration.

In general, the input and output variables include both static (dc) and time-variable components. In the absence of an incremental input signal, the

D.O. Pederson and K. Mayaram, *Analog Integrated Circuits for Communication*, DOI 10.1007/978-0-387-68030-9_3,
© 2008 Springer Science+Business Media, LLC

state of the amplifier is referred to as the quiescent situation, and the input and output variables are at their quiescent values. For the development of this section, interest is primarily in the incremental input, $v_i$, and the incremental response, $v_o$, about the quiescent operating point, $V_I$ and $V_O$, respectively. For completeness, however, we start the development with total variables. The output can be written as a function of the input as follows:

$$V_O + v_o = F(V_I + v_i) \tag{3.1}$$
$$= a_o + a_1 v_i + a_2 v_i^2 + a_3 v_i^3 + \dots$$

where $a_o = F(V_I) = V_O(V_I)$ is the quiescent output voltage which is a function of the quiescent input voltage. The incremental output is

$$v_o = a_1 v_i + a_2 v_i^2 + a_3 v_i^3 + \dots \tag{3.2}$$

The number of terms of the power series which must be retained for an adequate description of the amplifier depends upon the amplifier itself and the amplitude of the input. Whatever the case, ultimately the higher-order terms must become insignificant if a truncated series is to be used.

We now assume that the incremental input voltage is a pure sinusoidal tone,

$$v_i = V_{iA} \cos \omega_1 t \tag{3.3}$$

where the amplitude (zero-to-peak value) is $V_{iA}$ and the frequency of the tone is $f_1 = \omega_1 / 2\pi$. The procedure to establish the distortion in the output is to insert this input into the various terms and to use trigonometric identities to establish the output voltage as a sum of terms of the harmonics.

$$v_o = b_0 + b_1 \cos \omega_1 t + b_2 \cos 2\omega_1 t + \dots \tag{3.4}$$

This is a form of a Fourier expansion of $v_o$. (For the cosine input, and with no energy storage within the amplifier, only cosine terms appear in the output $v_o$. Had the input been assumed to be a sine function, the Fourier expansion of $v_o$ has alternating sines and cosines.) The values of the $b_i$ in terms of the amplifier coefficients $a_i$ and the sinusoidal input magnitude for terms through the third harmonic are

$$b_0 = \frac{a_2}{2} V_{iA}^2 + \dots \tag{3.5}$$

$$b_1 = a_1 V_{iA} + \frac{3}{4} a_3 V_{iA}^3 + \dots \approx a_1 V_{iA} + \dots$$

$$b_2 = \frac{a_2}{2} V_{iA}^2 + \dots$$

$$b_3 = \frac{a_3}{4} V_{iA}^3 + \dots$$

$$\dots$$

For each term in (3.5), only the initial term from the power-series expansion is usually included.

Notice that a dc term, $b_0$, is present even though the quiescent value, $a_o = V_O$, has been eliminated. This term represents a dynamic shift in the operating point of the amplifier due to distortion generation. Contributions to $b_0$ are produced by each even-powered term of (3.2). The expression for $b_0$ in (3.5) is due to $a_2 v_i^2$. In the developments in Section 2.5, the term $b_0'$ is the label for the complete dc value, $V_O + b_0$.

From $a_3 v_i^3$, an expansion/compression term in $b_1$, the fundamental, can be added as shown in (3.5) depending on the sign of $a_3$. If higher-order terms are included in (3.2), new components are added to each $b_i$ above, as symbolized with the notation, $+ \ldots$.

The harmonic distortion factors can again be introduced as is done in the previous chapters. If only the leading terms in (3.5) are used,

$$HD_2 = \frac{|b_2|}{|b_1|} \tag{3.6}$$

$$\approx \frac{1}{2}\frac{a_2}{a_1}V_{iA}$$

$$HD_3 = \frac{|b_3|}{|b_1|} \tag{3.7}$$

$$\approx \frac{1}{4}\frac{a_3}{a_1}V_{iA}^2$$

$$\ldots$$

A general expression for THD is usually too cumbersome for easy use.

## 3.2 Common-Emitter Stage Example

For an example in the use of the technique above, consider the simple common-emitter stage shown in Figure 3.2a. This circuit is simplified to concentrate on the essential nonlinear aspects by first identifying the bias elements and replacing them with pure sources. Finally, the 'ground point' is shifted to the emitter and a Thevenin equivalent is used at the output to obtain the 'essential' circuit configuration shown in Figure 3.2b. In the latter, it is to be noted that the input voltage is of the form

$$V_1 = V_{BB} + v_1 \tag{3.8}$$

The dc component, $V_{BB}$, is equal to 0.774 V for a collector current of 1 mA and for $I_s = 10^{-16}$ A. The signal source resistance is assumed to be very small;

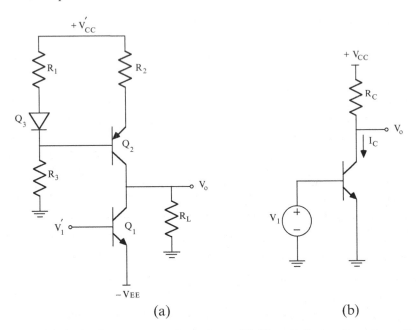

(a)                                              (b)

**Fig. 3.2.** (a) A simple common-emitter stage. (b) 'Essential' circuit configuration.

therefore, $V_{be} = V_1$. In the next chapter the effects of the source resistance on the distortion generation are considered.

A circuit model for the transistor is next introduced as shown in Figure 3.3. The transistor parameter beta is assumed constant, $\beta = \beta_{ac}$. Once again it

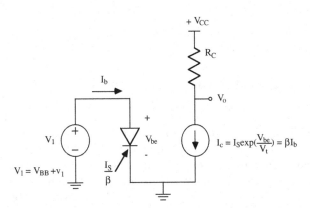

**Fig. 3.3.** Equivalent circuit of CE stage with simple device models.

is assumed that the transistor never enters saturation; thus, only a single nonlinearity (pn junction) is included. The output voltage is

$$V_o = V_{CC} - R_C I_c \qquad (3.9)$$

where

$$I_c = I_C + i_c = I_S \exp\left(\frac{V_{be}}{V_t}\right) = I_S \exp\left(\frac{V_{BB} + v_1}{V_t}\right) \qquad (3.10)$$

$$= \left[I_S \exp\left(\frac{V_{BB}}{V_t}\right)\right]\left[\exp\left(\frac{v_1}{V_t}\right)\right]$$

$$= I_{CA} \exp\left(\frac{v_1}{V_t}\right)$$

The quiescent value of the collector current, $I_C$, is labeled $I_{CA}$ for emphasis.

$$I_{CA} = I_S \exp\left(\frac{V_{BB}}{V_t}\right) \qquad (3.11)$$

A power series expansion of $\exp\left(\frac{v_1}{V_t}\right)$ is next introduced using $\exp(x) = 1 + x + \frac{1}{2}x^2 + \frac{1}{6}x^3 + \ldots$.

$$I_c = I_{CA}\left[1 + \left(\frac{v_1}{V_t}\right) + \frac{1}{2}\left(\frac{v_1}{V_t}\right)^2 + \frac{1}{6}\left(\frac{v_1}{V_t}\right)^3 + \ldots\right]$$

$$I_c = I_{CA}\left[1 + a_1' v_1 + a_2' v_1^2 + \ldots\right] \qquad (3.12)$$

Primed variables are used above, in relation to Equation (3.1), since $I_{CA}$ is a multiplier of the series expansion, $a_i = I_{CA}a_i'$. Also it is convenient to use $I_c$ as the output variable rather than the output voltage. The two variables are simply related from (3.9). The coefficients of the power series for $I_c$ are

$$a_0 = I_c(V_{BB}) = I_{CA} \qquad (3.13)$$

$$a_1 = a_1' I_{CA} = I_{CA}\left(\frac{1}{V_t}\right)$$

$$a_2 = a_2' I_{CA} = \frac{1}{2} I_{CA}\left(\frac{1}{V_t}\right)^2$$

$$a_3 = a_3' I_{CA} = \frac{1}{6} I_{CA}\left(\frac{1}{V_t}\right)^3$$

The input voltage variation is next taken to be

$$v_1 = V_{1A} \cos \omega_1 t \qquad (3.14)$$

From the results of the last section, the coefficients of the Fourier expansion of the output, including only the leading term of each, are

$$b_0 \approx \frac{I_{CA}}{4} \left( \frac{V_{1A}}{V_t} \right)^2 \tag{3.15}$$

$$b_1 \approx I_{CA} \left( \frac{V_{1A}}{V_t} \right)$$

$$b_2 \approx \frac{I_{CA}}{4} \left( \frac{V_{1A}}{V_t} \right)^2$$

$$b_3 \approx \frac{I_{CA}}{24} \left( \frac{V_{1A}}{V_t} \right)^3$$

Note that the nonzero value for $b_0$ indicates that a shift in the bias value of $V_O$ or $I_{CA}$ occurs. In the examples of the previous chapter the $b_0$ term was not present because the voltage transfer characteristic of the balanced ECP or SCP is antisymmetrical about the operating point for $V_1 = V_2$.

The approximate expressions for the distortion factors are

$$HD_2 \approx \frac{1}{4} \left( \frac{V_{1A}}{V_t} \right) \tag{3.16}$$

$$HD_3 \approx \frac{1}{24} \left( \frac{V_{1A}}{V_t} \right)^2$$

For a numerical example, let $V_{BB} = 0.774$ V. For $I_S = 1 \times 10^{-16}$ A and operation at 300° K,

$$V_t = 25.85 \text{ mV and } I_{CA} = 1.0 \text{ mA} \tag{3.17}$$

For $V_{1A} = 10$ mV, the Fourier components of the collector current are found from Equation (3.15). The corresponding amplitude values for the output voltage are obtained by multiplying by $-R_C = -5$ k$\Omega$. For the latter, and neglecting the sign change,

$$b_0 = 0.19 \text{ V} \tag{3.18}$$
$$b_1 = 1.93 \text{ V}$$
$$b_2 = 0.19 \text{ V}$$
$$b_3 = 0.012 \text{ V}$$

The estimated harmonic distortion factors are

$$HD_2 = 9.8\% \tag{3.19}$$
$$HD_3 = 0.62\%$$

The input file for this circuit for a Spice simulation is given in Figure 3.4a. For this example, $V_{CC} = 15$ V and $R_C = 5$ kΩ. The input bias voltage $V_{BB}$ is included in the $V_1$ specifications. The node voltages and the small-signal parameters for the quiescent state are given also in Figure 3.4a. The dc voltage transfer characteristic is given in Figure 3.4b. The output voltage waveform for a 10 mV sinusoidal input is shown in Figure 3.4c, and the harmonic outputs are given in Figure 3.4d. The output coefficients are

$$b_0 = 10.05 - 9.855 = 0.185 \text{ V} \tag{3.20}$$
$$b_1 = 1.942 \text{ V}$$
$$b_2 = 0.1835 \text{ V}$$
$$b_3 = 0.0114 \text{ V}$$

(In $b_0$, the actual collector current is 0.99 mA. Therefore, the estimated quiescent value of $V(3)$ is $15 - (0.99 \text{ mA})(5 \text{ kΩ}) \approx 10.05$ V.) The distortion factors from the Spice2 output are $HD_2 = 9.5\%$ and $HD_3 = 0.59\%$. The comparison of simulated and estimated values is very close.

As an exercise for the reader, the single-ended, single-device, ac-coupled, and bypassed stage of Figure 3.5 should be analyzed. In particular, the bias elements, including the average (dc) values of the coupling and bypass capacitors as voltage sources, should be combined to obtain the 'essential' stage configuration of Figure 3.2b. Again, a shift of ground point is necessary. The conclusion can be drawn that it is usually possible to reduce a practical implementation of a stage into the essential configuration.

## 3.3 The Ideal CE Stage with a Large Sinusoidal Input Voltage

If the amplitude of the input sinusoid is not small with respect to $V_t$, a truncated portion of the power series expansion of $I_c$ is not adequate. However, the function $\exp(d\cos\omega_1 t)$ has a known expansion.

$$\exp(d\cos\omega_1 t) = I_0(d) + 2I_1(d)\cos\omega_1 t + 2I_2(d)\cos 2\omega_1 t + \ldots \tag{3.21}$$
$$+ 2I_n(d)\cos n\omega_1 t + \ldots$$

where the $I_n(d)$ are modified Bessel functions of order $n$ [20]. A table of values of these functions as well as normalized plots are given in Figures 3.6a and b. If we introduce the sinusoidal input of (3.14) in the collector current expression

```
CE STAGE, FIG 3.4
V1 1 0 0.774 SIN(0.774 10M 100KHZ)
.TF V(3) V1
 .DC V1 0.6 0.9 .01
 .PLOT DC V(3)
*.TRAN 0.5U 20U
*.PLOT TRAN V(3)
*.FOUR 100K V(3)
Q1 3 1 0 MOD1
RC 5 3 5K
VCC 5 0 15
.MODEL MOD1 NPN BF=100 IS=1.0E-16
.OPTIONS NOPAGE NOMOD
.OPTIONS RELTOL=1E-6
.WIDTH OUT=80
.END
```

| NODE | VOLTAGE | NODE | VOLTAGE | NODE | VOLTAGE |
|------|---------|------|---------|------|---------|
| ( 1) | 0.7740 | ( 3) | 10.0401 | ( 5) | 15.0000 |

```
****    SMALL-SIGNAL CHARACTERISTICS

        V(3)/V1                          = -1.918D+02
        INPUT RESISTANCE AT V1           =  2.607D+03
        OUTPUT RESISTANCE AT V(3)        =  5.000D+03
```

(a)

(b)

(c)

```
FOURIER COMPONENTS OF TRANSIENT RESPONSE V(3)
DC COMPONENT =    9.853D+00
```

| HARMONIC NO | FREQUENCY (HZ) | FOURIER COMPONENT | NORMALIZED COMPONENT | PHASE (DEG) | NORMALIZED PHASE (DEG) |
|---|---|---|---|---|---|
| 1 | 1.000D+05 | 1.943D+00 | 1.000000 | 179.996 | 0.000 |
| 2 | 2.000D+05 | 1.838D-01 | 0.094586 | 90.006 | -89.990 |
| 3 | 3.000D+05 | 1.147D-02 | 0.005900 | 0.583 | -179.413 |
| 4 | 4.000D+05 | 4.154D-04 | 0.000214 | -94.482 | -274.479 |
| 5 | 5.000D+05 | 1.292D-04 | 0.000066 | 119.519 | -60.477 |
| 6 | 6.000D+05 | 1.177D-04 | 0.000061 | 114.839 | -65.157 |
| 7 | 7.000D+05 | 1.162D-04 | 0.000060 | 118.313 | -61.683 |
| 8 | 8.000D+05 | 1.161D-04 | 0.000060 | 122.005 | -57.992 |
| 9 | 9.000D+05 | 1.153D-04 | 0.000059 | 126.664 | -53.332 |

```
        TOTAL HARMONIC DISTORTION =     9.477058  PERCENT
```

(d)

**Fig. 3.4.** (a) Spice input file for CE stage, dc operating point and small-signal characteristics. (b) Dc voltage transfer characteristic of CE stage. (c) Transient output voltage waveform of CE stage. (d) Fourier components of output voltage.

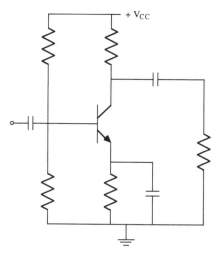

**Fig. 3.5.** A single-ended, single device, ac bypassed CE stage.

in (3.10) and use the Bessel function expansion of (3.21), the Fourier expansion of $I_c$ is obtained directly.

$$I_c = I_{CA}[I_0(d) + 2I_1(d)\cos\omega_1 t + \ldots + 2I_n(d)\cos n\omega_1 t + \ldots] \qquad (3.22)$$

$$= I_{CA}I_0(d)\left[1 + \frac{2I_1(d)}{I_0(d)}\cos\omega_1 t + \ldots + \frac{2I_n(d)}{I_0(d)}\cos n\omega_1 t + \ldots\right]$$

$$= I_{dc}\left[1 + \frac{2I_1(d)}{I_0(d)}\cos\omega_1 t + \ldots + \frac{2I_n(d)}{I_0(d)}\cos n\omega_1 t + \ldots\right]$$

where $d = \frac{V_{1A}}{V_t}$ is the normalized amplitude of the input sinusoid.

In the last expression, the dc collector current under any drive amplitude is identified.

$$I_{dc} = I_{CA}I_0(d) \qquad (3.23)$$

$$= I_S\left[\exp\left(\frac{V_{BB}}{V_t}\right)\right]I_0(d)$$

As mentioned above, all even-ordered terms of the power-series expansion of $I_c$ contribute to the dc shift of the quiescent bias point and are included in $I_0(d)$.

Expression (3.22) can be equated to the Fourier expansion of (3.4). The Fourier coefficients of the collector current are

$$I_{CA} + b_0 = I_{dc} \qquad (3.24)$$

| x | $I_0(x)$ | $I_1(x)$ | $I_2(x)$ | $I_3(x)$ | $\dfrac{I_1(x)}{I_0(x)}$ | $\dfrac{I_2(x)}{I_0(x)}$ | $\dfrac{I_3(x)}{I_0(x)}$ |
|---|---|---|---|---|---|---|---|
| 0.0 | 1.0000 | 0.0000 | 0.0000 | 0.0000 | 0.000 | 0.000 | 0.000 |
| 0.1 | 1.0025 | 0.0501 | 0.0013 | 0.0000 | 0.050 | 0.001 | 0.000 |
| 0.2 | 1.0100 | 0.1005 | 0.0050 | 0.0002 | 0.100 | 0.005 | 0.000 |
| 0.4 | 1.0404 | 0.2040 | 0.0203 | 0.0013 | 0.196 | 0.020 | 0.001 |
| 0.6 | 1.0920 | 0.3137 | 0.0464 | 0.0046 | 0.287 | 0.043 | 0.004 |
| 0.8 | 1.1665 | 0.4329 | 0.0844 | 0.0111 | 0.371 | 0.072 | 0.010 |
| 1.0 | 1.2661 | 0.5652 | 0.1357 | 0.0222 | 0.446 | 0.107 | 0.018 |
| 1.2 | 1.3937 | 0.7147 | 0.2026 | 0.0394 | 0.513 | 0.145 | 0.028 |
| 1.4 | 1.5534 | 0.8861 | 0.2875 | 0.0645 | 0.570 | 0.185 | 0.041 |
| 1.6 | 1.7500 | 1.0848 | 0.3940 | 0.0999 | 0.620 | 0.225 | 0.057 |
| 1.8 | 1.9896 | 1.3172 | 0.5260 | 0.1482 | 0.662 | 0.264 | 0.074 |
| 2.0 | 2.2796 | 1.5906 | 0.6889 | 0.2127 | 0.698 | 0.302 | 0.093 |
| 2.2 | 2.6291 | 1.9141 | 0.8891 | 0.2976 | 0.728 | 0.338 | 0.113 |
| 2.4 | 3.0493 | 2.2981 | 1.1342 | 0.4079 | 0.754 | 0.372 | 0.134 |
| 2.6 | 3.5533 | 2.7554 | 1.4337 | 0.5496 | 0.776 | 0.404 | 0.155 |
| 2.8 | 4.1573 | 3.3011 | 1.7994 | 0.7305 | 0.794 | 0.433 | 0.176 |
| 3.0 | 4.8808 | 3.9534 | 2.2452 | 0.9598 | 0.810 | 0.460 | 0.197 |
| 3.2 | 5.7472 | 4.7343 | 2.7883 | 1.2489 | 0.824 | 0.485 | 0.217 |
| 3.4 | 6.7848 | 5.6701 | 3.4495 | 1.6119 | 0.836 | 0.508 | 0.238 |
| 3.6 | 8.0277 | 6.7927 | 4.2540 | 2.0661 | 0.846 | 0.530 | 0.257 |
| 3.8 | 9.5169 | 8.1404 | 5.2325 | 2.6326 | 0.855 | 0.550 | 0.277 |
| 4.0 | 11.302 | 9.7595 | 6.4222 | 3.3373 | 0.864 | 0.568 | 0.295 |
| 4.2 | 13.442 | 11.706 | 7.8684 | 4.2120 | 0.871 | 0.585 | 0.313 |
| 4.4 | 16.010 | 14.046 | 9.6258 | 5.2955 | 0.877 | 0.601 | 0.331 |
| 4.6 | 19.093 | 16.863 | 11.761 | 6.6355 | 0.883 | 0.616 | 0.347 |
| 4.8 | 22.794 | 20.253 | 14.355 | 8.2903 | 0.889 | 0.630 | 0.364 |
| 5.0 | 27.240 | 24.336 | 17.506 | 10.331 | 0.893 | 0.643 | 0.379 |
| 5.2 | 32.584 | 29.254 | 21.332 | 12.845 | 0.898 | 0.655 | 0.394 |
| 5.4 | 39.009 | 35.182 | 25.978 | 15.939 | 0.902 | 0.666 | 0.409 |
| 5.6 | 46.738 | 42.328 | 31.620 | 19.742 | 0.906 | 0.677 | 0.422 |
| 5.8 | 56.038 | 50.946 | 38.470 | 24.415 | 0.909 | 0.687 | 0.436 |
| 6.0 | 67.234 | 61.342 | 46.787 | 30.151 | 0.912 | 0.696 | 0.448 |
| 6.2 | 80.718 | 73.886 | 56.884 | 37.187 | 0.915 | 0.705 | 0.461 |
| 6.4 | 96.962 | 89.026 | 69.141 | 45.813 | 0.918 | 0.713 | 0.472 |
| 6.6 | 116.54 | 107.30 | 84.021 | 56.383 | 0.921 | 0.721 | 0.484 |
| 6.8 | 140.14 | 129.38 | 102.08 | 69.328 | 0.923 | 0.729 | 0.495 |
| 7.0 | 168.59 | 156.04 | 124.01 | 85.175 | 0.926 | 0.736 | 0.505 |
| 7.2 | 202.92 | 188.25 | 150.63 | 104.57 | 0.928 | 0.742 | 0.515 |
| 7.4 | 244.34 | 227.17 | 182.94 | 128.29 | 0.930 | 0.749 | 0.525 |
| 7.6 | 294.33 | 274.22 | 222.17 | 157.29 | 0.932 | 0.755 | 0.534 |
| 7.8 | 354.68 | 331.10 | 269.79 | 192.75 | 0.933 | 0.761 | 0.543 |
| 8.0 | 427.56 | 399.87 | 327.60 | 236.08 | 0.935 | 0.766 | 0.552 |

Equation for higher order terms: $I_{n+1}(x) = I_{n-1}(x) - \dfrac{2n}{x} I_n(x)$

**Fig. 3.6.** (a) Values of modified Bessel functions for n = 1, 2, 3.

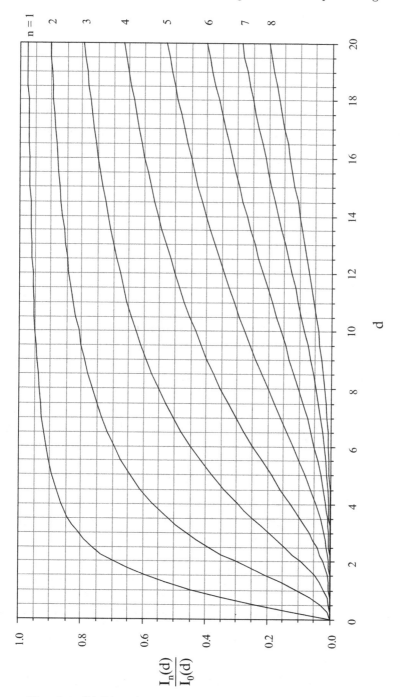

**Fig. 3.6.** (b) Plot of normalized Bessel functions for n = 1 to 8.

$$b_1 = I_{dc} \left[ \frac{2I_1(d)}{I_0(d)} \right]$$

$$b_2 = I_{dc} \left[ \frac{2I_2(d)}{I_0(d)} \right]$$

$$b_n = I_{dc} \left[ \frac{2I_n(d)}{I_0(d)} \right]$$

Note that the quiescent value $I_{CA}$ is added to the $b_0$ term to obtain $I_{dc}$. The harmonic distortion factors are

$$HD_2 = \frac{|b_2|}{|b_1|} = \frac{I_2(d)}{I_1(d)} \tag{3.25}$$

$$HD_3 = \frac{|b_3|}{|b_1|} = \frac{I_3(d)}{I_1(d)}$$

In the table of Figure 3.6a, values for the $I_n(x)(= I_n(d))$ are given for $n = 1$, 2 and 3. Values are also given for the ratios of $\frac{I_1(d)}{I_0(d)}$, $\frac{I_2(d)}{I_0(d)}$, and $\frac{I_3(d)}{I_0(d)}$. In Figure 3.6b, plots of the ratios $\frac{I_n(d)}{I_0(d)}$ are given for $n$ up to 8. The largest value of $d \approx 20$ corresponds to an input sinusoidal amplitude of approximately 0.5 V.

For a numerical example, let $d = 1$, corresponding to an input amplitude of approximately 26 mV. From Figure 3.6a, $\frac{I_1(1)}{I_0(1)} \approx 0.45$ and $\frac{I_2(1)}{I_0(1)} \approx 0.11$. Therefore, from (3.25), $HD_2 = 24\%$. From (3.16), the results from the simple power series expansion, $HD_2$ for $d = 1$ is estimated to be 25%. From the curves of Figure 3.6b, the $HD_3$ is estimated from the ratio of the third harmonic to the fundamental for $d = 1$. The result is $HD_3 \approx \frac{0.02}{0.48} = 4\%$ . The value from (3.16) is 4.1%.

The Bessel functions have simple asymptotic values. For d very small,

$$I_0(d) \mid_{d \ll 1} = 1 + 0.25d^2 \tag{3.26}$$

$$I_1(d) \mid_{d \ll 1} = 0.5d$$

$$I_2(d) \mid_{d \ll 1} = 0.12d^2$$

These approximations lead to an expression for $HD_2$ approximately equal to that of (3.16).

From the curves of Figure 3.6b, it is evident that $\frac{I_1(d)}{I_0(d)}$ approaches a value of 0.97 for large $d$.

## 3.4 Power Series and Fourier Series Characterizations

In Section 3.1, a general power-series expansion is proposed. In the example of the bipolar, common-emitter stage, the coefficients of the power series are obtained from a circuit analysis using simplified models for the transistor. In the general situation, such an approach may not be possible or appropriate. In that circumstance, a transfer characteristic may be available, say from measurements with an actual amplifier. The coefficients of the truncated power series (a polynomial) can be obtained from a solution of a set of simultaneous equations.

A transfer characteristic is illustrated in Figure 3.7a. A polynomial to approximate this curve can be established starting with a general polynomial of a desired degree. For a cubic,

$$v_o = a_1 v_i + a_2 v_i^2 + a_3 v_i^3 \tag{3.27}$$

Note that variations about the quiescent operating point are taken as the variables rather than the total input and output variables. For a third-order polynomial, the three coefficients are not known. Therefore, we choose three different values of the variational input voltage, $(v_{i1}, v_{i2}, v_{i3})$, and determine from the transfer curve the corresponding variational output voltage values $(v_{o1}, v_{o2}, v_{o3})$. These values are successively used in the polynomial.

$$v_{o1} = a_1 v_{i1} + a_2 v_{i1}^2 + a_3 v_{i1}^3 \tag{3.28}$$
$$v_{o2} = a_1 v_{i2} + a_2 v_{i2}^2 + a_3 v_{i2}^3$$
$$v_{o3} = a_1 v_{i3} + a_2 v_{i3}^2 + a_3 v_{i3}^3$$

We now have three linear equations in the three unknowns, $a_1$, $a_2$, $a_3$. The solution for the coefficients is straightforward, albeit tedious.

Another technique to obtain the values of the harmonic components of the output waveform is to accomplish a simplified form of Fourier analysis, again starting with a set of input voltage values and the corresponding output voltage values from the transfer characteristic. We do not find the $a_i$ but obtain the $b_i$ directly; and in particular, we choose a convenient set of input voltages. Again the input is assumed to be a sinusoid of zero-to-peak amplitude $V_{iA}$. Choose for the set of input voltages the maximum value $V_{iA}$ (for a cosine input this occurs at $\omega_1 t = 0$), the half maximum (at $\omega_1 t = 60°$), the zero value (at $\omega_1 t = 90°$), the half minimum (at $\omega_1 t = 120°$), and the input minimum (at $\omega_1 t = 180°$). These choices are illustrated in Figure 3.7b. From the transfer characteristic, the corresponding set of output voltages, $V_2$, $V_1$, $V_0$, $V_3$, and $V_4$ can be obtained. (These values may all be total values or incremental, variational values.) The output waveform with these values of time (really $\omega_1 t$) are plotted in Figure 3.7c.

We now propose that the output waveform be expressed as a truncated Fourier series of cosine terms only.

$$V_O + v_o = (V_O + b_0) + b_1 \cos \omega_1 t + b_2 \cos 2\omega_1 t + b_3 \cos 3\omega_1 t + \dots \tag{3.29}$$

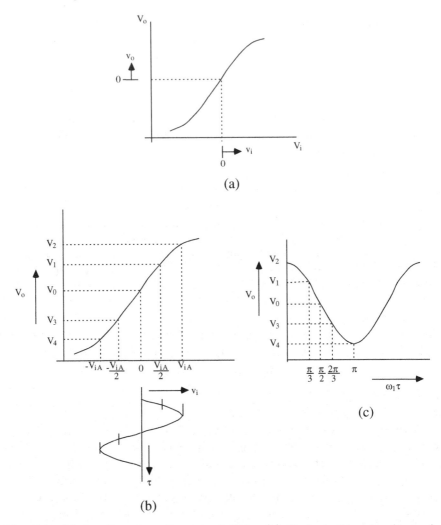

**Fig. 3.7.** (a) A voltage transfer characteristic. (b) Choices of input voltages for calculating harmonic components. (c) Output voltage waveform and the values of output voltage for the chosen time instants.

where for completeness the total output variable is included, i.e., the quiescent output voltage $V_O$ is introduced. Each value of output voltage is inserted into the equation with the corresponding value of $\omega_1 t$. The result is a set of equations in the unknown values of the Fourier coefficients.

The above choices of the input for a third-order situation lead to a five-point analysis; the five input time values described above lead to a corresponding set of output values, $V_2$, $V_1$, $V_0$, $V_3$, and $V_4$ as shown in Figure 3.7b. The results for the solution for the $b_i$ are

$$V_O + b_0 = \frac{1}{6}\left(V_2 + 2V_1 + 2V_3 + V_4\right) \tag{3.30}$$

$$b_1 = \frac{1}{3}\left(V_2 + V_1 - V_3 - V_4\right)$$

$$b_2 = \frac{1}{4}\left(V_2 - 2V_0 + V_4\right)$$

$$b_3 = \frac{1}{6}\left(V_2 - 2V_1 + 2V_3 - V_4\right)$$

$$b_4 = \frac{1}{12}\left(V_2 - 4V_1 + 6V_0 - 4V_3 + V_4\right)$$

For a numerical example, the following output voltage values have been obtained from the transfer characteristic of an output stage for the five input times.

$$V_2 = 11.09 \text{ V} \tag{3.31}$$

$$V_1 = 5.29 \text{ V}$$

$$V_0 = -0.27 \text{ V}$$

$$V_3 = -5.84 \text{ V}$$

$$V_4 = -10.37 \text{ V}$$

The Fourier coefficients are found from Equation (3.30).

$$V_O + b_0 = -0.063 \text{ V} \tag{3.32}$$

$$b_1 = 10.86 \text{ V}$$

$$b_2 = 0.315 \text{ V}$$

$$b_3 = -0.13 \text{ V}$$

$$b_4 = 0.12 \text{ V}$$

The harmonic distortion factors are

$$HD_2 = 2.9\% \tag{3.33}$$

$$HD_3 = 1.2\%$$

$$\cdots$$

$$THD \approx 3.3\%$$

The transfer characteristic used in the above example was obtained from a Spice simulation of an amplifier. A transient run which included a Fourier analysis yielded THD = 3.1%.

A simpler analysis can be obtained if only three input points are used. The three input time points are usually taken to be the $\omega_1 t$ values of 0, 90°, and 180°. The corresponding output voltage values are $V_2$, $V_0$, and $V_4$ and are the same as shown in Figure 3.7. The solution for the $b_i$ leads to

$$V_O + b_0 = \frac{1}{4}(V_2 + 2V_0 + V_4) \tag{3.34}$$

$$b_1 = \frac{1}{2}(V_2 - V_4)$$

$$b_2 = \frac{1}{4}(V_2 - 2V_0 + V_4)$$

$$HD_2 = \frac{b_2}{b_1} = \frac{1}{2}\left(\frac{V_2 - 2V_0 + V_4}{V_2 - V_4}\right) \tag{3.35}$$

From the voltage values in Equation (3.31),

$$V_O + b_0 = 0.045 \text{ V} \tag{3.36}$$

$$b_1 = 10.73 \text{ V}$$

$$b_2 = 0.315 \text{ V}$$

$$HD_2 = 2.9\% \text{ V}$$

These values compare closely with the results from the five-point analysis above except for the dc term, $V_O + b_0$.

Another example illustrates that care must be taken in using the three-point estimation. Three output voltage values obtained from the transfer characteristic of a MOS amplifier stage are

$$V_2 = 4.234 \text{ V} \tag{3.37}$$

$$V_0 = 2.5 \text{ V}$$

$$V_4 = 0.49 \text{ V}$$

The Fourier coefficients from the three-point analysis are

$$V_O + b_0 = 2.431 \text{ V} \tag{3.38}$$

$$b_0 = -0.069 \text{ V}$$

$$b_1 = 1.87 \text{ V}$$

$$b_2 = -0.07 \text{ V}$$

The estimated value of the second harmonic factor is $HD_2 = b_2/b_1 = 3.7\%$. From a Spice2 simulation of the same circuit for which the transfer characteristic is obtained, the Fourier outputs from a transient run for an input amplitude which reaches the proper extremes are

$$b_0 = -0.04 \text{ V} \tag{3.39}$$
$$b_1 = 2.016 \text{ V}$$
$$b_2 = 0.084 \text{ V}$$
$$b_3 = 0.125 \text{ V}$$
$$b_4 = 0.039 \text{ V}$$

The harmonic distortion factors are

$$HD_2 = 4.18\% \tag{3.40}$$
$$HD_3 = 6.20\%$$
$$\cdots$$
$$THD = 7.90\%$$

For this example, clearly the three-point analysis is not sufficient even if it predicts adequately $HD_2$.

## 3.5 The Common-Source MOS Amplifier Stage

For circuits involving MOS devices, it is useful to duplicate the developments of Section 3.2 for the single-stage MOS circuit. In the circuit of Figure 3.8a, the driver or 'inverter' transistor is an NMOS device, and the load element is a PMOSFET. The simplified, 'essential' circuit is shown in Figure 3.8b and simply replaces the load device with a resistor $R_D$. A simple circuit model for the driver device next is introduced which substitutes a dependent current source $I_D$. The three drain current equations for the dependent current source are given in Section 1.4.2, and the equation for operation in the saturation region is repeated below.

Assume now that operation is restricted to an operating point in the normal active region (MOS saturation) and that the amplitude of the sinusoidal input voltage is sufficiently small so that operation is restricted to the saturation region. The total input voltage is $V_1 = V_{GS} = V_{GG} + v_1$. $V_{GG}$ is the quiescent value of $V_{GS}$ and in this case is also the value of the dc bias source voltage at the gate node. The output drain current is

$$I_d = \frac{k'}{2} \frac{W}{L} (V_{GS} - V_T)^2 \tag{3.41}$$
$$I_d = \left[ \frac{k'}{2} \frac{W}{L} (V_{GG} - V_T)^2 \right] \left[ 1 + \frac{v_1}{V_{GG} - V_T} \right]^2$$

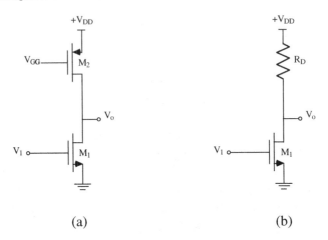

(a)                                    (b)

```
MOS CS STAGES, FIG 3.8
V1 1 0 1.5 SIN(1.5 0.05 100K)
.TF V(2) V1
.DC V1 1.1 1.8 0.025
.PLOT DC V(2) (0,5)
*.TRAN .1U 20U
*.PLOT TRAN V(2) (0,5)
*.FOUR 100K V(2)
VDD 3 0 5
VSS 5 0 -5
M1 2 1 0 5 MOD1 W=80U L=8U
RD 3 2 66.7K
.MODEL MOD1 NMOS VTO=1 KP=30U
.OPTIONS NOPAGE NOMOD RELTOL=1E-6
.WIDTH OUT=80
.END
```

(c)

```
NODE    VOLTAGE    NODE    VOLTAGE    NODE   VOLTAGE    NODE    VOLTAGE

( 1)    1.5000    ( 2)    2.4987    ( 3)   5.0000    ( 5)   -5.0000


****       SMALL-SIGNAL CHARACTERISTICS

        V(2)/V1                                  = -1.000D+01
        INPUT RESISTANCE AT V1                   =  1.000D+20
        OUTPUT RESISTANCE AT V(2)                =  6.670D+04
```

(d)

**Fig. 3.8.** (a) Common-source stage with PMOSFET load. (b) Simplified 'essential' circuit of CS stage. (c) Spice input file. (d) Dc operating point and small-signal characteristics.

The first term in brackets is the quiescent drain current which we denote $I_{DA}$. If the square of the second term is taken, the result has the form of a quadratic polynomial:

$$I_d = I_{DA}\left[1 + 2\frac{v_1}{V_{GG} - V_T} + \left(\frac{v_1}{V_{GG} - V_T}\right)^2\right] \tag{3.42}$$

$$= I_{DA}\left(a_0' + a_1'v_1 + a_2'v_1^2\right)$$

where

$$
\begin{aligned}
a_0 &= I_{DA}a_0' \tag{3.43}\\
&= I_{DA}\\
a_1 &= I_{DA}a_1'\\
&= I_{DA}\frac{2}{V_{GG} - V_T}\\
a_2 &= I_{DA}a_2'\\
&= \frac{I_{DA}}{(V_{GG} - V_T)^2}
\end{aligned}
$$

The input voltage variation is now assumed to have a sinusoidal variation

$$v_1 = V_{1A}\cos\omega_1 t \tag{3.44}$$

and this is introduced into the drain current expression to obtain the coefficients of the Fourier series description of the output current. However, we have done this earlier and can use the results directly. The coefficients of the Fourier series are

$$
\begin{aligned}
b_0 &= \frac{1}{2}\left[\frac{I_{DA}}{(V_{GG} - V_T)^2}\right]V_{1A}^2 \tag{3.45}\\
b_1 &= 2\left[\frac{I_{DA}}{(V_{GG} - V_T)}\right]V_{1A}\\
b_2 &= \frac{1}{2}\left[\frac{I_{DA}}{(V_{GG} - V_T)^2}\right]V_{1A}^2
\end{aligned}
$$

The second-harmonic distortion factor is

$$HD_2 = \frac{|b_2|}{|b_1|} = \frac{1}{4}\frac{V_{1A}}{V_{GG} - V_T} \tag{3.46}$$

There are no higher-order coefficients since the mathematical model of the transfer characteristic is a quadratic.

For a numerical example, the device parameters and circuit element values are shown in Figure 3.8c, the input file for a Spice run. For the basic circuit, the circuit values are $V_{DD} = 5$ V, $R_D = 66.7$ k$\Omega$, and $V_{GG} = 1.5$ V. The input bias voltage is included in the $V_1$ specification. The sinusoidal input amplitude is 50 mV. The estimated dc state, for $V_{GG} = 1.5$ V, is $I_D = 37.5$ $\mu$A and $V_O = V(2) = V_{DD} - I_D R_D = 2.5$ V. From the Spice data of Figure 3.8d, $V_O = 2.5$ V. Values of magnitudes of the $b_i$ for the output voltage are obtained by multiplying the expressions of Equation (3.45) by $-R_D$. These estimates, neglecting the minus signs, are

$$b_0 = 0.0125 \text{ V} \tag{3.47}$$
$$b_1 = 0.50 \text{ V}$$
$$b_2 = 0.0125 \text{ V}$$
$$HD_2 = \frac{|b_2|}{|b_1|} = 2.5\%$$

The dc transfer characteristics of the stage is presented in Figure 3.9a. The transient output waveform for $V_{1A} = 50$ mV, is shown in Figure 3.9b and the Fourier outputs are shown in Figure 3.10. For the basic stage, the Fourier coefficients and the harmonic factors are

$$b_0 = 2.486 - 2.50 = -0.014 \text{ V} \tag{3.48}$$
$$b_1 = 0.500 \text{ V}$$
$$b_2 = 0.0122 \text{ V}$$
$$HD_2 = 2.46\%$$
$$\cdots$$
$$THD = 2.46\% \tag{3.49}$$

The comparison with the estimates is close, as expected, since the Level-1 MOS model in Spice for saturated operation has the same square-law characteristic as used above.

## 3.6 Intermodulation Distortion

To this point, the variational input has been restricted to a single sinusoidal tone. The distortion produced by the nonlinearities of the circuit are harmonics of the input. Additional distortion components are produced when two or more different sinusoids are present in the input. Intermodulation ($IM$) is defined to occur when two or more sinusoidal signals are applied to a nonlinear circuit. In addition to the fundamentals of the input signals as well as their harmonics, the development which follows shows that a great number of other

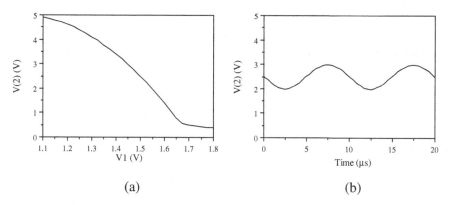

**Fig. 3.9.** (a) Dc transfer characteristics of CS stage. (b) Transient output voltage waveform.

```
FOURIER COMPONENTS OF TRANSIENT RESPONSE V(6)      Basic Circuit (T_step = 0.1 μs)
DC COMPONENT =    2.486D+00
HARMONIC    FREQUENCY      FOURIER      NORMALIZED      PHASE      NORMALIZED
  NO          (HZ)        COMPONENT     COMPONENT       (DEG)     PHASE (DEG)

   1       1.000D+05     4.976D-01     1.000000      179.999       0.000
   2       2.000D+05     1.224D-02     0.024600       90.012      -89.987
   3       3.000D+05     9.620D-06     0.000019       87.905      -92.094
   4       4.000D+05     1.164D-05     0.000023      102.248      -77.751
   5       5.000D+05     1.083D-05     0.000022      114.173      -65.826
   6       6.000D+05     9.718D-06     0.000020      116.985      -63.014
   7       7.000D+05     9.650D-06     0.000019      117.157      -62.842
   8       8.000D+05     1.003D-05     0.000020      120.667      -59.332
   9       9.000D+05     1.006D-05     0.000020      126.858      -53.141

        TOTAL HARMONIC DISTORTION =     2.460027  PERCENT


FOURIER COMPONENTS OF TRANSIENT RESPONSE V(2)      Typical Circuit (T_step = 0.1 μs)
DC COMPONENT =    2.460D+00
HARMONIC    FREQUENCY      FOURIER      NORMALIZED      PHASE      NORMALIZED
  NO          (HZ)        COMPONENT     COMPONENT       (DEG)     PHASE (DEG)

   1       1.000D+05     2.016D+00     1.000000     -179.988       0.000
   2       2.000D+05     8.432D-02     0.041819       89.726     269.714
   3       3.000D+05     1.250D-01     0.061973      179.820     359.807
   4       4.000D+05     3.853D-02     0.019111      -89.767      90.221
   5       5.000D+05     2.632D-02     0.013052        0.658     180.646
   6       6.000D+05     1.723D-02     0.008547      -91.597      88.391
   7       7.000D+05     7.343D-03     0.003642       -2.471     177.517
   8       8.000D+05     1.212D-02     0.006012       90.894     270.882
   9       9.000D+05     7.516D-04     0.000373      -22.342     157.646

        TOTAL HARMONIC DISTORTION =     7.904223  PERCENT
```

**Fig. 3.10.** Fourier components of output voltage for the basic MOS common-source stage.

spurious components appear in the output. The output frequency spectrum is illustrated in Figure 3.11, where for the plot shown, $\omega_2 > \omega_1 > \frac{2}{3}\omega_2$. The new components are the result of the original signals and their harmonics 'beating' with each other.

Let the bias state of an amplifier support appropriate active region operation of all devices. The circuit input and output variables of the amplifier are taken to be the variations about the bias state. Let the output variable be labeled $v_o$, and the input $v_i$. The transfer characteristic of the amplifier is described by a power series.

$$v_o = a_1 v_i + a_2 v_i^2 + a_3 v_i^3 + \dots \tag{3.50}$$

The variational input is assumed to consist of two sinusoids of different frequencies.

$$v_i = V_{1A} \cos \omega_1 t + V_{2A} \cos \omega_2 t \tag{3.51}$$

Inserting this expression into the output series yields

$$
\begin{aligned}
v_o = &\, a_1 \left(V_{1A} \cos \omega_1 t + V_{2A} \cos \omega_2 t\right) + \\
&\, a_2 \left(V_{1A} \cos \omega_1 t + V_{2A} \cos \omega_2 t\right)^2 + \\
&\, a_3 \left(V_{1A} \cos \omega_1 t + V_{2A} \cos \omega_2 t\right)^3 + \\
&\, \dots
\end{aligned}
\tag{3.52}
$$

For the moment, we consider only the second-order term

$$a_2 v_i^2 = a_2 \left(V_{1A} \cos \omega_1 t + V_{2A} \cos \omega_2 t\right)^2 \tag{3.53}$$

$$
\begin{aligned}
= &\, a_2 V_{1A}^2 \cos^2 \omega_1 t + a_2 V_{2A}^2 \cos^2 \omega_2 t \\
&+ 2a_2 V_{1A} V_{2A} \cos \omega_1 t \cos \omega_2 t
\end{aligned}
\tag{3.54}
$$

$$
\begin{aligned}
= &\, \frac{1}{2} a_2 \left(V_{1A}^2 + V_{2A}^2\right) + \frac{1}{2} a_2 V_{1A}^2 \cos 2\omega_1 t + \frac{1}{2} a_2 V_{2A}^2 \cos 2\omega_2 t \\
&+ a_2 V_{1A} V_{2A} \cos(\omega_1 + \omega_2)t + a_2 V_{1A} V_{2A} \cos(\omega_1 - \omega_2)t
\end{aligned}
\tag{3.55}
$$

The final result shows that this second-order term produces in the output dc (bias) shift factors, the harmonics of the two input signals, and sum and difference frequency sinusoidal terms. The last are often referred to as the 'beat' terms. These sum and difference terms are also defined as the second-order intermodulation ($IM_2$) terms. In the general case where higher-order terms are included in the power-series expression, $IM_2$ contributions are also produced by each even term of the power series. For low input levels, second-order $IM$ is produced primarily by the second-order term.

The distortion factor $IM_2$ is taken as the ratio of the amplitude of one of the beat terms to the amplitude of the fundamental output component.

$$IM_2 = \frac{a_2 V_{1A} V_{2A}}{a_1 V_{1A}} = \frac{a_2}{a_1} V_{2A} \qquad (3.56)$$

Notice that $IM_2$ is directly proportional to the signal level for low distortion values. For simplicity it is often assumed in defining $IM_2$ that the magnitudes of the two sinusoidal inputs are equal. In this case, $V_{2A}$ can be replaced by the common amplitude $V_{1A}$.

$$V_{1A} = V_{2A} \qquad (3.57)$$

Notice from a comparison of Equations (3.6) and (3.56) using (3.57) that $IM_2$ in this case is twice the value of $HD_2$ for the same amplifier.

$IM_2$ can also be expressed in terms of the output fundamental component. Let the amplitude of the fundamental component be labeled $V_{oA}$. For small distortion $V_{oA} \approx a_1 V_{1A}$. Therefore, again for equal input signal amplitudes,

$$IM_2 = \frac{a_2}{a_1^2} V_{oA} \qquad (3.58)$$

We next turn to the third-order terms in the power series of (3.50) and investigate and define third-order intermodulation distortion ($IM_3$). The third-order term in the power-series expansion is

$$
\begin{aligned}
a_3 v_i^3 &= a_3 \left( V_{1A} \cos \omega_1 t + V_{2A} \cos \omega_2 t \right)^3 \qquad (3.59) \\
&= a_3 V_{1A}^3 \cos^3 \omega_1 t + 3 a_3 V_{1A} V_{2A}^2 \cos \omega_1 t \cos^2 \omega_2 t \\
&\quad + 3 a_3 V_{1A}^2 V_{2A} \cos^2 \omega_1 t \cos \omega_2 t + a_3 V_{2A}^3 \cos^3 \omega_2 t
\end{aligned}
$$

It is clear that the output contains the third harmonics of the fundamentals as well as expansion/contraction components at the fundamental frequency of each. In addition, third-order $IM_3$ (beat) terms are produced at the sum and difference frequencies of the fundamental of each with the second harmonic of the other signal. In the following, only the cross products of (3.59) are included.

$$a_3 v_i^3 = \ldots a_{32} \cos(\omega_1 t \pm 2\omega_2 t) + a_{33} \cos(2\omega_1 t \pm \omega_2 t) \qquad (3.60)$$

where $a_{32}$ and $a_{33}$ represent the constants below.

$$a_{32} = \frac{3}{4} a_3 V_{1A} V_{2A}^2 \qquad (3.61)$$

$$a_{33} = \frac{3}{4} a_3 V_{1A}^2 V_{2A}$$

The method of specifying $IM_3$ depends upon the application. For broadband amplifiers, a common test condition is to adjust for equal output amplitudes at $\omega_1$ and $\omega_2$. For this case, $V_{1A} = V_{2A}$ if compression and/or contraction of the fundamentals are negligible. In this case, the amplitude for the third-order $IM_3$ components is

$$V_o(IM_3) = \frac{3}{4}a_3V_{1A}^3 \tag{3.62}$$

The expression for $IM_3$ in terms of the input signal level is

$$IM_3 = \frac{3}{4}\frac{a_3V_{1A}^3}{a_1V_{1A}} = \frac{3}{4}\frac{a_3}{a_1}V_{1A}^2 \tag{3.63}$$

In terms of the amplitude of the output fundamental, $V_{oA}$, and for small distortion,

$$IM_3 = \frac{3}{4}\frac{a_3}{a_1^3}V_{oA}^2 \tag{3.64}$$

Similar to $HD_3$, $IM_3$ varies as the square of the signal level for low distortion. Note that $IM_3 = 3HD_3$ from (3.7).

The distortion products produced by the nonlinearity of the circuit with two input signals are summarized in Figure 3.11 for $\omega_2 > \omega_1 > \frac{2}{3}\omega_2$. Note that the third-order $IM$ components in the output, i.e., those with frequencies of $2\omega_1 - \omega_2$ and $2\omega_2 - \omega_1$, may fall close to the fundamentals if $\omega_1$ is close to $\omega_2$. This commonly occurs with broadcast receiver applications. The $IM_3$ terms may fall in the passband of the receiver and cannot be filtered out.

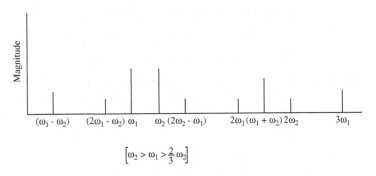

$$\left[\omega_2 > \omega_1 > \frac{2}{3}\omega_2\right]$$

**Fig. 3.11.** Output frequency spectrum.

For receivers, in contrast to broadband amplifiers, a common test condition for $IM_3$ is to apply a small desired signal with a frequency $\omega_1$ and a larger interfering signal at a suspect $\omega_2$. The amplitude of the interfering signal is increased until the $IM_3$ component reaches a predetermined fraction of the desired signal, often 1%. The human ear can usually detect the presence of the interfering signal above 1%. The amplitude, $V_{2A}$, of the interfering signal at $\omega_2$ is a measure of $IM_3$. For this case,

$$IM_3' = \frac{3}{4}\frac{a_3 V_{1A} V_{2A}^2}{a_1 V_{1A}} = \frac{3}{4}\frac{a_3}{a_1}V_{2A}^2 \tag{3.65}$$

$$IM_3'' = \frac{3}{4}\frac{a_3 V_{1A}^2 V_{2A}}{a_1 V_{1A}} = \frac{3}{4}\frac{a_3}{a_1}V_{1A}V_{2A}$$

The $IM_3'$ specification above is independent of the magnitude of the desired signal. (The largest $IM_3$ component of (3.65) is usually the one of interest.) An example of the use of this type of $IM_3$ specification is presented in Chapter 9.

The Spice .FOUR command can also be used to simulate intermodulation distortion with a proper choice of the fundamental frequency. For Fourier analysis, the signals must be periodic, implying that the tones must be commensurate.[1] Intermodulation is determined by simulating the circuit with two closely spaced tones at frequencies $f_1$ and $f_2 = f_1 + \Delta f$. For example, consider $f_1 = 50$ kHz and $f_2 = 51$ kHz. The third-order IM terms are at $2f_1 - f_2 = 49$ kHz and $2f_2 - f_1 = 52$ kHz. The slowest varying component is at $f_2 - f_1 = 1$ kHz, while another component of interest is at a much higher frequency of $2f_2 - f_1 = 52$ kHz. In this case, the ratio of the fastest (52 kHz) to slowest (1 kHz) frequencies is 52:1. Using a conventional transient analysis, at least one period of the slow 1 kHz component must be simulated. To resolve the fastest signal, a sufficient number of time points must be chosen in one period of the 52 kHz third-order IM term. The fundamental frequency for the .FOUR command is 1 kHz since all frequencies of interest are exactly divisible by this frequency. Thus, the two IM products are the 49th and 52nd harmonics of the fundamental frequency, respectively. Since Spice2 only provides the first nine harmonics of a fundamental frequency, these intermodulation terms cannot be directly calculated from Spice2 and Spice3 must be used. Simulators based on the harmonic-balance method [14], [15] are ideal for intermodulation calculations.

When energy storage effects can be ignored, the analysis frequency is not important. In this case, the intermodulation terms can be calculated from Spice2 by selecting appropriate frequencies for the two input tones. For the above example, one can choose $f_1 = 7$ kHz and $f_2 = 8$ kHz in Spice2. The two third-order IM terms are at $2f_1 - f_2 = 6$ kHz and $2f_2 - f_1 = 9$ kHz. The slowest varying component is still at $f_2 - f_1 = 1$ kHz while the highest frequency signal is at 9 kHz. With this selection of input frequencies and a fundamental frequency of 1 kHz for the .FOUR analysis, the third-order intermodulation terms are the 7th and 9th harmonics of the fundamental frequency.

---

[1] All frequencies must be exactly divisible by a single common frequency.

When three or more input signals are present at the input of an amplifier, more and more beats are obtained. For the case of three input signals, a 'triple beat' results. These components of the output are

$$V_O(TB) = \frac{3}{2}a_3 V_{1A} V_{2A} V_{3A} \cos{(\omega_1 \pm \omega_2 \pm \omega_3)} t \qquad (3.66)$$

A total of four terms are generated by Equation (3.66). The triple beat distortion factor TB is defined with equal input amplitudes.

$$TB = \frac{3}{2}\frac{a_3}{a_1} V_{1A}^2 \qquad (3.67)$$

Of particular importance is that this value is twice that of $IM_3$. Note that if the three frequencies are close, the triple beats also lie close to the fundamentals and may be difficult to filter out.

Another form of distortion is *cross modulation*. For this case, one of the two inputs is assumed to be amplitude modulated. Because of the nonlinearity of the amplifier, the carrier and sidebands of the modulated signal beat with the other signal, and the modulation of the carrier is transferred to the other signal. This problem is also brought up in Chapter 9.

## 3.7 Compression and Intercept Points

For high-frequency circuits distortion is specified in terms of compression and intercept points using extrapolated small-signal output power levels. The 1 dB compression point is defined as the input power for which the fundamental output power is 1 dB below the extrapolated small-signal fundamental power as shown in Figure 3.12a. Considering the power series expansion in (3.5) the fundamental term is

$$b_1 \approx a_1 V_{iA} + \frac{3}{4}a_3 V_{iA}^3 \qquad (3.68)$$

where gain compression occurs for $a_3 < 0$.

The 1 dB compression point can be calculated from the power series expansion. The fundamental output power ($P_{fund}$), in dBs, is given by

$$P_{fund} = 20\log\left(a_1 V_{iA} - \frac{3}{4}|a_3| V_{iA}^3\right) \qquad (3.69)$$

whereas the extrapolated small-signal output power, $P_{ss}$, (in dBs) is

$$P_{ss} = 20\log{(a_1 V_{iA})} \qquad (3.70)$$

From the definition of the 1 dB compression point, $P_{fund} = P_{ss} - 1$ dB, we have

$$20 \log \left( a_1 V_{iA} - \frac{3}{4} |a_3| V_{iA}^3 \right) = 20 \log \left( a_1 V_{iA} \right) - 1 \text{ dB} \qquad (3.71)$$

$$= 20 \log \left( \frac{a_1 V_{iA}}{1.12} \right)$$

Upon simplification, it can be shown than the input voltage amplitude corresponding to the 1 dB compression point is

$$V_{iA} = \sqrt{0.145 \left| \frac{a_1}{a_3} \right|} \qquad (3.72)$$

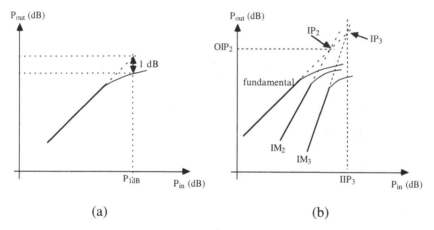

(a)     (b)

**Fig. 3.12.** Output power as a function of the input power for a nonlinear circuit. (a) 1 dB compression point. (b) Second- and third-order intercept points.

The intercept points $(IP)$ are defined as the intercept of the extrapolated small-signal power of the intermodulation terms with the fundamental as shown in Figure 3.12b. An important specification for narrow-band systems is the third-order intercept $(TOI$ or $IP_3)$. This is the point at which the extrapolated small-signal power of the fundamental and the third-order intermodulation term are identical. In general, the nth-order intercept point, $IP_n$, for $n \geq 2$, is the output power at which the extrapolated small-signal power of the fundamental and the $n$th intermodulation term intersect. The intercept points can be defined in terms of input power $(IIP_n)$ or output power $(OIP_n)$.

## Problems

**3.1.** The voltage transfer characteristic for an output stage is shown in Figure 3.13. For a bias input of 2 V and a sinusoidal input voltage of 4 V, estimate the value of THD in the output voltage waveform.

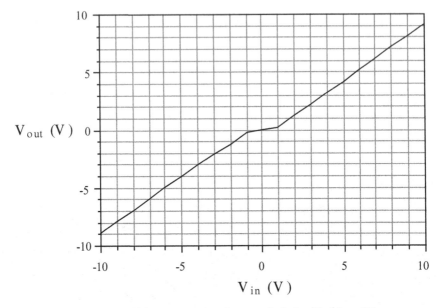

Fig. 3.13. Voltage transfer characteristic for Problem 3.1.

**3.2.** A MOS stage is shown in Figure 3.14.

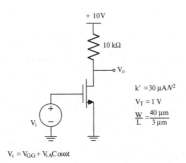

Fig. 3.14. MOS stage for Problem 3.2.

(a) Use Spice to establish the dc transfer characteristic of the stage.
(b) Choose a quiescent input bias voltage and an input sinusoidal amplitude to achieve 'just clipping' of the peaks of the output voltage.
(c) For the conditions of Part (b), use the three-point distortion analysis to estimate the value of HD2. Verify with Spice.

**3.3.** A MOS stage is shown in Figure 3.15.

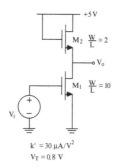

+5 V

$M_2$  $\frac{W}{L} = 2$

$V_o$

$M_1$  $\frac{W}{L} = 10$

$V_i$

$k' = 30 \ \mu A/V^2$
$V_T = 0.8 \ V$

**Fig. 3.15.** MOS stage for Problem 3.3.

(a) Use Spice to establish the dc transfer characteristic of the stage.
(b) Choose a load resistor to approximate the actual stage loading and achieve an idealized, basic stage.
(c) Choose a quiescent input bias voltage and an input sinusoidal amplitude to achieve 'just clipping' of the peaks of the output voltage. Estimate the values of HD2 and HD3. Verify with Spice.
(d) For the input conditions of (c) using the dc transfer characteristic of (a), use a three-point distortion analysis to estimate HD2, and a five-point distortion analysis to estimate HD2 and HD3. Verify your estimates with Spice.

**3.4.** A common-emitter amplifier circuit is shown in Figure 3.16.

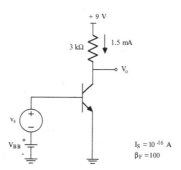

+ 9 V

3 kΩ       1.5 mA

$V_o$

$V_s$

$V_{BB}$

$I_S = 10^{-16} \ A$
$\beta_F = 100$

**Fig. 3.16.** Common-emitter amplifier for Problem 3.4.

(a) Determine the value of the input bias voltage.
(b) Estimate the amplitude of the input signal for an output voltage amplitude of 1.75V.
(c) Estimate the values of HD2 and HD3 for the input of Part (b) using the power series expansion method. Verify your results with Spice.
(d) If the input amplitude is doubled, what is the output voltage amplitude?

(e) Estimate the values of HD2 and HD3 for the input of Part (d) and verify your results with Spice.

(f) Determine the value of IM3 for the condition of Part (b) assuming that the corrupting signal has an input amplitude of one half of that of the desired input signal. Verify your result with Spice.

**3.5.** For the emitter-follower output circuit shown in Figure 3.17,

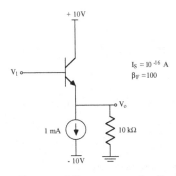

**Fig. 3.17.** Emitter-follower stage for Problem 3.5.

(a) Determine the bias state of the circuit.

(b) For $V_1 = 4.4\cos(2\pi 10^4 t)$, estimate the large-signal output response for $V_o(t)$.

(c) Estimate the harmonic distortion of $V_o(t)$ for the condition of Part (b) using the five-point method for distortion analysis.

(d) Verify the results of Parts (a) - (c) with Spice.

**3.6.** For the circuit in Problem 2.3 estimate the harmonic distortion of $V_{o2}(t)$ and $V_o(t)$ using the five-point method for an input amplitude of 20 mV. Use the information from the large-signal output response estimated in Problem 2.3 (c).

**3.7.** For the circuit in Problem 2.4 estimate the harmonic distortion of $V_o(t)$ using the five-point method for an input amplitude of 1 V. Use the information from the voltage transfer characteristic determined in Problem 2.4 (a).

**3.8.** Derive Equation (3.72) and show that the input voltage corresponding to the third-order intercept point is given by $V_{iA} = \sqrt{\dfrac{4}{3}\left|\dfrac{a_1}{a_3}\right|}$

# 4

# Distortion Generation with Source Resistance and Nonlinear Beta

## 4.1 Linearization of a Bipolar Stage Due to Source Resistance

For a basic MOS stage, whether 'single-ended' or differential, the input resistance at low frequencies is extremely large because of the insulated- gate structure of the device. Therefore, the presence of a finite resistance of a signal source has no effect on the performance of the stage. Only at very high frequencies is there an interaction between the signal source and the input capacitance.

This is not the case, however, for bipolar stages. The distortion generated by the nonlinear characteristic of base-emitter junctions of these devices is modified significantly by the presence of a source resistance. To illustrate this, assume that the signal source is a pure current source with an infinitely large signal-source resistance. The exponential nonlinearity of the input does not enter the scene, since the variational collector current output is beta times the input current. To the extent that beta is a constant, no distortion is present in the output. In Section 4.4, we consider the case where beta is not constant with the operating point and this results in distortion.

The linearization of the transfer characteristic of the stage due to the source resistance is easily appreciated by constructing a series of input V-I characteristics. A simple, common-emitter BJT stage including a signal source resistance, $R_S$, is shown in Figure 4.1a. A portion of the exponential $I_B - V_{BE}$ characteristic of the transistor is shown in Figure 4.1b. We wish to construct the nonlinear characteristic presented to the Thevenin-equivalent, open-circuit voltage source, $V_s$, of Figure 4.1a. From this combined characteristic, the input current to the device can be obtained, which multiplied by beta provides the output collector current.

We first sketch the inverse input characteristic of the transistor, the $V_{BE} - I_B$ curve shown in Figure 4.1c. The V-I characteristic of the source resistance, $R_S$, is the straight line shown in Figure 4.1d. The total V-I characteristic

D.O. Pederson and K. Mayaram, *Analog Integrated Circuits for Communication*, DOI 10.1007/978-0-387-68030-9_4,
© 2008 Springer Science+Business Media, LLC

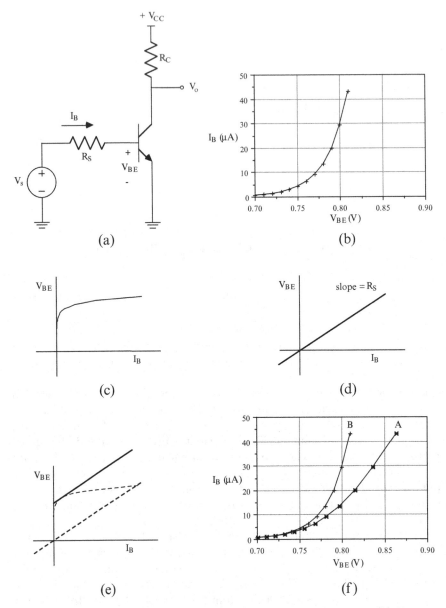

**Fig. 4.1.** (a) A simple CE stage including a signal-source resistance. (b) Exponential input characteristics of the transistor. (c) Inverse input characteristics of the transistor. (d) V-I characteristic of the source resistance. (e) The sum of the two components (c) and (d). (f) The inverse total characteristics shows linearization of the circuit.

presented to $V_s$ is the addition of these last two characteristics, since the current $I_B$ is common to both. The sum of the two components is illustrated in Figure 4.1e. Because the voltage source is assumed to be the independent variable, the inverse of the total characteristic is used, i.e., the $I_B - V_s$ plot as shown by Curve A in Figure 4.1f. For reference, the original $I_B - V_{BE}$ curve is also included as Curve B. From Figure 4.1e or Figure 4.1f, it is clear that a 'linearization' of the device characteristic is produced. The output collector current is the input current multiplied by beta. For a constant beta, there is less distortion generated for the same input signal amplitude in relation to the situation for $R_S = 0$. Of course, the incremental voltage gain of the stage, $v_o/v_s$, is reduced by the presence of $R_S$. However, as brought out below, for the same output level, there is less distortion generated when $R_S$ is present.

To illustrate the linearization of a BJT stage with $R_S$, a numerical example is used with the circuit shown in Figure 4.2a. The Spice input file is given in Figure 4.2b, in which the transistor parameters are also included. Different possible values for $R_S$ are indicated by the use of comment lines with a leading *. For $R_S = 1$ $\Omega$ and for a dc, quiescent input voltage of 0.78 V to produce $I_C = 1.25$ mA, the output voltage is 10 V $- 4$ k$\Omega$(1.25 mA) $= 5.0$ V. The predicted voltage gain of the stage is 193, and for an input amplitude of 15 mV, $HD_2 = 14.5\%$ using (3.16). Values of the node voltages for the quiescent bias state are given in the Figure 4.2c. The waveform of $V_o = V(3)$ is given in Figure 4.2d. The small-signal voltage gain of the stage is given by the results from a .TF run in Figure 4.2c. The large-signal gain and the output harmonic components with a sinusoidal excitation are obtained from the Fourier outputs given in Figure 4.2e. The Spice results are $V_o = 4.57$ V, a voltage gain of $3.01/.015 = 201$, $HD_2 = 14.1\%$, and $HD_3 = 1.3\%$. Notice that $I_C$ has changed to $(10 - 4.57)/4$ k$\Omega = 1.36$ mA due to large input drive conditions ($b_0$ effects).

For $R_S = 2$ k$\Omega$, the dc input voltage must be increased to 0.805 V ($= 0.78$ V $+ I_B R_S$) to maintain the same $V_{BE} = 0.78$ V. For $R_S = 2$ k$\Omega$, the Fourier output components are listed in Figure 4.2f. Notice that the voltage gain of the stage has decreased from 201 to 98. The predicted gain is 98. The second-harmonic distortion has decreased from 14.1\% to 3.69\%, by approximately a factor of four.

## 4.2 Distortion Reduction with Source Resistance

To obtain a measure of the linearization of a bipolar stage due to $R_S > 0$, we start with the basic stage shown in Figure 4.2a. It is assumed that operation is restricted to the normal active region. Therefore, an elementary circuit model for the BJT can be introduced as in Figure 4.3a. The no-signal, quiescent transistor state is

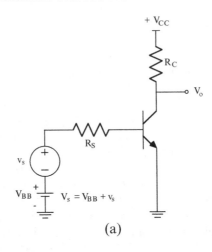

(a)

```
CE STAGE, FIGURE 4.2                                 MODEL      MOD1
VS 1 0 SIN(0.78 15MV 100KHZ)                         IB         1.25E-05
*VS 1 0 SIN(0.805 15M 100K)                          IC         1.25E-03
*VS 1 0 SIN(0.805 30.8M 100KHZ )                     VBE        .780
.TF V(3) VS                                          VBC        -4.219
*.TRAN 0.1U 20U                                      VCE        4.999
*.PLOT TRAN V(3)                                     BETADC     100.000
*.PLOT TRAN V(2)                                     GM         4.83E-02
*.FOURIER 100K V(3) V(2)                             RPI        2.07E+03
RS 1 2 1                                             RX         0.00E-01
*RS 1 2 2K                                           RO         1.00E+12
Q1 3 2 0 MOD1                                        CPI        0.00E-01
RC 5 3 4K                                            CMU        0.00E-01
VCC 5 0 10                                           CBX        0.00E-01
.MODEL MOD1 NPN BF=100 IS=1.0E-16                    CCS        0.00E-01
.OPTIONS NOPAGE NOMOD                                BETAAC     100.000
.OPTIONS RELTOL=1E-6                                 FT         7.69E+17
.WIDTH OUT=80
.END
```

(b)

| NODE | VOLTAGE | NODE | VOLTAGE | NODE | VOLTAGE | NODE | VOLTAGE |
|------|---------|------|---------|------|---------|------|---------|
| ( 1) | 0.7800 | ( 2) | 0.7800 | ( 3) | 4.9985 | ( 5) | 10.0000 |

```
****        SMALL-SIGNAL CHARACTERISTICS

        V(3)/VS                                 = -1.933D+02
        INPUT RESISTANCE AT VS                  =  2.070D+03
        OUTPUT RESISTANCE AT V(3)               =  4.000D+03
```

(c)

**Fig. 4.2.** (a) Circuit to illustrate linearization due to $R_S$. (b) Spice input file. (c) Dc operating point and small-signal characteristics.

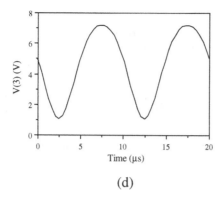

(d)

FOURIER COMPONENTS OF TRANSIENT RESPONSE V(3)    $R_S = 1\ \Omega$, $V_{SA} = 15$ mV ($T_{step} = 0.1\ \mu s$)
DC COMPONENT =    4.570D+00

| HARMONIC NO | FREQUENCY (HZ) | FOURIER COMPONENT | NORMALIZED COMPONENT | PHASE (DEG) | NORMALIZED PHASE (DEG) |
|---|---|---|---|---|---|
| 1 | 1.000D+05 | 3.007D+00 | 1.000000 | 179.995 | 0.000 |
| 2 | 2.000D+05 | 4.229D-01 | 0.140667 | 90.006 | -89.989 |
| 3 | 3.000D+05 | 3.936D-02 | 0.013092 | 0.374 | -179.621 |
| 4 | 4.000D+05 | 2.473D-03 | 0.000822 | -91.676 | -271.671 |
| 5 | 5.000D+05 | 3.447D-04 | 0.000115 | 134.658 | -45.337 |
| 6 | 6.000D+05 | 2.632D-04 | 0.000088 | 114.203 | -65.792 |
| 7 | 7.000D+05 | 2.546D-04 | 0.000085 | 118.293 | -61.702 |
| 8 | 8.000D+05 | 2.536D-04 | 0.000084 | 122.294 | -57.700 |
| 9 | 9.000D+05 | 2.510D-04 | 0.000083 | 126.573 | -53.422 |

TOTAL HARMONIC DISTORTION =    14.127785  PERCENT

(e)

FOURIER COMPONENTS OF TRANSIENT RESPONSE V(3)    $R_S = 2$ K$\Omega$, $V_{SA} = 15$ mV ($T_{step} = 0.1\ \mu s$)
DC COMPONENT =    4.943D+00

| HARMONIC NO | FREQUENCY (HZ) | FOURIER COMPONENT | NORMALIZED COMPONENT | PHASE (DEG) | NORMALIZED PHASE (DEG) |
|---|---|---|---|---|---|
| 1 | 1.000D+05 | 1.463D+00 | 1.000000 | 179.998 | 0.000 |
| 2 | 2.000D+05 | 5.400D-02 | 0.036903 | 90.009 | -89.989 |
| 3 | 3.000D+05 | 1.211D-03 | 0.000828 | 178.136 | -1.862 |
| 4 | 4.000D+05 | 7.187D-05 | 0.000049 | 98.416 | -81.582 |
| 5 | 5.000D+05 | 4.468D-05 | 0.000031 | 117.818 | -62.180 |
| 6 | 6.000D+05 | 3.955D-05 | 0.000027 | 116.330 | -63.668 |
| 7 | 7.000D+05 | 3.952D-05 | 0.000027 | 117.485 | -62.513 |
| 8 | 8.000D+05 | 4.029D-05 | 0.000028 | 121.362 | -58.636 |
| 9 | 9.000D+05 | 4.022D-05 | 0.000027 | 126.515 | -53.484 |

TOTAL HARMONIC DISTORTION =    3.691198  PERCENT

(f)

**Fig. 4.2.** (d) Transient output voltage waveform. (e) Fourier components of the output voltage for $R_S = 1\Omega$. (f) Fourier components of the output voltage for $R_S = 2$ k$\Omega$.

$$V_{BE} = V_B \tag{4.1}$$
$$V_{BS} = R_S I_B + V_B \tag{4.2}$$

where $I_B$ is the dc value of the base current at the operating point, $V_B$ is the dc value of the base voltage, and $V_{BS}$ is the value of the bias source voltage. It is assumed in the following development that the bias point is maintained constant even if $R_S$ is changed. Therefore, with changes of $R_S$, $V_{BS}$ will have to be changed to maintain $I_B$ and $V_B$ constant. The total input and output circuit variables, including both bias and incremental values, are written as follows:

$$I_b = I_B + i_b \tag{4.3}$$
$$V_b = V_B + v_b$$
$$V_o = V_O + v_o = V_{CC} - R_C I_c$$
$$I_c = I_C + i_c$$

From the circuit model of Figure 4.3a,

$$I_c = \beta I_b \tag{4.4}$$
$$= I_S \exp\left(\frac{V_b}{V_t}\right)$$

In terms of the input diode of the circuit model,

$$I_b = \frac{I_S}{\beta} \exp\left(\frac{V_b}{V_t}\right) \tag{4.5}$$

Note that $\beta$ is defined as the dc beta, cf., Section 1.4.1. The analysis procedure is to write the KVL equation at the input, to introduce a series expansion for the logarithmic relation of $V_b$ and $I_b$, and to employ a technique called harmonic balance to obtain expressions for the harmonics generated in $I_b$ and $V_b$ when the signal input is a sinusoid. At the input

$$v_s + V_{BS} = (I_B + i_b)R_S + V_B + v_b \tag{4.6}$$

For no input signal, $V_{BS} = I_B R_S + V_B$. Therefore,

$$v_s = i_b R_S + v_b \tag{4.7}$$

$v_b$ is logarithmically related to $i_b$. For the total variables,

$$I_B + i_b = \frac{I_S}{\beta} \exp\left(\frac{V_B + v_b}{V_t}\right) \tag{4.8}$$

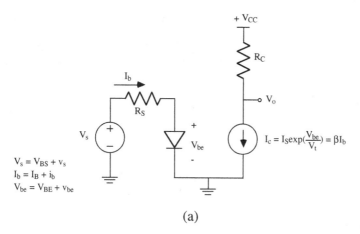

$V_s = V_{BS} + v_s$
$I_b = I_B + i_b$
$V_{be} = V_{BE} + v_{be}$

(a)

FOURIER COMPONENTS OF TRANSIENT RESPONSE V(3) $R_S = 2$ KΩ, $V_{SA} = 30.8$ mV ($T_{step} = 0.1$ μs)
DC COMPONENT =    4.767D+00

| HARMONIC NO | FREQUENCY (HZ) | FOURIER COMPONENT | NORMALIZED COMPONENT | PHASE (DEG) | NORMALIZED PHASE (DEG) |
|---|---|---|---|---|---|
| 1 | 1.000D+05 | 2.981D+00 | 1.000000 | 179.997 | 0.000 |
| 2 | 2.000D+05 | 2.260D-01 | 0.075824 | 90.007 | -89.990 |
| 3 | 3.000D+05 | 9.871D-03 | 0.003312 | 179.112 | -0.885 |
| 4 | 4.000D+05 | 7.003D-04 | 0.000235 | 93.512 | -86.485 |
| 5 | 5.000D+05 | 2.520D-04 | 0.000085 | 144.005 | -35.992 |
| 6 | 6.000D+05 | 1.452D-04 | 0.000049 | 116.785 | -63.212 |
| 7 | 7.000D+05 | 1.527D-04 | 0.000051 | 118.427 | -61.570 |
| 8 | 8.000D+05 | 1.525D-04 | 0.000051 | 122.084 | -57.913 |
| 9 | 9.000D+05 | 1.514D-04 | 0.000051 | 126.589 | -53.408 |

TOTAL HARMONIC DISTORTION =        7.589709  PERCENT

(b)

FOURIER COMPONENTS OF TRANSIENT RESPONSE V(2) $R_S = 2$ KΩ, $V_{SA} = 30.8$ mV ($T_{step} = 0.1$ μs)
DC COMPONENT =    7.788D-01

| HARMONIC NO | FREQUENCY (HZ) | FOURIER COMPONENT | NORMALIZED COMPONENT | PHASE (DEG) | NORMALIZED PHASE (DEG) |
|---|---|---|---|---|---|
| 1 | 1.000D+05 | 1.573D-02 | 1.000000 | 0.002 | 0.000 |
| 2 | 2.000D+05 | 1.131D-03 | 0.071864 | 90.003 | 90.000 |
| 3 | 3.000D+05 | 4.955D-05 | 0.003149 | 179.267 | 179.265 |
| 4 | 4.000D+05 | 3.226D-06 | 0.000205 | 93.428 | 93.426 |
| 5 | 5.000D+05 | 1.057D-06 | 0.000067 | 147.990 | 147.988 |
| 6 | 6.000D+05 | 5.867D-07 | 0.000037 | 114.121 | 114.119 |
| 7 | 7.000D+05 | 6.233D-07 | 0.000040 | 119.930 | 119.927 |
| 8 | 8.000D+05 | 5.894D-07 | 0.000037 | 124.055 | 124.053 |
| 9 | 9.000D+05 | 5.765D-07 | 0.000037 | 126.356 | 126.354 |

TOTAL HARMONIC DISTORTION =        7.193342  PERCENT

(c)

**Fig. 4.3.** (a) Equivalent circuit of CE stage with simple BJT model. (b) Fourier components of the output voltage. (c) Fourier components of the transistor input voltage.

$$= \left[ \frac{I_S}{\beta} \exp \left( \frac{V_B}{V_t} \right) \right] \exp \left( \frac{v_b}{V_t} \right)$$

$$= I_B \exp \left( \frac{v_b}{V_t} \right)$$

Dividing through by $I_B$, we obtain

$$1 + \frac{i_b}{I_B} = \exp \left( \frac{v_b}{V_t} \right) \tag{4.9}$$

$$v_b = V_t \ln \left( 1 + \frac{i_b}{I_B} \right)$$

This expression is used in the (variational) circuit equation (4.7).

$$v_s = i_b R_S + V_t \ln \left( 1 + \frac{i_b}{I_B} \right) \tag{4.10}$$

It is next assumed that the variational current change is less than the dc value, $i_b \ll I_B$. A series expansion of the ln function has the form

$$\ln(1 + x) = x - \frac{1}{2} x^2 + \frac{1}{3} x^3 - \ldots \tag{4.11}$$

This leads to

$$v_s = i_b R_S + V_t \left[ \frac{i_b}{I_B} - \frac{1}{2} \left( \frac{i_b}{I_B} \right)^2 + \frac{1}{3} \left( \frac{i_b}{I_B} \right)^3 - \ldots \right] \tag{4.12}$$

The variational signal source is assumed to be a single-tone sinusoid with an amplitude $V_{sA}$ and a radial frequency $\omega_1$.

$$v_s = V_{sA} \cos \omega_1 t \tag{4.13}$$

We now express $i_b$ as a sum of the Fourier harmonic components. For (4.13), since no energy storage is included, only cosine terms are needed.

$$i_b = b_{i1} \cos \omega_1 t + b_{i2} \cos 2\omega_1 t + b_{i3} \cos 3\omega_1 t + \ldots \tag{4.14}$$

The above series expression for $i_b$ is used in the circuit equation, Equation (4.12), and the required squaring, cubing, etc. is accomplished. The principle of harmonic balance states that each harmonic must satisfy the circuit equation independently. Thus, a set of equations is obtained from the circuit equation, one for each harmonic.

For the fundamental, if contributions from higher-order terms are neglected,

$$V_{sA} \cos \omega_1 t = R_S b_{i1} \cos \omega_1 t + V_t \frac{b_{i1}}{I_B} \cos \omega_1 t \tag{4.15}$$

This equation yields

$$b_{i1} = \frac{V_{sA}}{R_S + \frac{V_t}{I_B}} \tag{4.16}$$

$$= \frac{I_B}{1 + E} \frac{V_{sA}}{V_t} \tag{4.17}$$

A parameter $E$ is introduced.

$$E = R_S \frac{I_B}{V_t} = R_S \frac{g_m}{\beta} = \frac{R_S}{r_\pi} \tag{4.18}$$

where $g_m = I_C/V_t$ and $r_\pi = \frac{\beta}{g_m}$. Note that beta is assumed constant; therefore, $\beta_{ac} = \beta$. The parameter $E$ is related to the voltage gain from $v_s$ to $v_o$.

$$a_v = \frac{v_o}{v_s} = -\frac{g_m R_C}{1 + E} \tag{4.19}$$

$E$ is also comparable to the loop-gain parameter of feedback amplifiers. This is brought out in the next chapter.

For the second harmonic, the left-hand side of Equation (4.12) is zero since the excitation of the stage is a single sinusoid.

$$0 = \left(R_S + \frac{V_t}{I_B}\right) b_{i2} \cos 2\omega_1 t - \frac{1}{2}\frac{V_t}{I_B^2} (b_{i1} \cos \omega_1 t + \ldots)^2 + \ldots \tag{4.20}$$

The $\cos^2 \omega_1 t$ term, on the basis of $\cos^2 x = \frac{1}{2}\cos 2x + \frac{1}{2}$, produces a second-harmonic term as well as a dc term that can be neglected in this equation for only second harmonics. Higher-order even terms also produce second-harmonic contributions. It is assumed that the input excitation and the variational response is small relative to the dc state; therefore, these higher-order terms can be neglected for this analysis. However, in actual practice the higher-order terms do produce significant contributions. (In general, the harmonic equations must be solved simultaneously.) In the example below, an inspection of the dc and fundamental components of output voltages and currents in a Spice simulation shows when the higher-order terms are appreciable.

The solution of Equation (4.20), using (4.17) for $b_{i1}$, yields

$$b_{i2} = \frac{1}{4}\frac{I_B}{(1 + E)^3} \left(\frac{V_{sA}}{V_t}\right)^2 \tag{4.21}$$

For the third harmonics, contributions come from the first term directly, from the beat sum in the second term and from the cube of the fundamental in the third. Again, the contributions for higher-order terms are neglected.

(It can be asked why this beat term is not neglected along with the other higher-order terms. In general, $b_{i1} > b_{i2} > b_{i3} \cdots$. Therefore, the first beat term must be included.) From the circuit equation,

$$0 = \left(R_S + \frac{V_t}{I_B}\right) b_{i3} \cos 3\omega_1 t \tag{4.22}$$

$$-\frac{V_t}{2} \frac{1}{I_B^2} b_{i1} b_{i2} \cos(\omega_1 t + 2\omega_1 t) \tag{4.23}$$

$$+\frac{V_t}{3} \frac{1}{I_B^3} b_{i1}^3 \frac{1}{4} \cos 3\omega_1 t \tag{4.24}$$

The result using (4.17) and (4.21),

$$b_{i3} = \frac{1}{24} \frac{(1-2E)I_B}{(1+E)^5} \left(\frac{V_{sA}}{V_t}\right)^3 \tag{4.25}$$

From the expressions above, the harmonic-distortion parameters for the CE stage with $R_S > 0$ can be found.

$$HD_2 = \frac{|b_{i2}|}{|b_{i1}|} = \frac{1}{4} \frac{1}{(1+E)^2} \frac{V_{sA}}{V_t} \tag{4.26}$$

$$= \frac{1}{(1+E)^2} HD_2 \bigg|_{R_S=0}$$

$$HD_3 = \frac{|b_{i3}|}{|b_{i1}|} = \frac{1}{24} \frac{|1-2E|}{(1+E)^4} \left(\frac{V_{sA}}{V_t}\right)^2 \tag{4.27}$$

$$= \frac{|1-2E|}{(1+E)^4} HD_3 \bigg|_{R_S=0}$$

The possibility of a cancellation of $HD_3$ for the CE stage should be noted if $E = \frac{1}{2}$.

The harmonic components of the variational collector current, $i_c$, are simply those above for $i_b$ multiplied by $\beta$. The components of the output voltage, $v_o$, are found by multiplying by $-\beta R_C$.

At times it is necessary to inspect the harmonics of the transistor input voltage. This is simply accomplished using the input circuit equation.

$$v_b = v_s - i_b R_S \tag{4.28}$$

For the fundamental component, $v_{b1}$,

$$v_{b1} = V_{sA} - \frac{R_S I_B \frac{V_{sA}}{V_t}}{1 + E} \tag{4.29}$$

$$= \frac{V_{sA}}{1 + E}$$

For the second harmonic, $v_{b2}$,

$$v_{b2} = -R_S b_{i2} \tag{4.30}$$

$$= -\frac{1}{4} V_t \frac{E}{(1 + E)^3} \left(\frac{V_{sA}}{V_t}\right)^2$$

The second-harmonic factor for the base voltage is

$$HD_2 = \frac{1}{4} \frac{E}{(1 + E)^2} \frac{V_{sA}}{V_t} \tag{4.31}$$

Notice that this is $E$ different than the corresponding value for the input current and the output variables.

Similarly, the third harmonic follows from $v_{b3} = -R_S b_{i3}$. The third harmonic factor for $v_b$ is that of the input current, $i_b$, multiplied by $E$.

We return now to the circuit example of the last section and Figure 4.2a. For $R_S = 2$ k$\Omega$, $E = 0.967$, the predicted voltage gain of the stage is 98. The predicted value of $HD_2$ from Equation (4.26) is 3.75%. The simulation-value from Figure 4.2f is 3.69%. The predicted value of $HD_3$ is 0.087%, while the simulation yields 0.083%.

On the basis of (4.19), $V_{sA}$ should be increased by $(1 + E) = 1.967$ to produce the same fundamental output voltage with $R_S = 2$ k$\Omega$ relative to the case for $R_S = 1$ $\Omega$. Therefore, $V_{sA}$ should be 29.5 mV. This value, however, only produces a fundamental output amplitude of 2.87 V, rather than the 3.02 V value for $R_S = 1$ $\Omega$, due to compression effects of higher-order terms.

If the value of the sinusoidal input is increased to 30.8 mV, approximately the same output level is attained as for $R_S = 1$. The Fourier components of the output voltage are given in Figure 4.3b. Notice that the distortion factors are reduced by approximately $(1 + E)$ in relation to the values for $R_S = 1$ $\Omega$ and at the same output level. Clearly, the voltage gain has been reduced, as has the distortion. The linearization of the stage has been accomplished at the expense of voltage gain.

Also notice in Figure 4.3b that the values of the dc component of the output voltage has changed from 4.57 V for $R_S = 1$ $\Omega$ to 4.77 V. This effect is the result of higher-order terms in the nonlinearity introducing expansion and contraction components in the lower harmonics.

For completeness, the harmonic components for the transistor input voltage are given in Figure 4.3c for 30.8 mV drive and $R_S = 2$ k$\Omega$.

In the next chapter, the effects of negative feedback on distortion are considered. It is pointed out that the same reduction of distortion is obtained with reduction of voltage gain whether the reduction is due to the presence of $R_S$ or the application of negative feedback. In fact, approximately the same expressions for $HD_2$ and $HD_3$ are obtained with the proper substitution of a modified source resistance for the feedback resistance.

## 4.3 The ECP with Source Resistance

If an ECP is presented with a signal-source resistance, the small-signal voltage gain of the stage from the open-circuit voltage source to the output voltage is reduced. In addition, a linearization of the transfer characteristic is also produced with an attendant lowering of harmonic-distortion generation. The expressions for distortion reduction developed in the last section cannot be used directly because the even terms of the power series expansion are not present if the ECP is biased in a symmetric manner, i.e., if common-mode biasing is used.

The development of the expressions for the distortion components and $HD_3$ again use a series expansion involving logarithm expansions. Initially, $R_S$ is assumed equal to zero. For the ECP of Figure 4.4a, the (differential-mode) input voltage, is

$$V_1 = V_{BE1} - V_{BE2} \tag{4.32}$$

The base-emitter voltages have the general form

$$V_{BE} = V_{BB} + v_{be1} \tag{4.33}$$

where $V_{BB}$ is the (common-mode) quiescent bias voltage and $v_{be}$ is the incremental variable. Using this form in (4.32), we obtain

$$V_1 = v_{be1} - v_{be2} \tag{4.34}$$

The exponential base-current, emitter-base voltage characteristic of the BJT is now introduced.

$$
\begin{aligned}
I_b &= I_B + i_b \tag{4.35}\\
&= \frac{I_S}{\beta} \exp\left(\frac{V_{BE}}{V_t}\right)\\
&= \frac{I_S}{\beta} \exp\left(\frac{V_{BB} + v_{be}}{V_t}\right)\\
&= \left[\frac{I_S}{\beta} \exp\left(\frac{V_{BB}}{V_t}\right)\right] \exp\left(\frac{v_{be}}{V_t}\right)\\
&= I_B \exp\left(\frac{v_{be}}{V_t}\right)
\end{aligned}
$$

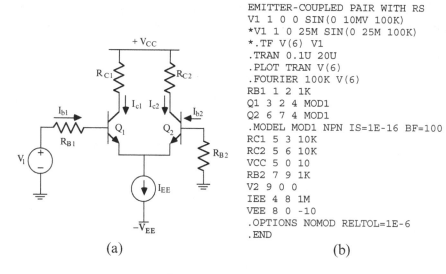

```
EMITTER-COUPLED PAIR WITH RS
V1 1 0 0 SIN(0 10MV 100K)
*V1 1 0 25M SIN(0 25M 100K)
*.TF V(6) V1
.TRAN 0.1U 20U
.PLOT TRAN V(6)
.FOURIER 100K V(6)
RB1 1 2 1K
Q1 3 2 4 MOD1
Q2 6 7 4 MOD1
.MODEL MOD1 NPN IS=1E-16 BF=100
RC1 5 3 10K
RC2 5 6 10K
VCC 5 0 10
RB2 7 9 1K
V2 9 0 0
IEE 4 8 1M
VEE 8 0 -10
.OPTIONS NOMOD RELTOL=1E-6
.END
```

(a)                                                                    (b)

```
FOURIER COMPONENTS OF TRANSIENT RESPONSE V(6)        V_1A = 10 mV (T_step = 0.1 μs)
DC COMPONENT =    5.049D+00
HARMONIC   FREQUENCY    FOURIER      NORMALIZED    PHASE      NORMALIZED
  NO         (HZ)      COMPONENT     COMPONENT     (DEG)     PHASE (DEG)

   1      1.000D+05    7.947D-01     1.000000      0.000        0.000
   2      2.000D+05    1.734D-05     0.000022     82.154       82.154
   3      3.000D+05    1.394D-03     0.001755     -0.149       -0.148
   4      4.000D+05    7.041D-06     0.000009    -84.741      -84.740
   5      5.000D+05    7.131D-06     0.000009    -43.007      -43.006
   6      6.000D+05    3.932D-06     0.000005    -53.342      -53.342
   7      7.000D+05    3.891D-06     0.000005    -67.600      -67.600
   8      8.000D+05    4.729D-06     0.000006    -64.133      -64.133
   9      9.000D+05    4.884D-06     0.000006    -52.731      -52.731

   TOTAL HARMONIC DISTORTION =          0.175482   PERCENT
```

(c)

```
FOURIER COMPONENTS OF TRANSIENT RESPONSE V(6)        V_1A = 25 mV (T_step = 0.1 μs)
DC COMPONENT =    5.049D+00
HARMONIC   FREQUENCY    FOURIER      NORMALIZED    PHASE      NORMALIZED
  NO         (HZ)      COMPONENT     COMPONENT     (DEG)     PHASE (DEG)

   1      1.000D+05    1.931D+00     1.000000      0.000        0.000
   2      2.000D+05    3.550D-05     0.000018     79.693       79.694
   3      3.000D+05    2.096D-02     0.010857     -0.032       -0.031
   4      4.000D+05    1.599D-05     0.000008    -80.371      -80.371
   5      5.000D+05    2.010D-04     0.000104     -4.088       -4.088
   6      6.000D+05    1.255D-05     0.000007    -57.390      -57.390
   7      7.000D+05    1.278D-05     0.000007    -62.019      -62.018
   8      8.000D+05    1.415D-05     0.000007    -62.294      -62.294
   9      9.000D+05    1.448D-05     0.000008    -53.010      -53.009

   TOTAL HARMONIC DISTORTION =          1.085716   PERCENT
```

(d)

**Fig. 4.4.** (a) An EC pair with source resistance. (b) Spice input file. (c) and (d) Fourier components of output voltage.

where $I_b$ is the total value of the base current, $I_B$ is the quiescent value and $i_b$ is the incremental value. The first and last expressions can be manipulated to obtain

$$1 + \frac{i_b}{I_B} = \exp\left(\frac{v_{be}}{V_t}\right) \tag{4.36}$$

$$v_{be} = V_t \ln\left(1 + \frac{i_b}{I_B}\right) \tag{4.37}$$

The ln function can be expanded and only the first few terms retained for $\frac{i_b}{I_B} \ll 1$.

$$\ln(1 + x) = x - \frac{1}{2}x^2 + \frac{1}{3}x^3 - \dots \tag{4.38}$$

Using this in (4.37) and (4.34), we obtain

$$V_1 = V_t\left[\frac{(i_{b1} - i_{b2})}{I_B} - \frac{1}{2}\frac{(i_{b1}^2 - i_{b2}^2)}{I_B^2} + \frac{1}{3}\frac{(i_{b1}^3 - i_{b2}^3)}{I_B^3} + \dots\right] \tag{4.39}$$

We now note that $i_{b1} = -i_{b2}$. This is verified from

$$I_{c1} + I_{c2} = (I_{C1} + i_{c1}) + (I_{C2} + i_{c2}) = I_{EE} \tag{4.40}$$
$$i_{c1} + i_{c2} = \beta i_{b1} + \beta i_{b2} = 0$$
$$i_{b1} = -i_{b2}$$

The result is

$$V_1 = V_t\left[2\left(\frac{i_{b1}}{I_B}\right) + \frac{2}{3}\left(\frac{i_{b1}}{I_B}\right)^3 + \dots\right] \tag{4.41}$$

Note that the even terms of the two series cancel. Only the odd terms remain.

For $R_S > 0$, the differential-mode, input voltage of the ECP of Figure 4.4a must now include the voltage drops across the two source resistances, $R_S$.

$$V_1 = R_S(i_{b1} - i_{b2}) + V_t \ln\left(1 + \frac{i_{b1}}{I_B}\right) - V_t \ln\left(1 + \frac{i_{b2}}{I_B}\right) \tag{4.42}$$

As in the earlier development, series expansions of the ln functions are introduced. For common-mode biasing, $i_{b1} = -i_{b2}$, and the even terms of the series cancel. The result has the form

$$V_1 = 2\left(R_S + \frac{V_t}{I_B}\right)i_{b1} + 0 + \frac{2}{3}V_t\left(\frac{i_{b1}}{I_B}\right)^3 + \dots \tag{4.43}$$

A pure input sinusoid is assumed and the base current is expressed as a Fourier series.

$$V_1 = V_{1A} \cos \omega_1 t \tag{4.44}$$

$$i_{b1} = b_{i1} \cos \omega_1 t + b_{i2} \cos 2\omega_1 t + b_{i3} \cos 3\omega_1 t + \dots \tag{4.45}$$

The harmonic-balance procedure is next used starting with the fundamental terms. Compression and expansion terms from the cubic and higher-order odd terms are neglected.

$$b_{i1} = \frac{1}{2} I_B \frac{\frac{V_{1A}}{V_t}}{1+E} \tag{4.46}$$

where the earlier-defined E parameter is again used.

$$E = R_S \frac{I_B}{V_t} \tag{4.47}$$

$$= \frac{R_S}{r_\pi}$$

It can be seen by inspection of (4.43) that all even-order terms are not present. When $V_1$ from (4.44) is incorporated into (4.43), higher-order odd terms produce potential even components. However, these components equate to zero. In particular, $b_{i2} = 0$.

The third-order term from (4.43) and (4.44) is

$$b_{i3} = \frac{1}{96} I_B \frac{\left(\frac{V_{1A}}{V_t}\right)^3}{(1+E)^4} \tag{4.48}$$

$HD_3$ for the ECP with $R_S > 0$ is

$$HD_3 = \frac{|b_{i3}|}{|b_{i1}|} \tag{4.49}$$

$$= \frac{1}{48} \frac{1}{(1+E)^3} \left(\frac{V_{iA}}{V_t}\right)^2$$

$$= \frac{1}{(1+E)^3} HD_3 \Big|_{R_S=0}$$

In the last expression, $HD_3$ from (2.22) without a compression term is used. The expressions of (4.49) should be compared with (4.27) for the CE stage with $R_S$.

Often symmetrical signal-source resistances are not used. That is, a second resistance equal to the signal-source resistance is not present or added to the 'other' side of the ECP. If the transistors are unilateral with no external or internal feedback, the above results still hold. The source resistance can be assumed to be symmetrically divided into two equal resistances, $\frac{R_S}{2}$. The value of the $E$ parameter is found from $E = \frac{\frac{R_S}{2}}{r_\pi}$. In effect, the input circuit-equation loop consists of $V_1$, two external resistances of $\frac{R_S}{2}$ and the ECP differential-mode input.

For a numerical example, circuit and device values for the ECP of Figure 4.4a are given in the Spice input file of Figure 4.4b. The amplitude of the input sinusoid is 10 mV or 25 mV, and the value of the two equal source (base) resistances, $R_{B1}$ and $R_{B2}$, is 1 k$\Omega$. For the 10 mV input, the Fourier components of the output voltage are given in Figure 4.4c. $HD_3$ is equal to 0.18%. For the circuit, the value of the $E$ parameter is 0.19. The estimated value of $HD_3$ from (4.49) is 0.18%. The nonzero values of the even harmonics are due to 'numerical noise' of the time integration in Spice.

If the amplitude of the input is increased to 25 mV, the estimated value of $HD_3$ is 1.15%. The Fourier components for this drive are given in Figure 4.4d. $HD_3$ is 1.09%. Contributions from the higher-order terms are now present in the value of the fundamental. This can be seen from a comparison of the ratio of the input amplitudes and the amplitudes of the output fundamentals, i.e., from the values of the large-signal gain.[1]

## 4.4 Nonlinear Beta and Distortion

As brought out in Section 4.1, output distortion due to the exponential nonlinearity of the idealized bipolar transistor decreases as the resistance of the source increases. In the limit, the source becomes a pure current source, and harmonic distortion due to the input nonlinearity disappears. Unfortunately, another nonlinearity is present in bipolar transistors. The short-circuit current gain (beta) of the transistor is not constant. Therefore, for a pure current source input, the output current and output voltage excursions are not amplified versions of the input excursions.

In Figure 4.5a, two plots of $I_C$ versus $I_B$ are shown. If beta is a constant, $\beta = \beta_0$, the collector current is a linear relation of the input current, as illustrated by the straight Line A in the figure. For an actual transistor, the current transfer characteristic is more like Curve B in the figure. In Figure 4.5b, the difference between the two curves is shown. It is clear that for Curve B, an input signal with equal positive and negative excursions about an operating point produces an output with unequal excursions. Corresponding plots of the dc beta with collector current are given in Figures 4.5c and d. In Figure 4.5c, log (beta) versus log ($I_C$) is plotted. In Figure 4.5d, beta versus log ($I_C$) is plotted with a reduced range for $I_C$.

The above distortion situation is readily studied with a Spice simulation. The circuit is shown in Figure 4.6a. The input circuit file is also given in

---

[1]    The large-signal gain is defined as the ratio of the amplitude of the fundamental component of the output to the amplitude of the sinusoidal input signal.

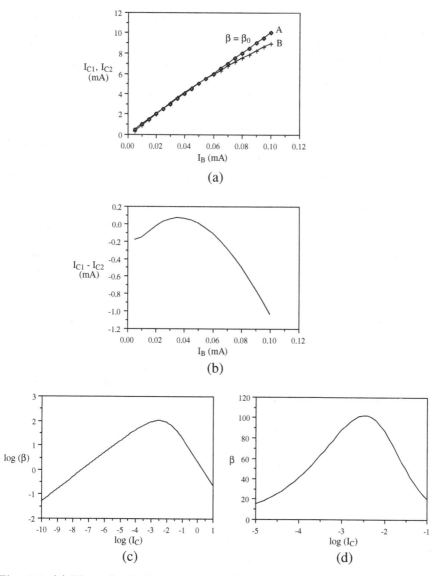

**Fig. 4.5.** (a) Plots of collector current as a function of base current. (b) Difference of the two collector currents A and B. (c) Log-log plot of the dc beta as a function of the collector current. (d) Plot of dc beta as a function of the collector current (log scale).

Figure 4.6a. The circuit is a parallel combination of two simple common-emitter stages each with a current-controlled, current source input consisting of a dc component and a sinusoidal variation. For transistor $Q_1$, the .MODEL definition includes several new parameters. These parameters, described below, introduce a nonlinear modeling of beta. The dc current transfer characteristic for $Q_1$ is that presented as Curve B of Figure 4.5a. The transfer characteristic of transistor $Q_2$, which has a constant $\beta$, is that of Curve A in Figure 4.5a. In Figure 4.6b, the harmonic components of the output current of $Q_1$ are given for a dc bias input current of 50 $\mu$A and a sinusoidal input current with an amplitude of 40 $\mu$A. The harmonic distortion, 6.2%, is certainly not negligible. Further details of this example are introduced in Section 4.5.

Many factors can produce the nonlinear behavior of beta. Two of the major ones are introduced in this section to illustrate the situation. In Figure 4.7, a semilog plot is given of $I_B$ and $I_C$ versus $V_{BE}$ of a small-sized bipolar transistor, in the normal operating region (reverse-biased collector-base junction and forward-biased emitter-base junction). For an idealized situation with beta constant, both curves are linear and have the same slope of value $1/V_t$. For an actual device, the base current breaks away from the linear region at a value of $I_C \approx I_L$ and approaches with ever lower current a new linear curve with an asymptote slope of $1/(2V_t)$. This is due to the dominance at low base currents of a component due to hole-electron recombination in the emitter-base junction itself [7]. At high currents, high-level injection effects at the emitter-base junction cause the collector current to break away from the linear region at a value of $I_C \approx I_K$ and approaches a new linear curve with a slope of $1/(2V_t)$ [6], [7]. The combination of these effects leads to the curves of beta versus $I_C$ as shown in Figure 4.5c or Figure 4.5d. (Note that a log scaling is used for $I_C$.)

The base current can be modeled as having two components.

$$I_B = I_{B1} + I_{B2} \qquad (4.50)$$
$$= \frac{I_S}{\beta_0} \exp\left(\frac{V_{BE}}{V_t}\right) + I_{SE} \exp\left(\frac{V_{BE}}{N_E V_t}\right)$$

$I_{B1}$ represents the base current that flows to support the hole-electron recombination in the base region, outside of the junction regions, and $\beta_0$ is assumed to be a constant. The second term represents the junction recombination mentioned above. The latter has a factor of $N_E$ in the exponent factor. In this book, $N_E = 2$ is used for convenience. Another parameter, $I_{SE}$, is also used to characterize this component. In Spice, the corresponding parameters are denoted $NE$ and $ISE$. Both Spice model parameters are used in the input file of Figure 4.6a for the MOD1 model.

At high currents, $I_{B1}$ is dominant, while at low currents $I_{B2}$ is dominant. The two base-current components are defined to be equal at a value of $I_C = I_L$, which leads to the following relation.

```
NONLINEAR BETA, FIGURE 4.6
I1 0 1 50U SIN( 50U 40U 100K)
.TRAN 0.1U 20U
.PLOT TRAN I(VC1) I(VC2)
.FOUR 100K I(VC1)
*.DC I1 5U 0.1MA 5UA
*.PLOT DC I(VC1)  I(VC2)  (0,10M)
*.PLOT DC V(8,2)  (-1,1)
V1 1 0 0
F1 0 3 V1 1
Q1 2 3 0 MOD1
.MODEL MOD1 NPN IS=1E-16 BF=230
+ISE=2E-12 NE=2
+IKF=10M
RC1 4 2 .5K
VC1 5 4 0
VCC 5 0 10
F2 0 6 V1 1
Q2 8 6 0 MOD2
VC2 5 7 0
RC2 7 8 0.5K
.MODEL MOD2 NPN BF=100 IS=1E-16
.OPTIONS NOMOD NOPAGE
.OPTIONS RELTOL=1E-6
.WIDTH OUT=80
.END
```

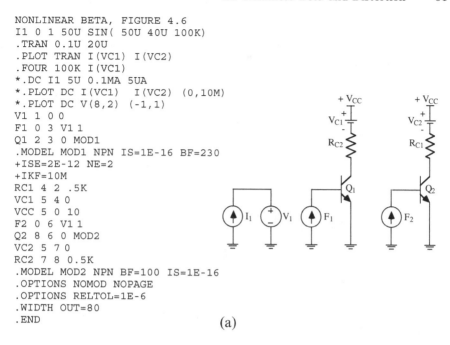

(a)

```
FOURIER COMPONENTS OF TRANSIENT RESPONSE I(VC1)
DC COMPONENT =   4.771D-03
HARMONIC   FREQUENCY     FOURIER      NORMALIZED      PHASE     NORMALIZED
   NO        (HZ)       COMPONENT     COMPONENT      (DEG)     PHASE (DEG)

    1     1.000D+05    3.674D-03     1.000000       0.002       0.000
    2     2.000D+05    2.278D-04     0.061996      90.002      90.001
    3     3.000D+05    1.505D-05     0.004097     179.507     179.505
    4     4.000D+05    3.973D-06     0.001081      90.573      90.571
    5     5.000D+05    3.099D-06     0.000843     177.901     177.899
    6     6.000D+05    1.317D-06     0.000358     -92.111     -92.113
    7     7.000D+05    5.313D-07     0.000145      11.930      11.928
    8     8.000D+05    3.275D-07     0.000089     101.821     101.819
    9     9.000D+05    1.772D-07     0.000048     147.876     147.874

    TOTAL HARMONIC DISTORTION =        6.214777  PERCENT
```

(b)

**Fig. 4.6.** (a) Circuit and Spice input file for studying distortion due to nonlinear beta. (b) Fourier components of the output current.

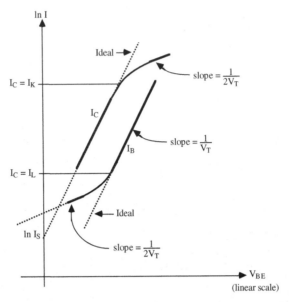

**Fig. 4.7.** Plot of collector and base currents (log scale) as a function of the base-emitter voltage (linear scale).

$$I_L = \frac{(\beta_0 I_{SE})^2}{I_S} \tag{4.51}$$

$\beta_0$ is the value of beta well above $I_L$.

The general value of dc beta is labeled $\beta$. If only the base current and the normal collector current are considered, $\beta$ can be written as

$$\beta = \frac{I_C}{I_B} \tag{4.52}$$

$$= \frac{I_C}{I_{B1} + I_{B2}}$$

$$= \frac{I_S \exp\left(\frac{V_{BE}}{V_t}\right)}{\frac{I_S}{\beta_0} \exp\left(\frac{V_{BE}}{V_t}\right) + I_{SE} \exp\left(\frac{V_{BE}}{2V_t}\right)}$$

This can be manipulated to obtain

$$\beta = \frac{\beta_0}{1 + \sqrt{\frac{I_L}{I_C}}} \tag{4.53}$$

where $I_L$ is as given above in Equation (4.51). A plot of the log of Equation (4.53) versus the log of $I_C$, obtained from a Spice simulation, is shown in

Figure 4.8a. Also shown are the low-current and high-current asymptotes of (4.53). Notice that because of the square-root dependency with $I_C$, the beta curve does not approach quickly the low-current asymptote and constant beta (horizontal) asymptote for currents well removed from $I_L$. This is in marked contrast to the situation for the bode plots of circuit analysis where a square-law relation for the magnitude functions is present and the asymptotes are approached quickly as the variable departs from the 'corner frequencies'. At $I_L$, beta is down by a factor of two from the maximum value (not the square root of two for bode plots).

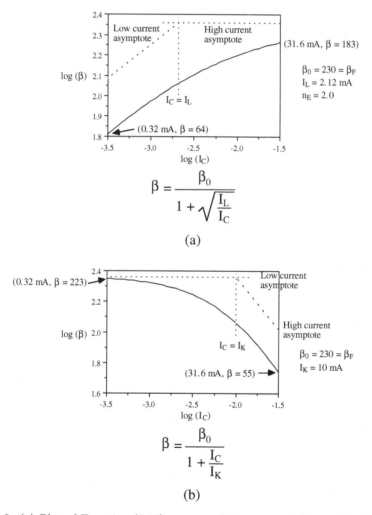

$$\beta = \frac{\beta_0}{1 + \sqrt{\dfrac{I_L}{I_C}}}$$

(a)

$$\beta = \frac{\beta_0}{1 + \dfrac{I_C}{I_K}}$$

(b)

**Fig. 4.8.** (a) Plot of Equation (4.53) versus collector current (log scale). (b) Plot of Equation (4.56) versus collector current (log scale).

For large collector currents, $I_C$ breaks away from the linear variation of lower currents as shown in Figure 4.7. The mid-current variation has a linear asymptote with a slope of $1/V_t$. However, the curve for the new high current region approaches a linear asymptote (for a semilog plot) with a slope of $1/(2V_t)$. As mentioned earlier, the phenomena which produces this effect is high-level injection into the base region from the emitter region. For large injection, the minority carrier density can no longer be considered negligible with respect to the majority carriers. In effect, the diffusion constant for minority carriers is reduced by two. Hence, there is a change in the slope relation to $1/(2V_t)$.

It is not possible to model the collector current as the sum of two components as is done for the base current. Instead we note that for large currents, $I_C$ and $I_B$ have the following asymptotic behavior:

$$I_C \sim \exp\left(\frac{V_{BE}}{2V_t}\right) \tag{4.54}$$

$$I_B \sim \exp\left(\frac{V_{BE}}{V_t}\right)$$

From Equation (4.54), it is seen that at large currents $\beta = I_C/I_B$ is inversely proportional to $I_C$.

$$\beta \sim \frac{\exp\left(\frac{V_{BE}}{2V_t}\right)}{\exp\left(\frac{V_{BE}}{V_t}\right)} \sim \frac{1}{\exp\left(\frac{V_{BE}}{2V_t}\right)} \sim \frac{1}{I_C} \tag{4.55}$$

Therefore, the overall variation of beta considering only the high current effects for $I_C$ can be modeled as

$$\beta = \frac{\beta_0}{1 + \frac{I_C}{I_K}} \tag{4.56}$$

where $\beta_0$ is the value of $\beta$ for values of $I_C$ well below $I_K$, and $I_K$ is a constant. A Spice plot of the dc beta with the log of $I_C$ for (4.56) is shown in Figure 4.8b, (only the mid-current component for $I_B$, $I_{B1}$, is included.) As illustrated, $I_K$ is defined as the collector current where the high-current asymptote and constant-beta (horizontal) asymptote meet. ($IKF$ is the corresponding beta parameter in Spice, cf., Figure 4.6a). Because of the inverse linear relation in Equation (4.55), the actual curve approaches the asymptotes quicker with variation of $I_C$ away from the 'corner current,' $I_K$, than is the case for the low-current situation considered above.

When both base current and collector effects are present, it can be shown that [8]

$$\beta = \frac{\beta_0}{1 + \frac{I_L}{I_C}^{\frac{(N_E-1)}{N_E}}\left(1 + \frac{I_C}{I_K}\right)^{\frac{1}{N_E}} + \frac{I_C}{I_K}} \tag{4.57}$$

For $N_E = 2$,

$$\beta = \frac{\beta_0}{1 + \sqrt{\frac{I_L}{I_C} + \frac{I_L}{I_K} + \frac{I_C}{I_K}}} \tag{4.58}$$

$$\approx \frac{\beta_0}{1 + \sqrt{\frac{I_L}{I_C} + \frac{I_C}{I_K}}}$$

A Spice plot of the log of beta with the log of $I_C$ is given in Figure 4.8c. Note that there are now three regions: one for the low-current region, one for the middle region where beta may be almost flat with $I_C$, and one for the high-current region. When $I_L$ and $I_K$ are at least several decades of magnitude apart, a flat beta region is obtained, and the maximum value of beta is unambiguous, $\beta_{\max} = BF$; however, when $I_L$ and $I_K$ are close, as is the situation in Figure 4.8c, the midregion asymptote is not clear, and appropriate values for $BF$, $ISE$, and $IKF$ for a Spice simulation are not easily established. The usual procedure is to obtain, for a transistor type, curves of dc beta with $I_C$ plotted on log-log coordinates. The low- and high-current regions are investigated relative to possible asymptotic slopes. If a substantial flat portion of the beta curve exists, the value of $\beta_{max} = BF$ is apparent.

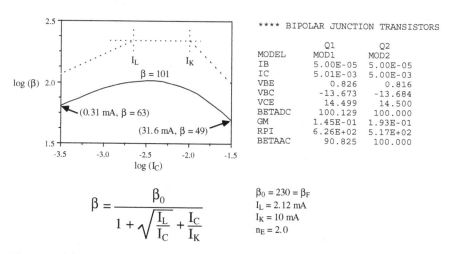

**Fig. 4.8.** (c) Plot of Equation (4.58) versus collector current (log scale) and model parameter values determined by Spice.

If the middle portion of the beta curve is rounded, a trial choice of a flat portion, with a value of $BF$ greater than $\beta_{max}$, the maximum value of beta, is made. Note from the example above that the maximum value of $\beta$ is approximately 100 while the value of $BF$ is 230. At a low collector current value, well in the lower region, the value of $\beta$ is noted and Equation (4.53) and then (4.51) are used to determine $I_L$ and $I_{SE}$. Similarly, a value well in the upper region is chosen, and Equation (4.56) is used to obtain a value of $I_K = IKF$. A Spice run can then be used to determine if $\beta_{max}$ is obtained. If not, an iteration of the above procedure is necessary.

The Spice BJT parameters that were used to generate the curve of Figure 4.8c are $BF = 230$, $IKF = 10$ mA, $IS = 1 \times 10^{-16}$ A, $NE = 2.0$, and $ISE = 2 \times 10^{-12}$ A ($I_L = 2.12$ mA). Also shown in the upper right-hand corner of the figure are the model values determined by Spice. Note that for the transistor of interest, $Q_1$, the dc beta equals 100 at the operating point, while the incremental or ac beta equals 90.8. Transistor $Q_2$ has a constant beta of 100.

## 4.5 Example of Distortion due to Beta($I_C$)

A single-stage amplifier is used as an example to illustrate the generation of distortion due to beta variations. The circuit is a simple CE stage with a current source input. The circuit values and device parameters are given in the earlier-used Spice input file, Figure 4.6a. The parameters of beta are those which pertain to Figure 4.8c: $BF = 230$, $IKF = 10$ mA, $NE = 2.0$, and $ISE = 2 \times 10^{-12}$ A ($I_L = 2.12$ mA). The plot of the collector current versus base current is given in Figure 4.9.

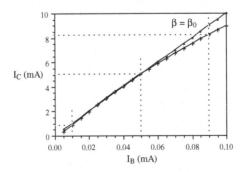

**Fig. 4.9.** Plot of collector currents versus base current with and without nonlinear beta. The values used for distortion calculation are shown by the dotted lines.

The signal-source conductance is assumed very small and can be neglected. The dc input current is $I_B = 50$ $\mu$A. From Figure 4.9, the corresponding value of $I_C$ is 5.0 mA. For a sinusoidal input signal amplitude of 40 $\mu$A, the collector current excursion is very large. The Spice output harmonics are given in Figure 4.6b. From these results, $HD_2 = 6.2\%$ and $HD_3 = 0.41\%$.

It is interesting to compare the Spice results with a simple three-point distortion analysis. From (3.34) in Section 3.4, the second-harmonic distortion factor is

$$HD_2 = \frac{1}{2}\left(\frac{I_{C2} - 2I_{C0} + I_{C4}}{I_{C2} - I_{C4}}\right) \qquad (4.59)$$

where a translation to collector-current maximum, mid, and minimum values has been made. For the sinusoidal input amplitude of 40 $\mu$A, the $I_C$ values from Figure 4.9 are 8.24 mA, 5.01 mA, and 0.85 mA. The estimated value for $HD_2$ is 6.2%.

If only the $\log(\beta)$ versus $\log(I_C)$ curve is available, in contrast to a beta-versus $I_B$ curve, a trial-and-error procedure usually is necessary to estimate the values of collector current at the extremes of the input drive.

The upper harmonics are not available from the simplified analysis used above. An alternative analysis, the differential-error analysis technique is introduced in Chapter 6. It is also very simple, uses small-signal values in contrast to dc (total) values, and also provides an estimate of $HD_3$.

## Problems

**4.1.** A common-emitter stage with a finite source resistance is shown in Figure 4.10. The bias state of $I_C = 2$ mA is maintained for any value of $R_S$.

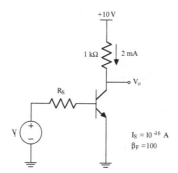

**Fig. 4.10.** Common-emitter stage for Problem 4.1.

(a) The input voltage is $V_1 = V_{BB} + V_{1A}\cos\omega_1 t$. $V_{BB}$ establishes the given quiescent collector current. For $V_{1A} = 30$ mV, determine the value of $R_S$ to achieve a fundamental output voltage amplitude of 2 V.

(b) Determine HD2 for the conditions of Part (a).
(c) For $R_S = 1$ k$\Omega$ and a fundamental output voltage amplitude of 1 V, determine the values of $V_{1A}$ and HD2.

**4.2.** A bipolar stage is shown in Figure 4.11.

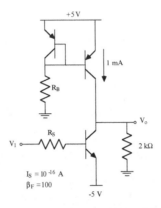

**Fig. 4.11.** Bipolar stage for Problem 4.2.

(a) Develop for the stage an idealized, basic configuration.
(b) For $R_S = 0.5$ $\Omega$, estimate HD2 and HD3 for a sinusoidal input amplitude of 15 mV. The input bias voltage must be chosen to achieve the specified bias state of the stage, $I_C = 1$ mA. Verify with Spice.
(c) For $R_S = 0.5$ $\Omega$, what is the practical maximum of input amplitude to achieve a reasonable output voltage waveform.
(d) For $R_S = 1.5$ k$\Omega$, estimate HD2 and HD3 with an input amplitude to achieve the same fundamental output voltage as in (b). Compare results. Note that the input bias voltage must be also adjusted to maintain the same quiescent collector current.
(e) For $R_S = 100$ k$\Omega$, and with $I_{KF} = 8$ mA and $I_L = 0.04$ mA, determine the THD for the stage if the input amplitude is adjusted to achieve 'just clipping'. The quiescent bias state of the circuit should remain the same as that of Part (b).

**4.3.** A common-emitter stage including a signal-source resistance is shown in Figure 4.12.
(a) For $R_S = 100$ k$\Omega$ and with $V_s = V_{BB} + V_{sA} \cos \omega_1 t$, determine HD2 and HD3 if the quiescent bias state is $I_C = 0.5$ mA and the fundamental output voltage amplitude is 4 V. Use Spice as appropriate.
(b) For $R_S = 10$ k$\Omega$, re-establish the same quiescent bias state and determine the necessary input drive to achieve the same fundamental output voltage amplitude. Compare the values of HD2 and HD3 for the transistor parameters of Figure 4.12 and for the case where beta is constant at the value of $\beta_F$.

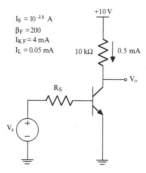

$I_S = 10^{-16}$ A
$\beta_F = 200$
$I_{KF} = 4$ mA
$I_L = 0.05$ mA

**Fig. 4.12.** Common-emitter stage for Problem 4.3.

**4.4.** A common-emitter amplifier with a signal-source resistance is shown in Figure 4.13.

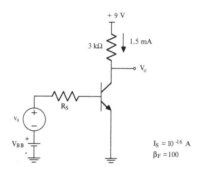

**Fig. 4.13.** Common-emitter amplifier for Problem 4.4.

(a) For $R_S = 2$ k$\Omega$ estimate HD2 and HD3 for a sinusoidal input amplitude that produces an output voltage amplitude of 3.5 V. Note that the input bias must be selected to provide the quiescent bias state of $I_C = 1.5$ mA. Use Spice to verify your results.

(b) For $R_S = 100$ k$\Omega$, and with $I_{KF} = 6$ mA and $I_L = 0.05$ mA, use Spice to determine the THD with an input amplitude to achieve 'just clipping'. The quiescent bias state of the circuit should remain as that in Part (a).

**4.5.** An emitter-coupled pair with source resistances is shown in Figure 4.14.

(a) For $V_1 = 0 + 0.1\cos\omega_1 t$, determine the value of $R_{S1} = R_{S2}$ to produce a HD3 level of 1%.

(b) If the input level of $V_1$ is doubled to that of Part (a), what is the value of HD3?

(c) With $I_{KF} = 6$ mA and $I_L = 0.05$ mA, use Spice to determine HD3 for the conditions of Part (a) and (b).

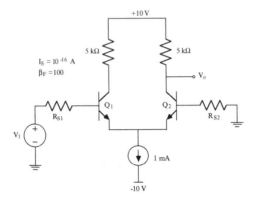

**Fig. 4.14.** Emitter-coupled pair for Problem 4.5.

# 5

# Distortion in Feedback Amplifiers

## 5.1 Effects of Negative Feedback

Negative feedback applied to an amplifier linearizes the transfer characteristic of the amplifier and reduces the distortion of the input signal that is generated by the nonlinearity. The gain of the amplifier at an operating point is also reduced accordingly. These aspects are illustrated using the amplifier block diagram of Figure 5.1a. The transfer characteristic of the amplifier without feedback is taken to be that of Curve A in Figure 5.1b. Notice that the output variable of the plot is $-v_o$. Initially, we are interested in negative feedback. Either the amplifier or the feedback must provide a net phase reversal. For plotting convenience here, $-v_o$ is used. For Curve A there is a significant nonlinearity. When feedback is applied, the slope at each point along the transfer characteristic is reduced by the corresponding amount of the 'loop gain,' i.e., the feedback, at that point. The slope of Curve B, at any input value in the figure, is the closed-loop gain at this point and tends to remain constant, even though the amount of feedback may be reduced at this operating point of interest when there is a falloff of gain of the original amplifier.

The above aspects are brought out algebraically by inspecting the closed-loop gain expression which is developed from the block diagram of Figure 5.1a. The gain of the original amplifier is

$$a = \frac{v_o}{v_i} \tag{5.1}$$

With feedback, the amplifier input voltage is

$$v_i = v_s - v_{fb} = v_s - f v_o \tag{5.2}$$

where $f$ is the fraction of the output voltage, $v_o$, which is fed back to the input. It is to be emphasized that, in Figure 5.1a and in (5.2), the input to the basic amplifier is the *difference* of the source voltage and the feedback voltage. Using this in Equation (5.1), we obtain the closed-loop gain function A.

D.O. Pederson and K. Mayaram, *Analog Integrated Circuits for Communication*, DOI 10.1007/978-0-387-68030-9_5,
© 2008 Springer Science+Business Media, LLC

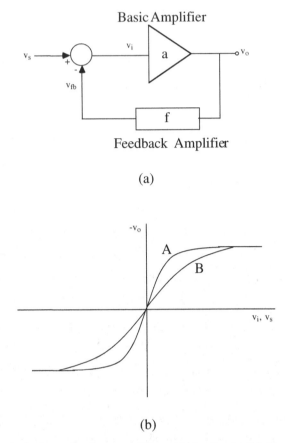

**Fig. 5.1.** (a) Block diagram of an amplifier with feedback. (b) Transfer characteristics of amplifier without (A) and with (B) feedback.

$$A = \frac{v_o}{v_s} = \frac{a}{1 + af} \tag{5.3}$$

'$af$' is the amount of feedback, i.e., the amount of the 'loop gain.' For a large magnitude of '$af$', the closed-loop gain expression can be expanded as follows, retaining only the first few terms.

$$\frac{v_o}{v_s} = \frac{1}{f}\left(1 - \frac{1}{af} + \dots\right) \tag{5.4}$$

As long as the magnitude of '$af$' is large and for $f$ constant, the slope of the closed-loop transfer characteristic is almost the constant, $1/f$. Therefore, a linearization of the amplifier is achieved. Consequently, distortion in the amplifier output is reduced with negative feedback.

To establish expressions for the amount of distortion that is produced in an amplifier with negative feedback, we again start with the idealized block diagram situation shown in Figure 5.1a. With no feedback, the transfer characteristic of the basic amplifier can be characterized by the usual power series in terms of the incremental values of the input and output variables.

$$v_o = a_1 v_i + a_2 v_i^2 + \ldots \tag{5.5}$$

where the expansion is taken about an operating point and the quiescent term of the power series is not included. With feedback, the amplifier input voltage is given in Equation (5.2). The output voltage becomes,

$$v_o = a_1(v_s - f v_o) + a_2(v_s - f v_o)^2 + \ldots \tag{5.6}$$

We next define a new power series for the closed-loop amplifier in terms of the source voltage.

$$v_o = a_1' v_s + a_2' v_s^2 + \ldots \tag{5.7}$$

Expressions for the $a_i'$ are obtained by interpreting the power series as a Taylor's series

$$v_o = \left.\frac{dv_o}{dv_s}\right|_{0,0} v_s + \frac{1}{2}\left.\frac{d^2 v_o}{dv_s^2}\right|_{0,0} v_s^2 + \ldots \tag{5.8}$$

where the necessary derivatives are evaluated at the operating point, $v_s = 0, v_o = 0$. This leads to the following expressions:

$$\left.\frac{dv_o}{dv_s}\right|_{0,0} = a_1'$$

$$\left.\frac{d^2 v_o}{dv_s^2}\right|_{0,0} = 2a_2'$$

$$\left.\frac{d^3 v_o}{dv_s^3}\right|_{0,0} = 6a_3'$$

Taking the derivatives of (5.6) and using the fact that $v_o = 0$ when $v_s = 0$, we obtain

$$a_1' = \frac{a_1}{(1 + a_1 f)} \tag{5.9}$$

$$a_2' = \frac{a_2}{(1 + a_1 f)^3}$$

$$a_3' = \frac{a_3(1 + a_1 f) - 2a_2^2 f}{(1 + a_1 f)^5}$$

From Chapter 3, the harmonic distortion factors for the open-loop amplifier are

$$HD_2 \approx \frac{1}{2}\frac{a_2}{a_1}V_{sA} \tag{5.10}$$

$$HD_3 \approx \frac{1}{4}\frac{a_3}{a_1}V_{sA}^2$$

where contributions from higher-order terms of the series are neglected, for simplicity, and where $V_{sA}$ is equal to the amplitude of the input sinusoid. Similarly, for the closed-loop amplifier,

$$HD_2 \approx \frac{1}{2}\frac{a_2'}{a_1'}V_{sA} = \frac{1}{2}\frac{a_2}{a_1}V_{sA}\frac{1}{(1+a_1 f)^2} \tag{5.11}$$

$$\approx \frac{1}{(1+a_1 f)^2}\,HD_2\big|_{\text{w/o fb}}$$

$$HD_3 \approx \frac{1}{4}\frac{a_3'}{a_1'}V_{sA}^2 = \frac{1}{4}\frac{a_3}{a_1}V_{sA}^2\frac{\left|1-\frac{2a_2^2 f}{a_3(1+a_1 f)}\right|}{(1+a_1 f)^3}$$

$$\approx \frac{\left|1-\frac{2a_2^2 f}{a_3(1+a_1 f)}\right|}{(1+a_1 f)^3}\,HD_3\big|_{\text{w/o fb}}$$

Magnitude signs are added to the last expression to conform to the magnitude definition of the distortion factors in the previous chapters, e.g., $HD_3 = |b_3|/|b_1|$. An alternate form for $HD_3$ for a simple CE stage with local feedback is given in Section 5.4.

As pointed out in Chapter 3, it often is desired to establish the distortion of an amplifier when the amplitude of the output fundamental is held constant. The small-signal gain expression of the original amplifier is used for the fundamental frequency component,

$$v_o = a_1 v_i \tag{5.12}$$

The resulting expressions for $HD_2$ and $HD_3$ for the open-loop amplifier where $V_{oA}$ is the amplitude of the output fundamental are

$$HD_2 \approx \frac{1}{2}\frac{a_2}{a_1^2}V_{oA} \tag{5.13}$$

$$HD_3 \approx \frac{1}{4}\frac{a_3}{a_1^3}V_{oA}^2$$

For the closed-loop amplifier,

$$HD_2 \approx \frac{1}{2}\frac{a_2'}{a_1'^2}V_{oA} = \frac{1}{2}\frac{a_2}{a_1^2}V_{oA}\frac{1}{1+a_1f} \tag{5.14}$$

$$\approx \frac{1}{1+a_1f}\, HD_2|_{\text{w/o fb}}$$

$$HD_3 \approx \frac{1}{4}\frac{a_3'}{a_1'^3}V_{oA}^2 = \frac{1}{4}\frac{a_3}{a_1^3}V_{oA}^2\frac{\left|1-\frac{2a_2^2f}{a_3(1+a_1f)}\right|}{1+a_1f}$$

$$\approx \frac{\left|1-\frac{2a_2^2f}{a_3(1+a_1f)}\right|}{1+a_1f}\, HD_3|_{\text{w/o fb}}$$

Notice that $HD_2$ is reduced by $(1+a_1f)^2$ when the input voltage is held constant and is reduced by $(1+a_1f)$ when the output voltage is held constant. In effect, for the former, $HD_2$ is reduced once by the reduction of the output voltage and reduced again by the linearization of the transfer characteristic. In the latter case, only the linearization is present.

In the $HD_3$ expressions a subtraction appears possible. Cancellation of the third-order distortion occurs if

$$\frac{2a_2^2f}{a_3(1+a_1f)} = 1 \tag{5.15}$$

This cancellation can be very important when $IM_3$ needs to be eliminated (reduced) in high-performance broadband amplifiers.

If no $a_2$ term in the amplifier power series exists, as is the case for a balanced emitter-coupled pair, no second-order distortion is present, and $HD_3$ is reduced by $(1+a_1f)^3$ when the input is held constant and is reduced by $(1+a_1f)$ when the output is held constant.

## 5.2 Feedback for a General Amplifier

The block diagram approach used in the last section to establish the effects of feedback is quite simple and leads to appropriate closed-loop expressions for the closed-loop gain, the amount of linearization, desensitization, etc. However, problems can arise in the general case in introducing properly the combining (subtracting) of signals at the input, in defining and working with the proper gain functions of the basic amplifier and the overall combination, and in establishing the correct 'open-loop' gain. Other potential problems occur with the loading of the source and load resistances on the basic amplifier as well as of the feedback elements.

Sections 8.5 and 8.6 of [6], and others have addressed these problems by treating the closed-loop amplifier as a combination of two-port networks. As brought out in [6], this technique is possible for most practical feedback configurations. In this section, one amplifier configuration is studied to illustrate the problems and a solution. Another two-port combination is studied in Section 5.4.

As an introduction to the two-port approach, a simple evaluation is first made of the typical amplifier shown in Figure 5.2a. It is assumed that the 'original' amplifier supplies both voltage and current gain and a phase reversal, i.e., a negative gain. As brought out below, it is helpful to replace the signal source with its Norton equivalent, as done in Figure 5.2b. The current through $R_f$ from the input to the output is labeled $i_f$ and subtracts from $i_s$, the short-circuit available current from the signal source. Therefore, the current in the amplifier is $i_a = i_s - i_f$ analogous to the voltage notation of Figure 5.1a.

In Figure 5.2c, a circuit model for the original amplifier is introduced. Note that its input resistance is $r_i$, its output resistance is $r_o$, and its transfer function is characteristized by a transconductance $G_m$. (The open-circuit voltage gain of the original amplifier is $-G_m r_o$. However, the actual gain of the amplifier is modified by the loading at its input and output due to $R_S$, $R_L$ and $R_f$.)

The 'basic' amplifier is defined as the original amplifier including the loading of $R_S$ in shunt at the input and of $R_L$ in shunt at the output as shown in Figure 5.2c. Finally, we assume for now that the combined input resistance at the input, $R_S \| r_i$, where the symbol $\|$ denotes the parallel combination, is much smaller than the feedback resistor. Therefore, loading effects of $R_f$ at the input can be neglected. However, the loading at the output by $R_f$ is added to the basic amplifier as shown in the figure. The total configuration is labeled the 'basic open-loop amplifier' and can be considered the $a$ block of the feedback configuration.

In general, the current through $R_f$ is $i_f = (v_1 - v_2)/R_f$. For the assumed small input resistance of the total amplifier, this is simply

$$i_f \approx -\frac{v_2}{R_f} \tag{5.16}$$

This current through $R_f$ is identified in Figure 5.2c.

The gain of the overall feedback configuration can now be developed similar to that obtained for the block diagram of Figure 5.1a. The output voltage can be written

$$v_2 = z_{21}^a (i_s - i_f) \tag{5.17}$$

Remember that $R_S \| r_i$ is assumed to be very small with respect to $R_f$. In (5.17), $z_{21}^a$ is the transfer impedance (resistance) of the basic open-loop amplifier of Figure 5.2c.

$$z_{21}^a = -(R_S \| r_i) G_m (r_o \| R_L \| R_f) \tag{5.18}$$

Including (5.16) in (5.17), we obtain an expression for the closed-loop gain

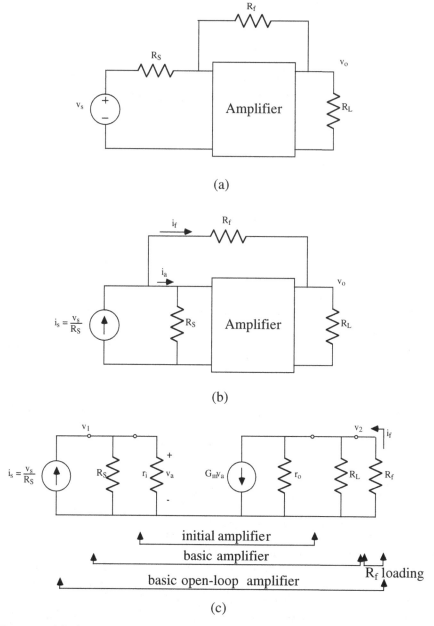

(a)

(b)

(c)

**Fig. 5.2.** (a) A typical amplifier with feedback. (b) Amplifier with signal source replaced with its Norton equivalent. (c) Circuit model of the initial and basic amplifiers.

$$\frac{v_2}{i_s} = \frac{z_{21}^a}{1 - \frac{z_{21}^a}{R_f}} \tag{5.19}$$

$$= z_{21}^T$$

Notice that the closed-loop 'gain' expression is not a voltage or current gain but an overall, total transfer impedance, $z_{21}^T$. Expression (5.19) has the desired closed-loop feedback form of the last section.

$$A = \frac{a}{1 + af} \tag{5.20}$$

Comparing (5.19) and (5.20), we can identify

$$A = A_z = z_{21}^T \tag{5.21}$$
$$a = a_z = z_{21}^a$$
$$f = f_y = \frac{-1}{R_f}$$

Thus, $A_z$ and $a_z$ are not gains and $f_y$ is not a voltage or current ratio.

$a_L$ is defined as the *loop-gain* of this configuration, i.e., the gain of the loop which has been opened. It can be identified as the ratio of the current, $i_f$, and the (Norton-equivalent) current from the signal source, $i_s$, as identified in Figure 5.2c.

$$a_L = \frac{i_f}{i_s} \tag{5.22}$$

$$= \frac{v_2}{i_s} \frac{i_f}{v_2}$$

$$= z_{21}^a \left(\frac{-1}{R_f}\right)$$

$$= a_z f_y$$

The loop gain, $a_L$, can be expressed as the product of $a_z f_y$, i.e., the product of the basic amplifier 'gain' and the feedback function. The loop gain, $a_L$, is a current gain for this configuration. However, it is not the product of two current gains. Rather it is the product of a transimpedance and a conductance.

If the basic amplifier provides a phase reversal, as in the case for Figure 5.2c when $G_m$ is positive, $z_{21}^a$ is negative and $a_L$ is a positive number. Conversely, if $a_L < 0$, the feedback is positive. The value of $(1 + a_L)$ determines the reduction of gain of the overall amplifier as feedback is applied, and also the reduction of distortion and the amount of linearization, and the desensitivity of the amplifier.

It is often convenient to deal directly with the voltage gain of the basic and overall amplifiers rather than with the transimpedances. The open-circuit available voltage from the signal source (the Thevenin-equivalent voltage), $v_s$, can be introduced, using

$$v_s = i_s R_S$$

The closed-loop voltage gain can then be written

$$A_v = \frac{v_2}{v_s} = \frac{A_z}{R_S} = \frac{\left.\frac{v_2}{v_s}\right|_a}{1 + a_L} = \frac{a_v}{1 + a_L} \tag{5.23}$$

where the notation $\left.\frac{v_2}{v_s}\right|_a$ pertains to the gain function of the $a$ block and where the open-loop voltage gain $a_v$ and the loop gain $a_L$ can be defined as

$$a_v = \left.\frac{v_2}{v_s}\right|_a = \frac{v_2^a}{i_s R_S} = \frac{z_{21}^a}{R_S} \tag{5.24}$$

$$a_L = \frac{-z_{21}^a}{R_f} = -a_v \frac{R_S}{R_f}$$

Note that $a_v$ is the voltage gain from $v_s$ (not $v_{in}$ of the amplifier) to the output voltage including the loading effects of the source, the load and feedback resistances. The loop gain, $a_L$, can now be identified as the product of a gain and a loss function. Again, for this feedback amplifier configuration, the basic gain without feedback must provide a phase reversal to produce a positive value of $a_L$ and negative feedback.

To extend these results to avoid the initial input resistance assumption, cf. (5.16), a shunt resistance equal to $R_f$ must be included at the input and output of the amplifier without feedback when determining $z_{21}^a$ or $a_v = v_2/v_s|_a$. This aspect is brought out in the following.

The configuration of Figure 5.2a is called a shunt-shunt feedback amplifier since the feedback resistance $R_f$ can be considered a two-port network as shown in Figure 5.3a. The overall configuration can be treated as two two-ports in parallel since the input and output voltages of the ports are common. The appropriate two-port parameters to describe this situation are the short-circuit admittance parameters, $y_{ij}$, where,

$$i_1 = y_{11}v_1 + y_{12}v_2 \tag{5.25}$$
$$i_2 = y_{21}v_1 + y_{22}v_2$$

An equivalent circuit for the short-circuit admittance description is shown in Figure 5.3b.

The 'natural' source description of Figure 5.3a is a Norton-equivalent current source, $i_s$. The 'natural' overall gain description is then the transimpedance of the combination, i.e., the ratio of the common output voltage, $v_2$, to the current source input which drives the common input node pair.

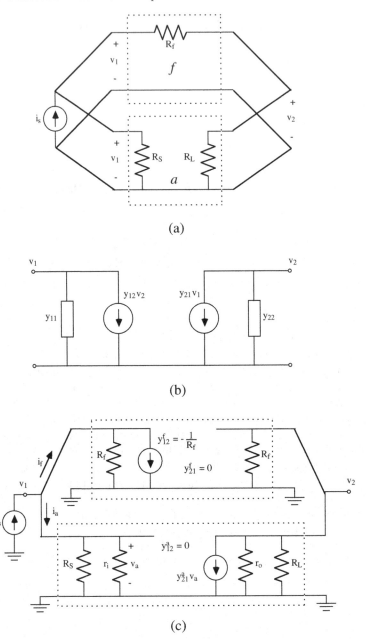

**Fig. 5.3.** (a) A two-port representation of the amplifier with shunt-shunt feedback. (b) Equivalent circuit for the short-circuit admittance parameters. (c) Loading effects of the source and load resistances.

$$\frac{v_2}{i_s} = z_{21}^T \tag{5.26}$$

$$= \frac{-y_{21}^T}{y_{11}^T y_{22}^T - y_{21}^T y_{12}^T}$$

where the short-circuit two-port parameters of the amplifier and the feedback circuit are added to provide the two-port parameters for the combination.

$$y_{ij}^T = y_{ij}^a + y_{ij}^f \tag{5.27}$$

The loading effects of the source conductance $1/R_S$ and the load conductance $1/R_L$ are included in $y_{11}^a$ and $y_{22}^a$ of the basic amplifier (the $a$ block) as illustrated in Figure 5.3c. If the numerator and denominator of (5.26) are divided by $y_{11}^T y_{22}^T$, the resulting function has the general form of a closed-loop gain function.

$$A_z = z_{21}^T = \frac{\frac{-y_{21}^T}{y_{11}^T y_{22}^T}}{1 + \left(\frac{-y_{21}^T}{y_{11}^T y_{22}^T}\right) y_{12}^T} \tag{5.28}$$

$$= \frac{a_z}{1 + a_z f_y}$$

With two approximations, appropriate gain and feedback circuits and functions can be defined for this general situation. The first approximation is that the forward transmission through the feedback resistance is much less than that for the amplifier. The second is that the reverse transmission through the amplifier is much less than that through the feedback resistance. In terms of the two-port parameters:

$$y_{21}^a \gg y_{21}^f \text{ and } y_{12}^a \ll y_{12}^f \tag{5.29}$$

For most amplifiers employing feedback deliberately, these criteria are usually well satisfied.

In Figure 5.3d, a circuit model for the $a$ block is shown. Note that it is assumed that $y_{12}^a = 0$. In the $f$ block, an equivalent circuit is used where $y_{21}^f$ is assumed equal to zero.

The loading effects of the feedback resistance (circuit) are included in the expression $a_z$, the 'gain' expression for the basic amplifier.

$$a_z = \left.\frac{v_2}{i_s}\right|_{\text{no fb}} = z_{21}^a = -\frac{y_{21}^a}{y_{11}^T y_{22}^T} \tag{5.30}$$

For the example of Figure 5.3c,

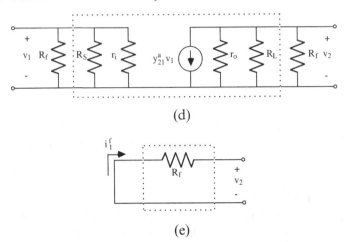

(d)

(e)

**Fig. 5.3.** (d) Circuit model for the $a$ block. (e) Definition of the short circuit parameter $y_{12}$.

$$y_{21}^T = y_{21}^a + y_{21}^f \approx y_{21}^a \qquad (5.31)$$
$$y_{11}^T = y_{11}^a + y_{11}^f$$
$$= \frac{1}{R_S \| r_i \| R_f}$$
$$y_{22}^T = \frac{1}{r_o \| R_L \| R_f}$$

From (5.28), the feedback function is

$$f_y = y_{12}^T \approx y_{12}^f \approx -\frac{1}{R_f} \qquad (5.32)$$

In terms of the definition of the short-circuit two-port parameter, $y_{12}$ is the ratio of the input current, $i_1^f$, through a short circuit and the output voltage, $v_2$, as shown in Figure 5.3e. The defined positive feedback current is the flow in $R_f$ from the input to the output.

The loop-gain function, $a_L$, is defined as

$$a_L = a_z f_y = \left( \frac{-y_{21}^a}{y_{11}^T y_{22}^T} \right) y_{12}^f \qquad (5.33)$$

Therefore, we have now a set of identifiable $a$, $f$, and $a_L$ functions in terms of particular circuit configurations. Care must be taken with respect to the signs of $a_z$, $f_y$, and $a_L$. The closed-loop gain function appears as Equation (5.28). The denominator of the gain function is $(1 + a_z f_y) = (1 + a_L)$, and the loop-gain function is $a_L = a_z f_y$. The effects of feedback are calculated from $(1 + a_L)$, a positive number for a negative feedback situation.

We return now to the nonlinear situation. It must first be emphasized that the above developments strictly hold only for a linear amplifier. Therefore, we must restrict attention to only reasonably small variations about a quiescent operating point. For the total open-loop amplifier, including necessary loading conditions of $R_f$, $R_S$, and $R_L$, a power series can be used to describe the transfer characteristic. For a shunt-shunt configuration, the output variable, $v_2$, is the output voltage of the combination as shown in Figure 5.3c, while the input variable, $i_s$, is the signal-source current (the short-circuit available current). The I/O variables are increments about the operating point, $V_2$ and $I_s$.

## 5.3 A CE Stage with Shunt Feedback

As an example to illustrate the material of the last section, consider the simple one-stage, common-emitter amplifier shown in Figure 5.4a. The frequency of interest is assumed to be sufficiently high so that the effects of the coupling capacitor $C_f$ can be neglected. A bias state of 1 mA is established by the dc (quiescent) value of the input source voltage. A circuit to determine the small-signal, 'open-loop' gain is shown in Figure 5.4b. Note that the signal-source resistance, the load resistance, and the feedback resistance are included. For the values given in the Spice input file of Figure 5.5,

$$a_v = \frac{-g_m R_L}{1 + \frac{R_S}{r_\pi \| R_f}} \left(1 + \frac{R_L}{R_f}\right)^{-1} \approx -87 \tag{5.34}$$

where the symbol $\|$ indicates the parallel combination of two resistances, and $g_m = I_C/V_t = (1/25.85)$ ℧ for the assumed operating point value of 1 mA for $I_C$. The value of the loop gain is obtained from Equation (5.34), by multiplying by $-R_S/R_f$, cf. (5.24).

$$a_L = -(-87)\frac{1 \text{ k}\Omega}{10 \text{ k}\Omega} = +8.7 \tag{5.35}$$

$1 + a_L = 9.7$ is the amount by which the gain is reduced with feedback, e.g., $A_v = -87/9.7 = -8.97$ and the amount by which the $HD_2$ is reduced for constant output amplitude, as described in the Section 5.1. These values of gain and distortion reduction compare favorably with results obtained using a Spice simulation.

For the Spice input file given in Figure 5.5, three stages are included in the simulation. The first is the closed-loop amplifier, the second is an open-loop stage including the proper loading of $R_f$ as above, and the third is for an open-loop stage without feedback loading. The Fourier output harmonic voltages are given in Figures 5.6a, b, and c. From the values of Figure 5.6b for the basic amplifier without feedback but with $R_f$ loading,

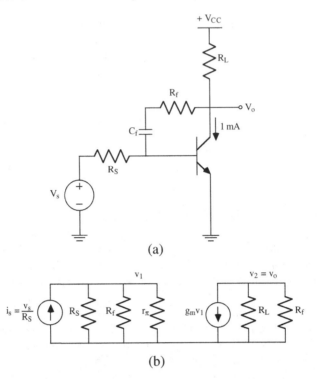

**Fig. 5.4.** (a) A simple one-stage common-emitter amplifier. (b) Circuit to determine the small-signal 'open-loop' gain.

$$| \, a_v \, | = 85.98 \tag{5.36}$$
$$HD_2 = 4.73\%$$

From Figure 5.6a for the stage with feedback,

$$| \, A_v \, | = 8.91 \tag{5.37}$$
$$HD_2 = 0.050\%$$

The gain is reduced by a factor of 9.65 by the feedback. The distortion is reduced by $95.3 = (9.76)^2$, which is approximately the square of the gain reduction, since the input amplitude is held constant.

From Figure 5.6c for the stage without $R_f$ loading, the open-loop gain is a factor of 15.5 larger than the closed-loop gain. The value of $HD_2$ for the open-loop stage without $R_f$ loading is 4.9% and is larger than the corresponding value of Figure 5.6b primarily because of the removal of the loading of $R_f$ at the input. The predicted level of $HD_2$ for the closed-loop stage on the basis of the gain reduction of 15.5 is 0.02%. Thus, the third stage does not provide a correct estimate of the loop-gain.

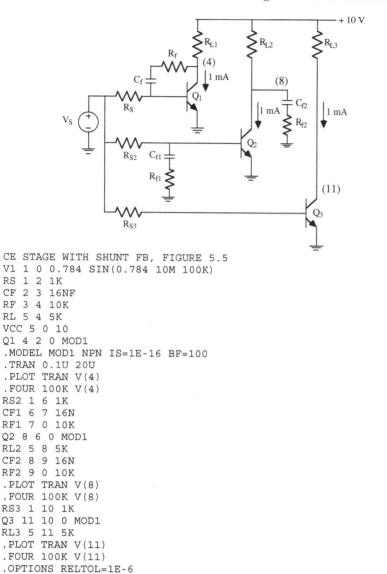

```
CE STAGE WITH SHUNT FB, FIGURE 5.5
V1 1 0 0.784 SIN(0.784 10M 100K)
RS 1 2 1K
CF 2 3 16NF
RF 3 4 10K
RL 5 4 5K
VCC 5 0 10
Q1 4 2 0 MOD1
.MODEL MOD1 NPN IS=1E-16 BF=100
.TRAN 0.1U 20U
.PLOT TRAN V(4)
.FOUR 100K V(4)
RS2 1 6 1K
CF1 6 7 16N
RF1 7 0 10K
Q2 8 6 0 MOD1
RL2 5 8 5K
CF2 8 9 16N
RF2 9 0 10K
.PLOT TRAN V(8)
.FOUR 100K V(8)
RS3 1 10 1K
Q3 11 10 0 MOD1
RL3 5 11 5K
.PLOT TRAN V(11)
.FOUR 100K V(11)
.OPTIONS RELTOL=1E-6
.WIDTH OUT=80
.END
```

| NODE | VOLTAGE | NODE | VOLTAGE | NODE | VOLTAGE | NODE | VOLTAGE |
|------|---------|------|---------|------|---------|------|---------|
| ( 1) | 0.7840 | ( 2) | 0.7741 | ( 3) | 5.0290 | ( 4) | 5.0290 |
| ( 5) | 10.0000 | ( 6) | 0.7741 | ( 7) | 0.0000 | ( 8) | 5.0290 |
| ( 9) | 0.0000 | ( 10) | 0.7741 | ( 11) | 5.0290 | | |

**Fig. 5.5.** Circuit, Spice input file, and dc operating point of three amplifier stages.

```
FOURIER COMPONENTS OF TRANSIENT RESPONSE V(4)
DC COMPONENT =   5.028D+00
HARMONIC   FREQUENCY    FOURIER     NORMALIZED    PHASE      NORMALIZED
  NO         (HZ)      COMPONENT    COMPONENT     (DEG)     PHASE (DEG)

   1      1.000D+05   8.913D-02    1.000000     179.468       0.000
   2      2.000D+05   4.416D-05    0.000496      89.616     -89.852
   3      3.000D+05   5.394D-07    0.000006      80.662     -98.806
   4      4.000D+05   9.548D-07    0.000011     100.561     -78.907

     TOTAL HARMONIC DISTORTION =     .049595  PERCENT
```

(a)

```
FOURIER COMPONENTS OF TRANSIENT RESPONSE V(8)
DC COMPONENT =   4.983D+00
HARMONIC   FREQUENCY    FOURIER     NORMALIZED    PHASE      NORMALIZED
  NO         (HZ)      COMPONENT    COMPONENT     (DEG)     PHASE (DEG)

   1      1.000D+05   8.598D-01    1.000000     179.771       0.000
   2      2.000D+05   4.066D-02    0.047295      89.706     -90.064
   3      3.000D+05   4.459D-04    0.000519       4.132    -175.639
   4      4.000D+05   7.797D-05    0.000091      57.615    -122.155

     TOTAL HARMONIC DISTORTION =    4.729808  PERCENT
```

(b)

```
FOURIER COMPONENTS OF TRANSIENT RESPONSE V(11)
DC COMPONENT =   4.959D+00
HARMONIC   FREQUENCY    FOURIER     NORMALIZED    PHASE      NORMALIZED
  NO         (HZ)      COMPONENT    COMPONENT     (DEG)     PHASE (DEG)

   1      1.000D+05   1.383D+00    1.000000     179.998       0.000
   2      2.000D+05   6.829D-02    0.049393      90.008     -89.990
   3      3.000D+05   5.127D-04    0.000371       5.318    -174.680
   4      4.000D+05   1.156D-04    0.000084      96.468     -83.530

     TOTAL HARMONIC DISTORTION =    4.939443  PERCENT
```

(c)

**Fig. 5.6.** Fourier components of the output voltage for (a) closed-loop amplifier, (b) open-loop amplifier including feedback loading, and (c) open-loop amplifier without feedback loading.

In Section 9.1, another feedback circuit is studied using the parallel two-port configuration.

## 5.4 The CE Stage With Emitter Feedback

In Figure 5.7a, a common-emitter stage is shown including an external resistor $R_e$ in series with the emitter. Negative feedback is produced by the series (feedback) resistor. As pointed out in [6], Section 8.6.1, this simple feedback circuit can be recognized as a degenerate form of a series-series combination of an amplifier two-port and a feedback resistive two-port. The two-port combination is developed by redrawing the stage including floating dc voltage sources as in Figure 5.7b. When a small-signal circuit model is included, the result is that of Figure 5.7c. Note that $i_1^T = i_1^a = i_1^f$ and that $v_1 = v_1^a + v_1^f$ or $v_1^a = v_1 - v_1^f$.

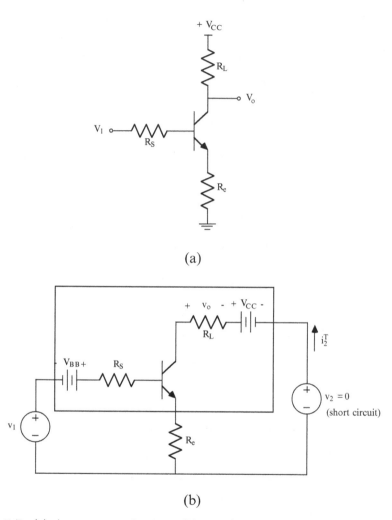

(a)

(b)

**Fig. 5.7.** (a) A common-emitter amplifier with emitter feedback. (b) Two-port representation of the amplifier.

The natural two-port description of the series-series combination is the mathematical dual of that of the shunt-shunt combination used in Section 5.2. The open-circuit impedance parameters are now used in place of the short-circuit admittance parameters for the shunt-shunt combination.

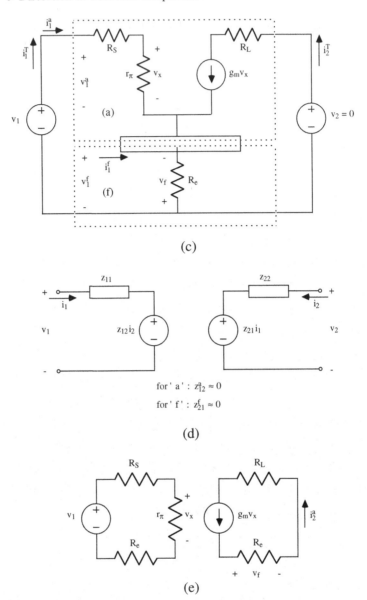

**Fig. 5.7.** (c) Small-signal two-port network representation. (d) An equivalent circuit for the open-circuit impedance parameters. (e) Open-loop small-signal circuit used for calculations.

$$v_1 = z_{11}i_1 + z_{12}i_2 \tag{5.38}$$
$$v_2 = z_{21}i_1 + z_{22}i_2$$

An equivalent circuit is given in Figure 5.7d.

As shown in Figures 5.7b and c, if the amplifier load resistance $R_L$ is included in the basic amplifier, the load of the combination is an equivalent short circuit and $v_2 = v_2^T = 0$. The output variable is taken as the short-circuit output current $i_2 = i_2^T$. Similarly, the signal-source resistance is included in the basic amplifier and $v_1$ is the input variable. The desired input voltage-to-output current transfer function is $y_{21}^T$.

$$\frac{i_2}{v_1} = y_{21}^T = \frac{-\frac{z_{21}^T}{z_{11}^T z_{22}^T}}{1 + \left(-\frac{z_{21}^T}{z_{11}^T z_{22}^T}\right) z_{12}^T} \tag{5.39}$$

$$\approx \frac{a_y}{1 + a_y f_z}$$

$$\approx \frac{a_y}{1 + a_L}$$

where

$$a_y = \frac{i_2^a}{v_1} = y_{21}^a = \frac{-z_{21}^a}{z_{11}^T z_{22}^T} \tag{5.40}$$

$$f_z = \frac{v_f}{i_2^a} = z_{12}^f = R_e$$

$$a_L = a_y f_z$$

$$= y_{21}^a z_{12}^f$$

Similar to the shunt-shunt case, note that $f$ is not a fraction of the output variable. The necessary conditions to identify these functions simply are:

$$z_{21}^a \gg z_{21}^f = z_{12}^f \tag{5.41}$$

$$z_{12}^a \ll z_{12}^f$$

The open-loop, small-signal circuit with which to calculate the functions in (5.40) is shown in Figure 5.7e. Note that the loading of the feedback resistor is included in both the input and output segments of the circuit. The 'open-loop' feedback voltage, $v_f$, appears across the $R_e$ in the output segment of the circuit as shown in the figure. If the transistor does not provide a finite output resistance, i.e., if $r_o = V_A/I_C = \infty$, the loading effects of $R_e$ on the output segment are absent. However, the voltage drop across this $R_e$ is $v_f$ as mentioned above.

It is often convenient to work with the voltage-gain function, $\frac{v_o}{v_s}$. This is easily obtained noting in Figure 5.7b that $v_o = -R_L i_2$.

$$A_v = \frac{v_o}{v_s} = \frac{-R_L i_2}{v_1} \tag{5.42}$$

$$= \frac{a_v}{1 + a_L}$$

where

$$a_L = a_v f' \tag{5.43}$$

$$a_v = -a_y R_L$$

$$f' = \frac{z_{12}^f}{R_L} = \frac{R_e}{R_L}$$

where $f'$ is the original $f_z$ divided by $R_L$ and can be identified as an output voltage ratio. From Figure 5.7e,

$$a_v = -g_m \frac{R_L}{\left(1 + \frac{R_S + R_e}{r_\pi}\right)} \tag{5.44}$$

$$f' = \frac{R_e}{R_L}$$

$$a_L = g_m \frac{R_e}{\left(1 + \frac{R_S + R_e}{r_\pi}\right)}$$

For $(R_S + R_e) \ll r_\pi$, the loop gain is $g_m R_e$ as expected.

As brought out in Section 5.1, $HD_3$ can be eliminated in some cases. For the CE stage with an emitter resistance, the condition is very simple if $(R_S + R_e) \ll r_\pi$ and $a_L \approx g_m R_e$. For this circuit, the incremental collector current is first expressed as a power series. From Chapter 3,

$$i_c = \left(\frac{I_{CA}}{V_t}\right) v_i + \left(\frac{I_{CA}}{2V_t^2}\right) v_i^2 \tag{5.45}$$

$$+ \left(\frac{I_{CA}}{6V_t^3}\right) v_i^3 + \dots$$

$$= a_1 v_i + a_2 v_i^2 + a_3 v_i^3 + \dots$$

Substitution of these coefficients into (5.11) yields

$$HD_3\big|_{\text{w fb}} = \frac{|1 - 2a_L|}{(1 + a_L)^4} HD_3\big|_{\text{w/o fb}} \tag{5.46}$$

The coefficients of (5.45) for the BJT lead to a very simple form for $HD_3$. The condition for the cancellation of $HD_3$ is

$$a_L = \frac{1}{2} \approx g_m R_e \qquad (5.47)$$

Equation (5.46) for $HD_3$ has the same form as (4.27) for the CE stage with $R_S$. Because of the similarity of (5.46) and (5.27), the condition to eliminate $HD_3$ is similar, $a_L = \frac{1}{2}$ and $E = \frac{1}{2}$. The equivalence in terms of circuit parameters is particularly simple for the situation where the loop gain is approximately equal to $g_m R_e$. The equivalent resistance of the signal source to reduce the voltage gain by $(1+g_m R_e)$ is $R_S = (\beta+1)R_e$. Using this relation in the expressions for $HD_2$ and $HD_3$, we find that the two stages are the same if $\beta_{ac} = \beta$.

For a numerical example of the $HD_3$ elimination; the reference CE stage without feedback or without a source resistance is that associated with $Q_1$ in the Spice input file of Figure 5.8a. The quiescent collector current is approximately 1 mA for the bias voltage of 0.774 V. Three other stages are also included in Figure 5.8a. Associated with $Q_2$ is the emitter resistor, $R_{e2} = \frac{1}{2g_m} = 12.92\ \Omega$. For this stage $a_L = \frac{1}{2}$ and $HD_3$ should be eliminated. Associated with $Q_3$ is a source resistance of $R_{S3} = (\beta+1)R_{e2} = 1.305$ k$\Omega$. As noted above, and from Chapter 4, this stage should also eliminate $HD_3$. For $Q_4$, $R_{e4}$ and $R_{S4}$ are one-half of the values of $R_{e2}$ and $R_{S3}$. Again, $HD_3$ should be eliminated.

Care must be taken in using (4.27) and (5.46) simultaneously. Both the E parameter and the loop gain $a_L$ depend in general on $R_S$ and $R_e$. The best procedure is to use one or the other by reflecting $R_e$ to an equivalent addition to $R_S$ or by reflecting $R_S$ to an equivalent addition to $R_e$, i.e., use either an equivalent total E or an equivalent total $a_L$.

For a 10 mV sinusoidal input amplitude, the Spice waveforms of the output voltage are virtually identical. The Spice Fourier outputs for all four stages are given in Figure 5.8b. $HD_3$ has been significantly reduced for the last three configurations. The $HD_3$ values for the last three cases are in the 'numerical-noise level' of the simulation. Note that $HD_2$ has been reduced in each of the last three cases by $(1+0.5)^2$, as expected.

As mentioned in Chapter 1, great care must be taken when requesting very small numbers from the Spice programs. An adequate number of points per period must be chosen for harmonic distortion analyses. Further, the value of RELTOL, the accuracy parameter in Spice, must be made small.

To return to the example, if a 20 mV drive is used, the reduction of $HD_3$ is not as large as for 10 mV, only about 140 instead of 197. Contributions from higher-order terms account for this difference.

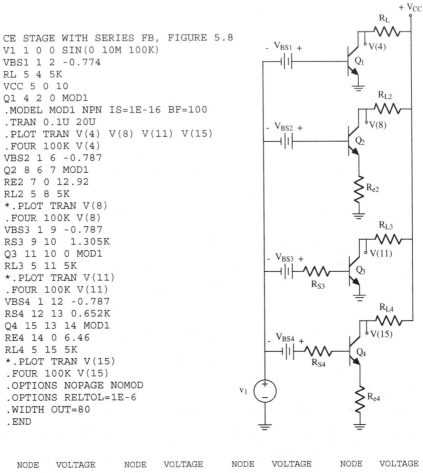

```
CE STAGE WITH SERIES FB, FIGURE 5.8
V1 1 0 0 SIN(0 10M 100K)
VBS1 1 2 -0.774
RL 5 4 5K
VCC 5 0 10
Q1 4 2 0 MOD1
.MODEL MOD1 NPN IS=1E-16 BF=100
.TRAN 0.1U 20U
.PLOT TRAN V(4) V(8) V(11) V(15)
.FOUR 100K V(4)
VBS2 1 6 -0.787
Q2 8 6 7 MOD1
RE2 7 0 12.92
RL2 5 8 5K
*.PLOT TRAN V(8)
.FOUR 100K V(8)
VBS3 1 9 -0.787
RS3 9 10  1.305K
Q3 11 10 0 MOD1
RL3 5 11 5K
*.PLOT TRAN V(11)
.FOUR 100K V(11)
VBS4 1 12 -0.787
RS4 12 13 0.652K
Q4 15 13 14 MOD1
RE4 14 0 6.46
RL4 5 15 5K
*.PLOT TRAN V(15)
.FOUR 100K V(15)
.OPTIONS NOPAGE NOMOD
.OPTIONS RELTOL=1E-6
.WIDTH OUT=80
.END
```

| NODE | VOLTAGE | NODE | VOLTAGE | NODE | VOLTAGE | NODE | VOLTAGE |
|------|---------|------|---------|------|---------|------|---------|
| ( 1) | 0.0000 | ( 2) | 0.7740 | ( 4) | 5.0401 | ( 5) | 10.0000 |
| ( 6) | 0.7870 | ( 7) | 0.0130 | ( 8) | 5.0330 | ( 9) | 0.7870 |
| ( 10) | 0.7740 | ( 11) | 5.0331 | ( 12) | 0.7870 | ( 13) | 0.7805 |
| ( 14) | 0.0065 | ( 15) | 5.0324 | | | | |

**Fig. 5.8.** (a) Circuit, Spice input file, and dc operating point for common-emitter amplifier with series feedback.

## 5.5 Alternative Loop-Gain Calculations

It is to be emphasized that the combined two-port technique is not a necessary method of solution. But it does provide the setting to establish the necessary and fundamental assumptions to employ the feedback equation as well as to include properly the loading effects of the feedback network.

If the loop gain is of primary interest, it is sometimes possible to 'break' or open the loop at a location convenient for a simple circuit evaluation. As

```
FOURIER COMPONENTS OF TRANSIENT RESPONSE V(4)
DC COMPONENT =    4.853D+00
```

| HARMONIC NO | FREQUENCY (HZ) | FOURIER COMPONENT | NORMALIZED COMPONENT | PHASE (DEG) | NORMALIZED PHASE (DEG) |
|---|---|---|---|---|---|
| 1 | 1.000D+05 | 1.943D+00 | 1.000000 | 179.996 | 0.000 |
| 2 | 2.000D+05 | 1.838D-01 | 0.094586 | 90.007 | -89.990 |
| 3 | 3.000D+05 | 1.147D-02 | 0.005900 | 0.584 | -179.412 |
| 4 | 4.000D+05 | 4.156D-04 | 0.000214 | -94.510 | -274.506 |

```
TOTAL HARMONIC DISTORTION =    9.477044  PERCENT
```

$$R_S = R_e = 0$$

```
FOURIER COMPONENTS OF TRANSIENT RESPONSE V(8)
DC COMPONENT =    4.978D+00
```

| HARMONIC NO | FREQUENCY (HZ) | FOURIER COMPONENT | NORMALIZED COMPONENT | PHASE (DEG) | NORMALIZED PHASE (DEG) |
|---|---|---|---|---|---|
| 1 | 1.000D+05 | 1.273D+00 | 1.000000 | 179.998 | 0.000 |
| 2 | 2.000D+05 | 5.353D-02 | 0.042066 | 90.009 | -89.989 |
| 3 | 3.000D+05 | 3.830D-05 | 0.000030 | 95.737 | -84.261 |
| 4 | 4.000D+05 | 8.810D-05 | 0.000069 | 96.754 | -83.244 |

```
TOTAL HARMONIC DISTORTION =    4.206633  PERCENT
```

$$R_e > 0$$

```
FOURIER COMPONENTS OF TRANSIENT RESPONSE V(11)
DC COMPONENT =    4.978D+00
```

| HARMONIC NO | FREQUENCY (HZ) | FOURIER COMPONENT | NORMALIZED COMPONENT | PHASE (DEG) | NORMALIZED PHASE (DEG) |
|---|---|---|---|---|---|
| 1 | 1.000D+05 | 1.272D+00 | 1.000000 | 179.998 | 0.000 |
| 2 | 2.000D+05 | 5.353D-02 | 0.042065 | 90.009 | -89.989 |
| 3 | 3.000D+05 | 3.830D-05 | 0.000030 | 95.828 | -84.170 |
| 4 | 4.000D+05 | 8.809D-05 | 0.000069 | 96.754 | -83.244 |

```
TOTAL HARMONIC DISTORTION =    4.206517  PERCENT
```

$$R_S > 0$$

```
FOURIER COMPONENTS OF TRANSIENT RESPONSE V(15)
DC COMPONENT =    4.978D+00
```

| HARMONIC NO | FREQUENCY (HZ) | FOURIER COMPONENT | NORMALIZED COMPONENT | PHASE (DEG) | NORMALIZED PHASE (DEG) |
|---|---|---|---|---|---|
| 1 | 1.000D+05 | 1.273D+00 | 1.000000 | 179.998 | 0.000 |
| 2 | 2.000D+05 | 5.355D-02 | 0.042073 | 90.009 | -89.989 |
| 3 | 3.000D+05 | 3.828D-05 | 0.000030 | 95.221 | -84.777 |
| 4 | 4.000D+05 | 8.813D-05 | 0.000069 | 96.754 | -83.245 |

```
TOTAL HARMONIC DISTORTION =    4.207286  PERCENT
```

$$R_S, R_e > 0$$

**Fig. 5.8.** (b) Fourier components of the output voltages.

an example, consider the feedback configuration of Figure 5.7c. The output is driven by the dependent current source, $g_m v_x$. The feedback loop can be opened by returning $g_m v_x$ to ac ground (or to a dc bias voltage source) as shown in Figure 5.9a. The current to ground is the loop response current, labeled $i_o$. The loop can be excited by an independent current source $i_i$ as shown in the figure. Since the circuit relations and interactions are unchanged, the loop gain is the negative of the ratio of $i_o/i_i$.

$$a_L = -\frac{i_o}{i_i} = \frac{g_m R_e}{\left(1 + \frac{R_S + R_e}{r_\pi}\right)} \tag{5.48}$$

This is the same as that obtained for the two-port approach, (5.44).

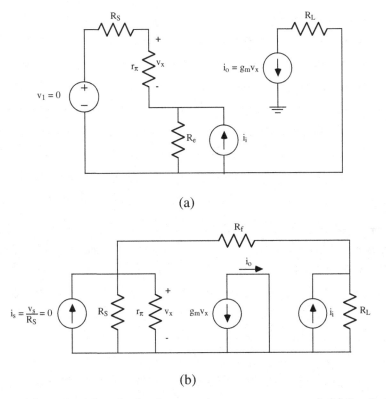

(a)

(b)

**Fig. 5.9.** (a) Feedback loop broken by returning $g_m v_x$ to ac ground. (b) Small-signal circuit for common-emitter amplifier with shunt-shunt feedback.

As a second example, consider the CE stage with shunt-shunt feedback, as in Figure 5.4a. The small-signal model is given in Figure 5.9b. Note that the feedback loop is broken at the $g_m v_x$ dependent current source by returning it to ac ground. The loop is excited by a new independent current source, $i_i$. The loop gain from simple circuit analysis is:

$$a_L = -\frac{i_o}{i_i} = \frac{R_i}{R_L + R_f + R_i} g_m R_L \tag{5.49}$$

The loop gain obtained from the two-port approach can be determined for Figure 5.4b.

$$a_L = \frac{R_f}{(R_i + R_f)} \frac{R_i}{(R_L + R_f)} g_m R_L \tag{5.50}$$

The difference between (5.49) and (5.50) is due to the forward transmission to the output through $R_f$, which is neglected in the two-port approach. The two expressions are approximately equal if $\frac{R_L R_i}{R_f} < (R_i + R_L + R_f)$ which is usually satisfied for a practical amplifier stage.

## 5.6 Emitter Feedback in the ECP

Negative feedback can be introduced into the emitter-coupled pair by adding in each emitter lead a series resistor, as shown in Figure 5.10a. The expected effects of negative feedback are realized, e.g., linearization of the transfer characteristic as shown in Figure 5.10b and the reduction of distortion at the expense of reduced overall gain. At the input,

$$V_1 = V_{BE1} - V_{BE2} + (I_{E1} - I_{E2})R_e \tag{5.51}$$

Once again it is assumed that the transistor parameter $\beta$ is large and that the base currents can be neglected in relation to the collector currents. This leads to

$$I_{C1} \approx \mid I_{E1} \mid \tag{5.52}$$
$$I_{C2} \approx \mid I_{E2} \mid$$
$$I_{C1} + I_{C2} \approx I_{EE}$$

$$V_1 = V_t \ln \frac{I_{C1}}{I_{C2}} + (2I_{C1} - I_{EE})R_e \tag{5.53}$$

The first term of (5.53) is the earlier result of Chapter 2. The second term brings in the effects of feedback. For large values of $R_e$,

$$V_1 \approx (2I_{C1} - I_{EE})R_e \tag{5.54}$$

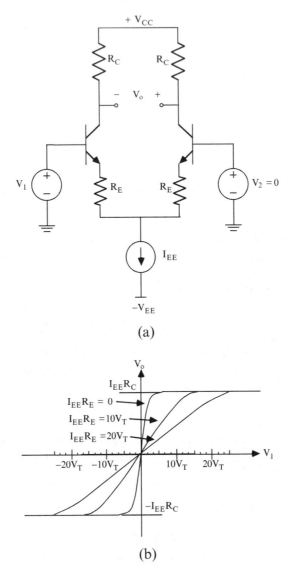

**Fig. 5.10.** (a) EC pair with series resistance in each emitter lead. (b) Linearization of the transfer characteristics of the EC pair by use of emitter resistances.

$$I_{C1} = \frac{1}{2}\left(\frac{V_1}{R_e} + I_{EE}\right)$$

$$I_{C2} = \frac{1}{2}\left(-\frac{V_1}{R_e} + I_{EE}\right)$$

$I_{C1}$ and $I_{C2}$ are now approximate linear functions of the input voltage for current values near the quiescent state as shown in Figure 5.10b. However, for increasing values of $I_{C1}$ (decreasing values of $I_{C2}$) the 'ln' term in Equation (5.53) takes over. Since the output voltages and thus the voltage gains are linearly proportional to the collector currents, it is clear that as the slope of the transfer characteristic is reduced by $R_e$, the voltage gain is reduced with the feedback, as expected.

The distortion reduction due to the emitter feedback can be estimated using the results obtained in Section 4.3. For the ECP with emitter feedback, the input voltage expression can be written

$$V_1 = V_{BE1} + R_e(\beta+1)I_{B1} - R_e(\beta+1)I_{B2} - V_{BE2} \tag{5.55}$$

Expressing the variables in terms of the dc and incremental components yields

$$v_1 = (\beta+1)R_e(i_{b1} - i_{b2}) + v_{be1} - v_{be2} \tag{5.56}$$

This is the same as (4.42) if $R_S$ is replaced with $(\beta+1)R_e$. Thus, the developments of Section 4.3 can be used directly after making the resistance substitution. For the present case

$$a_L = E = (\beta+1)\frac{R_e}{r_\pi} \tag{5.57}$$

$$= \frac{\beta+1}{\beta}g_m R_e$$

$$\approx g_m R_e$$

$$HD_3 = \frac{1}{48}\frac{1}{(1+a_L)^3}\left(\frac{V_{1A}}{V_t}\right)^2 \tag{5.58}$$

$$= \frac{1}{(1+a_L)^3}HD_3\bigg|_{R_e=0}$$

The input file for a Spice simulation is shown in Figure 5.10c. The circuit values include $I_{EE} = 2$ mA, $V_{CC} = 10$ V, $R_e = 25.85\ \Omega$, $R_{C1} = R_{C2} = 5$ k$\Omega$. The amplitude of the input sinusoid is chosen to be 10 mV. (A second ECP is also included, which does not have emitter feedback, to provide reference values.) The estimated gain values are

$$g_m = \frac{1}{25.85} \tag{5.59}$$
$$g_m R_e = 1$$
$$1 + a_L = 2$$
$$g_m R_C = 192$$
$$A_v = 48$$

The expected amplitude of the output excursion is $(10 \text{ mV})(48) = \pm 0.48$ V. From the values of Figure 5.10d, $HD_3$ without feedback is 0.295%. With feedback, the estimated value of $HD_3$ should be $2^3 = 8$ times less than the value without feedback, or $0.295\% \div 8 = 0.037\%$. From Figure 5.10e, $HD_3 = .037\%$. As shown in the outputs from Spice in Figures 5.10d and e, the amplitude of the fundamental output component without feedback is 0.943 V and 0.476 V with emitter feedback, a decrease of approximately 2.

## 5.7 Internal Feedback in the ECP and the SCP

Inherent, internal feedback is present in the ECP and the SCP due to several mechanisms. In terms of the circuit elements of the small-signal circuit model of a BJT, feedback arises due to resistive elements such as $r_b'$ and $r_o$ as well as to the capacitive elements such as $C_{jc}$. In this section, attention is given to the feedback effects due to basewidth modulation in bipolar transistors and channel-length modulation in MOS devices. These effects are modeled in Spice by the parameters, VA and LAMBDA, respectively [3]. In circuit terms, basewidth or channel-length modulation introduces an output resistance, $r_o$, in the circuit model of the transistor; for the BJT this element is connected from the collector to emitter, as shown in Figure 5.11a.

The effects of the feedback for several different situations involving different load and feedback resistances from Spice simulation data are given in Table 5.1. The device and circuit values are given in the Spice input file of Figure 5.11b. For the moment, actual resistors $R_{o1}$ and $R_{o2}$ are used rather than the internal $r_o$ of the transistors in order to keep separate the components of the collector current.

In the first row of the table are the values from a .TF run of the voltage gain, input and output resistance for an ideal ECP, i.e., with the $R_{oi}$ absent. In Row 2 are the data from a .TF run for an ECP with $R_{o1} = R_{o2} = 50 \text{ k}\Omega$, corresponding approximately to including $V_A = 50$ V in the transistor models. Because of the change in collectors currents due to the addition of the resistors, normalizations of the .TF values are made with respect to $I_C$-dependent parameters, as shown in the headings of the last three columns. Note that for this balanced situation the normalized voltage gain drops by 10%, the normalized input resistance remains unchanged, and the normalized output

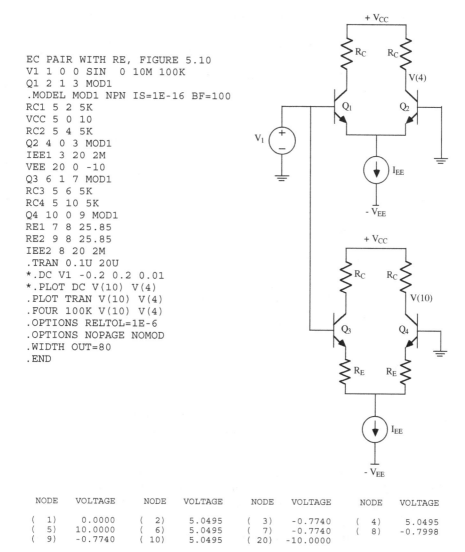

```
EC PAIR WITH RE, FIGURE 5.10
V1 1 0 0 SIN  0 10M 100K
Q1 2 1 3 MOD1
.MODEL MOD1 NPN IS=1E-16 BF=100
RC1 5 2 5K
VCC 5 0 10
RC2 5 4 5K
Q2 4 0 3 MOD1
IEE1 3 20 2M
VEE 20 0 -10
Q3 6 1 7 MOD1
RC3 5 6 5K
RC4 5 10 5K
Q4 10 0 9 MOD1
RE1 7 8 25.85
RE2 9 8 25.85
IEE2 8 20 2M
.TRAN 0.1U 20U
*.DC V1 -0.2 0.2 0.01
*.PLOT DC V(10) V(4)
.PLOT TRAN V(10) V(4)
.FOUR 100K V(10) V(4)
.OPTIONS RELTOL=1E-6
.OPTIONS NOPAGE NOMOD
.WIDTH OUT=80
.END
```

| NODE | VOLTAGE | NODE | VOLTAGE | NODE | VOLTAGE | NODE | VOLTAGE |
|------|---------|------|---------|------|---------|------|---------|
| ( 1) | 0.0000 | ( 2) | 5.0495 | ( 3) | -0.7740 | ( 4) | 5.0495 |
| ( 5) | 10.0000 | ( 6) | 5.0495 | ( 7) | -0.7740 | ( 8) | -0.7998 |
| ( 9) | -0.7740 | ( 10) | 5.0495 | ( 20) | -10.0000 | | |

**Fig. 5.10.** (c) Circuit and Spice input file for EC pairs with and without emitter resistances.

resistance drops by 5%. For a balanced ECP, the output voltages, $V_{o1}$ and $V_{o2}$, are out of phase; thus, for equal $r_o$ for the two transistors, the feedback currents through the resistors into the common emitter node are equal and opposite. The feedback due to the $r_o$ is negative for $Q_1$ and positive for $Q_2$. On a normalized basis, the input resistance should remain unchanged because of the negative and positive feedback introduced to the common-emitter node. The overall voltage gain of the ECP is reduced by $\left(1 + \frac{R_L}{r_o}\right)$. To appreciate

```
FOURIER COMPONENTS OF TRANSIENT RESPONSE V(4)
DC COMPONENT =    5.049D+00
HARMONIC   FREQUENCY     FOURIER     NORMALIZED      PHASE     NORMALIZED
   NO        (HZ)       COMPONENT    COMPONENT      (DEG)     PHASE (DEG)

   1      1.000D+05    9.432D-01     1.000000      0.000        0.000
   2      2.000D+05    2.016D-05     0.000021     81.879       81.879
   3      3.000D+05    2.784D-03     0.002951     -0.092       -0.092
   4      4.000D+05    8.276D-06     0.000009    -84.203      -84.203
   5      5.000D+05    1.454D-05     0.000015    -24.079      -24.078
   6      6.000D+05    4.857D-06     0.000005    -54.029      -54.029
   7      7.000D+05    4.817D-06     0.000005    -66.910      -66.909
   8      8.000D+05    5.779D-06     0.000006    -63.846      -63.846
   9      9.000D+05    5.961D-06     0.000006    -52.771      -52.771

      TOTAL HARMONIC DISTORTION =          0.295136  PERCENT
```

(d)

```
FOURIER COMPONENTS OF TRANSIENT RESPONSE V(10)
DC COMPONENT =    5.049D+00
HARMONIC   FREQUENCY     FOURIER     NORMALIZED      PHASE     NORMALIZED
   NO        (HZ)       COMPONENT    COMPONENT      (DEG)     PHASE (DEG)

   1      1.000D+05    4.756D-01     1.000000      0.000        0.000
   2      2.000D+05    1.063D-05     0.000022     82.456       82.457
   3      3.000D+05    1.750D-04     0.000368     -0.676       -0.675
   4      4.000D+05    4.265D-06     0.000009    -85.345      -85.345
   5      5.000D+05    3.385D-06     0.000007    -56.439      -56.438
   6      6.000D+05    2.243D-06     0.000005    -52.454      -52.454
   7      7.000D+05    2.222D-06     0.000005    -68.084      -68.083
   8      8.000D+05    2.735D-06     0.000006    -64.485      -64.484
   9      9.000D+05    2.829D-06     0.000006    -52.681      -52.681

      TOTAL HARMONIC DISTORTION =          0.036889  PERCENT
```

(e)

**Fig. 5.10.** (d) Fourier components of output voltage without emitter resistances. (e) Fourier components of output voltage with emitter resistances.

the latter, recognize that with both feedback paths present, a balanced pair is obtained. The gain can then be calculated from a differential half-circuit. The resulting $R_L$ is loaded by $r_o$. There is a 5% decrease in the normalized value of $R_{out}$.

If $R_{C1}$ is reduced to a very low value, say 0.1 $\Omega$ , the negative feedback due to $R_{o1}$ is negligible and only the positive feedback effects, due to $R_{o2}$, remain. This circumstance often occurs in bandpass amplifier stages where high-frequency performance is critical, cf., Chapter 9. Notice that in relation to the balanced circuit, the normalized voltage gain in Row 3 of the table increases by about 5% as does the input resistance. The normalized output resistance is unchanged in relation to the balanced case. An evaluation of this feedback situation is given below. In the last row of the table, data is given for the case where $R_{o1}$ is removed. The same relative data is obtained as that of Row 3; however, the actual values of $A_v$, etc., are different because of the change of quiescent bias-current levels.

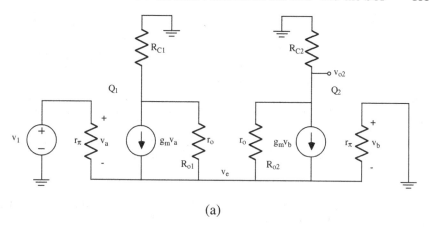

(a)

FIGURE 5.11
V1 1 0 0 SIN 0 10M 100K
.TF V(4) V1
Q1 2 1 3 MOD1
.MODEL MOD1 NPN IS=1E-16 BF=100
*+VA=50
RC1 5 2 5K
*RC1 5 2 0.1
RO1 2 3 50K
VCC 5 0 10
RC2 5 4 5K
Q2 4 0 3 MOD2
.MODEL MOD2 NPN IS=1E-16 BF=100
*+VA=50
RO2 4 3 50K
IEE1 3 20 2M
VEE 20 0 -10
*.TRAN 0.5U 20U
*.DC V1 -0.2 0.2 0.01
.PLOT DC V(4)
.PLOT TRAN V(4)
.OPTIONS NOPAGE NOMOD
.OPTIONS RELTOL=1E-6
.WIDTH OUT=80
.END

| NODE | VOLTAGE | NODE | VOLTAGE | NODE | VOLTAGE | NODE | VOLTAGE |
|------|---------|------|---------|------|---------|------|---------|
| ( 1) | 0.0000 | ( 2) | 5.0437 | ( 3) | -0.7708 | ( 4) | 5.0437 |
| ( 5) | 10.0000 | ( 20) | -10.0000 | | | | |

(b)

**Fig. 5.11.** (a) Small-signal circuit of EC pair with $r_o$. (b) Spice input file and dc operating point of EC pair with $r_o$.

| $R_{C1}$ | $R_{C2}$ | $R_{o1}$ | $R_{o2}$ | $I_{C2}$ | $A_v$ | $R_{in}$ | $R_{out}$ | $\frac{A_v}{0.5 g_m R_{C2}}$ | $\frac{R_{in}}{2 r_\pi}$ | $\frac{R_{out}}{R_{C2}}$ |
|---|---|---|---|---|---|---|---|---|---|---|
| 5 | 5 | - | - | 0.990 | 95.7 | 5.23 | 5 | 1.00 | 1.00 | 1.00 |
| 5 | 5 | 50 | 50 | 0.875 | 76.9 | 5.91 | 4.77 | 0.91 | 1.00 | 0.95 |
| 0.1 | 5 | 50 | 50 | 0.824 | 75.8 | 6.59 | 4.76 | 0.95 | 1.05 | 0.95 |
| 0.1 | 5 | - | 50 | 0.935 | 86.0 | 5.80 | 4.76 | 0.95 | 1.05 | 0.95 |

**Table 5.1.** Circuit element values, bias current, and small-signal circuit parameters showing the effect of feedback. All values are in k$\Omega$ or mA, as appropriate.

It is helpful to inspect the collector currents as $V_A$ is introduced ($R_o$ in the examples above). For the simple Ebers-Moll model of the BJT, the collector current including base-width modulation is [6]

$$I_C = \left[ I_S \exp \left( \frac{V_{BE}}{V_t} \right) \right] \left( 1 + \frac{V_{CE}}{V_A} \right) \qquad (5.60)$$
$$= I_C' \left( 1 + \frac{V_{CE}}{V_A} \right)$$

where $I_C'$ is the current with $V_A$ effects not present. Notice that $I_C$ increases as $V_A$ is introduced if $V_{BE}$ is kept constant. In an ECP, however, as $V_A$ is included, the total collector current is kept constant by the common-emitter current source. Therefore, $I_C'$ and $V_{BE}$ must be reduced. This is observed in the data of Row 2 in Table 5.1.

The reciprocal of the value of the output resistor, $r_o$, is found from (5.60) by taking the differential with respect to $V_{CE}$.

$$\frac{1}{r_o} = \frac{I_C'}{V_A} \qquad (5.61)$$
$$= \frac{\frac{I_C}{V_A}}{1 + \frac{V_{CE}}{V_A}}$$

For a MOS device, the corresponding drain current and output resistance expressions are for the Level-1 MOSFET (Shichman-Hodges) model [6]:

$$I_D = I_D' (1 + \lambda V_{DS}) \qquad (5.62)$$
$$I_D' = \frac{k'}{2} \frac{W}{L} (V_{GS} - V_T)^2$$
$$\frac{1}{r_o} = \frac{\lambda I_D}{1 + \lambda V_{DS}}$$

where $\lambda$ is the Spice model parameter LAMBDA.

The change of ECP or SCP performance with $r_o$ can be estimated from both an ordinary circuit analysis and by using the feedback approach of this chapter. First, we include the output resistance, $r_o$, of only the right-hand transistor. (If the output resistance of $Q_1$, on the left, is included for $R_{C1} = 0$, its $r_o$ is effectively in shunt with the common-emitter current source. The small-signal, node-to-ground resistance across the current source is very low because of the emitters of the two transistors.)

In the ECP circuit model of Figure 5.11a, $v_1$, $v_e$ and $v_{o2}$ are all in phase. Therefore, the $r_o$ of $Q_2$ introduces positive feedback. To set up the feedback situation more clearly, the first transistor can be viewed as an emitter follower feeding a common-base stage with shunt-shunt positive feedback. On the basis of the conclusions made earlier in this chapter, we can expect that the input and output resistances as well as the voltage gain should increase with $\frac{1}{(1-|a_L|)}$ since $a_L$ has a negative value for positive feedback. The feedback analysis is given below.

For a conventional circuit analysis, the incremental circuit model of the ECP of Figure 5.11a can be reduced to that of Figure 5.12a by introducing a Thevenin equivalent for $Q_1$ acting as an emitter follower. In Figure 5.12b, a common-base model is introduced for $Q_2$. For the resulting circuit, the conventional circuit analysis leads to

$$A_v = \frac{v_o}{v_i} \approx \frac{1}{2} g_m R_L \frac{1}{\left(1 + \frac{R_L}{2r_o}\right)} \tag{5.63}$$

where the approximation is made that $g_m \gg \frac{1}{r_o}$.

The input resistance looking into the emitter of $Q_2$ is

$$R'_{in} \approx \frac{1}{g_{m2}} \left(1 + \frac{R_L}{r_o}\right) \tag{5.64}$$

The input resistance looking in the base of $Q_1$ is

$$R_{in} \approx 2r_\pi \left(1 + \frac{R_L}{2r_o}\right) \tag{5.65}$$

Notice that the voltage gain of the ECP is reduced by $(1 + \frac{R_L}{2r_o})$ relative to the value for $r_o = \infty$. The input resistance of the ECP is increased by the same amount. The output resistance looking back into the circuit from $R_{C2}$ can be determined directly from Figure 5.12b. However, it is simpler to use the approximate configuration of Figure 5.12c. From the figure, the feedback current $i_a$ splits into two equal parts of $\frac{i_a}{2}$ each at the common-emitter node. Since $Q_2$ can be considered to be a common-base stage for an input at its emitter with a current gain of approximately unity, the current out of the collector of $Q_2$ contains the component, $\frac{i_a}{2}$, as shown in the figure. Summing currents at $v_{o2}$, we obtain

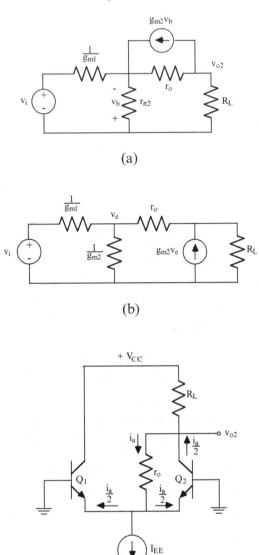

(a)

(b)

(c)

**Fig. 5.12.** (a) Equivalent small-signal circuit for the EC pair. (b) Modified circuit for (a) using a common-base model for $Q_2$. (c) Circuit for calculation of output resistance of EC pair.

$$i_o = i_{R_L} + i_a - \frac{i_a}{2} \qquad (5.66)$$

where

$$i_{R_L} = \frac{v_o}{R_L}$$

$$i_a = \frac{v_o - v_e}{r_o} \approx \frac{v_o}{r_o}$$

The resistance seen by $v_o$ is

$$\frac{v_o}{i_o} = R_L \| R_{out} \qquad (5.67)$$

where $R_{out} = 2r_o$

If the ECP has a finite source resistance of $R_S$ at the base input of $Q_1$, an additional factor in $R_{out}$ appears.

$$R_{out} \approx 2r_o \left( 1 + \frac{R_S}{2r_\pi} \right) \qquad (5.68)$$

The output resistance of the ECP is modified for the active-load, current-mirror situation shown in Figure 5.13. This load arrangement is often referred to as a balance-to-unbalance converter. The voltage gain of an ECP from the base of $Q_1$ to the output at $Q_2$ with this type of load is twice that obtained with equal collector resistance loads.

The output resistance can be estimated using an extension of the feedback current technique leading to (5.66) above. As in Figure 5.12c, the feedback current due to $r_{on}$ of $Q_1$ can be neglected because of the very small value of the 'load' resistance presented to $Q_1$ due to the diode-connected $Q_3$ in Figure 5.13. The feedback current from $r_{on}$ of $Q_2$, labeled $i_a$, is again approximately, $i_a \approx v_o/r_{on}$. This current divides into two equal components, as shown in the figure. The emitter input to $Q_2$ provides a component of $i_a/2$ into the output node. The emitter input into $Q_1$ becomes the input to $Q_3$, which is mirrored and becomes an output current of $i_a/2$ out of the $v_o$ node and into $Q_4$. Finally, due to the presence of $r_{op}$ in $Q_4$, another current component from the output node flows into $Q_4$ with a value $i_b = v_o/r_{op}$

The total output current is

$$i_o = i_a - \frac{i_a}{2} + \frac{i_a}{2} + i_b = i_a + i_b = \frac{v_o}{r_{on}} + \frac{v_o}{r_{op}} \qquad (5.69)$$

from which the output resistance is

$$R_{out} = \frac{v_o}{i_o} = r_{on} \| r_{op} \qquad (5.70)$$

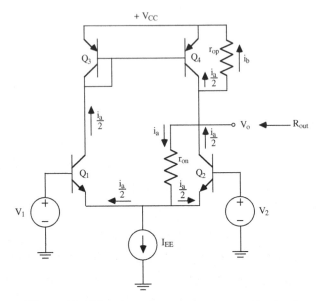

**Fig. 5.13.** EC pair with current-mirror active load.

## Problems

**5.1.** A common-emitter stage with series feedback and with a finite source resistance is shown in Figure 5.14.

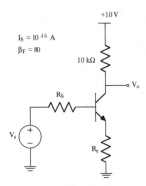

**Fig. 5.14.** Common-emitter stage for Problem 5.1.

(a) For $R_S = 0$ and with $V_{BB}$ chosen to produce $I_C = 0.5$ mA, estimate the value of $R_e$ needed to eliminate HD3. Verify with Spice for the case where the input signal amplitude is 20 mV.
(b) What value of $R_S$ with $R_e = 0$ provides the same estimated cancellation at the same quiescent bias state? Compare Spice results with those of (a).

**5.2.** A common-emitter stage is shown in Figure 5.15.

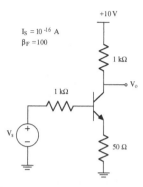

**Fig. 5.15.** Common-emitter stage for Problem 5.2.

(a) Determine the dc input voltage to achieve a bias collector current of 1 mA.
(b) For a sinusoidal input voltage that produces a fundamental output voltage of 0.5 V, estimate the value of HD2.
(c) Verify the results of (b) with Spice.

**5.3.** A MOS stage is shown in Figure 5.16.

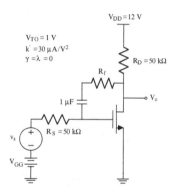

**Fig. 5.16.** MOS stage for Problem 5.3.

(a) Use Spice to establish the dc transfer characteristic of the stage.
(b) Confirm that the values of $V_{GG} = 1.7$ V and $V_{sA} = 0.4$ V are suitable values at the input.
(c) For $R_f = \infty$, estimate the value of HD2 for the drive of (b). Confirm your estimates with Spice.
(d) For $R_f = 200$ k$\Omega$, estimate THD for the same drive. Confirm with Spice.
(e) For $R_f = 200$ k$\Omega$ and $R_S = 0$, what is HD2?

**5.4.** A MOS circuit is shown in Figure 5.17.

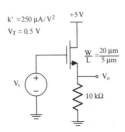

**Fig. 5.17.** MOS circuit for Problem 5.4.

(a) Using feedback ideas calculate HD2 for a sinusoidal input of 1 V. The input bias voltage establishes a drain current of 200 $\mu$A.
(b) Verify the result of (a) with Spice.
(c) Will there be any HD3 for this circuit? Explain your answer and estimate the value of HD3.
(d) How does the result in (c) compare with Spice?

**5.5.** A emitter-coupled pair with emitter resistances is shown in Figure 5.18. Determine HD3 for $V_1 = 0 + 50$ mV$cos\omega_1 t$.

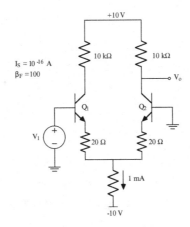

**Fig. 5.18.** Emitter-coupled pair for Problem 5.5.

# 6

# Basic IC Output Stages

## 6.1 Requirements for an Output Stage

In this chapter, the performance is studied of typical output stages for analog integrated circuits. In general, the output stage should be able to drive heavy loads, resistive and capacitive. This implies that the output stage should have a very low output resistance. At the input, the input resistance should be large so that the loading on a previous stage is as small as possible. Voltage gain, per se, is usually provided by the preceding 'gain' stages and is not necessary in the output stage. However, power gain (and thus current gain) is required. Of course, the transfer characteristic of the stage should be as linear as possible in order not to introduce harmonic distortion. All of these considerations point to the use of a feedback amplifier. In particular, a series-shunt combination is appropriate for the high input resistance, the low output resistance, the linearity, and reduced voltage gain. The simplest form of such amplifiers are the 'follower' stages, e.g., an emitter follower (EF) for bipolar transistors and the source follower (SF) for MOS devices. (The follower stages can be considered as forms of two two-port combinations of the series-shunt variety. Note that a CE or CS stage with resistance in the common input-output lead is a series-series configuration.) Finally, any output stage should convert the dc power from the bias sources into ac power as efficiently as possible. Therefore, power-conversion efficiency must be given attention.

The first basic output stage to be investigated is the emitter follower. In later sections of this chapter, the source follower and various push-pull, parallel combinations of emitter followers and of source followers are introduced. A new method of estimating distortion of very linear stages is also presented.

## 6.2 The Emitter Follower

The schematic diagram of a typical emitter follower as used in an analog integrated circuit is shown in Figure 6.1a. Devices $Q_a$ and $Q_b$ together with

D.O. Pederson and K. Mayaram, *Analog Integrated Circuits for Communication*, DOI 10.1007/978-0-387-68030-9_6,
© 2008 Springer Science+Business Media, LLC

resistors $R_a$, $R_b$, $R_e$ are part of the bias circuitry. As usual, we initially replace these elements with an ideal source element, in this case a current source, $I_{EE}$, as shown in Figure 6.1b. The symbol, $(-V_{EE})$, is included to keep in mind the actual source of Figure 6.1a. For the idealized circuit, a small-signal

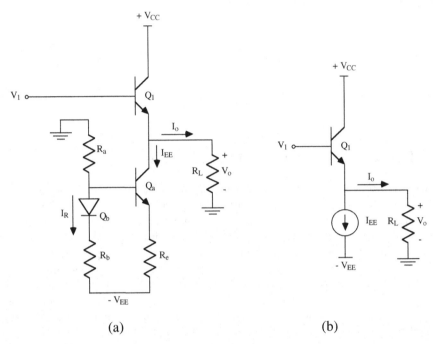

(a)                                                (b)

**Fig. 6.1.** (a) Schematic diagram of a typical emitter follower. (b) Idealized circuit for the emitter follower.

equivalent circuit is shown in Figure 6.2a, where a quiescent operating point in the normal active region of the transistor is assumed. At the input, the small-signal input resistance is

$$R_{in} = r_\pi + (\beta_{ac} + 1)R_L \approx \beta_{ac}\left(\frac{1}{g_m} + R_L\right) \tag{6.1}$$

At the output, a Thevenin equivalent circuit is shown in Figure 6.2b. The small-signal output resistance is

$$R_o \approx \frac{r_\pi + R_s}{\beta_{ac}} = \frac{1}{g_m} + \frac{R_s}{\beta_{ac}} \tag{6.2}$$

where $R_s$ is the value of the signal-source resistance presented to the follower. From the figure, it is clear that the open-circuit, small-signal voltage gain is unity. For any reasonable value of the load resistance, the voltage gain is close to one.

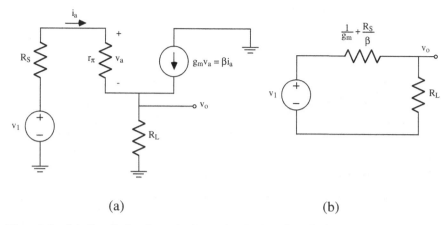

(a)                                            (b)

**Fig. 6.2.** (a) Small-signal equivalent circuit for the idealized emitter follower. (b) Thevenin equivalent circuit at the output.

For the large-signal case, the above results must be used with caution. However, we know from Chapter 5 that there is significant negative feedback for operation in the normal operating region. Thus the small-signal gain and resistance features of the follower should be achieved over a large input and output operating range.

An expression relating the input voltage and the output voltage is readily obtained using the idealized large-signal equivalent circuit and equations for the BJT. However, the result is not that useful.

$$V_1 = V_t \ln \left[ \frac{\frac{\beta}{\beta+1} \left( I_{EE} + \frac{V_o}{R_L} \right)}{I_S} \right] + V_o \tag{6.3}$$

where $\beta$ is the dc beta. This expression can be plotted. However, it is simpler to use the Spice circuit simulator to produce the dc transfer voltage characteristic. To obtain this characteristic from the circuit of Figure 6.1b, the following values are chosen: $V_{CC} = 10$V, $R_L = 10$ k$\Omega$ and $I_{EE} = 1$ mA. For $I_S = 10^{-16}$ A and $BF = \beta = \beta_{ac} = 100$, the quiescent input voltage should be 0.774V. The circuit schematic diagram and the Spice input file are given in Figure 6.3a. Note that three circuits are included to permit simulation with three different values of $R_L$, simultaneously.

The dc transfer characteristics are shown in Figure 6.3b. The linearity of the input-output relation is evident, and for $R_L \geq 10$ k$\Omega$ the linearity holds over almost the output range of $V_{CC} = +10$V down to $-V_{EE} = -10$V. The slope of the transfer characteristic is approximately unity. The transfer characteristic for $R_L = 20$ k$\Omega$ is similar to that for 10 k$\Omega$. For $R_L = 5$ k$\Omega$, the minimum value of output voltage is $-5$V. The emitter current of the BJT is equal to zero and the transistor is cutoff when $I_{EE} = -V_o/R_L$. (For the

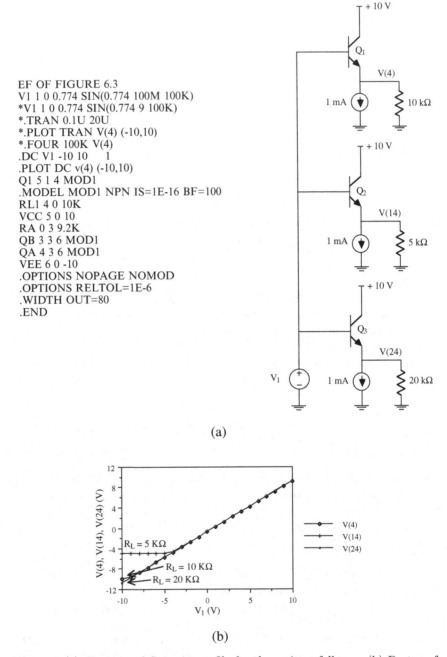

EF OF FIGURE 6.3
V1 1 0 0.774 SIN(0.774 100M 100K)
*V1 1 0 0.774 SIN(0.774 9 100K)
*.TRAN 0.1U 20U
*.PLOT TRAN V(4) (-10,10)
*.FOUR 100K V(4)
.DC V1 -10 10    1
.PLOT DC v(4) (-10,10)
Q1 5 1 4 MOD1
.MODEL MOD1 NPN IS=1E-16 BF=100
RL1 4 0 10K
VCC 5 0 10
RA 0 3 9.2K
QB 3 3 6 MOD1
QA 4 3 6 MOD1
VEE 6 0 -10
.OPTIONS NOPAGE NOMOD
.OPTIONS RELTOL=1E-6
.WIDTH OUT=80
.END

(a)

(b)

**Fig. 6.3.** (a) Circuit and Spice input file for the emitter follower. (b) Dc transfer characteristics of the emitter follower.

actual stage of Figure 6.1a, one must also consider when the device $Q_a$ enters saturation. This can occur above $-I_{EE}R_L$.)

As a check on the distortion components that are generated by the transistor, Spice runs were made for input sinusoids with amplitudes of 100 mV and 9V, the latter is shown in the commented line of the Spice input file of Figure 6.3a. For $R_L = 10$ k$\Omega$, the output Fourier responses from Spice2 are given in Figures 6.4a and b. The emitter follower is clearly a low distortion stage. In the next section, methods to estimate the low values of distortion are presented.

```
FOURIER COMPONENTS OF TRANSIENT RESPONSE V(4)      V1A = 100 mV
DC COMPONENT =    4.677D-05
HARMONIC   FREQUENCY    FOURIER     NORMALIZED     PHASE      NORMALIZED
  NO        (HZ)       COMPONENT    COMPONENT      (DEG)     PHASE (DEG)

   1      1.000D+05    9.922D-02    1.000000      0.000        0.000
   2      2.000D+05    1.612D-06    0.000016     79.597       79.597
   3      3.000D+05    7.696D-07    0.000008   -146.478     -146.478

     TOTAL HARMONIC DISTORTION =           0.002380   PERCENT
```

(a)

```
FOURIER COMPONENTS OF TRANSIENT RESPONSE V(4)      V1A = 9 V
DC COMPONENT =    7.404D-03
HARMONIC   FREQUENCY    FOURIER     NORMALIZED     PHASE      NORMALIZED
  NO        (HZ)       COMPONENT    COMPONENT      (DEG)     PHASE (DEG)

   1      1.000D+05    8.922D+00    1.000000     -0.008        0.000
   2      2.000D+05    1.007D-02    0.001129    -81.314      -81.306
   3      3.000D+05    5.344D-03    0.000599      4.617        4.625

     TOTAL HARMONIC DISTORTION =           0.137456   PERCENT
```

(b)

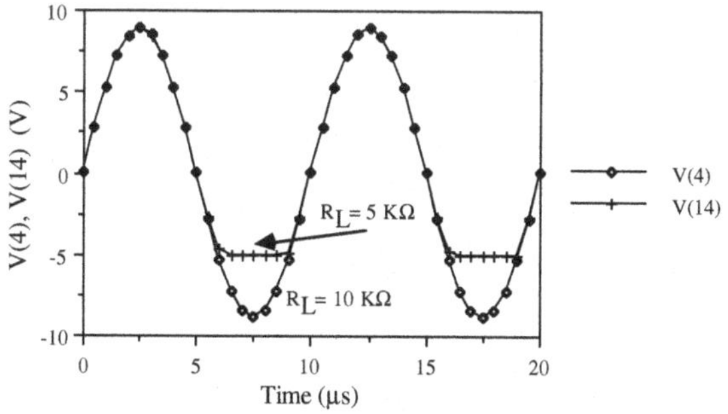

(c)

**Fig. 6.4.** Fourier components of output voltage for (a) $V_{1A} = 100$ mV, (b) $V_{1A} = 9$ V. (c) Output voltage waveforms for a sinusoidal input of 9 V.

The output voltage waveforms for a sinusoidal input of 9V and with $R_L = 5$ k$\Omega$ and 10 k$\Omega$ are shown in Figure 6.4c. For $R_L = 5$ k$\Omega$, the output voltage waveform exhibits severe clipping due to the earlier turnoff of the transistor. For optimum results, the design of the EF must include the proper choice and relations between the input amplitude, $V_{CC}$, $R_L$, $I_{EE}$, and $V_{EE}$.

For reference, the input file, the dc transfer characteristic, the output voltage waveform and the output voltage harmonics for the original circuit of Figure 6.1a are given in Figure 6.5 for $V_{1A} = 9$ V and $R_L = 10$ k$\Omega$.

```
EF OF FIGURE 6.1
V1 1 0 0.774 SIN(0.774 9 100K)
.TRAN 0.1U 20U
.PLOT TRAN V(4)  (-10,10)
.FOUR 100K V(4)
*.DC V1 -10 10 1
*.PLOT DC v(4)  (-10,10)
Q1 5 1 4 MOD1
.MODEL MOD1 NPN IS=1E-16 BF=100
RL1 4 0 10K
VCC 5 0 10
RA 0 3 9.2K
QB 3 3 6 MOD1
QA 4 3 6 MOD1
VEE 6 0 -10
.OPTIONS NOPAGE NOMOD
.OPTIONS RELTOL=1E-6
.WIDTH OUT=80
.END
```

(a)

| NODE | VOLTAGE | NODE | VOLTAGE | NODE | VOLTAGE | NODE | VOLTAGE |
|------|---------|------|---------|------|---------|------|---------|
| ( 1) | 0.7740  | ( 3) | -9.2262 | ( 4) | 0.0005  | ( 5) | 10.0000 |
| ( 6) | -10.0000|      |         |      |         |      |         |

(b)

**Fig. 6.5.** (a) Spice input file for the emitter follower. (b) Dc transfer characteristics and dc operating point of the emitter follower.

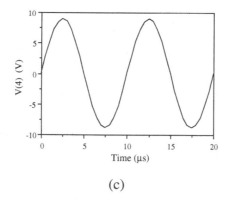

(c)

```
FOURIER COMPONENTS OF TRANSIENT RESPONSE V(4)
DC COMPONENT =    9.179D-03
HARMONIC    FREQUENCY      FOURIER      NORMALIZED      PHASE       NORMALIZED
   NO         (HZ)        COMPONENT     COMPONENT      (DEG)       PHASE (DEG)

    1       1.000D+05     8.920D+00     1.000000       0.000         0.000
    2       2.000D+05     1.024D-02     0.001148     -89.851       -89.851
    3       3.000D+05     4.270D-03     0.000479      -0.559        -0.558

    TOTAL HARMONIC DISTORTION =          0.126817   PERCENT
```

(d)

**Fig. 6.5.** (c) Output voltage waveform for $V_{1A} = 9$ V and $R_L = 10$ k$\Omega$. (b) Fourier components of output voltage.

## 6.3 Distortion Calculation using Differential Error

The excellent linearity of the emitter follower, as with other negative feedback amplifiers, makes difficult the calculation and estimation of the harmonic distortion which is generated by the stage. The method of *Differential Error* which is introduced in this section is a technique to estimate $HD_2$ and $HD_3$ for these small-distortion situations.

In the differential-error approach, we find the expression for the small-signal gain at the quiescent operating point and at input/output levels corresponding to the positive and negative-peak excursions. These determinations include the effects of the different instantaneous operating-point conditions. A pair of new parameters is defined from which $HD_2$ and $HD_3$ can be determined.

At an operating point, express the output voltage as a power series, omitting the quiescent terms,

$$v_o = a_1 v_i + a_2 v_i^2 + a_3 v_i^3 + \dots \tag{6.4}$$

The derivative of the variational output voltage at this operating point is

$$a = \frac{dv_o}{dv_i} = a_1 + 2a_2v_i + 3a_3v_i^2 + \ldots \tag{6.5}$$

Note that the coefficient $a_1$ is the small-signal voltage gain. Let the input sinusoidal variation be

$$v_i = V_{iA} \cos \omega_1 t \tag{6.6}$$

The amplitude, $V_{iA}$, is the zero-to-peak input excursion. The negative maximum excursion is $-V_{iA}$. These two values are used in the equations above to obtain the coefficients $a_2$ and $a_3$.

$$a^+ = a_1 + 2a_2V_{iA} + 3a_3V_{iA}^2 + \ldots \tag{6.7}$$
$$a^- = a_1 + 2a_2(-V_{iA}) + 3a_3(-V_{iA})^2 + \ldots$$

Notice that $a^+$ and $a^-$ are used to denote the output derivative expressions at the extremes of the input variation. We next take the difference between these expressions and the gain for a linear amplifier, $a_1$, and normalize this difference with respect to the linear gain. Two new coefficients are defined as follows:

$$E^+ = \frac{a^+ - a_1}{a_1} \tag{6.8}$$
$$= \frac{2a_2V_{iA} + 3a_3V_{iA}^2 + \ldots}{a_1}$$
$$E^- = \frac{a^- - a_1}{a_1}$$
$$= \frac{-2a_2V_{iA} + 3a_3V_{iA}^2 + \ldots}{a_1}$$

The sum and difference of these new quantities isolates either quantities containing only $a_2$ or $a_3$.

$$E^+ - E^- \approx \frac{4a_2V_{iA}}{a_1} \tag{6.9}$$
$$E^+ + E^- \approx \frac{6a_3V_{iA}^2}{a_1}$$

where the higher-order terms have been neglected. The coefficients $a_2$ and $a_3$ are

$$a_2 = \frac{a_1(E^+ - E^-)}{4V_{iA}} \tag{6.10}$$
$$a_3 = \frac{a_1(E^+ + E^-)}{6V_{iA}^2}$$

To relate these constants to the harmonic-distortion factors, remember that

$$HD_2 \approx \frac{1}{2}\frac{a_2}{a_1}V_{iA} \tag{6.11}$$

$$HD_3 \approx \frac{1}{4}\frac{a_3}{a_1}V_{iA}^2$$

where contributions from the higher-order terms of the power-series are neglected. Therefore,

$$HD_2 = \frac{E^+ - E^-}{8} \tag{6.12}$$

$$HD_3 = \frac{E^+ + E^-}{24}$$

For a numerical example, the emitter follower shown in Figure 6.6a is used. The values of the circuit components and the transistor parameters are given in the Spice input file of Figure 6.6b. At the quiescent state, $V_o = V(4) = 0V$ and $V_i = V_1 = 0.80V$ for $I_C = 2.86$ mA. The input is taken to be a 10V sinusoid. The small-signal voltage gain at any operating point in the normal active region is

$$a_v = \frac{1}{1 + \frac{\beta}{\beta+1}\frac{1}{g_m R_L}} = \frac{1}{1 + \frac{V_T}{|I_E| R_L}} \tag{6.13}$$

Because of accuracy considerations, $\beta + 1$ is retained and not approximated as $\beta$, and $|I_E|$ is used rather than $I_C$.

For the circuit and device parameter values of Figure 6.6,

$$a_1 = 0.99910 \tag{6.14}$$

Note that several digits of accuracy are used. (For this accuracy, the appropriate value of $V_{BE}$ for $I_{EE} = 2.86$ mA is 0.80095V. To illustrate the procedure, iteration for $V_{BE}$ accuracy is omitted). At the positive peak of the input, the output voltage is

$$V_o \approx V_1 - V_{BE} = 10.8 - 0.8 = 10.0 \text{ V} \tag{6.15}$$

For this example, a correction is not made for a change of $V_{BE}$. The corresponding value of $|I_E|$ is 2.86 mA $+ \frac{10 \text{ V}}{10 \text{ k}\Omega} = 3.86$ mA. The voltage gain from (6.13) is

$$a_v^+ = 0.99933 \tag{6.16}$$

At the negative peak input,

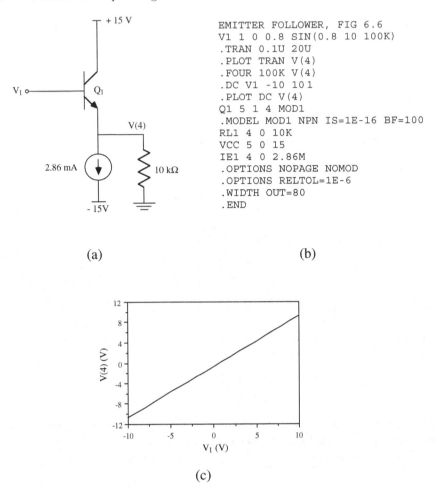

EMITTER FOLLOWER, FIG 6.6
V1 1 0 0.8 SIN(0.8 10 100K)
.TRAN 0.1U 20U
.PLOT TRAN V(4)
.FOUR 100K V(4)
.DC V1 -10 10 1
.PLOT DC V(4)
Q1 5 1 4 MOD1
.MODEL MOD1 NPN IS=1E-16 BF=100
RL1 4 0 10K
VCC 5 0 15
IE1 4 0 2.86M
.OPTIONS NOPAGE NOMOD
.OPTIONS RELTOL=1E-6
.WIDTH OUT=80
.END

(a)                                    (b)

(c)

```
FOURIER COMPONENTS OF TRANSIENT RESPONSE V(4)
DC COMPONENT =  -5.751D-04
HARMONIC    FREQUENCY      FOURIER      NORMALIZED      PHASE      NORMALIZED
   NO         (HZ)        COMPONENT     COMPONENT      (DEG)      PHASE (DEG)

    1       1.000D+05     9.938D+00     1.000000      0.000        0.000
    2       2.000D+05     6.025D-04     0.000061     -87.232      -87.231
    3       3.000D+05     5.357D-05     0.000005     -53.428      -53.427

     TOTAL HARMONIC DISTORTION =       0.006267    PERCENT
```

(d)

**Fig. 6.6.** (a) Emitter-follower circuit. (b) Spice input file. (c) Dc transfer characteristics. (d) Fourier components of the output voltage.

$$V_o \approx -10.0 \text{ V} \tag{6.17}$$

$$|I_E| = 2.86 \text{ mA} - \frac{10 \text{ V}}{10 \text{ k}\Omega} = 1.86 \text{ mA}$$

$$a_v^- = 0.99861$$

The $E$ parameters are

$$E^+ = 23 \times 10^{-5} \tag{6.18}$$

$$E^- = -49 \times 10^{-5}$$

These lead to the estimations

$$HD_2 = 0.0090\% \tag{6.19}$$

$$HD_3 = 0.0011\%$$

Surprisingly, if iteration of $V_{BE}$ and $|I_E|$ is used to obtain greater accuracy, almost the same results are produced. The results from a Spice run with the input file of Figure 6.6b are given in Figures 6.6c and d: $HD_2 = 0.0061\%$ and $HD_3 = 0.0005\%$.

The differential-error technique can also be used in many other circumstances. As a second example of this technique, the simple CE stage of Section 4.5 is used which includes nonlinear beta for the transistor. The circuit diagram is repeated as Figure 6.7a. The circuit values and transistor parameters are given in the Spice input file of Figure 6.7b. Notice that a second transistor with a constant beta is also included, for reference. In Figure 6.8, the current transfer characteristics are shown for the two transistors. For the stage containing $Q_1$, the chosen problem is to estimate the harmonic distortion with a quiescent operating condition of $I_C = 5$ mA and a sinusoidal input amplitude of 40 $\mu$A, zero-to-peak. The source conductance is assumed small enough to ignore. As developed in Chapter 4, the distortion is due primarily to the nonlinearity of the current transfer characteristic.

At the desired quiescent operating point of $I_C = 5$ mA, the base-current input from the transfer characteristic is approximately $I_B = 50$ $\mu$A. For the sinusoidal input excursion of 40 $\mu$A, the input current extremes are 10 $\mu$A and 90 $\mu$A.

The small-signal current gains can be estimated at the quiescent and extreme operating points. These values can be obtained by taking values in Figure 6.8a of output currents from adjacent points about the specified input values. At the quiescent point the small-signal current gain is

$$a_1 = \frac{(5.187 - 4.824) \times 10^{-3}}{(5.2 - 4.8) \times 10^{-5}} = 90.75 \tag{6.20}$$

The corresponding values at the input extremes are

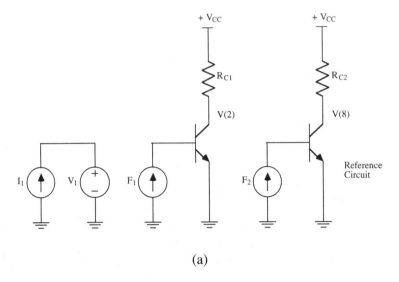

(a)

FIGURE 6.7

```
I1 0 1 50U SIN( 50U 40U 100K)
*.TRAN 0.1U 20U 0 0.1U
*.PLOT TRAN  I(VC1)
*.FOUR 100K I(VC1)
.DC I1 0 0.1MA 2UA
.PLOT DC I(VC1) I(VC2)  (0,8M)
V1 1 0 0
F1 0 3 V1 1
Q1 2 3 0 MOD1
.MODEL MOD1 NPN IS=1E-16 BF=230
+ISE=2E-12 NE=2
+IKF=10M
RC1 4 2 .1K
VC1 5 4 0
VCC 5 0 15
F2 0 6 V1 1
Q2 8 6 0 MOD2
VC2 5 7 0
RC2 7 8 0.1K
.MODEL MOD2 NPN BF=100 IS=1E-16
.OPTIONS RELTOL=1E-6
.OPTIONS NOPAGE
.WIDTH OUT=80
.END
```

|  | Q1 | Q2 |
|---|---|---|
| MODEL | MOD1 | MOD2 |
| IB | 5.00E-05 | 5.00E-05 |
| IC | 5.01E-03 | 5.00E-03 |
| VBE | 0.826 | 0.816 |
| VBC | -13.673 | -13.684 |
| VCE | 14.499 | 14.500 |
| BETADC | 100.129 | 100.000 |
| GM | 1.45E-01 | 1.93E-01 |
| RPI | 6.26E+02 | 5.17E+02 |
| RX | 0.00E-01 | 0.00E-01 |
| RO | 1.50E+12 | 1.00E+12 |
| CPI | 0.00E-01 | 0.00E-01 |
| CMU | 0.00E-01 | 0.00E-01 |
| CBX | 0.00E-01 | 0.00E-01 |
| CCS | 0.00E-01 | 0.00E-01 |
| BETAAC | 90.825 | 100.000 |
| FT | 2.31E+18 | 3.08E+18 |

(b)

**Fig. 6.7.** (a) Simple common-emitter stage. (b) Spice input file and transistor parameters with nonlinear beta.

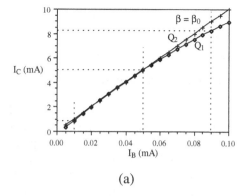

(a)

```
FOURIER COMPONENTS OF TRANSIENT RESPONSE I(VC1)
DC COMPONENT =    4.771D-03
HARMONIC   FREQUENCY     FOURIER     NORMALIZED     PHASE     NORMALIZED
   NO        (HZ)       COMPONENT    COMPONENT      (DEG)    PHASE (DEG)

    1      1.000D+05    3.674D-03    1.000000       0.002      0.000
    2      2.000D+05    2.278D-04    0.061996      90.002     90.001
    3      3.000D+05    1.505D-05    0.004097     179.507    179.505

    TOTAL HARMONIC DISTORTION =         6.214777   PERCENT
```

(b)

**Fig. 6.8.** (a) Current transfer characteristics for transistors with and without non-linear beta. (b) Fourier components of the collector current show distortion due to nonlinear beta.

$$a_i^+ = \frac{(8.389 - 8.097) \times 10^{-3}}{(9.192 - 8.792) \times 10^{-5}} = 73.00 \tag{6.21}$$

$$a_i^- = \frac{(1.075 - 0.630) \times 10^{-3}}{(1.208 - 0.808) \times 10^{-5}} = 111.30$$

The $E$ parameters are

$$E^+ = -0.1956 \tag{6.22}$$
$$E^- = +0.2264$$

The harmonic distortion factors are

$$HD_2 = 5.3\% \tag{6.23}$$
$$HD_3 = 0.13\%$$

The values obtained from a Spice2 run are $HD_2 = 6.2\%$, $HD_3 = 0.41\%$. From Section 4.5, the result from a three-point analysis is $HD_2 = 6.2\%$. Clearly, for large distortion situations, the differential-error technique must be used with caution.

## 6.4 Power-Conversion Efficiency

As mentioned in Section 6.1, the output stage must deliver power to a given load. This power must come primarily from the dc power supplies, and it is of interest to establish the efficiency of the conversion process of dc to ac power. It is assumed that the output stage has a large ac power gain and that the ac power input from the preceding stage can be neglected. For an emitter follower, the voltage gain is approximately unity while the ac current gain is approximately $\beta_{ac}$. Thus, the ac power gain can be 100 or more.

For the EF of Figure 6.1b, repeated in Figure 6.9a, and for maximum unclipped output, the results from Section 6.2 indicate that the values of the supply voltage, the current source and the load resistance should obey the rule

$$V_{CC} = I_{EE}R_L \tag{6.24}$$

where $V_{CEsat} \approx 0$. The maximum output voltage excursion is then $\pm V_{CC}$. Therefore, the rms value of the maximum output voltage, assuming that it is a pure sine wave, is

$$V_o|_{rms} = \frac{V_{CC}}{\sqrt{2}} \tag{6.25}$$

The ac power developed in $R_L$ is then

$$P_{ac} = \frac{1}{2}\frac{V_{CC}^2}{R_L} = \frac{1}{2}V_{CC}I_{EE}. \tag{6.26}$$

The dc power for the EF of Figure 6.9a is supplied by sources, $V_1, V_{CC}$, and $I_{EE}$. The dc power for the 'input' signal source $V_1$ for $V_o = 0$ is $V_1 I_B = \frac{V_{BE}I_C}{\beta}$. For $V_{BE} \ll V_{CC}$ and/or $\beta \gg 1$, this dc input power can be neglected. For the circuit of Figure 6.9a, with $V_o = 0$, $I_{EE}$ has zero dc volts across it and supplies no dc power to the circuit. (See below for additional comments on this dc power.) The voltage source $V_{CC}$ supplies the dc power $V_{CC}I_C$. For $I_C \approx I_{EE}$, the dc power supplied to the circuit is

$$P_{dc} \approx V_{CC}I_{EE} \tag{6.27}$$

The conversion efficiency is the ratio of the ac power developed in $R_L$ under maximum output excursion to the dc power.

$$\eta_{max} = \frac{P_{ac}}{P_{dc}} = \frac{1}{2} = 50\% \tag{6.28}$$

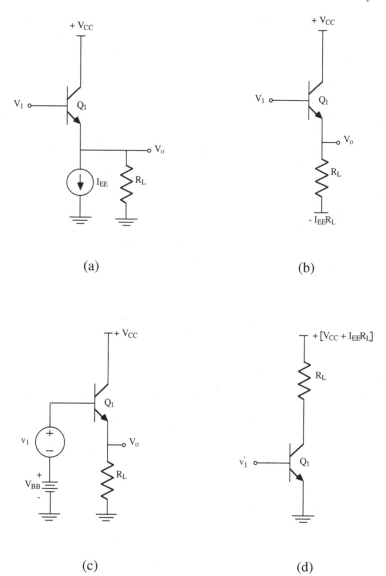

**Fig. 6.9.** (a) Emitter-follower circuit. (b) Emitter-follower circuit with Thevenin equivalent of $I_{EE}$ and $R_L$. (c) Emitter-follower circuit with $V_{CC}$ and $V_{BB}$ as the only dc sources. (d) Ground point of emitter-follower circuit moved to obtain a common-emitter configuration.

Equations (6.27) and (6.28) only apply to Figure 6.9a and can be misleading. For the circuit of Figure 6.1a, the $I_{EE}$ source is realized with a transistor current source. The dc power required to produce $I_{EE}$ is $V_{EE}(I_R + I_{EE}) \approx 2I_{EE}V_{EE}$ for $I_R = I_{EE}$. For $V_{CC} = V_{EE}$ and $V_{EE} = I_{EE}R_L, P_{dc} = 3V_{CC}I_{EE}$ and $\eta_{max} = 16.7\%$. Finally, note that if a Thevenin equivalent of $I_{EE}$ and $R_L$ in Figure 6.9a is used to produce Figure 6.9b, the dc power supplied by the equivalent voltage source $I_{EE}R_L(= V_{EE})$ is $V_{EE}I_C$. Therefore, $P_{dc} = 2V_{CC}I_{EE}$ and $\eta_{max} = 25\%$.

Since in Figure 6.1a, the current defining elements $R_a, R_b$ and $Q_b$ may be shared with other stages, an appropriate value for $\eta_{max}$ is approximately 20% for an IC emitter follower.

Finally, consider the circuit of Figure 6.9c where $V_{CC}$ and the bias value of $V_1$ are the only dc sources. For $V_{CEsat} \approx 0$, the quiescent value of $V_o$ is chosen to be $V_{CC}/2$, one half of the collector supply voltage. The quiescent value of $I_C$ is $V_{CC}/2R_L$ and the corresponding quiescent value of $V_1$ is $V_{BB} + V_{CC}/2$. $V_{BB}$ is the needed value of $V_{BE}$ to support $I_C$.

The total dc input power is

$$P_{dc} = \frac{V_{CC}^2}{2R_L} + \frac{V_{CC}^2}{4R_L\beta}\left(1 + \frac{2V_{BB}}{V_{CC}}\right)$$

For large $\beta$, the second term can be neglected. The maximum limiting output excursion occurs for $V_{CE} = V_{CEsat}$ or $V_{CE} = V_{CC}$. For $V_{CEsat} \approx 0$, the amplitude of the input sinusoid can be $V_1A = V_{CC}/2$. But this leads to a peak value of $V_1$ of $V_{CC} + V_{BB}$. It is common to limit the peak value of $V_1$ to $V_{CC}$. Therefore, $V_{1A} = (V_{CC}/2 - V_{BB})$. The ac power developed in $R_L$ is

$$P_{ac} = \frac{1}{2}\frac{(V_{CC}/2)^2}{R_L}\left(1 - \frac{2V_{BB}}{V_{CC}}\right)^2$$

The maximum conversion efficiency is

$$\eta = \frac{P_{ac}}{P_{dc}} = 25\%\left(1 - \frac{2V_{BB}}{V_{CC}}\right)^2$$

For the previous example where $V_{BB} = 0.8$ V and $V_{CC} = 10$V, $\eta = 18\%$.

Next we look at the power dissipated in the EF device. To do this, the $I_C - V_{CE}$ characteristic of the transistor is helpful. For convenience, move the ground point of the emitter follower in Figure 6.9b to obtain the common-emitter configuration of Figure 6.9d. The plot of the load line for this arrangement is shown in Figure 6.10. Let the value of input voltage $V_1$ of Figure 6.9a or b be such that for quiescent conditions the output voltage of the EF is 0. The quiescent operating point in Figure 6.10 is then $I_{EE}$ and $V_{CC}$, as shown. This point is independent of the value of the load, and the load line passes through the point. From an inspection of several load lines, it is obvious if $V_{CEsat} \approx 0$ that the maximum excursions of $V_{CE}$ and $I_C$ are obtained with $V_{CC} = I_{EE}R_L$.

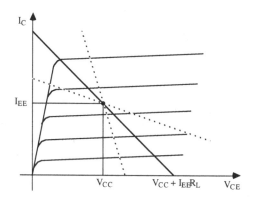

**Fig. 6.10.** Transistor output characteristics and dc load lines.

The dc power dissipated in the transistor is $V_{CC}I_{EE}$ at the quiescent condition. We now show that the quiescent state is the condition for maximum dissipation in the transistor. With a sinusoidal input signal, both the collector current and the collector-emitter voltage, ideally, will also vary sinusoidally. For maximum input, $V_{CE}(t) = V_{CC}(1 + \sin \omega_1 t)$ and $I_C = I_{EE}(1 - \sin \omega_1 t)$. The instantaneous power in the transistor is

$$
\begin{aligned}
P_{dev} &= V_{CE}(t)I_C(t) \\
&= V_{CC}(1 + \sin \omega_1 t)I_{EE}(1 - \sin \omega_1 t) \\
&= \frac{V_{CC}I_{EE}}{2}(1 + \cos 2\omega_1 t)
\end{aligned}
\tag{6.29}
$$

The average value of $P_{dev}$ is $V_{CC}I_{EE}/2$. Therefore at maximum output, the power dissipated in the transistor is only one half of the amount at the quiescent state. A study of the instantaneous power in the device shows that the maximum power is developed in the device at the quiescent operating point and that instantaneous power is less the further away the instantaneous operating point is along the load line.

For a numerical example, we return to the circuit and values of Figures 6.6a and 6.6b. The load line for the original values of the circuit does not lead to optimum conversion or maximum power output. The limiting condition is the negative output voltage swing, cf., Figure 6.10. The maximum output voltage excursion is 15 V, and the maximum variation of the collector current excursion is $15/10$ k$\Omega = 1.5$ mA. The power developed in $R_L$ is then $P_{ac} = 1/2 \times 15^2/10$ k$\Omega = 11.25$ mW. The power supplied by the sources including the $-I_{EE}R_L$ source corresponding to Figure 6.9b is $2V_{CC}I_{EE} = 85.8$ mW. Therefore, the efficiency of conversion is only 13%.

If $R_L$ is lowered to $V_{CC}/I_{EE} = 5.25$ k$\Omega$, the optimum condition is reached. The maximum output power increases to 21.4 mW with a conversion efficiency of 25%. The power dissipated in the device under maximum output is 21.4 mW.

## 6.5 The Source Follower

As brought out earlier, an enhancement-mode MOS device is similar in operation and required biasing to a bipolar transistor of the same polarity. Just as the configuration of a source-coupled pair follows directly from an emitter-coupled pair, an emitter follower leads directly to the source follower of Figure 6.11a. For proper operation of the NMOS transistor, the bulk connection must be returned to the lowest dc potential in the circuit, $-V_{SS}$ in this example. The principal differences in performance result from the lower value of $g_m$ obtainable from the MOS device for the same current levels. For the followers, the output resistance and voltage gains are directly related to the value of $1/g_m$. For normal active-region operation with a collector current of 1 mA, the value of $1/g_m$ for a bipolar device is 25.85 $\Omega$. For the MOS device, from Section 1.4.2, and for operation in the normal active region, $g_m = \sqrt{2k'(W/L)I_D}$. For $k' = 30$ $\mu$A/V$^2$ and $W/L = 100$, the value of $1/g_m$ is $\approx 400$ $\Omega$ for a drain current of 1 mA.

The Thevenin equivalent of the SF is the same as that of the EF shown in Figure 6.2b. Therefore, with a large value of $1/g_m$ the small-signal voltage gain of the source follower is not close to unity for a small value of load resistance. Usually, then, a large value of $R_L$ must be used for MOS circuits.

Similarly, for the same load resistance and device current, the loop gain of the SF is not as great as that for the EF, and the improvement of the linearity of the source follower is less in comparison with the bipolar case. However, the original nonlinearity of the MOS device (square-law) is not as severe as that of the bipolar device (exponential). Finally, the maximum voltage swing with small distortion is less for the SF than for the EF since the 'resistive' region of operation, which corresponds to the saturation region of a BJT, occurs at larger values of device voltage ($V_{DS}$).

As an example, the circuit of Figure 6.11a is simplified to that of Figure 6.11b. In Figure 6.12a, circuit values and MOS parameters are included in a Spice input file. For the simplest MOS model in Spice, including only parameters $KP$ and $VTO$, the voltage transfer ratio of the circuit obtained from a Spice run is shown in Figure 6.12b. From this curve, it is seen that very linear operation is obtained for an input excursion from $V_1 = -4$ V to $V_1 = V_{DD} = +5$ V. The slope of the transfer characteristic in the linear region is $\approx 0.92$. The maximum unclipped output voltage is obtained with a

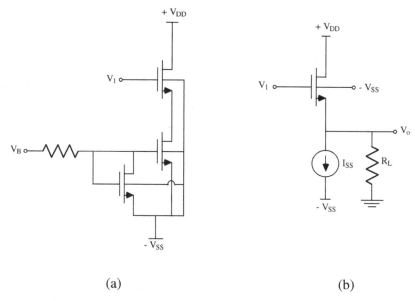

(a)                                                                (b)

**Fig. 6.11.** (a) A typical MOS source-follower circuit. (b) Simplified circuit for the source follower.

dc input bias of $V_1 = 0.5$ V with an input amplitude of 4.5 V. The output voltage excursion is then also approximately $\pm 4.5$ V centered at $-0.94$ V.

```
SOURCE FOLLOWER,    FIGURE 6.12
V1 1 0 0 SIN(0.5 4.5 100K)
.TRAN 0.1U 20U
.PLOT TRAN V(3) I(VDD)
.FOUR 100K V(3) I(VDD)
.DC V1 -5 5 0.5
.PLOT DC V(3)
M1 2 1 3 4 MOD1 W=800U L=8U
.MODEL MOD1 NMOS KP=30U VTO=0.7
RL 3 0 5K
VDD 2 0 5
ISS 3 0 1MA
VSS 4 0 -5
.OPTIONS RELTOL=1E-6
.OPTIONS NOPAGE NOMOD
.WIDTH OUT=80
.END
```

(a)                                                                (b)

**Fig. 6.12.** (a) Spice input file for source follower. (b) Voltage transfer characteristics of the source follower.

With a 4.5 V sinusoidal input voltage biased at 0.5 V, a transient analysis with Spice leads to the waveform and the distortion components listed in Figure 6.13. The large-signal voltage gain is obtained from the ratio of the amplitude of the fundamental output to the input amplitude.

$$a_v = \frac{b_1}{V_{1A}} = \frac{4.054}{4.5} = 0.90 \tag{6.30}$$

From the output listing, $HD_2 = 1.63\%$ and $THD = 1.72\%$.

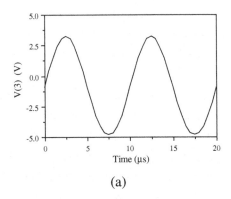

(a)

```
FOURIER COMPONENTS OF TRANSIENT RESPONSE V(3)
DC COMPONENT =   -8.763D-01
HARMONIC    FREQUENCY    FOURIER     NORMALIZED    PHASE      NORMALIZED
   NO         (HZ)      COMPONENT    COMPONENT     (DEG)      PHASE (DEG)

    1       1.000D+05   4.054D+00    1.000000     -0.001        0.000
    2       2.000D+05   6.591D-02    0.016258    -89.986      -89.985
    3       3.000D+05   2.106D-02    0.005195     -0.122       -0.121

      TOTAL HARMONIC DISTORTION =            1.722526   PERCENT
```

(b)

```
FOURIER COMPONENTS OF TRANSIENT RESPONSE I(VDD)
DC COMPONENT =   -8.247D-04
HARMONIC    FREQUENCY    FOURIER     NORMALIZED    PHASE      NORMALIZED
   NO         (HZ)      COMPONENT    COMPONENT     (DEG)      PHASE (DEG)

    1       1.000D+05   8.109D-04    1.000000    179.999        0.000
    2       2.000D+05   1.318D-05    0.016258     90.014      -89.985
    3       3.000D+05   4.213D-06    0.005195    179.878       -0.121

      TOTAL HARMONIC DISTORTION =            1.722526   PERCENT
```

(c)

**Fig. 6.13.** (a) Output voltage waveform for the source follower. (b) Fourier components of the output voltage. (c) Fourier components of the drain current.

Also from the output listings, the ac power in $R_L$ can be calculated.

$$P_{ac} = \frac{(4.054)^2}{2} \left( \frac{1}{5 \text{ k}\Omega} \right) = 1.64 \text{ mW} \tag{6.31}$$

The average input current from the voltage source $V_{DD}$ is obtained from the Spice runs, cf. Figure 6.13c, and equals 0.825 mA. The dc input power from both $V_{DD}$ and $I_{SS}$ is

$$P_{dc} = 0.825 \text{ mA} \times 5 \text{ V} + 1.0 \text{ mA} \times 0.88 \text{ V} = 5.0 \text{ mW} \tag{6.32}$$

where the dc value for $V_o$ is taken from the values of Figure 6.13b. The conversion efficiency for this SF is $\eta = 33\%$. As brought out in Section 6.4, this value of $\eta$ is an idealized value. For an actual SF, as in Figure 6.11a, the dc power input is larger by an approximate factor of two to three. Therefore, $\eta_{max} \approx 10 - 20\%$.

Other output stages using MOS devices are introduced in a later section of this chapter.

## 6.6 Push-Pull Emitter Followers

If one parallels devices in an emitter or source follower, the current drive capability into a low load resistance is improved. However, the distortion generation remains the same and the conversion efficiency is the same. Through the use of complementary devices, a 'push-pull' drive situation can be achieved in which distortion cancellation is obtained together with the possibility of better conversion efficiency.

In Figure 6.14, both npn and pnp transistors are used in a 'parallel' arrangement. Because of the opposite polarity of the devices, as $V_1$ is increased, the input voltage of $Q_1$, the npn unit, is increased while the input voltage is reduced for $Q_2$, the pnp unit. As is seen in the development below, while $Q_1$ pushes more current toward $R_L$ as it turns 'on,' $Q_2$ turns toward 'off' and rejects current, again forcing more current into $R_L$. At the positive input extreme, $Q_1$ is fully on and 'pushing' current into $R_L$. At the negative input extreme, $Q_2$ is fully on and 'pulling' current from $R_L$. The load current thus sustains a push-pull excitation.

To evaluate the performance of this circuit, it is helpful initially to assume that the load resistance is extremely small. Therefore, the inputs to the transistors are approximately

$$V_{BE1} \approx V_1 + V_{BB} \tag{6.33}$$
$$V_{BE2} \approx V_1 - V_{BB}$$

We next introduce simple nonlinear circuit models for the transistors, assuming operation only in the off and normal active regions. The circuit model for the push-pull follower is then that of Figure 6.15. The load current is

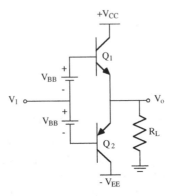

**Fig. 6.14.** A 'push-pull' circuit configuration.

the sum of the two emitter currents, which for devices with large betas is approximately the sum of the two collector currents.

$$I_L = -I_{E1} - I_{E2} \approx I_{C1} + I_{C2} \tag{6.34}$$

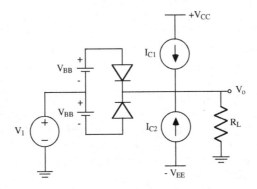

**Fig. 6.15.** Push-pull circuit with nonlinear transistor models.

Remember that all transistor currents are defined as positive into the transistor. The actual currents must carry their sign. The collector currents in terms of the input voltage are

$$I_{C1} = I_S \exp\left(\frac{V_{BE1}}{V_t}\right) \tag{6.35}$$

$$I_{C2} = -I_S \exp\left(\frac{-V_{BE2}}{V_t}\right)$$

Notice that both the coefficient and the exponential have negative signs for the pnp unit. Equal devices are assumed, for simplicity, except for the different polarity. Thus, the magnitudes of the device parameters are the same. At the quiescent operating point, $V_1 = 0, V_o = 0$, the current in each transistor is labeled $I_{CA}$ where $I_{CA} = I_S \exp\left(\frac{V_{BB}}{V_t}\right)$. The collector currents in terms of this value are

$$I_{C1} = I_{CA} \exp\left(\frac{V_1}{V_t}\right) \tag{6.36}$$

$$= I_{CA}\left[1 + \frac{V_1}{V_t} + \frac{1}{2}\left(\frac{V_1}{V_t}\right)^2 + \cdots\right]$$

$$I_{C2} = -I_{CA} \exp\left(\frac{-V_1}{V_t}\right)$$

$$= -I_{CA}\left[1 + \left(\frac{-V_1}{V_t}\right) + \frac{1}{2}\left(\frac{-V_1}{V_t}\right)^2 + \cdots\right]$$

$$= I_{CA}\left[-1 + \frac{V_1}{V_t} - \frac{1}{2}\left(\frac{V_1}{V_t}\right)^2 + \cdots\right]$$

In the above expressions, the exponentials have been expanded in power series, and it is to be noted that the signs of the even terms are opposite, while the signs of the odd powers are the same. Thus, in the expression for the load current, which is the algebraic sum of the two collector currents, the even powers cancel and do not appear.

$$I_L \approx I_{C1} + I_{C2} \tag{6.37}$$

$$= I_{CA}\left[0 + 2\left(\frac{V_1}{V_t}\right) + 0 + \frac{2}{6}\left(\frac{V_1}{V_t}\right)^3 + \cdots\right]$$

In particular, there is no dc power dissipated in the load resistance, and there are no even harmonics generated. For a sinusoidal input, the third-harmonic distortion factor for the output voltage remains the same as for one device, since both the fundamental and the third harmonic are increased by two.

For a nonzero value of $R_L$, we can expect the usual properties of negative feedback. The small-signal voltage gain approaches unity, and the harmonic distortion is reduced. If the output voltage is held constant, $HD_3$ is reduced by $(1 + a_L)$ where $a_L$ is the loop gain. Higher-order odd harmonics are also reduced significantly.

For a numerical example, the push-pull emitter follower of Figure 6.14 is evaluated using Spice. The input file is given in Figure 6.16. As shown in the schematic diagram, two circuits are included, one for the situation where the models for transistors, $Q_1$ and $Q_2$ are the same except for the

```
BJT CLASS AB, FIGURE 6.16
VIN 1 0 0 SIN (0 9.9 100K)
VB1 6 1 0.774
VB2 1 7 0.774
.TRAN .1U 20U
.PLOT TRAN V(3) V(13)
.PLOT TRAN I(VCC1) I(VEE1) (-10M,10M
.PLOT TRAN I(VCC2) I(VEE2) (-10M,10M
.FOUR 100K V(3)
.FOUR 100K I(VCC1)
.FOUR 100K I(VEE1)
.PLOT DC V(3) V(13)
.DC VIN -10 10 0.5
Q1 5 6 3 MOD1
Q2 4 7 3 MOD2
VCC1 5 0 10
VEE1 4 0 -10
RL 3 0   1K
.MODEL MOD1 NPN BF=100 IS=1E-16 RB=1
.MODEL MOD2 PNP BF=100 IS=1E-16 RB=1
.MODEL MOD3 PNP BF=20   IS=1E-15 RB=
VB3 16 1 0.774
VB4 1 17 0.714
Q3 15 16 13 MOD1
Q4 14 17 13 MOD3
RL2 13 0 1K
VCC2 15 0 10
VEE2 14 0 -10
.FOUR 100K V(13) I(VCC2) I(VEE2)
.OPTIONS RELTOL=1E-6
.END
```

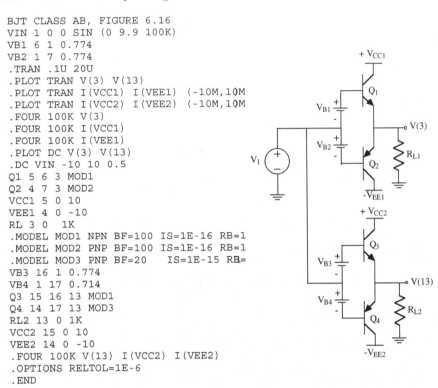

**Fig. 6.16.** Circuit and Spice input file for push-pull emitter follower.

npn or pnp designation; the other where the two transistors have different parameters, to check on the situation where the npn and pnp transistors differ. For the top circuit, the value of $V_{BB}$ is chosen to be 0.774 V corresponding to quiescent collector currents of 1 mA for $I_S = 10^{-16}$ A. Other choices are $V_{CC} = V_{EE} = 10$ V. The dc voltage transfer characteristic is shown in Figure 6.17. Excellent linearity and a near-unity voltage gain are achieved over the range $V_1 \pm V_{CC} = \pm 10$ V.

In the bottom circuit, the model for $Q_4$ has beta reduced from 100 to 20 and $I_S$ increased to $10^{-15}$ A. The dc transfer characteristic is coincident with the curve for equal parameters. Note that for this pnp transistor, $V_{BB} = 0.714$ V. The change in linearity is small because of the large value of negative feedback even for a weak pnp unit.

In order to check on the distortion generation, transient analyses of the circuits of Figure 6.16 have been made with an input sinusoidal amplitude of 9.9 V. The Fourier components of the two output voltages, $V(3)$ and $V(13)$, are given in Figure 6.18. As usual for cases of small distortion, extreme care must be taken in setting up the simulation. The parameter RELTOL is made

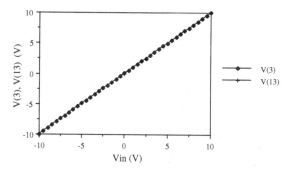

**Fig. 6.17.** Dc voltage transfer characteristics of BJT class-AB push-pull stage.

$1 \times 10^{-6}$. Notice in Figures 6.18a and b, the very small values of even-order harmonics in both cases.

```
FOURIER COMPONENTS OF TRANSIENT RESPONSE V(3)          Equal Transistors
DC COMPONENT =  -2.699D-04
HARMONIC     FREQUENCY       FOURIER      NORMALIZED      PHASE       NORMALIZED
   NO          (HZ)         COMPONENT     COMPONENT       (DEG)      PHASE (DEG)

    1       1.000D+05      9.782D+00      1.000000       0.000         0.000
    2       2.000D+05      2.289D-04      0.000023      83.476        83.477
    3       3.000D+05      8.400D-03      0.000859    -179.795      -179.795

      TOTAL HARMONIC DISTORTION =          0.089799   PERCENT
```

(a)

```
FOURIER COMPONENTS OF TRANSIENT RESPONSE V(13)       Unequal Transistors
DC COMPONENT =  -4.292D-04
HARMONIC     FREQUENCY       FOURIER      NORMALIZED      PHASE       NORMALIZED
   NO          (HZ)         COMPONENT     COMPONENT       (DEG)      PHASE (DEG)

    1       1.000D+05      9.782D+00      1.000000       0.000         0.000
    2       2.000D+05      1.558D-04      0.000016      80.349        80.349
    3       3.000D+05      8.313D-03      0.000850    -179.790      -179.790

      TOTAL HARMONIC DISTORTION =          0.088813   PERCENT
```

(b)

**Fig. 6.18.** Fourier components of output voltage with (a) equal transistors, and (b) unequal transistors.

For the situation of unequal transistor parameter values, the values of the odd harmonics are close to those of the equal-beta-magnitude case.

If we monitor the collector currents for the transient runs, the waveforms shown in Figures 6.19a and b are obtained. Note that the devices turn off during a portion of a cycle. This is called Class-AB operation. As is clear from the data of Figure 6.18, the output voltage waveform is very sinusoidal. The cancellation phenomenon is the explanation. The Fourier components of

the collector currents are shown in Figure 6.20. The richness in harmonics in
the collector current waveforms is evident.

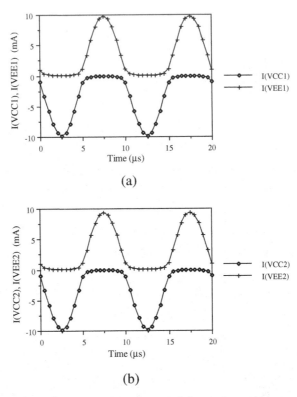

(a)

(b)

**Fig. 6.19.** Collector current waveforms with (a) equal transistors, and (b) unequal
transistors.

To check on the power-conversion efficiency for the case of equal beta
magnitudes, we look at the dc components of the two circuits. From the values
of Figure 6.20, the dc currents from the two voltage sources are $I_{C1} = 3.32$
mA and $I_{C2} = -3.32$ mA. Therefore, the dc power supplied by $V_{CC}$ and
$V_{EE}$ is $P_{dc} = 66.4$ mW. From the value of the output fundamental voltage
in Figure 6.18, 9.78 V, the fundamental output power developed in $R_L$ is
$(9.78)^2/2(1 \text{ k}\Omega) = 47.8$ mW. The conversion efficiency is then $\eta = 72\%$.
There is a significant improvement in conversion efficiency in relation to the
single-ended emitter follower.

For the case of unequal transistors, the corresponding numbers from Figures 6.18 and 6.20 lead to $P_{dc} = 65.2$ mW, $P_{ac} = 47.8$ mW and $\eta = 73\%$.

The fact that excellent performance can be achieved with distorted collector current waveforms leads us to push the limit. If we remove the bias sources,
$V_{BB}$, in Figure 6.14, both transistors are off for the quiescent state. The Spice

```
FOURIER COMPONENTS OF TRANSIENT RESPONSE I(VCC1)
DC COMPONENT =  -3.318D-03
HARMONIC   FREQUENCY      FOURIER      NORMALIZED     PHASE      NORMALIZED
   NO        (HZ)        COMPONENT     COMPONENT      (DEG)     PHASE (DEG)

    1      1.000D+05     4.841D-03     1.000000      179.963       0.000
    2      2.000D+05     1.836D-03     0.379234       90.015     -89.948
    3      3.000D+05     3.725D-06     0.000769       55.630    -124.333

      TOTAL HARMONIC DISTORTION =     38.373638  PERCENT


FOURIER COMPONENTS OF TRANSIENT RESPONSE I(VEE1)
DC COMPONENT =   3.319D-03
HARMONIC   FREQUENCY      FOURIER      NORMALIZED     PHASE      NORMALIZED
   NO        (HZ)        COMPONENT     COMPONENT      (DEG)     PHASE (DEG)

    1      1.000D+05     4.843D-03     1.000000     -179.964       0.000
    2      2.000D+05     1.838D-03     0.379453      -89.987      89.977
    3      3.000D+05     5.412D-06     0.001118      -34.003     145.961

      TOTAL HARMONIC DISTORTION =     38.391306  PERCENT


FOURIER COMPONENTS OF TRANSIENT RESPONSE I(VCC2)
DC COMPONENT =  -3.322D-03
HARMONIC   FREQUENCY      FOURIER      NORMALIZED     PHASE      NORMALIZED
   NO        (HZ)        COMPONENT     COMPONENT      (DEG)     PHASE (DEG)

    1      1.000D+05     4.842D-03     1.000000      179.963       0.000
    2      2.000D+05     1.833D-03     0.378566       90.015     -89.948
    3      3.000D+05     3.578D-06     0.000739       58.548    -121.415

      TOTAL HARMONIC DISTORTION =     38.302514  PERCENT


FOURIER COMPONENTS OF TRANSIENT RESPONSE I(VEE2)
DC COMPONENT =   3.196D-03
HARMONIC   FREQUENCY      FOURIER      NORMALIZED     PHASE      NORMALIZED
   NO        (HZ)        COMPONENT     COMPONENT      (DEG)     PHASE (DEG)

    1      1.000D+05     4.657D-03     1.000000     -179.964       0.000
    2      2.000D+05     1.765D-03     0.378895      -89.987      89.977
    3      3.000D+05     5.309D-06     0.001140      -32.989     146.975

      TOTAL HARMONIC DISTORTION =     38.331377  PERCENT
```

**Fig. 6.20.** Fourier components of collector currents for Class-AB push-pull stages with equal and unequal transistors.

input file is shown in Figure 6.21a. The voltage transfer characteristics for this situation are shown in Figure 6.21b for both equal and nonequal transistor parameter values. A distinct nonlinearity near zero input volts is seen. This produces the kink in the output voltage waveform as shown in Figure 6.22a which is called crossover distortion. However, the output voltage is quite low in harmonics as is shown in the Fourier outputs shown in Figure 6.22b; for the equal-beta case, $THD = 4.45\%$. The odd harmonics fall off very slowly, from $HD_3 = 3.6\%$ to $HD_9 = 0.9\%$. This harmonic content can be appreciated by looking at the waveform of the difference of the output voltage and the fundamental of the output. This is shown in Figure 6.23a. This difference waveform is clearly rich in harmonics.

The collector currents flow for less than half of the input cycle as is shown in Figure 6.23b. According to a strict definition, this corresponds to Class-C

```
BJT CLASS B, FIGURE 6.21
VIN 1 0 0 SIN(0  9.9 100K)
VB1 6 1 0
VB2 1 7 0
.TRAN .1U 20U
.PLOT TRAN V(3) V(13)
.PLOT TRAN I(VCC1) I(VEE1) (-10M,10M
.PLOT TRAN I(VCC2) I(VEE2) (-10M,10M
.FOUR 100K V(3)
.FOUR 100K I(VCC1)
.FOUR 100K I(VEE1)
.PLOT DC V(3) V(13)
.DC VIN -10 10 0.5
Q1 5 6 3 MOD1
Q2 4 7 3 MOD2
VCC1 5 0 10
VEE1 4 0 -10
RL 3 0  1K
.MODEL MOD1 NPN BF=100 IS=1E-16 RB=1
.MODEL MOD2 PNP BF=100 IS=1E-16 RB=1
.MODEL MOD3 PNP BF=20   IS=1E-15 RB=
VB3 16 1 0
VB4 1 17 0
Q3 15 16 13 MOD1
Q4 14 17 13 MOD3
RL2 13 0 1K
VCC2 15 0 10
VEE2 14 0 -10
.FOUR 100K V(13) I(VCC2) I(VEE2)
.OPTIONS RELTOL=1E-6
.END
```

(a)

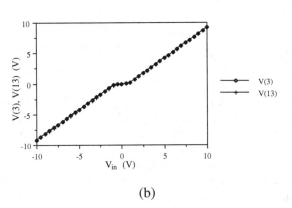

(b)

**Fig. 6.21.** (a) Spice input file for BJT Class-B stage. (b) Dc voltage transfer characteristics of Class-B stage.

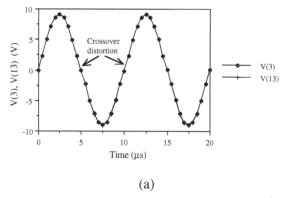

(a)

```
FOURIER COMPONENTS OF TRANSIENT RESPONSE V(3)
DC COMPONENT =    1.306D-03
HARMONIC    FREQUENCY       FOURIER      NORMALIZED     PHASE      NORMALIZED
  NO          (HZ)         COMPONENT     COMPONENT      (DEG)     PHASE (DEG)

   1        1.000D+05     8.805D+00      1.000000       0.017       0.000
   2        2.000D+05     4.035D-03      0.000458      32.309      32.292
   3        3.000D+05     3.203D-01      0.036380     179.653     179.636

       TOTAL HARMONIC DISTORTION =        4.452303  PERCENT


FOURIER COMPONENTS OF TRANSIENT RESPONSE V(13)
DC COMPONENT =   -2.719D-02
HARMONIC    FREQUENCY       FOURIER      NORMALIZED     PHASE      NORMALIZED
  NO          (HZ)         COMPONENT     COMPONENT      (DEG)     PHASE (DEG)

   1        1.000D+05     8.843D+00      1.000000       0.014       0.000
   2        2.000D+05     6.221D-03      0.000703      56.498      56.483
   3        3.000D+05     3.081D-01      0.034840     179.719     179.705

       TOTAL HARMONIC DISTORTION =        4.264578  PERCENT
```

(b)

**Fig. 6.22.** (a) Output voltage waveforms for Class-B stage. (b) Fourier components of output voltage with equal and unequal transistors.

operation. It is conventional, however, to refer to this case as Class B, where each device conducts for approximately one-half of an input cycle. The corresponding Fourier components of the collector current waveforms are given in Figures 6.23c and d.

From the dc values in the Fourier current outputs, the dc power supplied by $V_{CC}$ and $V_{EE}$ is $P_{dc} = 54.4$ mW for equal betas. The ac power developed in $R_L$ is $P_{ac} = 38.8$ mW. The conversion efficiency is 71%. Of course, the dc power from the $V_{BB}$ sources is now absent. For the unequal transistor example, $P_{dc} = 53.6$ mW, $P_{ac} = 39.1$ mW, and $\eta = 73\%$.

In practice, a push-pull emitter follower stage is designed to include a small voltage bias for the devices; equivalent values of $V_{BB}$ lead to quiescent values of collector current of 0.1 to 1 mA.

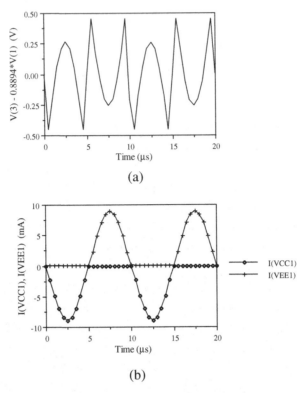

**Fig. 6.23.** (a) Harmonic content in the output voltage waveform. (b) Collector current waveforms.

In the output stage of many analog ICs, improved performance of the push-pull EF can be obtained by using composite transistors, as shown in Figure 6.24. Composite device $Q_1$ is the combination of a small-area npn transistor, $Q_{1a}$, driving a large-area pnp, $Q_{1b}$. Composite device $Q_2$ is the combination of a small-area pnp transistor driving a large-area npn transistor. From the input to the output, each of two parallel paths involve the effective cascade of two devices with complementary-polarity. In particular, the current gain at an operating point of each path is the product of a poor gain device and a high-gain device. Over the range of large-signal operation, the nonlinear beta characteristics are partially compensated by the other polarity device in its path and by the composite beta characteristic of the parallel path. Because of the adequate current gains, the output resistance of both paths is significantly smaller than that provided by a two-device, push-pull EF over the full input voltage variation.

```
FOURIER COMPONENTS OF TRANSIENT RESPONSE I(VCC1)
DC COMPONENT =   -2.718D-03
HARMONIC   FREQUENCY     FOURIER     NORMALIZED     PHASE     NORMALIZED
  NO         (HZ)       COMPONENT    COMPONENT      (DEG)    PHASE (DEG)

   1       1.000D+05    4.360D-03    1.000000     -179.867      0.000
   2       2.000D+05    2.023D-03    0.464091       90.076    269.943
   3       3.000D+05    1.572D-04    0.036056       -3.388    176.479

    TOTAL HARMONIC DISTORTION =     47.561413    PERCENT
```

```
FOURIER COMPONENTS OF TRANSIENT RESPONSE I(VEE1)
DC COMPONENT =    2.717D-03
HARMONIC   FREQUENCY     FOURIER     NORMALIZED     PHASE     NORMALIZED
  NO         (HZ)       COMPONENT    COMPONENT      (DEG)    PHASE (DEG)

   1       1.000D+05    4.358D-03    1.000000      179.901      0.000
   2       2.000D+05    2.025D-03    0.464706      -90.019   -269.921
   3       3.000D+05    1.604D-04    0.036804        2.633   -177.269

    TOTAL HARMONIC DISTORTION =     47.637512    PERCENT
```

(c)

```
FOURIER COMPONENTS OF TRANSIENT RESPONSE I(VCC2)
DC COMPONENT =   -2.718D-03
HARMONIC   FREQUENCY     FOURIER     NORMALIZED     PHASE     NORMALIZED
  NO         (HZ)       COMPONENT    COMPONENT      (DEG)    PHASE (DEG)

   1       1.000D+05    4.360D-03    1.000000     -179.867      0.000
   2       2.000D+05    2.023D-03    0.464091       90.076    269.943
   3       3.000D+05    1.572D-04    0.036056       -3.388    176.479

    TOTAL HARMONIC DISTORTION =     47.561413    PERCENT
```

```
FOURIER COMPONENTS OF TRANSIENT RESPONSE I(VEE2)
DC COMPONENT =    2.641D-03
HARMONIC   FREQUENCY     FOURIER     NORMALIZED     PHASE     NORMALIZED
  NO         (HZ)       COMPONENT    COMPONENT      (DEG)    PHASE (DEG)

   1       1.000D+05    4.229D-03    1.000000      179.897      0.000
   2       2.000D+05    1.951D-03    0.461405      -90.020   -269.917
   3       3.000D+05    1.427D-04    0.033743        3.012   -176.884

    TOTAL HARMONIC DISTORTION =     47.283488    PERCENT
```

(d)

**Fig. 6.23.** Fourier components of collector currents with (c) equal transistors, and (d) unequal transistors.

## 6.7 The Push-Pull Source Follower

A schematic diagram for a push-pull source follower is directly obtained from the BJT push-pull emitter follower by substituting n-channel and p-channel enhancement-mode MOS devices for the npn and pnp (enhancement-mode) transistors. The resulting circuit configuration is shown in Figure 6.25a. Of course, the values of the circuit parameters must be changed. The values of $V_{G1}$ and $V_{G2}$, shown in the Spice input file of Figure 6.25b, provide Class-AB performance. The voltage transfer characteristic is that of Figure 6.25c.

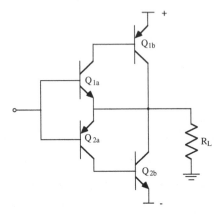

**Fig. 6.24.** Push-pull emitter follower with composite transistors.

There is some crossover distortion present, and the output range with respect to the values of the supply voltages is not as large as obtainable for a BJT configuration. The maximum input sinusoid to achieve just clipping at the output is 5 V zero-to-peak. However, the maximum gate potential of $M_1$ is 5.7 V. The harmonic components of the output voltage are given in Figure 6.26. Note that $HD_2 = 1.31\%, HD_3 = 2.85\%$ and $THD = 3.43\%$. These results are significantly worse than those for the BJT circuit.

The harmonic components of the supply voltage currents are given in Figures 6.27a and b. From the values of the dc components, the dc power supplied to the source followers is $P_{dc} = 2.5$ mW. The power delivered to the load is $P_{ac} = 1.6$ mW. The conversion efficiency is 64%. The results are not as good as those obtainable for the BJT.

## 6.8 Push-Pull, Single-Polarity Output Stages

Output stages have been developed for both analog MOS and bipolar circuits that employ devices of only one polarity and also have a push-pull action. In Figure 6.28a, a two-stage MOS configuration is present. The first stage is a normal inverter with an enhancement-mode load. The output of this stage drives the upper transistor in the final stage. This transistor may be either enhancement-mode (with a very small or zero value of $V_T$) or depletion-mode ($V_T < 0$) and operates as a source follower. The input signal is also fed directly to the lower final-stage enhancement-mode transistor. Both paths to the output involve a net phase reversal and the gain characteristic of a common-source stage.

For the circuit and device values given in the Spice input file of Figure 6.28b, the voltage transfer characteristic of the MOS output stage is shown in Figure 6.29. The input voltage range must be limited to less than the supply

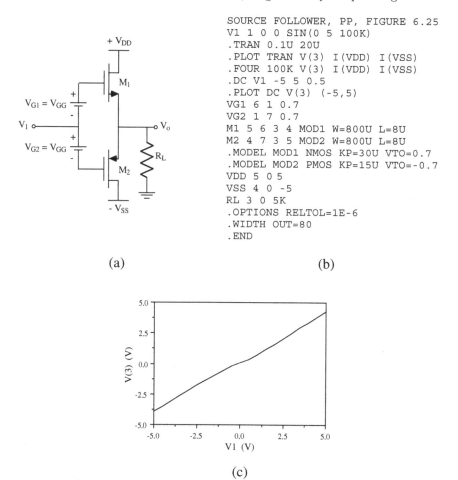

```
SOURCE FOLLOWER, PP, FIGURE 6.25
V1 1 0 0 SIN(0 5 100K)
.TRAN 0.1U 20U
.PLOT TRAN V(3) I(VDD) I(VSS)
.FOUR 100K V(3) I(VDD) I(VSS)
.DC V1 -5 5 0.5
.PLOT DC V(3) (-5,5)
VG1 6 1 0.7
VG2 1 7 0.7
M1 5 6 3 4 MOD1 W=800U L=8U
M2 4 7 3 5 MOD2 W=800U L=8U
.MODEL MOD1 NMOS KP=30U VTO=0.7
.MODEL MOD2 PMOS KP=15U VTO=-0.7
VDD 5 0 5
VSS 4 0 -5
RL 3 0 5K
.OPTIONS RELTOL=1E-6
.WIDTH OUT=80
.END
```

(a)                                    (b)

(c)

**Fig. 6.25.** (a) A push-pull source-follower circuit. (b) Spice input file. (c) Dc voltage transfer characteristics for the source follower.

```
FOURIER COMPONENTS OF TRANSIENT RESPONSE V(3)
DC COMPONENT =   9.722D-02
HARMONIC   FREQUENCY    FOURIER     NORMALIZED     PHASE    NORMALIZED
  NO         (HZ)       COMPONENT   COMPONENT      (DEG)    PHASE (DEG)

   1       1.000D+05   4.001D+00    1.000000      0.011      0.000
   2       2.000D+05   5.260D-02    0.013146    -90.170    -90.181
   3       3.000D+05   1.139D-01    0.028469    179.612    179.601

   TOTAL HARMONIC DISTORTION =       3.432367  PERCENT
```

**Fig. 6.26.** Fourier components of the output voltage.

```
FOURIER COMPONENTS OF TRANSIENT RESPONSE I(VDD)
DC COMPONENT =   -2.618D-04
HARMONIC    FREQUENCY      FOURIER     NORMALIZED     PHASE      NORMALIZED
   NO         (HZ)        COMPONENT    COMPONENT      (DEG)     PHASE (DEG)

    1       1.000D+05     4.149D-04    1.000000      179.939       0.000
    2       2.000D+05     1.846D-04    0.444848       90.027     -89.911
    3       3.000D+05     1.009D-05    0.024314        2.425    -177.514

        TOTAL HARMONIC DISTORTION =      45.368813   PERCENT
```

(a)

```
FOURIER COMPONENTS OF TRANSIENT RESPONSE I(VSS)
DC COMPONENT =    2.423D-04
HARMONIC    FREQUENCY      FOURIER     NORMALIZED     PHASE      NORMALIZED
   NO         (HZ)        COMPONENT    COMPONENT      (DEG)     PHASE (DEG)

    1       1.000D+05     3.853D-04    1.000000     -179.911       0.000
    2       2.000D+05     1.741D-04    0.451747      -89.961      89.950
    3       3.000D+05     1.272D-05    0.033000       -2.619     177.291

        TOTAL HARMONIC DISTORTION =      46.012688   PERCENT
```

(b)

**Fig. 6.27.** Fourier components of the supply currents. (a) Positive supply, and (b) negative supply.

voltage. The linearity of the characteristic is not good and distortion generation is significant. The Spice results for a sinusoidal input with an amplitude of 0.75 V and a bias value of 1.05 V are shown in Figure 6.30. The total harmonic distortion is 6.0%. Clearly, this stage should be used primarily within a feedback amplifier. Some improvement is obtained when a resistance is used in the source leads of $M_1$ and $M_3$.

From the Fourier outputs of the output voltage and current of Figure 6.30, $P_{ac} = 7.0$ mW, $P_{dc} = 16.74$ mW, and $\eta = 42\%$.

A bipolar push-pull circuit can also be developed with only one polarity-type of device. An example is the so-called totem-pole circuit shown in Figure 6.31a. This output stage was developed originally for digital logic circuits, in particular for TTL circuits. For digital circuits, a diode-connected transistor is also included in the emitter lead of $Q_2$, leading to a stack of three transistors, hence the name totem pole. The voltage transfer characteristic for this circuit is obtained from a Spice run. For the circuit values and device parameters shown in Figure 6.31b, the transfer characteristic is that shown in Figure 6.31c. The characteristic is quite linear, while transistors $Q_1$ and $Q_3$ are in the normal active region. Here, transistor $Q_1$ operates simultaneously as an emitter follower feeding $Q_3$ and a common-emitter stage (with $R_e = R_{B2}$ feedback) feeding the emitter follower, $Q_2$. The $Q_1$ stage is often called a phase splitter, since the two outputs are out of phase. Notice that both paths to the output load include a common emitter stage and an emitter follower. Near the bottom of the transfer curve, $Q_3$ begins to conduct significantly. Then $Q_1$ saturates and pulls up the output through $Q_2$. In design, the choice of the

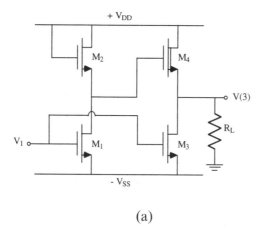

(a)

```
MOS TOTEM, FIGURE 6.28
V1 1 4 0 SIN(1.05 0.75 100K)
R1 1 0 1E9
C1 1 0 1P
.TRAN 0.1U 20U
.PLOT TRAN V(3) I(VDD) I(VSS)
.FOUR 100K V(3) I(VDD) I(VSS)
.DC V1 0 2 0.1
.PLOT DC V(3) (-5,5)
M1 2 1 4 4 MOD1 W=80U L=8U
M2 5 5 2 4 MOD1 W=18U L=80U
M3 3 1 4 4 MOD1 W=800U L=8U
M4 5 2 3 4 MOD2 W=800U L=8U
C2 2 0 1P
VDD 5 0 5
VSS 4 0 -5
RL 3 0 1K
.MODEL MOD1 NMOS KP=30U VTO=0.25 LAMBDA=0.01
.MODEL MOD2 NMOS KP=30U VTO=-1.5 LAMBDA=0.01
.OPTIONS RELTOL=1E-6
.WIDTH OUT=80
.END
```

(b)

**Fig. 6.28.** (a) A two-stage MOS output stage. (b) Spice input file for the MOS output stage.

resistors is critical to achieve a linear characteristic. In use, the input voltage excursion must be limited if significant distortion is to be avoided.

For a sinusoidal input voltage with an amplitude of 0.4 V, zero-to-peak, and a quiescent bias of 1.1 V, a transient Spice run provides the output waveform of Figure 6.32 and the Fourier output components given in Figure 6.33. The Fourier components of the separate voltage supply for the phase splitter are also given. The distortion of the output voltage, $V(6)$, is 3.9%. The power-conversion properties of the circuit can be estimated by inspecting the

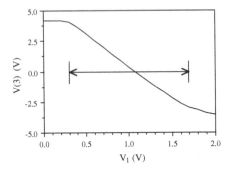

**Fig. 6.29.** Dc voltage transfer characteristics of the MOS output stage.

dc terms in the harmonic components of the current supplied by $V_{CC}$. The dc power supplied to the totem-pole stage is (5 V)(2.01 mA) $\approx$ 10.0 mW. The ac power delivered to $R_L$ is 1.24 mW. The conversion efficiency is only 12%. This is very low relative to what can be obtained from the push-pull emitter-follower output stage operated in near Class B.

## Problems

**6.1.** The dc current transfer characteristic for a BJT stage is shown in Figure 6.34. For a dc input which provides a collector current of 4 mA and an input sinusoidal voltage of 2.0 V, use the method of differential error to estimate HD2 in the output waveform.

**6.2.** An emitter-follower stage is shown in Figure 6.35.
(a) Estimate the maximum value of $V_{1A}$ for a reasonable output voltage waveform.
(b) For $V_{1A} = 9$ V, use Spice to establish THD. Check results with differential error estimation.
(c) Can feedback ideas and formulations be used to estimate the THD for the drive condition of (b)?
(d) For the drive of (b), determine the average power delivered to the load, the average power supplied by the dc voltage sources, and the power-conversion efficiency.

**6.3.** A source-follower stage is shown in Figure 6.36. The input voltage is $0 + 5\cos\omega t$. At the zero input, the positive peak input, and the negative peak input, the corresponding values of the drain current are 0.85 mA, 1.32 mA, and 0.38 mA, respectively. Use the differential error method to estimate HD2 in the output voltage.

**6.4.** A source-follower stage is shown in Figure 6.37.

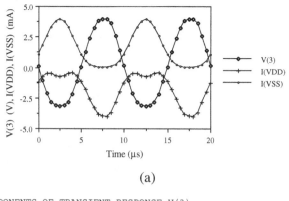

(a)

```
FOURIER COMPONENTS OF TRANSIENT RESPONSE V(3)
DC COMPONENT =   2.472D-01
HARMONIC   FREQUENCY    FOURIER     NORMALIZED     PHASE     NORMALIZED
   NO        (HZ)      COMPONENT    COMPONENT     (DEG)     PHASE (DEG)

    1      1.000D+05   3.753D+00    1.000000    178.777       0.000
    2      2.000D+05   1.745D-01    0.046504   -112.230    -291.007
    3      3.000D+05   1.289D-01    0.034339   -166.331    -345.108

    TOTAL HARMONIC DISTORTION =        5.985200  PERCENT
```

(b)

```
FOURIER COMPONENTS OF TRANSIENT RESPONSE I(VDD)
DC COMPONENT =  -1.797D-03
HARMONIC   FREQUENCY    FOURIER     NORMALIZED     PHASE     NORMALIZED
   NO        (HZ)      COMPONENT    COMPONENT     (DEG)     PHASE (DEG)

    1      1.000D+05   1.758D-03    1.000000     -2.537       0.000
    2      2.000D+05   6.007D-04    0.341630     83.753      86.290
    3      3.000D+05   1.111D-04    0.063162     16.087      18.624

    TOTAL HARMONIC DISTORTION =       34.880506  PERCENT
```

(c)

```
FOURIER COMPONENTS OF TRANSIENT RESPONSE I(VSS)
DC COMPONENT =   1.550D-03
HARMONIC   FREQUENCY    FOURIER     NORMALIZED     PHASE     NORMALIZED
   NO        (HZ)      COMPONENT    COMPONENT     (DEG)     PHASE (DEG)

    1      1.000D+05   1.995D-03    1.000000      0.002       0.000
    2      2.000D+05   4.355D-04    0.218250    -89.966     -89.968
    3      3.000D+05   1.842D-05    0.009230      0.113       0.111

    TOTAL HARMONIC DISTORTION =       21.847090  PERCENT
```

(d)

**Fig. 6.30.** (a) Output voltage and supply current waveforms. Fourier components of (b) output voltage, (c) positive supply current, and (d) negative supply current.

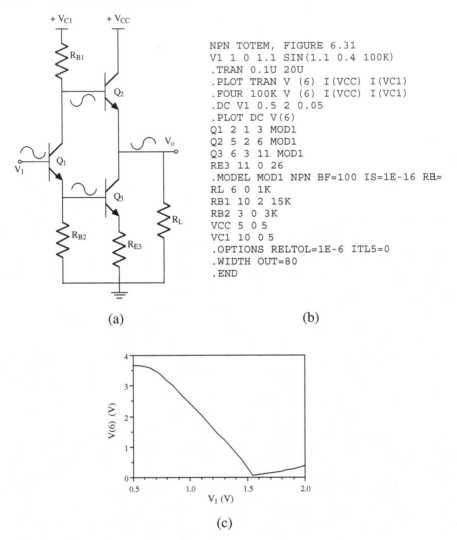

NPN TOTEM, FIGURE 6.31
```
V1 1 0 1.1 SIN(1.1 0.4 100K)
.TRAN 0.1U 20U
.PLOT TRAN V(6) I(VCC) I(VC1)
.FOUR 100K V(6) I(VCC) I(VC1)
.DC V1 0.5 2 0.05
.PLOT DC V(6)
Q1 2 1 3 MOD1
Q2 5 2 6 MOD1
Q3 6 3 11 MOD1
RE3 11 0 26
.MODEL MOD1 NPN BF=100 IS=1E-16 RB=
RL 6 0 1K
RB1 10 2 15K
RB2 3 0 3K
VCC 5 0 5
VC1 10 0 5
.OPTIONS RELTOL=1E-6 ITL5=0
.WIDTH OUT=80
.END
```

(a)                          (b)

(c)

**Fig. 6.31.** (a) A totem-pole bipolar npn output stage. (b) Spice input file for bipolar output stage. (c) Dc voltage transfer characteristics for the totem-pole output stage.

(a) Establish the value of $V_{GS}$ to produce $V_o = 0$ V.

(b) For $V_{sA} = 2$ V, estimate HD2 and HD3 using an analysis technique of your choice. Verify with Spice. What is the maximum value of $V_{sA}$ for adequate operation?

(c) For $V_{sA} = 2$ V, determine the average power delivered to the load, the average power supplied by the dc voltage sources, and the power-conversion efficiency.

**6.5.** A push-pull emitter-follower is shown in Figure 6.38.

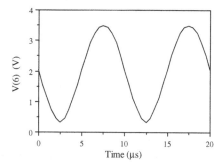

**Fig. 6.32.** Output voltage waveform for a sinusoidal input amplitude of 0.4 V.

```
FOURIER COMPONENTS OF TRANSIENT RESPONSE V(6)
DC COMPONENT =    1.982D+00
HARMONIC    FREQUENCY      FOURIER       NORMALIZED      PHASE      NORMALIZED
   NO         (HZ)        COMPONENT      COMPONENT       (DEG)     PHASE (DEG)

    1       1.000D+05     1.572D+00      1.000000      179.999       0.000
    2       2.000D+05     5.876D-02      0.037384       90.006      -89.993
    3       3.000D+05     7.965D-04      0.000507        1.492     -178.507

     TOTAL HARMONIC DISTORTION =         3.943662   PERCENT
```

(a)

```
FOURIER COMPONENTS OF TRANSIENT RESPONSE I(VCC)
DC COMPONENT =   -2.008D-03
HARMONIC    FREQUENCY      FOURIER       NORMALIZED      PHASE      NORMALIZED
   NO         (HZ)        COMPONENT      COMPONENT       (DEG)     PHASE (DEG)

    1       1.000D+05     1.468D-03      1.000000       -0.001       0.000
    2       2.000D+05     1.884D-05      0.012831       89.979      89.980
    3       3.000D+05     6.044D-05      0.041171       -0.020      -0.019

     TOTAL HARMONIC DISTORTION =         5.017887   PERCENT
```

(b)

```
FOURIER COMPONENTS OF TRANSIENT RESPONSE I(VC1)
DC COMPONENT =   -1.487D-04
HARMONIC    FREQUENCY      FOURIER       NORMALIZED      PHASE      NORMALIZED
   NO         (HZ)        COMPONENT      COMPONENT       (DEG)     PHASE (DEG)

    1       1.000D+05     1.063D-04      1.000000      179.999       0.000
    2       2.000D+05     4.178D-06      0.039314       90.006      -89.993
    3       3.000D+05     4.000D-08      0.000376        2.226     -177.773

     TOTAL HARMONIC DISTORTION =         4.100335   PERCENT
```

(c)

**Fig. 6.33.** Fourier components of (a) output voltage, (b) collector current of $Q_2$, and (c) collector current of $Q_1$.

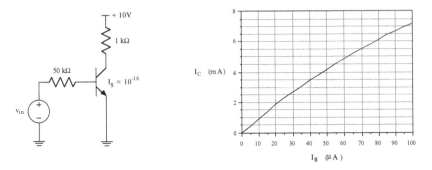

**Fig. 6.34.** Dc transfer characteristics for Problem 6.1.

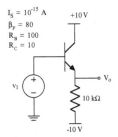

**Fig. 6.35.** Emitter-follower stage for Problem 6.2

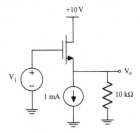

**Fig. 6.36.** Source-follower stage for Problem 6.3

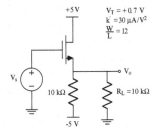

**Fig. 6.37.** Source-follower stage for Problem 6.4

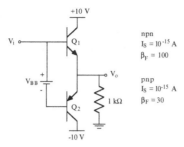

**Fig. 6.38.** Push-pull emitter-follower for Problem 6.5

(a) For $I_C = 0.5$ mA determine the value of $V_{BB}$ for quiescent operation.
(b) Determine the necessary zero-to-peak level of $V_i$ to achieve 'just clipping' of both the positive and negative peaks of $V_o$.
(c) Estimate the efficiency of power-supply conversion and the THD for the conditions of Part (b).
(d) Reduce $V_{BB}$ by 60 mV. Recalculate the estimates of Part (c).

**6.6.** A push-pull emitter-follower is shown in Figure 6.39.

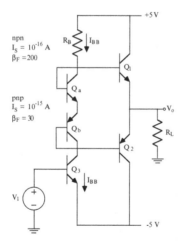

**Fig. 6.39.** Push-pull emitter-follower for Problem 6.6

(a) Replace transistor $Q_3$ with a voltage source, $V_s$, the quiescent value of which produces a dc current of 1 mA in the transistors. Note that all the transistors of the same type have the same $I_S$. Determine the necessary sinusoidal input amplitude to achieve 'just clipping' of the output voltage waveform.
(b) Use the method of differential error to estimate THD for $V_{sA} = 4.5$ V.
(c) Estimate the efficiency of power-supply conversion for the conditions of Part (b).

(d) Replace $V_s$ with $Q_3$ and $V_1$. Find the necessary value of the bias of the input voltage to produce currents of 1 mA in the transistors. Determine the necessary sinusoidal drive to achieve the output amplitude of (b). Comment on the difficulties of this request. Establish the distortion in the output voltage waveform for your choice of drive.

# 7

# Transformers

## 7.1 Introduction

Within an integrated circuit, active devices such as bipolar and MOS transistors can be realized readily along with certain passive devices such as resistors and capacitors. Inductors and transformers cannot be realized except for high-frequency applications [21], [22]. Although tuned, bandpass frequency responses can be realized with ICs containing feedback amplifiers, resistors and capacitors only, for many moderate-frequency applications, an inductor is needed and an element external to the IC must be utilized. Often, it is almost as convenient to achieve a two-winding 'coil' as one-winding; therefore, the transformer is available.

In the next two chapters, tuned circuits and bandpass amplifiers that utilize inductors and transformers are studied. Oscillators including these elements are studied later. In other chapters, transformers are often a necessary component of demodulation and rectifier circuits. Amplifier-$RC$ configurations to achieve the necessary bandpass responses are also briefly included in the next few chapters.

In this chapter, the basic low-frequency transformer is presented together with the circuit models and parameters which characterize its electrical performance. Not only is the transformer valuable and important in its own right as a circuit component, but also the concept of voltage, current, and resistance transformations are important tools and concepts in the design and evaluation of integrated circuit functions.

## 7.2 Elementary Coupled Coils

Figure 7.1a shows a sketch of a simple coil of wire wound about a closed magnetic medium, such as a ferrite torroid or a laminated set of iron-compound sheets, producing a continuous magnetic path, with a permeability, $\mu$, much larger than unity. Note that positive input current and voltage polarities are

D.O. Pederson and K. Mayaram, *Analog Integrated Circuits for Communication*, DOI 10.1007/978-0-387-68030-9_7,
© 2008 Springer Science+Business Media, LLC

defined. For such an arrangement, a magnetic flux $\phi$ is established in the 'core' for an input dc current. If the input current $I_1$ is increased, an induced voltage $V_1$ arises with the polarity shown in Figure 7.1a, in a direction as to oppose the change in flux linkage, $\lambda$, which produces it (Lenz's Law). The relation between the voltage and the changing flux linkage is Faraday's Law.

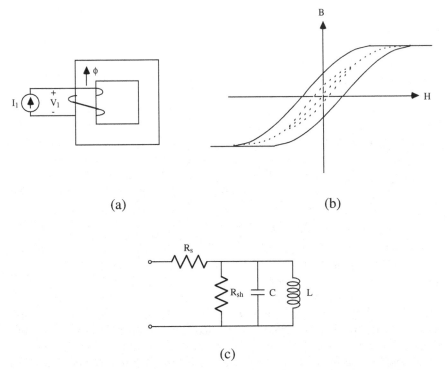

(a)                                    (b)

(c)

**Fig. 7.1.** (a) A simple coil of wire wound about a closed magnetic medium. (b) A typical BH curve for an iron core. (c) Circuit model representation for the coil.

$$V_1 = \frac{d\lambda}{dt} = \frac{dN\phi}{dt} \qquad (7.1)$$

where $\lambda$ is the magnetic flux linkage between the flux and the coil. For a simple situation, $\lambda$ can be considered to be the product $N\phi$ of the turns of the coil $N$ and the flux $\phi$ produced by the current $I_1$ flowing through the wire, i.e., a single turn. Again for a simple case, the flux produced by $I_1$ is

$$\phi = KNI_1 \qquad (7.2)$$

where $K$ is a constant. This is the equivalent of Ohm's Law for a simple magnetic element, particularly if one identifies $NI_1$ as the magnetomotive force, mmf. Using this expression, one obtains

$$V_1 = KN^2 \frac{dI_1}{dt} = L\frac{dI_1}{dt} \tag{7.3}$$

Note that the inductance of the coil is defined by

$$L = KN^2 \tag{7.4}$$

In general, and usually, the magnetic media is not linear. The relation between flux $\phi$ and the magnetomotive force, $NI_1$, is not a linear curve. It is more fundamental in studying the linearity relation to deal with the flux density $B$ and the magnetic field $H$ rather than with the flux and the mmf. A typical B-H curve for an iron core has the shape of the curves of Figure 7.1b for small and large driving levels. Note that both a hysteresis effect is present as well as a saturation phenomenon. The area within the hysteresis loop can be shown to represent an energy loss due to the flow of eddy currents in the magnetic media and is comparable to the loss in a resistor. Of course, for a dc input current there is no core loss. The loss problem is certainly present for a sinusoidal time excitation.

The saturation effects in a magnetic media lead to a nonlinear response in the circuit using the inductor. There is also a loss mechanism in the resistance of the wire. If the effects described above are combined into a circuit, the coil can be represented by the circuit model shown in Figure 7.1c. Notice that a shunt capacitor $C$ has been added across the inductance element. This represents the (distributed) electrostatic coupling of the turns of the coil with each other.

From the circuit model of Figure 7.1c, it is seen that at a particular frequency a parallel resonance occurs. These resonant circuits are studied in detail in the next chapter. For now, note that operation of the coil must be at frequencies well below the resonant frequency if it is to be considered a simple inductance.

If two coils are used as in Figure 7.2a, there are two induced voltages for a change of the flux. For the moment, assume that the right side, which is labeled for now the output, is open-circuited ($I_2 = 0$). The signal input is assumed to be applied to the left side, which is called the input side. For simplicity, the effects of the wire and core losses are neglected. In addition, it is assumed that all of the flux produced by the input current couples with the secondary winding, i.e., there is no 'leakage' flux. A current at the input establishes the magnetic flux. For a change of the input current, the induced voltage at the input is

$$V_1 = \frac{dN_1\phi}{dt} \tag{7.5}$$

where $N_1$ is the number of turns of the input winding. At the terminals of the output winding, the change of flux also produces an induced voltage.

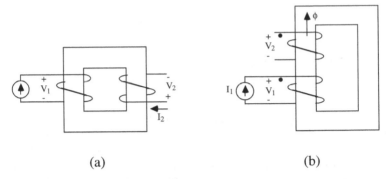

**Fig. 7.2.** (a) Two coils of wire wound on a core. (b) Two windings arranged on the same side of the core.

$$V_2 = \frac{dN_2\phi}{dt} \tag{7.6}$$

A different turns number $N_2$ is assumed for the output winding. The relation between the flux and the input current is next applied as in Equation (7.2) leading to

$$V_2 = KN_1N_2\frac{dI_1}{dt} \tag{7.7}$$

The product $KN_1N_2$ is defined as the mutual inductance $M$ of the coupled coils.

$$V_2 = M\frac{dI_1}{dt} \tag{7.8}$$

Note that the ratio of the input and output induced voltages is equal to the ratio of the turns of the input to output windings of the coils.

$$\frac{V_1}{V_2} = \frac{N_1}{N_2} = n \tag{7.9}$$

where $n$ is called the turns ratio of the coupled coils.

In Figure 7.2b, the two windings are arranged on the same side of the core. Notice that the two windings are shown to have the same winding sense. Using the 'right-hand rule' of basic electrical physics, one has an increasing flux in the up direction for an increasing input current. This produces a positive induced voltage for the defined port polarities of the figure for each coil. To denote the winding sense of the coupled coils, we usually *pole* the coupled coils with a dot notation. As shown in the figure, the dot is placed at the node of each pair which provides the same polarities of the two ports. If the two windings are in fact the same winding with a 'tap' connection, the arrangement is called an *autotransformer*. It is clear that since the two windings have the same winding sense, the voltage across the combination must have the same polarity as the two separate ports.

Next, we consider the case of three windings as illustrated in Figure 7.3. The two output windings have the opposite winding senses; the top coil has the same winding sense as the input, while the lower coil has the opposite sense. Notice that the location of the dot polarity notation is different for the two output coils. For an increasing input current, the two output voltages are of opposite polarity from the top node to bottom node of each node pair. Finally, we place a very large load resistor $R_L$ across each output. Because of the induced voltages, currents in these loads must flow for a change in the magnetic flux, i.e., a change in the input current. Each of these currents produces a 'back' flux which opposes the changing flux. In spite of the different winding senses of the two output coils, the direction of the back flux is the same as shown in the figure. This can be verified by using the right-hand rule for both coils for an assumed input current increase with resulting output voltages and load currents.

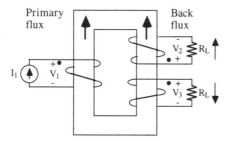

**Fig. 7.3.** Three windings on a core.

We return to a two coupled-coil situation as shown in Figure 7.4a, where the portrayal of the core and of the winding sense of the coils are not shown. It now is understood that a core material is present, which may not be an iron compound. The relative winding sense of the two coils is included with the dot polarity notation. The ratio of the number of turns of the two windings is introduced with the turns ratio, $n$. By definition, the winding associated with the parameter $n$ is called the *primary* side of the coupled coils and the other winding is referred to as the *secondary* side. This nomenclature holds even if $n < 1$, i.e., the primary winding has less turns than the secondary winding. Figure 7.4a is the schematic diagram of a two-winding transformer. We have seen above that the open-circuit input to output voltage ratio for a change of input current and with no leakage flux is $n$. The ratio of the currents in the two windings now is investigated. First, it is necessary to establish the voltage-current relations for this 'inductive two-port.' For an open-circuit output, the relations of the voltages to the input current are

$$V_1 = L_1 \frac{dI_1}{dt} \qquad (7.10)$$

$$V_2 = M \frac{dI_1}{dt}$$

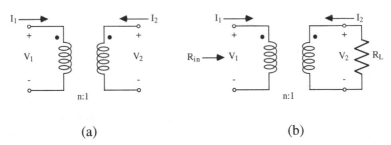

(a)                                        (b)

**Fig. 7.4.** (a) Schematic diagram of a two-winding transformer. (b) Transformer with a load resistance attached to the secondary side.

A new notation is now used. $L_1$ is defined as the self inductance of the input winding. It is proportional to $N_1^2$, as brought out above. The parameter $M$ is called the mutual inductance of the coil set. It is proportional to $N_1 N_2$, the product of the two turns. The ratio of $V_1$ to $V_2$ is equal to $n = N_1/N_2$ only if there is no leakage flux, i.e., if the total flux produced by $I_1$ couples with the secondary and vice versa. In general, $L_1$ includes a component of leakage. Therefore, the multiplier constant, $K$, for $L_1$ is not identical to the constant for $M$.

The next possibility is that an 'input' current is also present at the output (secondary). Note in Figure 7.4a that an assumed positive direction for this current is chosen in accordance to the usual network-theory practice. Positive node-pair voltages are also defined. The output current drive also produces flux. For an assumed linear core material, superposition of the flux due to the primary winding and due to the secondary can be used. At the input, the voltage is the sum of the self excitation and the mutual excitation from the secondary.

$$V_1 = L_1 \frac{dI_1}{dt} + M \frac{dI_2}{dt} \tag{7.11}$$

At the output node pair,

$$V_2 = M \frac{dI_1}{dt} + L_2 \frac{dI_2}{dt} \tag{7.12}$$

The mutual excitation parameter $M$ is the same for both sides. This is a fundamental property of passive, linear networks and can be derived either from network theory or directly from energy considerations [23]. A new self-inductance parameter, $L_2$, for the secondary is also used and, in general, includes leakage inductance for that winding.

The two equations above are examined in more detail shortly. For now, we establish the relations of the two currents $I_1$ and $I_2$ under special circumstances. In Figure 7.4b, a load resistance has been added at the secondary side of the transformer. The $V$–$I$ relation of the load resistance is

$$V_2 = -I_2 R_L \tag{7.13}$$

Using this, one obtains

$$0 = M\frac{dI_1}{dt} + L_2\frac{dI_2}{dt} + R_L I_2 \tag{7.14}$$

Consider now that either the values of the inductances, $L_2$ and $M$, are very large or that $R_L$ is very small. In either case, the $R_L$ term above can be neglected. The equation can be integrated to obtain

$$\frac{I_1}{I_2} = -\frac{L_2}{M} = -\frac{N_2}{N_1} = -\frac{1}{n} \tag{7.15}$$

where it is assumed that there is no leakage component in $L_2$. The current ratio of the primary to the secondary sides is the negative inverse of the turns ratio of the transformer.

It is convenient to define an ideal circuit element on the basis of the results and assumptions made in the above developments. The *ideal transformer* is defined to be a perfectly coupled set of coils without leakage, without winding or core losses, and with a mutual inductance which is infinitely large. The symbol for this element is shown in Figure 7.5, which also includes the defining voltage and current relations:

$$\frac{V_1}{V_2} = n \tag{7.16}$$

$$\frac{I_1}{I_2} = -\frac{1}{n}$$

$$L_m = \infty$$

$$k = 1$$

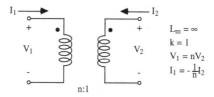

**Fig. 7.5.** An ideal transformer.

The magnetizing inductance $L_m$ and the coefficient of coupling $k$ are described below. If the ideal transformer is loaded at the output (secondary) with a load resistance $R_L$, the input resistance is

$$R_{in} = \frac{V_1}{I_1} = \frac{nV_2}{\frac{1}{n}I_2} = n^2 R_L \tag{7.17}$$

If the input side is loaded with a resistance $R_S$, the output resistance is

$$R_{out} = \frac{1}{n^2} R_S \tag{7.18}$$

Thus, the ideal transformer provides voltage, current, and resistance transformation.

The ideal transformer is used as a basic element of a circuit model for a general transformer as is developed in the next section.

For the main flux (and flux linkage) of the core without leakage, we note that only two parameters or elements are needed: the self inductance of the idealized coupled coils (without leakage and losses) and the mutual inductance, which are related simply.

$$L_1 = KN_1^2 \tag{7.19}$$
$$L_2 = KN_2^2$$
$$M = KN_1N_2$$
$$M^2 = L_1L_2$$

where now the constants $K$ are equal, since leakage is assumed zero.

The two parameters to characterize *ideally coupled coils* are the primary side inductance which is called the *magnetizing inductance* of the coupled coils, $L_m$, and the turns ratio of the idealized situation (without leakage.) A circuit model for ideally coupled coils is shown in Figure 7.6 and consists of an ideal transformer and a single inductor, $L_m$. (The magnetizing inductance and the turns ratio can also be taken from the 'low' side. Thus, care must be used in establishing a consistent pair of parameters.)

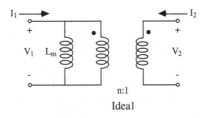

**Fig. 7.6.** Circuit model for ideally coupled coils.

## 7.3 Circuit Model for a Transformer

The ideal transformer and the several effects present in an actual transformer can be included in a circuit model of the transformer as shown in Figure 7.7a. The wire losses of the input and output windings are represented by $R_{s1}$ and $R_{s2}$, respectively. The input-side and output-side leakage inductances are denoted $L_{l1}$ and $L_{l2}$. The magnetizing inductance of the coupled coils is modeled by $L_m$, while the 'internal ideal transformer' has a turns ratio of $n_1$. A new notation for the turns ratio is introduced here to avoid confusion with the parameter for actual coupled coils. A polarity must also be assigned to the ideal transformer to include different connection possibilities. The secondary winding has a degree of freedom with respect to the output polarity of the induced voltage. Across $L_m$ is shown a shunt resistor, $R_{sh}$, which models the core losses due to eddy currents. Two capacitors are also included. $C_c$ models the electrostatic coupling of the two windings. $C_m$ represents the turn-to-turn electrostatic coupling of each winding to itself. Because of the presence of the ideal transformer, the shunt capacitances of the two windings can be combined into a single capacitor.

Simplifications can be made to the circuit model of Figure 7.7a. First, we restrict operation to frequencies below the open-circuit resonant frequency of $L_m$ and $C_m$. Further, we assume that the coupling effects of $C_c$ are small with respect to that of the ideal transformer. (The frequency restriction corresponds to a time-domain restriction to slow transients. This will be discussed later.) $C_m$ and $C_c$ can then be omitted. Next, the effects of wire loss and leakage inductance can be combined on one side of the transformer model. The new values are not just the sum of the two initial values. This aspect is treated later in this chapter. Nonetheless, the effects of wire loss and leakage can be included in a more gross modeling sense. Finally, for the reduced circuit model, assume that $R_{sh}$, $R_{s1}$, and $R_{s2}$ are removed from the model and incorporated in the source and load resistances presented to the transformer. (Of course, if leakage inductance is appreciable, the removal of $R_{sh}$ introduces inaccuracy.) The resulting inductive circuit model is that shown in Figure 7.7b. Notice that three elements or parameters describe completely this inductive 'two-port,' $L_i$, $L_m$ and $n_1$. Alternately, the two port can be described by $L_1, L_2$ and $M$, cf., (7.11) and (7.12).

## 7.4 Inductive Two-Port Parameters

Alternate, equivalent inductive two ports can be used in place of that of Figure 7.7b. These include the 'T' of inductors of Figure 7.7c and the 'pi' circuit shown in Figure 7.7d. The inductive two port can also be described in terms of open-circuit inductive parameters, $l_{ij}$, and short-circuit reciprocal parameters, $\Gamma_{ij}$. For the former, the two-port equations are

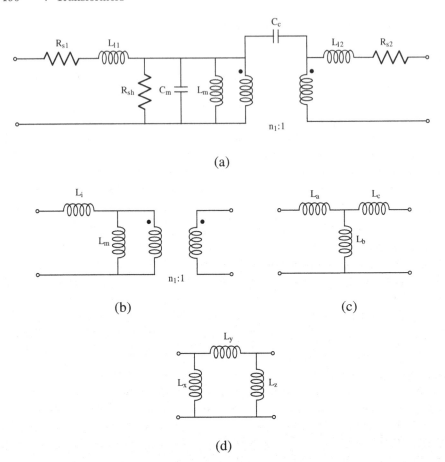

**Fig. 7.7.** (a) Circuit model for an actual transformer. (b) Simplified inductive model of the transformer. (c) 'T'-circuit equivalent of (b). (d) 'Pi'-circuit equivalent of (b).

$$v_1 = sl_{11}i_1 + sl_{12}i_2 \tag{7.20}$$
$$v_2 = sl_{12}i_1 + sl_{22}i_2$$

$$v_1 = j\omega l_{11}i_1 + j\omega l_{12}i_2$$
$$v_2 = j\omega l_{12}i_1 + j\omega l_{22}i_2$$

(Lower-case variables are now used since small-signal, steady-state variations about a quiescent state are required.) In the second equation the parameter $l_{21}$ is the proper constant, but for passive reciprocal devices $l_{12} = l_{21}$.

For the short-circuit description,

$$i_1 = \frac{\Gamma_{11}}{j\omega}v_1 + \frac{\Gamma_{12}}{j\omega}v_2 \tag{7.21}$$

$$i_2 = \frac{\Gamma_{12}}{j\omega}v_1 + \frac{\Gamma_{22}}{j\omega}v_2$$

The assumed polarities are shown in Figure 7.8. For the first set of equations above, if the output is open, $i_2$ is equal to zero, and the 'open-circuit' input impedance is

$$z_{11} = \frac{v_1}{i_1} = j\omega l_{11} \qquad (7.22)$$

$l_{11}$ is called the open-circuit input inductance. Similarly, $l_{22}$ is called the open-circuit output inductance. The transfer impedance from the input to the open-circuit output is

$$z_{21} = z_{12} = \frac{v_2}{i_1} = j\omega l_{12} \qquad (7.23)$$

Therefore, $l_{12}$ is referred to as the open-circuit transfer inductance. If sinusoidal variables are assumed in (7.11) and (7.12), the equations in the steady state become the same as (7.20) and $L_1 = l_{11}, L_2 = l_{22}, M = l_{12}$.

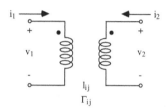

**Fig. 7.8.** Assumed polarities for the short-circuit description.

In terms of the parameters of Figure 7.7b, by inspection,

$$l_{11} = L_i + L_m \qquad (7.24)$$
$$l_{12} = \frac{L_m}{n_1}$$
$$l_{22} = \frac{L_m}{n_1^2}$$

In relation to the developments in Section 7.2, the turns ratio $n$ of the coupled coils is defined as $n = \sqrt{\frac{l_{11}}{l_{22}}}$. A parameter can be defined to portray the relative effects of the leakage inductance. This is the *coefficient of coupling*, $k$. By definition,

$$k = \frac{|l_{12}|}{\sqrt{l_{11} \times l_{22}}} \qquad (7.25)$$

In terms of the parameters of the circuit model of Figure 7.7b,

$$k = \frac{1}{\sqrt{\frac{L_m + L_i}{L_m}}} \tag{7.26}$$

If the leakage is zero, $L_i = 0$, cf., Figure 7.7b, $l_{12}^2 = l_{11}l_{22}$, and $k = 1$. This is another definition of *perfectly coupled coils*. For this case, $n = n_1$. The circuit model for this case is shown in Figure 7.6.

It is left as an exercise for the reader to repeat the developments for the other two inductive circuit models of Figures 7.7c and d to obtain

$$l_{11} = L_a + L_b \tag{7.27}$$
$$l_{12} = L_b$$
$$l_{22} = L_b + L_c$$
$$k = \frac{L_b}{\sqrt{(L_a + L_b)(L_b + L_c)}}$$

$$\Gamma_{11} = \frac{1}{L_x} + \frac{1}{L_y} \tag{7.28}$$
$$\Gamma_{12} = \frac{-1}{L_y}$$
$$\Gamma_{22} = \frac{1}{L_y} + \frac{1}{L_z}$$
$$k = \frac{|\Gamma_{12}|}{\sqrt{\Gamma_{11} \times \Gamma_{22}}}$$

If the coefficient of coupling is introduced, the inductive circuit model of the transformer can be represented as in Figure 7.9a. It is to be noted that the correct turns ratio for this circuit model is $n$, not $n_1$. For the model where leakage is included on only one side, as in Figure 7.7b, the appropriate model is that shown in Figure 7.9b. Note the presence of the $k^2$ parameter in the latter. Again, $n$ is used, not $n_1$.

As a final exercise, we investigate what happens in coupled coils with a linearly increasing input current. To solve this case, use the simple circuit model of Figure 7.10a where leakage and losses have been neglected. The input current is a ramp function with a slope value of $A$. If the output resistance is transformed to the primary side, the simpler circuit of Figure 7.10b is obtained. The circuit equation for this circuit is

$$v_1 = L_m \frac{d}{dt}\left(i_1 - \frac{v_1}{n^2 R_L}\right) \tag{7.29}$$

If one uses transform notation, the voltage function $v_1(s)$ is

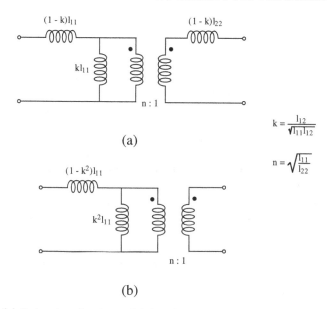

$$k = \frac{l_{12}}{\sqrt{l_{11}l_{12}}}$$

$$n = \sqrt{\frac{l_{11}}{l_{22}}}$$

**Fig. 7.9.** (a) Inductive circuit model for the transformer with leakage. (b) Circuit model with leakage included on only one side.

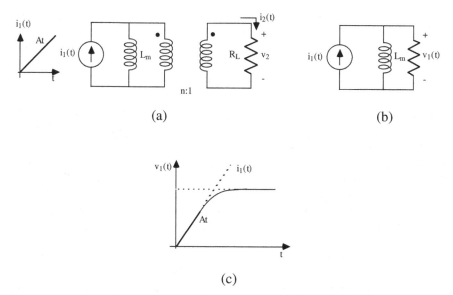

**Fig. 7.10.** (a) Circuit model with ramp input. (b) Simplified circuit. (c) Voltage response in the time domain.

$$v_1(s) = L_m \frac{s}{1 + \frac{s}{s_1}} i_1(s) \tag{7.30}$$

where $s_1 = n^2 R_L / L_m$.

The Laplace transform of the ramp function is $A/s^2$. Therefore, $v_1(s)$ can be written

$$v_1(s) = C_1 \frac{1}{s} + C_2 \frac{1}{s + s_1} \tag{7.31}$$

The voltage response in the time domain is shown in Figure 7.10c. This is also the time-response form of $v_2(t)$ and $i_2(t) = -i_L(t)$. For even a perfectly coupled set of coils, the input/output current ratio is not a constant even if the voltage ratio is $n$. Only for the ideal transformer are both the voltage and current ratios a constant.

In the next section, a transformer is used in a simple output stage. The different parameters and approximations above are illustrated with actual numerical values.

## 7.5 A Transformer-Coupled, Class-A BJT Output Stage

A bipolar output stage including an input transformer and an output transformer is shown in Figure 7.11a. Both transformers are assumed to have coefficients of coupling of $k \approx 1$. Because the input transformer usually has very low wire losses, relative to the input resistance of the diode-connected transistor $Q_b$, normal diode biasing of the transistor input occurs. If a restriction is made to input frequencies for which the inductive effects of the transformer are negligible, the equivalent input to the transistor base is that shown in Figure 7.11b. Note that at the secondary side of the transformer, the open-circuit, signal-source voltage is divided by $n_i$ while the source resistance is divided by $n_i^2$. The bias source $V_{BB}$, which is the quiescent value of $V_{BE}$, is the 'diode' voltage $V_D$ and is established by the current through $R_B$. A dashed-line inductor is placed across the reflected source resistance to remind us that for dc, a short-circuit path through the secondary-transformer winding is present. Therefore for dc,

$$I_D = \frac{V_{CC} - V_D}{R_B} \approx \frac{V_{CC}}{R_B} \tag{7.32}$$

The approximation is valid if $V_{CC} \gg V_D$. For convenience in the following, a dc voltage source, $V_{BB}$, is used in place of $R_B$ and the bias diode. A set of circuit and device values for Figure 7.11a is given in the Spice input file of Figure 7.12. Notice that each transformer is characterized by two inductances, the open-circuit input and output inductances, and a coefficient of coupling. The turns ratio of the transformer is the square root of the ratio of the inductances. For a desired value of quiescent collector current, say $I_{CA} = 1$ mA,

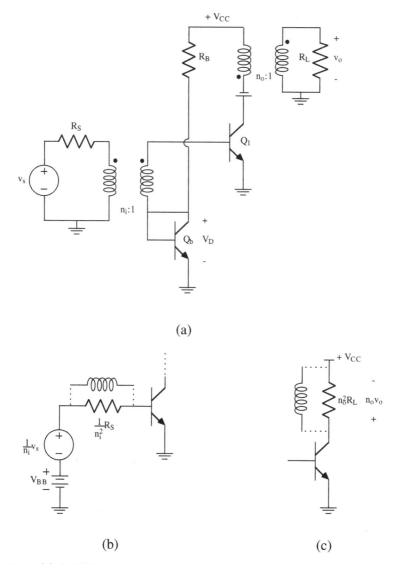

(a)

(b)                                    (c)

**Fig. 7.11.** (a) A BJT output stage with input and output transformers. (b) Equivalent input to the transistor base. (c) Equivalent circuit at the output of the transistor.

the base-bias voltage $V_{BB} = 0.774$ V. (Actually, $I_{CA} \approx 0.99$ mA due to the base currents.)

```
SINGLE STAGE XSFMR, FIGURE 7.12
VS 1 0 0 SIN(0 19M 1MEG)
*VS 1 0  0 SIN( 0 26.2M 1MEG)
.TRAN 0.01U 2U
.PLOT TRAN V(30)
.PLOT TRAN I(VC2)
.FOUR 1MEG V(30) I(VC2)
RS1 1 2 1
*RS1 1 2 1000
L1 2 0 .1
L2 3 4 .1
VBB 4 0 0.774
KA L1 L2 1
Q1 5 3 0 MOD1
VC2 5 8 0
LA 8 20 .1
LC 30 0 1M
KF LA LC 1
RL 30 0 100
VCC 20 0 10
.MODEL MOD1 NPN IS=1E-16 BF=100
.WIDTH OUT=80
.OPTIONS RELTOL=1E-6
.OPTIONS NOPAGE NOMOD
.END
```

**Fig. 7.12.** Spice input file for the transformer coupled BJT output stage.

At the transistor output, the load resistor $R_L$ can be transferred across to the transformer primary, as shown in Figure 7.11c by a resistor $n_o^2 R_L$. Note that the voltage across the reflected resistance is $n_o$ times the actual output voltage. Due to the polarity assumption for the output transformer, $n_o v_o$ is inverted relative to $v_o$. Again a dashed-line inductor is placed across the reflected load to illustrate the dc path in the primary winding from $V_{CC}$ to the collector. The load line for this situation is shown in Figure 7.13. For the quiescent condition (no signal input voltage), the operating point is $I_{CA}$ and $V_{CC}$. Remember there is a dc path across $n_o^2 R_L$. For frequencies sufficiently high so that the reactance of $L_m$ is very large, the slope of the load line through the operating point is $-1/n_o^2 R_L$. The x-axis intercept of the load line is $V_{CC} + I_{CA} n_o^2 R_L$. The y-axis intercept is $I_{CA} + V_{CC}/n_o^2 R_L$. Because of transistor saturation, the maximum negative voltage swing at the collector is $V_{CC} - V_{CEsat}$. Usually, $V_{CEsat}$ is small relative to $V_{CC}$ and can be neglected. The maximum positive voltage swing at the collector is $V_{CC} + I_{CA} n_o^2 R_L - V_{CC} = I_{CA} n_o^2 R_L$.

As is done in Chapter 6, assume that no distortion of the input signal is generated and that the collector current and collector voltage are both pure sinusoids for a sinusoidal input. For a maximum drive to the transistor input

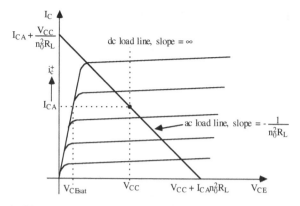

**Fig. 7.13.** Transistor output characteristics with dc and ac load lines.

without clipping of the collector voltage and current, i.e., with the transistor operating exclusively in the normal active region, maximum power is developed in $n_o^2 R_L$, for

$$n_o^2 R_L = \frac{V_{CC}}{I_{CA}} \qquad (7.33)$$

The ac output power is

$$P_{ac} = \frac{1}{2} V_{CC} I_{CA} \qquad (7.34)$$

The dc power delivered from the source to the output circuit is

$$P_{dc} = V_{CC} I_{CA} \qquad (7.35)$$

The ideal maximum conversion efficiency is $\eta = 50\%$

This value is significantly larger than can be obtained for the 'resistive' CE stage of Chapter 3. However, in common with a simple CE stage, the output voltage and current variables for maximum output excursions are very nonsinusoidal, and the harmonic distortion is not small. In addition, there may be new frequency and transient degradation effects that are introduced by the magnetizing inductance of the transformers. These aspects are brought out below.

The actual conversion efficiency obtainable and the effects of the magnetizing inductances can be brought out with Spice simulations. In the input Spice file of Figure 7.12, typical transformer parameters have been chosen. Note that for both transformers, the coefficient of coupling, $k$, is set equal to 1 for an assumed closely coupled situation. (For some Spice-type simulators, $k$ should be chosen to be 0.999.) The turns-ratio parameter is not used explicitly in the Spice input. Rather the open-circuit input and output inductances are used. As brought out in the last section, this specification is equivalent. Given $l_{11}$ and $l_{22}$, and for $k = 1$,

$$L_m = l_{11} \tag{7.36}$$

$$n = \sqrt{\frac{l_{11}}{l_{22}}}$$

$$l_{22} = \frac{l_{11}}{n^2} = \frac{L_m}{n^2}$$

For the input transformer of the circuit of Figure 7.11a, $n_i = 1$, and at the output, $n_o = 10$. The value for the open-circuit, primary input inductances (the magnetizing inductance values) of both transformers is chosen to be 0.1 H. This is a typical, small value for inexpensive coupling transformers. We see below that this value is often not large enough to permit adequate operation at low frequencies.

For a small value for $R_s$, the input voltage to the base of the transistor is the signal input, since the turns ratio is unity. We expect then the same harmonic generation as encountered with the simple resistive CE stage of Chapter 3, due to the exponential input and transfer characteristics of the transistor. The input signal-source voltage to achieve maximum output conditions can be estimated using the small-signal voltage gain of the stage at the quiescent state.

$$a_v = \frac{v_o}{v_1} = -\frac{1}{R_s + r_\pi} \beta n_o^2 \frac{R_L}{n_o} \tag{7.37}$$

For $R_s = 1\Omega$ and with $r_\pi = 2585\Omega$ (for an approximate 1 mA bias), $a_v = -38.5$. With $V_{CC} = 10$ V, a 10 V collector voltage swing is the largest possible value for just clipping. This corresponds to an output voltage with a zero-to-peak amplitude of 1 V and an input voltage amplitude of 26 mV. If this signal level is used in the Spice simulation, the output is severely distorted. Definite negative peak clipping is observed in the Spice output waveforms. In effect, the gain for the negative excursion is larger than that predicted by the small-signal calculation at the quiescent point.

A value of the maximum amplitude of a sinusoidal input drive to produce 'just clipping' can be obtained using the Bessel function curves of Chapter 3, Figure 3.6. An estimate can be made considering only the first two terms of the power series expansion for $I_C$.

$$i_c = I_C - I_{CA} \approx I_{CA} \left[ \frac{v_1}{V_t} + \frac{1}{2} \left( \frac{v_1}{V_t} \right)^2 \right] \tag{7.38}$$

where $I_{CA}$ is the quiescent value of $I_C$. For a sinusoidal input voltage with an amplitude $V_{1A}$

$$i_c = I_{CA} \left[ \frac{V_{1A}}{V_t} \sin \omega_1 t - \frac{1}{4} \left( \frac{V_{1A}}{V_t} \right)^2 \cos 2\omega_1 t + \frac{1}{4} \left( \frac{V_{1A}}{V_t} \right)^2 \right] \qquad (7.39)$$

Notice that the dc shift term is included. For a positive input excursion of $i_c$, the maximum of $i_c$ occurs at $\omega_1 t = 90°$. The peak value is

$$i_c{}^+ = I_{CA} \frac{V_{1A}}{V_t} \left[ 1 + \frac{2}{4} \frac{V_{1A}}{V_t} \right] \qquad (7.40)$$

The input amplitude is

$$V_{1A} = V_t \left[ -1 + \sqrt{1 + 2 \frac{i_c{}^+}{I_{CA}}} \right] \qquad (7.41)$$

For an example, refer first to the load line of Figure 7.13. If $n_o^2 R_L > V_{CC}/2I_{CA}$, $i_c{}^+ = V_{CC}/n_o^2 R_L - I_{CA}$. For $n_o^2 R_L < V_{CC}/I_{CA}$, 'off clipping' occurs before 'saturation clipping'. For the maximum excursion of $i_c$, $I_{CA} = V_{CC}/2R_C$, $i_c{}^+ = I_{CA}$.

$$V_{1A} = V_t \sqrt{-1 + 3}$$
$$= 18.9 \text{ mV}$$

The output-voltage and collector-current waveforms for $V_{sA} = 19$ mV are given in Figures 7.14a and b. Notice that the output voltage shows some compression at the positive peaks, which corresponds to low values of $V_{CE}$. The collector current, measured by voltage source $V_{C2}$ in the Spice input file, shows more clearly that transistor saturation is encountered. Harmonic outputs for $V_o$ and $I_c$ are given in Figure 7.15. The second harmonic distortion of $v_o$ is $HD_2 \approx 17\%$. The other harmonic components are small but not negligible. Note that the average value of the collector current is 1.13 mA. This is significantly larger than the quiescent value of 0.99 mA. Therefore, at this signal level, significant expansion/contraction of the dc and fundamental components are being produced by the higher-order terms of the power series of the nonlinearities.

From the two sets of output, the ac power in the load and the dc power from $V_{CC}$ are

$$P_{ac} = \frac{(0.774)^2}{2 \times 100} = 3.0 \text{ mW} \qquad (7.42)$$
$$P_{dc} = (1.13 \text{ mA})(10 \text{ V}) = 11.3 \text{ mW}$$

The conversion efficiency is $\eta = 27\%$. This is almost a factor of two lower than the 50% value of the idealized case.

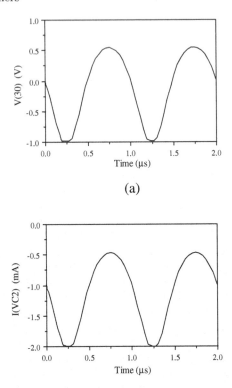

**Fig. 7.14.** Waveforms of (a) output voltage, and (b) collector current.

```
FOURIER COMPONENTS OF TRANSIENT RESPONSE V(30)          Rₛ = 1 Ω, Vₛₐ = 19 mV
DC COMPONENT = -1.053D-01
HARMONIC   FREQUENCY     FOURIER     NORMALIZED     PHASE      NORMALIZED
   NO        (HZ)       COMPONENT    COMPONENT      (DEG)     PHASE (DEG)

    1      1.000D+06    7.702D-01    1.000000    -179.087       0.000
    2      2.000D+06    1.296D-01    0.168221      91.200     270.287
    3      3.000D+06    8.214D-03    0.010665       0.239     179.326

    TOTAL HARMONIC DISTORTION =     16.903684  PERCENT


FOURIER COMPONENTS OF TRANSIENT RESPONSE I(VC2)
DC COMPONENT = -1.126D-03
HARMONIC   FREQUENCY     FOURIER     NORMALIZED     PHASE      NORMALIZED
   NO        (HZ)       COMPONENT    COMPONENT      (DEG)     PHASE (DEG)

    1      1.000D+06    7.669D-04    1.000000    -179.998       0.000
    2      2.000D+06    1.296D-04    0.168995      90.012     270.010
    3      3.000D+06    9.331D-06    0.012167       0.281     180.279

    TOTAL HARMONIC DISTORTION =     16.992872  PERCENT
```

**Fig. 7.15.** Fourier components of the output voltage and collector current.

For a larger value of $R_s$, say $R_s = 1$ k$\Omega$, the distortion is reduced due to the linearization of the input characteristic, as brought in Chapter 4. The output voltage is also reduced because of the voltage division between $R_s$ and $r_\pi$. The signal input voltage must be increased in relation to the last example to achieve just clipping of the output. (In this transformer-coupled stage, it is not necessary to adjust the bias input level to compensate for the bias effects of $R_s$. The secondary of the input transformer provides a constant dc path.)

From the results of Chapter 4, the value of the parameter $E = R_s/r_\pi = 1000/2585 = 0.38$. Therefore, the input should be increased by $(1+E) = 1.38$, i.e., $V_{1A} = 26.2$ mV. For a zero-to-peak input voltage of 26.2 mV, just clipping is obtained. (Because of expansion effects which affect the value of the dc collector current and therefore $r_\pi$, the small-signal calculation of $E$ may not provide exact results.) The Fourier components of the output voltage, across $R_L$, and the collector current are shown in Figure 7.16. The value of $HD_2$ is seen to be 12.5%. The estimated value of $HD_2$ from the results for $R_s = 1\Omega$ is $17\%/1.38 = 12.3\%$. For $Rs = 1$ k$\Omega$, the conversion efficiency is: $\eta = 24.7\%$.

Problems with a transformer occur when the reactance of the magnetizing inductance at a given frequency has a magnitude comparable to the resistances presented to the transformer. At the primary of the input transformer, the ac circuit is shown in Figure 7.17a. Notice that the input resistance of the transistor, $r_\pi$, is reflected across the input transformer by $n_i^2$. The parallel resistance presented across $L_{mi}$ is

$$R_i = R_s \parallel n_i^2 r_\pi \qquad (7.43)$$

where for the present example, $n_i = 1$. The reactance of $L_{mi}$ should be large relative to this value. If the input frequency is $\omega_i$,

$$\omega_i L_{mi} > R_s \parallel n_i^2 r_\pi \qquad (7.44)$$

In terms of the constraint on the input frequency,

$$f_i \geq \frac{R_s \parallel n_i^2 r_\pi}{2\pi L_{mi}} = f_{-3\text{dB}} \qquad (7.45)$$

As noted in (7.45), the $RL$ ratio is equal to the -3 dB corner frequency of the high-pass circuit of Figure 7.17a. The signal input frequency must be at least three times greater than this highpass corner frequency for no more than a 10% effect in the (linear) frequency response. The signal input frequency should be greater than 10 times for a negligible effect.

In Figure 7.12, two values of $R_s$ are given. For $R_s = 1\ \Omega$, $f_{-3\text{dB}} = 1.6$ Hz. For this small-signal-source resistance, little frequency effects should be encountered even at the low audio frequencies.

For $R_s = 1$ k$\Omega$, $f_{-3\text{dB}} = 1.1$ kHz. Therefore, for a reasonable value of source resistance and even with a reasonable value of magnetizing inductance, frequency effects due to the input transformer can be expected in the audio frequencies.

```
SINGLE STAGE XSFMR, FIG 7.16
*VS 1 0 0 SIN(0 19M 1MEG)
VS 1 0  0 SIN( 0 26.2M 1MEG)
.TRAN 0.01U 2U
.PLOT TRAN V(30)
.PLOT TRAN I(VC2)
.FOUR 1MEG V(30) I(VC2)
*RS1 1 2 1
RS1 1 2 1000
L1 2 0 .1
L2 3 4 .1
VBB 4 0 0.774
KA L1 L2 1
Q1 5 3 0 MOD1
VC2 5 8 0
LA 8 20 .1
LC 30 0 1M
KF LA LC 1
RL 30 0 100
VCC 20 0 10
.MODEL MOD1 NPN IS=1E-16
BF=100
.WIDTH OUT=80
.OPTIONS RELTOL=1E-6
.OPTIONS NOPAGE NOMOD
.END
```
                                            (a)

FOURIER COMPONENTS OF TRANSIENT RESPONSE V(30)    $R_S = 1 \text{ k}\Omega$, $V_{SA} = 26.2 \text{ mV}$
DC COMPONENT =  -7.073D-02

| HARMONIC NO | FREQUENCY (HZ) | FOURIER COMPONENT | NORMALIZED COMPONENT | PHASE (DEG) | NORMALIZED PHASE (DEG) |
|---|---|---|---|---|---|
| 1 | 1.000D+06 | 7.301D-01 | 1.000000 | -179.031 | 0.000 |
| 2 | 2.000D+06 | 9.159D-02 | 0.125447 | 91.248 | 270.279 |
| 3 | 3.000D+06 | 1.107D-03 | 0.001516 | 1.325 | 180.356 |

TOTAL HARMONIC DISTORTION =     12.546613   PERCENT

FOURIER COMPONENTS OF TRANSIENT RESPONSE I(VC2)
DC COMPONENT =  -1.086D-03

| HARMONIC NO | FREQUENCY (HZ) | FOURIER COMPONENT | NORMALIZED COMPONENT | PHASE (DEG) | NORMALIZED PHASE (DEG) |
|---|---|---|---|---|---|
| 1 | 1.000D+06 | 7.280D-04 | 1.000000 | -179.941 | 0.000 |
| 2 | 2.000D+06 | 9.163D-05 | 0.125862 | 90.098 | 270.039 |
| 3 | 3.000D+06 | 1.857D-06 | 0.002552 | 1.761 | 181.702 |

TOTAL HARMONIC DISTORTION =     12.589112   PERCENT

                                            (b)

**Fig. 7.16.** (a) Spice input file for transformer coupled stage. (b) Fourier components of the output voltage and collector current.

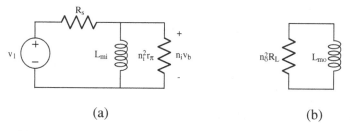

(a)                              (b)

**Fig. 7.17.** (a) Ac circuit at the primary of the input transformer. (b) Ac circuit at the output.

At the output side of the stage, the ac small-signal circuit is shown in Figure 7.17b. For small frequency effects due to $L_{mo}$,

$$\omega_i L_{mo} > n_o^2 R_L \tag{7.46}$$

The corresponding highpass corner frequency for the element values of the Spice input file of Figure 7.12 is

$$f_{-3dB} = \frac{n_o^2 R_L}{2\pi L_{mo}} = 15.9 \text{ kHz} \tag{7.47}$$

Again, the signal frequency must be at least three times this frequency for no more than 10% effect on the fundamental components. For the present example, the output side of the stage clearly determines the lowest frequency at which degradation occurs, even for $R_s = 1$ k$\Omega$.

If one must operate with an input frequency equal to or smaller than the corner frequencies above, severe frequency distortion will occur. For the present simulation example, an input frequency of 1 MHz is chosen as shown in the Spice input file, Figure 7.12, to avoid such problems. If the frequency of the input sinusoid is decreased, the reactances (energy storage) of the magnetizing inductances come into play. In Figure 7.18 are shown the harmonic distortion factors of the output voltage for input frequencies of 10 kHz, 100 kHz, and 1 MHz. The input signal amplitude remains at 26.2 mV and $R_{s1} = 1$ k$\Omega$. For $R_s = 1$ k$\Omega$, 10 kHz is below the high-pass corner frequency produced by the output transformer and above the input corner frequency. It is clear that a significant increase of $HD_2$ is introduced as the input frequency is lowered.

Of course, using transformers with larger magnetizing inductances is a possibility. But this involves greater component expense. Further, one must also be concerned about the fact that with a larger $L_m$, the self-resonant frequency of the transformer may be less, leading to another possible frequency degradation. Finally, as brought out below with larger values of $L_m$, nonlinearity of the transformers usually becomes a new source of distortion.

| $V_1 = 26.2$ mV | $R_S = 1$ kΩ | |
|---|---|---|
| Frequency | HD$_2$ | HD$_3$ |
| 1 MHz | 12.7 % | 0.17 % |
| 100 kHz | 14.4 % | 0.40 % |
| 10 kHz | 20.7 % | 0.75 % |

**Fig. 7.18.** Harmonic distortion factors for various input frequencies.

## 7.6 Maximum Power Transfer

For the transformer-coupled output stage, new problems may enter the scene primarily due to the output transformer. First, for the quiescent condition, a dc current flows through the primary which 'biases' the transformer on its B-H characteristic. Most practical transformers exhibit a nonlinear B-H curve; thus, more distortion is produced (for excursions about the operating point) than if the transformer is biased at the origin of the B-H characteristic. A Spice run will not show this unless a nonlinear model for the transformer is introduced.[1]

A second problem concerns the core losses of a transformer, which have not been included in the above examples. The reflected load resistance, $n_o^2 R_L$, is usually large. If the core loss modeled by $R_{sh}$ of the transformer is included, the efficiency results above must be corrected. Similarly, for large $n_o$, the output resistance of the transistor, $r_o$, must also be included.

In general, for maximum power transfer to the load, the reflected load resistance must equal the parallel combination of $R_{sh}$ of the transformer and $r_o$ of the transistor.

$$n_o^2 R_L = R_{sh} \parallel r_o \qquad (7.48)$$

As an example, let $r_o$ at the quiescent operating point be 10kΩ (corresponding to a collector current of 6 mA and $V_A = 60$ V). Let the value of core loss characterized at the transformer primary also be 10kΩ. The equivalent input resistance at the primary side presented to the reflected load resistance, $n_o^2 R_L$, is 5 kΩ. For a load resistance of 100Ω , the value of $n_o$ for maximum power transfer should be 7.07.

For the case where $n_o$ is chosen to obtain maximum power transfer, the efficiency of conversion is reduced by one half. This follows because at the primary side of the output transformer, the reflected load is equal to the loading supplied by $R_{sh}$ and $r_o$. Only one half of the ac power developed at

---

[1]   Several Spice-type simulators provide a nonlinear transformer model, cf., Pspice [24]. A model for the nonlinear transformer is introduced with the addition of a model name in the coefficient-of-coupling element and the inclusion of a new .MODEL line. The parameters of the .MODEL description can include transformer core material, shape, and size.

this point is transformed to $R_L$. Since the earlier ideal maximum is 50%, the idealized maximum for maximum power transfer is 25%. It can be expected that the maximum obtainable conversion is less than this.

## 7.7 Class-A Push-Pull Operation

Just as two windings, rather than one for the coil, leads to a transformer, so a center-tap connection can be introduced to obtain a more versatile transformer. The use of center-tapped transformers in an output stage is shown in the bipolar output stage shown in Figure 7.19. This is a push-pull, parallel situation comparable to the push-pull emitter-followers and source-followers of Chapter 6. For the push-pull stage, a common bias arrangement uses a bias resistance from $V_{CC}$ and a diode-connected transistor as shown in Figure 7.19b. In effect, this provides a simple voltage-bias source, $V_{BB}$, as used in the Figure 7.19a. $V_{BB}$ is then the quiescent value of $V_{BE}$. The input drive in this section is restricted to values for which the collector currents always flow, i.e., for Class-A operation. (The voltage sources, $V_{C1}$ and $V_{C2}$, are used to monitor the collector currents).

For the center-tapped input transformer, care must be taken with respect to labeling the primary and secondary sides. For convenience in this example, the turns ratio of the input transformer is given as the ratio of one-half of the right-hand-side (RHS) voltage to the left-hand-side (LHS) voltage. Each transistor input has an incremental base-emitter voltage of one-half of the total RHS voltage.

Similarly, for the output transformer, the turns ratio is taken as one-half of the LHS voltage to the RHS voltage, (which in this case is the output voltage $v_o$).

Assume for the moment that $R_s$ is small. Because of the center-tap in the RHS of the input transformer which is at ac ground, equal and opposite voltage signals are delivered across the transformer to the inputs of the two transistors. The RHS winding of the input transformer is continuous with the same sense of winding. The incremental transferred voltage from point a to b is equal to that from b to c, i.e., the voltage from c to b is the negative of that from a to b. The two transistors have the same base-bias voltage; thus, the same quiescent collector currents. For an increase of input voltage, $v_1$, the collector current of the top transistor, $Q_1$, increases while that of the bottom transistor, $Q_2$, decreases. The variations of the two collector currents are 'out of phase.'

Notice also that there is no net dc flux in the input transformer. The base currents flowing to each transistor from $V_{BB}$ produce equal and opposite dc fluxes. Thus the quiescent bias point for the input transformer occurs at the origin of its B-H characteristic. Similarly, at the output, the LHS, the primary, of the output transformer is assumed to be a center-tapped single winding. For the same bias, there is no net primary flux for the quiescent condition,

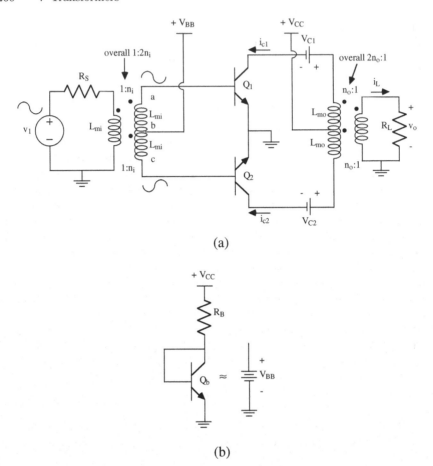

(a)

(b)

**Fig. 7.19.** (a) A bipolar push-pull output stage with center-tapped transformer. (b) Bias arrangement for the push-pull stage.

since the collector currents flow 'out' of the primary windings at the top and the bottom. Consequently, this transformer is also 'biased' at the origin of the B-H characteristic. (The above description presupposes that the base currents are equal as are the collector currents. For an ideal transformer situation, an inconsistency exists if this is not true. The circuit equations then become indeterminate and the solution 'blows up.' In practice, the resistive properties of the transistors and transformers provide a circuit equilibrium.)

For the increase of input signal voltage, the increase of collector current of $Q_1$ increases the transformer flux in one direction. Since the transistor current of $Q_2$ is decreasing but in the opposite direction, the change in flux due to it is in the same direction as the change in flux due to the top transistor. The change of flux then is increased by two and effective, parallel operation is obtained. Since the dc currents in the primary of the transformers cancel and the incrementals, or fundamentals, add, we expect in this push-pull configuration for all even harmonics of the output to cancel and all odd harmonics to add.

A harmonic generation analysis can be made to establish the harmonic generation and cancellation, such as done in Chapter 6. This is quite straightforward for the case where the signal-source resistance is very small. The variation of the collector current of the upper transistor for the case where the reflected source resistance is very small is

$$i_{c1} = I_{c1} - I_{CA} = I_S \exp\left(\frac{V_1}{V_t}\right) - I_{CA} \tag{7.49}$$

$$= I_{CA}\left[\frac{v_1}{V_t} + \frac{1}{2}\left(\frac{v_1}{V_t}\right)^2 + \frac{1}{6}\left(\frac{v_1}{V_t}\right)^3 + \ldots\right]$$

where $V_1 = V_{BB} + v_1$ and $I_{CA} = I_S \exp(V_{BB}/V_t)$ is the quiescent value of $I_{c1}$ and $I_{c2}$. Notice that a series expansion is introduced for the exponential relation of the collector current to the transistor input voltage. For the lower transistor

$$i_{c2} = I_{c2} - I_{CA} = I_S \exp\left(\frac{V_2}{V_t}\right) - I_{CA} \tag{7.50}$$

$$= I_{CA}\left[\frac{v_2}{V_t} + \frac{1}{2}\left(\frac{v_2}{V_t}\right)^2 + \frac{1}{6}\left(\frac{v_2}{V_t}\right)^3 + \ldots\right]$$

For $Q_2$ the incremental input voltage is $v_2 = -v_1$. If this is inserted in Equation (7.50), and if a comparison is made with Equation (7.49), it is seen that the even terms of the two series expansions have the same sign.

In the transformer, the net flux is produced by the difference of the two currents. In terms of incremental currents, $i_{c1}$ pulls current out of the top node of the primary while $i_{c2}$ pulls current out of the bottom node. The output current in the secondary of the transformer can be obtained from superposition of these two 'input' currents, since we assume a linear circuit. The load current in the output load resistance is

$$i_L = -n_o i_{c1} + n_o i_{c2} = -n_o(i_{c1} - i_{c2}) \tag{7.51}$$

Using $v_1 = -v_2$ in (7.50) and using (7.49) and (7.50) in (7.51), we can observe that all even harmonics in $i_L$ cancel. For the odd harmonics, and in particular for the fundamental,

$$i_L = -2n_o i_{c1} \tag{7.52}$$

$HD_2$ in $i_L$ and $v_o$ is thus eliminated, while $HD_3$ is the same as for a single-ended stage. (Again, if the collector currents are not exactly equal in magnitude, an inconsistency exists unless there is resistance in parallel with the outputs of each transistor.)

For a numerical example, the Spice input file for the circuit of Figure 7.19a is given in Figure 7.20a. The signal-source resistance is $1\,\Omega$ for the first example. The maximum input signal amplitude for the push-pull situation is not the same as that of the single-ended stage.

```
BJT PP, FIGURE 7.20
V1 1 0 0 SIN(0 9.5M 1MEG)
*V1 1 0 0 SIN(0 12.8M 1MEG)
.TRAN 0.05U 2U 0 0.02U
.PLOT TRAN V(12)
.PLOT TRAN I(VC1) I(VC2)
.FOUR 1MEG V(12) I(VC1) I(VC2)
RS 1 2 1
*RS 1 2 1K
LA 2 0 0.1
LB 3 4 0.1
LC 4 5 0.1
KA LA LB 1
KB LA LC 1
KC LB LC 1
VBB 4 0 0.774
Q1 6 3 0 MOD1
Q2 7 5 0 MOD1
.MODEL MOD1 NPN BF=100 IS=1E-16
VC1 8 6 0
VC2 9 7 0
LD 8 10 0.1
LE 10 9 0.1
LF 12 0 1M
KD LD LE 1
KE LD LF 1
KF LE LF 1
VCC 10 0 10
RL 12 0 100
.OPTIONS RELTOL=1E-6
.OPTIONS NOPAGE NOMOD
.WIDTH OUT=80
.END
```

**Fig. 7.20.** (a) Spice input file for the circuit of Figure 7.19a.

In order to estimate the maximum input signal amplitude to provide just clipping of the collector voltage, the load resistance presented to the transistors is needed. For $Q_1$, this is the input resistance looking into the top half of the primary (LHS) of the output transformer. From Equation (7.52), for the fundamental,

$$R_{in} = \frac{n_o v_o}{\frac{1}{2n_o}i_L} = 2n_o^2 R_L \tag{7.53}$$

This is one-half of the total resistance appearing across the primary, $(2n_o)^2 R_L$. Note that the resistance presented to $Q_1$, $R_{in}$, is twice the value pertaining to the single-ended stage having the same turns ratio as each transformer half, cf., Figures 7.11 and 7.12. Therefore, the required current drive to produce a given voltage excursion for the Class-A push-pull stage is only one-half of that for the single-ended stage. 19 mV is the voltage drive for the single-ended stage of Figure 7.11 to obtain an output amplitude of 0.74 V. We need only 9.5 mV for the present push-pull stage of Figure 7.20a to obtain the same output excursion. (Because of the cancellation of the second harmonics for the push-pull stage, a larger drive can be used before severe distortion is encountered. See the example below.)

The output Fourier components from a Spice simulation with a sinusoidal input amplitude of 9.5 mV are given in Figure 7.20b. The even harmonics are very small as expected. For this example, the value of the output fundamental voltage is 0.74 V, $HD_2 \approx 0\%$, and $HD_3 = 0.57\%$. The conversion efficiency is $\eta = 13\%$. The waveforms of the output voltage and collector currents are shown in Figures 7.20c and d.

To produce a fundamental output voltage of 1 V, corresponding to a collector voltage excursion of 10 V, the input drive should be increased at $(1.0/0.74)(9.5 \text{ mV}) = 12.8$ mV. The output voltage and collector current waveforms with this drive are shown in Figure 7.20e and the harmonic components are given in Figure 7.20f. Note that $V_{oA} = 1.003$ V, $HD_2 = 0.4\%$, $HD_3 = 0.31\%$, and $THD = 1.1\%$. The ac power output is 5.3mW and the conversion efficiency is 25%. By achieving full output, we have regained the power conversion efficiency of the single-ended stage. (Remember that the dc power of the push-pull stage is twice that of the single-ended stage.)

Significantly higher efficiency is obtained using Class-AB operation, as described in the next section, but at the expense of higher distortion.

For the case where the signal-source resistance is not zero, linearization of the transistor exponential characteristics occur. However, the use of the linearization formulas of Chapter 4 to the remaining odd terms of the power series of Equations (7.49) and (7.50) cannot be used directly to estimate the odd harmonic distortion. There is a new interaction at the right-hand side of the input transformer (the inputs of the transistors). Because of the finite reflected load at this side of the input transformer, distortion components generated in the base currents of the two transistors develop new distortion voltages across the transistor inputs. These are amplified by the transistors and appear in the collector and load currents.

The estimation of the distortion components of $v_o$ is not as easy as for the single-ended stage. As a first guess, we ignore the harmonics in the base voltages. The input voltages are reduced from the signal-source voltage because of the voltage division between $(2n_i)^2 R_s$ and the series combination of both $r_\pi$.

```
FOURIER COMPONENTS OF TRANSIENT RESPONSE V(12)        R_S = 1 Ω, V_SA = 9.5 mV
DC COMPONENT =    1.012D-02
HARMONIC    FREQUENCY    FOURIER      NORMALIZED     PHASE      NORMALIZED
  NO          (HZ)      COMPONENT     COMPONENT      (DEG)     PHASE (DEG)

   1       1.000D+06    7.396D-01     1.000000     -179.093       0.000
   2       2.000D+06    5.766D-05     0.000078      154.538     333.631
   3       3.000D+06    4.207D-03     0.005688        0.298     179.391
   4       4.000D+06    8.272D-05     0.000112       24.191     203.284

       TOTAL HARMONIC DISTORTION =      0.569323  PERCENT


FOURIER COMPONENTS OF TRANSIENT RESPONSE I(VC1)       R_S = 1 Ω, V_SA = 9.5 mV
DC COMPONENT =    1.026D-03
HARMONIC    FREQUENCY    FOURIER      NORMALIZED     PHASE      NORMALIZED
  NO          (HZ)      COMPONENT     COMPONENT      (DEG)     PHASE (DEG)

   1       1.000D+06    3.700D-04     1.000000       -0.009       0.000
   2       2.000D+06    3.369D-05     0.091060      -89.818     -89.809
   3       3.000D+06    2.054D-06     0.005551     -179.270    -179.261
   4       4.000D+06    9.192D-08     0.000248       89.101      89.110

       TOTAL HARMONIC DISTORTION =      9.122958  PERCENT


FOURIER COMPONENTS OF TRANSIENT RESPONSE I(VC2)
DC COMPONENT =    1.026D-03
HARMONIC    FREQUENCY    FOURIER      NORMALIZED     PHASE      NORMALIZED
  NO          (HZ)      COMPONENT     COMPONENT      (DEG)     PHASE (DEG)

   1       1.000D+06    3.700D-04     1.000000     -179.996       0.000
   2       2.000D+06    3.369D-05     0.091050      -90.177      89.819
   3       3.000D+06    2.048D-06     0.005534       -1.455     178.541
   4       4.000D+06    1.007D-07     0.000272       90.402     270.398

       TOTAL HARMONIC DISTORTION =      9.121806  PERCENT
```

(b)

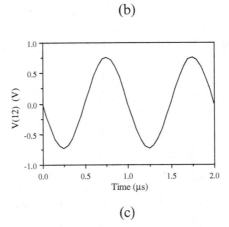

(c)

**Fig. 7.20.** (b) Fourier components of the output voltage and collector currents. (c) Output voltage waveform for $V_{1A} = 9.5$ mV.

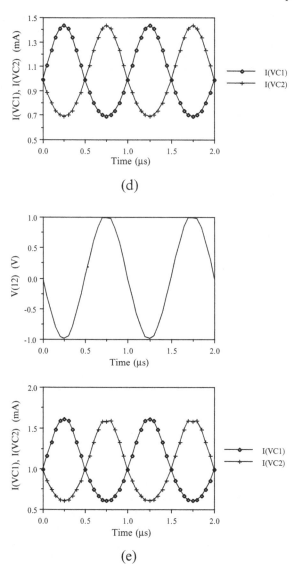

**Fig. 7.20.** (d) Collector current waveforms for $V_{1A} = 9.5$ mV. (e) Output voltage and collector current waveforms for $V_{1A} = 12.8$ mV.

```
FOURIER COMPONENTS OF TRANSIENT RESPONSE V(12)          R_S = 1 Ω, V_SA = 12.8 mV
DC COMPONENT =  1.193D-02
HARMONIC   FREQUENCY    FOURIER     NORMALIZED     PHASE     NORMALIZED
   NO        (HZ)      COMPONENT    COMPONENT     (DEG)     PHASE (DEG)

    1      1.000D+06   1.003D+00    1.000000    -179.100       0.000
    2      2.000D+06   4.001D-03    0.003991      92.795     271.895
    3      3.000D+06   3.103D-03    0.003095      -6.581     172.520
    4      4.000D+06   3.350D-03    0.003341     -85.082      94.018

         TOTAL HARMONIC DISTORTION =      1.107031  PERCENT


FOURIER COMPONENTS OF TRANSIENT RESPONSE I(VC1)
DC COMPONENT =  1.053D-03
HARMONIC   FREQUENCY    FOURIER     NORMALIZED     PHASE     NORMALIZED
   NO        (HZ)      COMPONENT    COMPONENT     (DEG)     PHASE (DEG)

    1      1.000D+06   5.035D-04    1.000000      -0.015       0.000
    2      2.000D+06   6.004D-05    0.119232     -89.948     -89.933
    3      3.000D+06   3.287D-06    0.006528     179.260     179.275
    4      4.000D+06   1.301D-06    0.002584     -84.436     -84.421

         TOTAL HARMONIC DISTORTION =     11.960218  PERCENT


FOURIER COMPONENTS OF TRANSIENT RESPONSE I(VC2)  ·
DC COMPONENT =  1.051D-03
HARMONIC   FREQUENCY    FOURIER     NORMALIZED     PHASE     NORMALIZED
   NO        (HZ)      COMPONENT    COMPONENT     (DEG)     PHASE (DEG)

    1      1.000D+06   4.996D-04    1.000000     179.995       0.000
    2      2.000D+06   5.607D-05    0.112232     -90.303    -270.299
    3      3.000D+06   4.888D-07    0.000979    -132.604    -312.600
    4      4.000D+06   4.673D-06    0.009353     -86.220    -266.215

         TOTAL HARMONIC DISTORTION =     11.375026  PERCENT
```

**Fig. 7.20.** (f) Fourier components of the output voltage and collector currents.

$$\frac{v_b}{v_1} = n_i \frac{1}{1 + 2E} \tag{7.54}$$

where $E = n_i^2 R_s / r_\pi$. In effect, $2E$ appears since $n_i^2 R_s$ 'sees' $r_\pi/2$, not $r_\pi$.

As an example to illustrate these aspects, a source resistance of 1 kΩ is included in the circuit of Figure 7.19a. (Again as for the earlier, single-ended transformer-coupled case, the bias voltage does not change as $R_s$ is added.) The Spice input file of Figure 7.20a includes this resistance as a comment line. The new input file is given in Figure 7.21. For the quiescent collector current of approximately 1 mA, $r_\pi \approx 2585\Omega$ and the value of $2E$ is 0.76. The input voltage to maintain the same output voltage excursion as in the earlier example of this section should be $1.76 \times 12.8$ mV = 22.5 mV. This input level leads to some clipping of the output voltage and the collector currents as shown in Figures 7.22a and b. The output harmonic generation is listed in Figure 7.22c. A series of simulations of both the single-ended stage and the push-pull stage with equivalent drive for unclipped output voltage shows that, in addition to the cancellation of the even harmonics for the push-pull stage, there is also significant harmonic interaction in this stage at the transistor inputs.

```
BJT PP, FIGURE 7.21
*V1 1 0 0 SIN(0 12.8M 1MEG)
V1 1 0 0 SIN(0 22.5M 1MEG)
.TRAN 0.05U 10U 8U 0.02U
.PLOT TRAN V(12)
.PLOT TRAN I(VC1) I(VC2)
.FOUR 1MEG V(12) I(VC1) I(VC2)
*RS 1 2 1
RS 1 2 1K
LA 2 0 0.1
LB 3 4 0.1
LC 4 5 0.1
KA LA LB 1
KB LA LC 1
KC LB LC 1
VBB 4 0 0.774
Q1 6 3 0 MOD1
Q2 7 5 0 MOD1
.MODEL MOD1 NPN BF=100 IS=1E-16
VC1 8 6 0
VC2 9 7 0
LD 8 10 0.1
LE 10 9 0.1
LF 12 0 1M
KD LD LE 1
KE LD LF 1
KF LE LF 1
VCC 10 0 10
RL 12 0 100
.OPTIONS RELTOL=1E-6
.OPTIONS NOPAGE NOMOD
.WIDTH OUT=80
.END
```

**Fig. 7.21.** Spice input file with $R_S = 1$ k$\Omega$.

## 7.8 Class-AB Operation

Class-AB operation of the push-pull, transformer-coupled output stage of Figure 7.19a is also possible by lowering the quiescent operating bias voltage, $V_{BB}$, relative to the input signal level and the maximum peak collector currents. An improvement in conversion efficiency is obtained, while the cancellation of the even harmonics is maintained. It is seen below that the price paid is of greater distortion.

To illustrate the results that can be obtained with AB operation, the circuit of Figure 7.19a is used with the Spice input file of Figure 7.23a. A bias voltage of $V_{BB} = 0.753$ V is used. The input voltage was varied to determine the level to obtain just clipping of the collector voltage waveforms. This value is approximately 33 mV, as given in Figure 7.23a. The output voltage waveform is shown in Figure 7.23b. The collector current waveforms are shown in Figure 7.23c. The harmonic components of $V_o$ and the two collector currents are given in Figure 7.23d. Although cancellation of the even harmonics is obtained, the

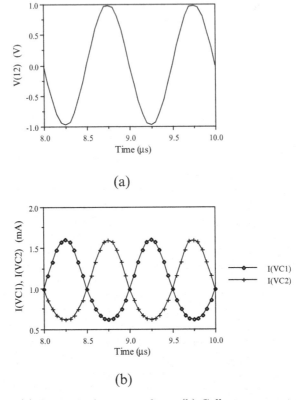

**Fig. 7.22.** (a) Output voltage waveform. (b) Collector current waveforms.

distortion level is not as low that for Class-A operation. From the values of Figure 7.23d, $HD_3 = 2.4\%$, $P_{dc} = 10.8\text{mW}$, $P_{ac} = 4.1\text{mW}$, and $\eta = 38\%$.

Although estimates of performance for the AB stage can be developed, the accuracy is limited because of the new nonlinearities which are involved. Results from a series of Spice simulations can quickly provide the necessary design information.

## Problems

**7.1.** Derive Equations (7.27) and (7.28).

**7.2.** A transformer-coupled amplifier is shown in Figure 7.24. The transformer turns ratios are to be determined. Both transformers have a primary-side magnetizing inductance of 1 H and a core loss modeled by a shunt 10 k$\Omega$ resistance at the primary. The coefficient of coupling for each unit is approximately one. (a) For no feedback, $R_f = \infty$, design the circuit to achieve the desired dc state. Specifically state the quiescent operating point of each transistor.

```
FOURIER COMPONENTS OF TRANSIENT RESPONSE V(12)          Rₛ = 1 kΩ, V₁ₐ = 22.5 mV
DC COMPONENT =    6.376D-03
HARMONIC   FREQUENCY     FOURIER      NORMALIZED     PHASE      NORMALIZED
   NO        (HZ)       COMPONENT     COMPONENT      (DEG)     PHASE (DEG)

    1      1.000D+06    9.886D-01     1.000000     -179.042      0.000
    2      2.000D+06    1.486D-03     0.001503       96.027    275.069
    3      3.000D+06    2.073D-03     0.002097       -1.549    177.492
    4      4.000D+06    1.369D-03     0.001384      -85.585     93.456

    TOTAL HARMONIC DISTORTION =          0.590674  PERCENT


FOURIER COMPONENTS OF TRANSIENT RESPONSE I(VC1)
DC COMPONENT =    1.051D-03
HARMONIC   FREQUENCY     FOURIER      NORMALIZED     PHASE      NORMALIZED
   NO        (HZ)       COMPONENT     COMPONENT      (DEG)     PHASE (DEG)

    1      1.000D+06    4.946D-04     1.000000        0.040      0.000
    2      2.000D+06    5.903D-05     0.119339      -89.790    -89.830
    3      3.000D+06    1.351D-06     0.002732     -178.619   -178.659
    4      4.000D+06    2.725D-07     0.000551      -87.390    -87.430

    TOTAL HARMONIC DISTORTION =         11.943904  PERCENT


FOURIER COMPONENTS OF TRANSIENT RESPONSE I(VC2)
DC COMPONENT =    1.051D-03
HARMONIC   FREQUENCY     FOURIER      NORMALIZED     PHASE      NORMALIZED
   NO        (HZ)       COMPONENT     COMPONENT      (DEG)     PHASE (DEG)

    1      1.000D+06    4.943D-04     1.000000     -179.944      0.000
    2      2.000D+06    5.756D-05     0.116437      -90.027     89.917
    3      3.000D+06    6.643D-07     0.001344      -10.118    169.826
    4      4.000D+06    1.654D-06     0.003345      -87.781     92.163

    TOTAL HARMONIC DISTORTION =         11.668364  PERCENT
```

**Fig. 7.22.** (c) Fourier components of the output voltage and collector currents.

(b) For $R_f = \infty$, determine the turns ratio of the transformers to achieve maximum power transfer.

(c) For an approximate maximum unclipped output voltage, estimate $\eta$, the power-conversion efficiency, and HD2 of the output voltage.

(d) If the turns ratio of the output transformer is adjusted, can a better conversion efficiency be obtained?

(e) Above what input frequency is proper high-pass operation produced?

(f) Choose $R_f$ so that a loop gain of 20 dB is obtained for the design conditions of (b). How are the values of $\eta$ and HD2 changed if the fundamental of the output voltage is maintained?

**7.3.** A MOS transformer-coupled amplifier is shown in Figure 7.25.

(a) For no feedback, $R_f = \infty$, estimate the maximum unclipped output voltage for a sinusoidal input voltage.

(b) What is the input voltage for the condition of (a)?

(c) Calculate the value of output voltage HD2 for the input voltage of (b)

(d) For $R_f = 10$ kΩ, and with the same input voltage level as in (b), what is the value of HD2 in the output voltage?

```
BJT PP, CLASS AB, FIGURE 7.23
V1 1 0 0 SIN(0 33M 1MEG)
.TRAN 0.05U 2U 0 0.01U
.PLOT TRAN V(12)
.PLOT TRAN I(VC1) I(VC2)
.FOUR 1MEG V(12) I(VC1) I(VC2)
RS 1 2 1K
LA 2 0 0.1
LB 3 4 0.1
LC 4 5 0.1
KA LA LB 1
KB LA LC 1
KC LB LC 1
VBS 4 0 0.753
Q1 6 3 0 MOD1
Q2 7 5 0 MOD1
.MODEL MOD1 NPN BF=100 IS=1E-16
VC1 8 6 0
VC2 9 7 0
LD 8 10 0.1
LE 10 9 0.1
LF 12 0 1M
KD LD LE 1
KE LD LF 1
KF LE LF 1
VCC 10 0 10
RL 12 0 100
.OPTIONS RELTOL=1E-6
.OPTIONS NOPAGE NOMOD
.WIDTH OUT=80
.END
```

(a)

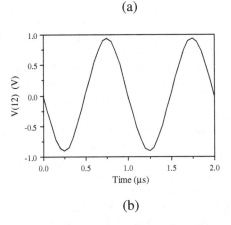

(b)

**Fig. 7.23.** (a) Spice input file for Class-AB push-pull stage. (b) Output voltage waveform.

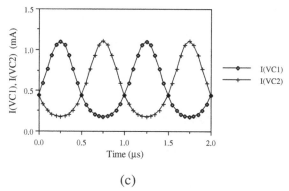

(c)

```
FOURIER COMPONENTS OF TRANSIENT RESPONSE V(12)        R_S = 1 k Ω, V_1A = 33 mV
DC COMPONENT =    1.327D-02
HARMONIC   FREQUENCY    FOURIER     NORMALIZED     PHASE      NORMALIZED
   NO        (HZ)      COMPONENT    COMPONENT      (DEG)     PHASE (DEG)

    1      1.000D+06    9.045D-01    1.000000    -179.025       0.000
    2      2.000D+06    2.263D-04    0.000250     -31.932     147.093
    3      3.000D+06    2.175D-02    0.024043       0.560     179.585
    4      4.000D+06    1.103D-04    0.000122      24.723     203.748

       TOTAL HARMONIC DISTORTION =      2.404959   PERCENT


FOURIER COMPONENTS OF TRANSIENT RESPONSE I(VC1)
DC COMPONENT =    5.394D-04
HARMONIC   FREQUENCY    FOURIER     NORMALIZED     PHASE      NORMALIZED
   NO        (HZ)      COMPONENT    COMPONENT      (DEG)     PHASE (DEG)

    1      1.000D+06    4.521D-04    1.000000       0.062       0.000
    2      2.000D+06    9.924D-05    0.219524     -89.863     -89.925
    3      3.000D+06    1.080D-05    0.023880    -179.714    -179.776
    4      4.000D+06    1.259D-07    0.000279     -88.803     -88.865

       TOTAL HARMONIC DISTORTION =     22.081917   PERCENT


FOURIER COMPONENTS OF TRANSIENT RESPONSE I(VC2)
DC COMPONENT =    5.405D-04
HARMONIC   FREQUENCY    FOURIER     NORMALIZED     PHASE      NORMALIZED
   NO        (HZ)      COMPONENT    COMPONENT      (DEG)     PHASE (DEG)

    1      1.000D+06    4.530D-04    1.000000    -179.931       0.000
    2      2.000D+06    9.939D-05    0.219419     -89.874      90.057
    3      3.000D+06    1.081D-05    0.023871       0.062     179.993
    4      4.000D+06    1.123D-07    0.000248     -89.874      90.057

       TOTAL HARMONIC DISTORTION =     22.071371   PERCENT
```

(d)

**Fig. 7.23.** (c) Collector current waveforms. (d) Fourier components of the output voltage and collector currents.

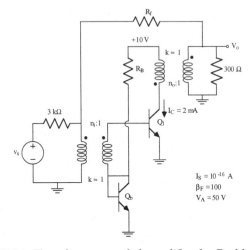

**Fig. 7.24.** Transformer-coupled amplifier for Problem 7.2.

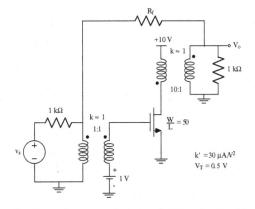

**Fig. 7.25.** Transformer-coupled amplifier for Problem 7.3.

**7.4.** A push-pull transformer-coupled stage employing MOS devices is shown in Figure 7.26.

(a) Determine the input voltage level to provide the maximum unclipped output voltage.

(b) What is the ac power output and the THD for the input of (a)?

**7.5.** A push-pull transformer-coupled stage employing BJT devices is shown in Figure 7.27. Assume $L_m$ is large for both transformers.

(a) For a low-frequency input voltage amplitude of 25 mV, determine the fundamental power output and THD.

(b) What is the maximum power output that can be obtained?

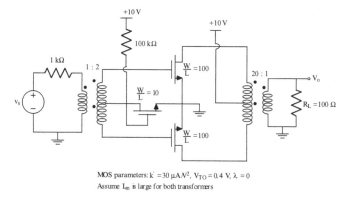

MOS parameters: $k' = 30\ \mu A/V^2$, $V_{TO} = 0.4$ V, $\lambda = 0$
Assume $L_m$ is large for both transformers

**Fig. 7.26.** A push-pull transformer-coupled amplifier for Problem 7.4.

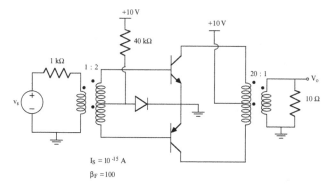

$I_S = 10^{-15}$ A

$\beta_F = 100$

**Fig. 7.27.** A push-pull transformer-coupled stage for Problem 7.5.

# 8

# Tuned Circuits in Bandpass Amplifiers

## 8.1 Introduction

Bandpass amplifiers can be operated both as small-signal circuits and as large-signal amplifiers, similar to the situation for lowpass amplifiers. For a small-signal amplifier, the desired overall gain (transfer) function should provide a magnitude versus frequency response such as that shown in Figure 8.1. The peak-magnitude response occurs at what is labeled the *center frequency*, $f_o$. By definition, the frequencies where the magnitude is 'down' from the peak value by a given amount are called the *passband* edges, and the interval between these is the *bandwidth* of the passband. Commonly, the bandedges are defined in terms of the $-3$ dB points and the bandwidth is then called the $-3$ dB bandwidth. These aspects are developed in greater detail in the next section.

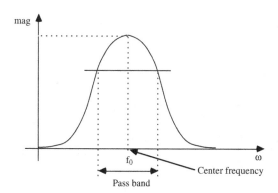

**Fig. 8.1.** Frequency response of the overall gain.

In this book, only a brief review of the small-signal situation for lowpass amplifiers is necessary because of the prerequisite courses. For bandpass amplifiers, it is necessary to treat in some detail the properties of the basic tuned

D.O. Pederson and K. Mayaram, *Analog Integrated Circuits for Communication*, DOI 10.1007/978-0-387-68030-9_8,
© 2008 Springer Science+Business Media, LLC

circuits, both for their proper use in bandpass amplifiers and to the extensive use of tuned circuits in many other nonlinear analog integrated circuits, such as those introduced in the following chapters of this book. In the remainder of this chapter, elementary circuits and analysis and evaluation techniques are introduced. In the next chapter, bandpass amplifiers are introduced which illustrate the use of these ideas in evaluation and design. The major emphasis is given to bandpass amplifiers employing inductive transformers. However, basic types of tuned amplifiers using only active elements (transistors and op amps) and resistors and capacitors are also included. The latter configurations are especially useful in their IC realization. Two examples of this type of tuned amplifier are given in Section 9.1.

## 8.2 The Single-Tuned Circuit

The parallel combination of an inductor $L$ and a capacitor $C$ fed from a signal source with a source resistance $R_s$ is shown in Figure 8.2a. If a Norton

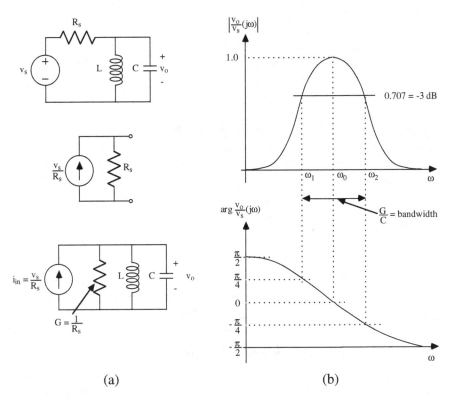

(a)     (b)

**Fig. 8.2.** (a) A parallel-tuned circuit. (b) Magnitude and phase as a function of frequency.

equivalent of the source is used, as shown in the figure, a parallel-tuned circuit results. For reasons that will be clear later, this circuit is referred to as a 'single-tuned' bandpass or resonant circuit. (A series resonant circuit has the same properties and is considered in the next section.) The voltage transfer function of this circuit is taken as the ratio of the output voltage to the signal-source voltage. For the Norton equivalent, the corresponding network function is the input impedance of the circuit. In the small-signal transform domain,

$$a_v(s) = \frac{v_o}{v_s} = \frac{\frac{G}{C}s}{s^2 + \frac{G}{C}s + \frac{1}{LC}} \tag{8.1}$$

$$= \frac{\frac{G}{C}s}{(s - s_1)(s - s_2)}$$

where $G = 1/R_s$. The zeros of $a_v(s)$ are located at the origin and at infinity. The poles of $a_v(s)$ are located at

$$s_1, s_2 = -\frac{G}{2C} \pm j\sqrt{\frac{1}{LC} - \left(\frac{G}{2C}\right)^2} \tag{8.2}$$

In the sinusoidal steady state, $s = j\omega$,

$$a_v(j\omega) = \frac{j\omega\omega_b}{(\omega_o^2 - \omega^2) + j\omega\omega_b} \tag{8.3}$$

where two constants are introduced for convenience:

$$\omega_b = \frac{G}{C} \tag{8.4}$$

$$\omega_o^2 = \frac{1}{LC}$$

An alternate form is

$$a_v(j\omega) = \frac{1}{1 - j\frac{\omega_o}{\omega_b}\left(\frac{\omega_o}{\omega} - \frac{\omega}{\omega_o}\right)} \tag{8.5}$$

The magnitude function is

$$|a_v(j\omega)| = \frac{\omega\omega_b}{\sqrt{(\omega_o^2 - \omega^2)^2 + (\omega\omega_b)^2}} \tag{8.6}$$

$$= \frac{1}{\sqrt{1 + \left(\frac{\omega_o}{\omega_b}\right)^2\left(\frac{\omega_o}{\omega} - \frac{\omega}{\omega_o}\right)^2}} \tag{8.7}$$

The phase function is

$$\angle a_v(j\omega) = \frac{\pi}{2} - \tan^{-1}\left(\frac{\omega\omega_b}{\omega_o^2 - \omega^2}\right) \tag{8.8}$$

The plots of the magnitude and phase functions with frequency are shown in Figure 8.2b. The peak magnitude response occurs at the resonant frequency of $L$ and $C$,

$$\omega_o = \frac{1}{\sqrt{LC}} \tag{8.9}$$

The value of the peak response of the magnitude function is one since the effects of $L$ and $C$ cancel at $\omega_o$. This is also the frequency where the phase response is zero.

The bandedges, $\omega_i$, of the passband are defined in terms of the frequencies where the magnitude response is down by 0.707 or $-3$ dB of the peak value, 1. Setting $|a_v(j\omega_i)| = 0.707$ and solving for $\omega_i$ yields

$$\omega_i = \omega_2, \omega_1 = \pm\frac{\omega_b}{2} + \sqrt{\omega_o^2 + \left(\frac{\omega_b}{2}\right)^2} \tag{8.10}$$

At the bandedge frequencies, the phase of the response function is $+\pi/4 = +45°$ at $\omega_1$ and $-\pi/4 = -45°$ at $\omega_2$. The difference of the two bandedges is the $-3$ dB bandwidth, $\omega_b$,

$$\omega_b = \omega_2 - \omega_1 = \frac{G}{C} = \frac{1}{R_sC} \tag{8.11}$$

Notice for this simple circuit that the geometric mean of the bandedge frequencies is the center frequency.

$$\sqrt{\omega_1\omega_2} = \omega_o \tag{8.12}$$

For a numerical example, choose $R_s = 2.43$ k$\Omega$, $C = 341.5$ pF, and $L = 0.66$ $\mu$H. These values provide

$$f_o = \frac{1}{2\pi}\omega_o = 10.6 \text{ MHz} \tag{8.13}$$

$$bw = \frac{1}{2\pi}\omega_b = 0.192 \text{ MHz}$$

Notice that the cyclic values are defined and given in contrast with radian values.

The transfer-function magnitude in Equation (8.7) is useful in determining quickly the response for frequencies other than the center and bandedge frequencies. The quantity $\omega_o/\omega_b$ is called the $Q$ (quality factor) of the tuned circuit.

$$Q = \frac{\omega_o}{\omega_b} = \frac{f_o}{bw} \tag{8.14}$$

$$= \frac{R_s}{\sqrt{L/C}}$$

An alternate expression for $Q$ for this RLC circuit is also given. For this single-tuned circuit, $Q$ is equal to $2\pi$ times the ratio of the energy stored at $\omega_o$ to the loss per cycle of the sinusoidal response. The larger the value of $Q$, the sharper is the magnitude response, i.e., the smaller the bandwidth of the response relative to the center frequency. In terms of $Q$, the magnitude response can be written as

$$|a_v(j\omega)| = \frac{1}{\sqrt{1 + Q^2 \left(\frac{\omega_o}{\omega} - \frac{\omega}{\omega_o}\right)^2}} \tag{8.15}$$

For a 'high-$Q$' situation, i.e., a narrowband situation, the magnitude response for frequencies well beyond the passband is approximately

$$|a_v(j\omega)| \approx \frac{1}{Q \left|\frac{\omega_o}{\omega} - \frac{\omega}{\omega_o}\right|} \tag{8.16}$$

$$\approx \frac{1}{2Q \left(\frac{\Delta\omega}{\omega_o}\right)}$$

where $\Delta\omega$ is the difference of the frequency of interest and $\omega_o$. To obtain the last approximation,

$$\frac{\omega_o}{\omega} - \frac{\omega}{\omega_o} = \frac{(\omega + \omega_o)(\omega - \omega_o)}{\omega\omega_o} \approx \frac{2\omega\Delta\omega}{\omega\omega_o} = \frac{2\Delta\omega}{\omega_o}$$

For a numerical example, let $f_o = 560$ kHz and $f_x = 610$ kHz, the center frequencies of two local AM radio stations. For a $Q$ of 50, the $-3$ dB bandwidth of the tuned circuit is 11.2 kHz. The relative response at $f_x$, which is 50 kHz away from the center frequency, is 0.112 ($-19$ dB) for the approximation and 0.116 from an exact calculation.

## 8.3 Lowpass Equivalents

Simple lowpass circuits, such as those shown in Figure 8.3, also can provide a bandpass response when properly designed. The responses are not identical to the response of the parallel $GLC$ circuit of the last section. However, the circuits provide in many circumstances an easy way to achieve a suitable bandpass approximation.

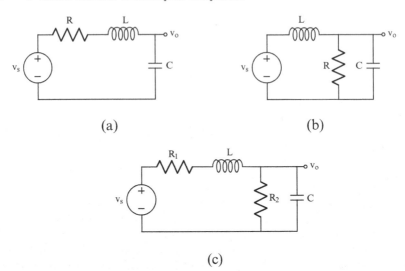

**Fig. 8.3.** (a) A series resonant RLC circuit. (b) Another simple lowpass circuit. (c) A lowpass circuit with two resistors.

The circuit of Figure 8.3a is just a series resonant $RLC$ circuit. A voltage signal source is used, and the capacitor is shown with one terminal connected to the reference (ground) potential. In circuits employing electronic devices, a shunt capacitance to ground is a usual occurrence. The output quantity for the circuit of Figure 8.3a is taken as the voltage across the capacitor. The transfer function of the circuit is of lowpass type, i.e., zero transmission to the output occurs only at infinite frequency.

$$a_v(s) = \frac{v_o}{v_s} = \frac{\frac{1}{LC}}{s^2 + \frac{R}{L}s + \frac{1}{LC}} \tag{8.17}$$

The transmission zeros of $a_v(s)$ are both at infinity. The pole-zero plot for Equation (8.17) is shown in Figure 8.4a. The lowpass nature of the circuit is clear. For the parallel circuit of the last section, the pole-zero plot based upon (8.1) and (8.2) is that of Figure 8.4b and has a zero at the origin. In the lowpass circuit, if the pole pair is located very close to the $j\omega$ axis relative to the distance to the origin, the magnitude responses of the two circuits near $\omega = \omega_o$ are very similar.

The response of (8.17) for the sinusoidal steady state is

$$a_v(j\omega) = \frac{\omega_o^2}{(\omega_o^2 - \omega^2) + j\omega\omega_b} \tag{8.18}$$

where it has been recognized that the peak response occurs at the resonance of $L$ and $C$, $\omega_o^2 = 1/LC$, and that the $-3$ dB bandwidth of the response is $\omega_b = R/L$, both values in radian measure. (Note that the bandwidth is not

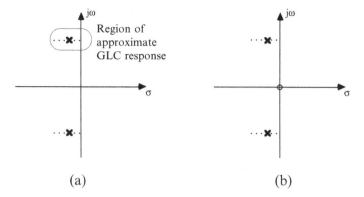

**Fig. 8.4.** (a) Pole-zero plot for Equation (8.17). (b) Pole-zero plot for a parallel-tuned circuit.

equal to $G/C$ as for the pure bandpass circuit of the last section.) The peak magnitude response at $\omega = \omega_o$ is

$$|a_v(j\omega_o)| = \frac{\omega_o}{\omega_b} = Q \tag{8.19}$$

The values of the phase responses for $\omega = \omega_o$ are quite different. From Equation (8.18), $\angle a_v(j\omega_o) = -90°$, while for the parallel circuit, $\angle a_v(j\omega_o) = 0°$.

It is often convenient to express the transfer function of the lowpass circuit in terms of the exact bandpass function multiplied by a correction function. For the lowpass example of Figure 8.3a, multiplication and division by the variable $s$ and arranging the constants leads to

$$a_v(s) = N_o(s)N_a(s) \tag{8.20}$$

where

$$N_o(s) = \frac{\omega_b s}{s^2 + \omega_b s + \omega_o^2} \tag{8.21}$$

$$N_a(s) = \frac{\omega_o^2}{\omega_b s} \tag{8.22}$$

For the sinusoidal response, $s = j\omega$,

$$N_a(j\omega) = \left|\frac{\omega_o^2}{\omega_b \omega}\right| \angle - 90° \tag{8.23}$$

$N_a(j\omega)$ introduces a phase shift and also a 'tilt' in the magnitude response relative to the true bandpass response, $N_o(s)$.

Another simple lowpass circuit is that of Figure 8.3b. The voltage transfer function for this arrangement is

$$a_v(s) = \frac{v_o}{v_1} = \frac{\omega_o^2}{s^2 + \omega_b s + \omega_o^2} \qquad (8.24)$$

where $\omega_b = G/C$ as in the pure bandpass circuit and $\omega_o^2 = 1/LC$. In the sinusoidal steady state, the peak-magnitude response occurs at $\omega = \omega_o$ and has the value $Q = \omega_o/\omega_b$. The correction function for this circuit in relation to the response for the pure bandpass case is

$$N_a(s) = \frac{\omega_o^2}{\omega_b s} \qquad (8.25)$$

This is the same function as for the series resonant circuit of Figure 8.3a.

A somewhat more complicated circuit is the lowpass circuit of Figure 8.3c where two resistors are present. The voltage transfer function for this circuit is

$$a_v = \frac{v_o}{v_s} = \frac{\frac{1}{LC}}{s^2 + \left(\frac{R_1}{L} + \frac{G_2}{C}\right)s + \frac{1+R_1 G_2}{LC}} \qquad (8.26)$$

where $G_2 = 1/R_2$. The center frequency of the peak-magnitude response is not $1/\sqrt{LC}$ but occurs at

$$\omega_o = \sqrt{\frac{1 + R_1 G_2}{LC}} \qquad (8.27)$$

The $-3$ dB bandwidth of the response is likewise changed.

$$\omega_b = \frac{R_1}{L} + \frac{G_2}{C} \qquad (8.28)$$

The correction function relative to the pure simple bandpass circuit is

$$N_a(s) = \frac{1}{1 + R_1 G_2} \frac{\omega_o^2}{\omega_b s} \qquad (8.29)$$

The center-frequency magnitude is $Q/(1+R_1 G_2)$ where $Q$ is defined as $\omega_o/\omega_b$. The values of $\omega_o$ and $\omega_b$ must be calculated from Equations (8.27) and (8.28).

An approximate technique for analyzing circuits of the type shown in Figures 8.3b and c is presented in Section 8.7.

## 8.4 Transformer-Coupled Single-Tuned Circuits

Since an inductor is used in the simple bandpass circuit of the Section 8.2, it is natural to consider a two-winding coil to implement the inductor and to achieve the transformation properties as well as the dc blocking of the transformer. For the case where the coefficient of coupling of the two windings is very close to unity, i.e., for the case of almost perfectly coupled coils, the resulting expressions for the transfer function, the center frequency, and the bandwidth are easily modified from the earlier results by introducing appropriately the turns ratio.

The circuit of Figure 8.5a is one example of the use of a transformer. The magnetizing inductance $L_m$ is assumed to be characterized on the primary side of the transformer, and the turns ratio from the primary to the secondary is $n$. The capacitor can be transferred across the ideal transformer from right to left as shown in Figure 8.5b, and the circuit configuration of Section 8.2 is obtained. The center frequency and the bandwidth expressions must now include $(1/n)^2 C$ rather than $C$. The center-frequency magnitude response at the center frequency is

$$|a_v(j\omega_o)| = \frac{1}{n} \tag{8.30}$$

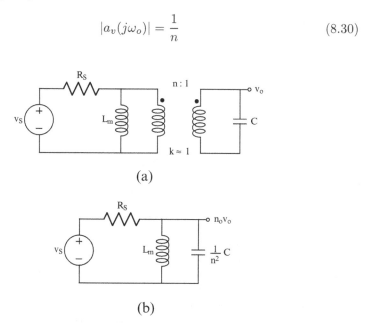

(a)

(b)

**Fig. 8.5.** (a) A transformer-coupled single-tuned circuit. (b) Circuit with capacitor transferred across the ideal transformer.

It is also possible to achieve the same results by transferring the primary-side elements across to the secondary. It is sometimes helpful in understanding the process first to use a Norton equivalent for $v_s$ and $R_s$. Each parallel element on the primary side is transferred across in turn. The current source moves as $n(v_s/R_s)$, the conductor as $n^2(1/R_s)$, and the inductor as $(1/n)^2 L_m$.

## 8.5 Single-Tuned, Bandpass Circuits with Loosely Coupled Transformers

Above audio frequencies, core materials for transformers in order to achieve closely coupled coils are increasingly expensive or not available. Often, then, the transformers in a tuned circuit have a coefficient of coupling less than unity. In this section, this situation is explored for approximate single tuning. In the next section, double tuning of these stages is briefly introduced. For the latter, a coefficient of coupling less than one is usually a necessary condition.

In Figure 8.6a, the circuit of a transformer-coupled tuned circuit is shown. Because $k$, the coefficient of coupling, is less than one, an equivalent with only one inductor cannot be used. One suitable equivalent circuit for the coupled coils is developed in Figure 7.9b of Section 7.4 and is included in the circuit of interest in Figure 8.6b. Note that the 'high' side, the primary side, of the transformer is taken as the left side. The values for the transformer elements are

$$L_1 = k^2 l_{11} \qquad (8.31)$$
$$L_2 = (1 - k^2) l_{11}$$

where $l_{11}$ is the open-circuit inductance of the primary side. The 'original' turns ratio of the transformer is

$$n = \sqrt{\frac{l_{11}}{l_{22}}} \qquad (8.32)$$

The new internal, equivalent turns ratio in Figure 8.6b is

$$n' = kn \qquad (8.33)$$

Our procedure is first to transfer the source elements across the transformer and then to inspect the value of $R_s$ relative to the reactances of the inductances at frequencies near the center frequency. The result is an approximate single-tuned equivalent circuit.

In Figure 8.6c, the signal voltage source, the source resistor, and the inductors have been brought across the internal ideal transformer.

$$R'_s = \frac{R_s}{(kn)^2} \qquad (8.34)$$
$$L'_2 = \frac{L_2}{(kn)^2}$$
$$L'_1 = \frac{L_1}{(kn)^2}$$

This circuit has three independent energy storage elements and thus has three natural frequencies. A possible set is illustrated in Figure 8.6d. In Section 8.3, it is shown that a lowpass circuit can approximate a single-tuned bandpass circuit near the peak-magnitude frequency if a complex pair of natural frequencies is designed to occur at the same locations as for a single-tuned,

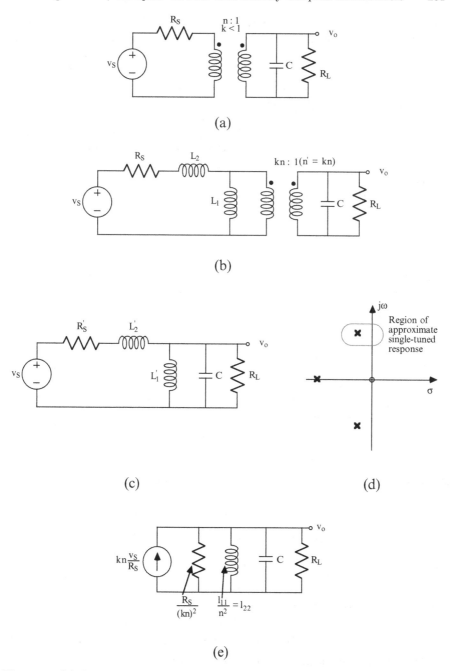

**Fig. 8.6.** (a) Single-tuned circuit with a loosely coupled transformer. (b) Single-tuned circuit with a circuit model for the transformer. (c) Signal source and resistor transferred across the ideal transformer. (d) A possible set of natural frequencies. (e) A parallel RLC circuit.

bandpass circuit. The present, more general band-pass response can also approximate a single-tuned response, as illustrated in Figure 8.6d.

For a high-Q resonant circuit and for an appropriate resistance matching, as described below, the reactance of $L'_2$ near $\omega_o$ is much less than the value of $R'_s$. The necessary condition is approximately $Q \gg 2(1 - k^2)n$. Thus, $L'_2$ can be neglected relative to $R'_s$, and we obtain a parallel $RLC$ circuit, as shown in Figure 8.6e, where a Norton equivalent has been used.

For a numerical example, we design a bandpass stage having the configuration of Figure 8.6a to achieve a center frequency of 1 MHz with a $Q$ of 20. The source resistance is 0.1 k$\Omega$, and the load resistance is 10 k$\Omega$. The coefficient of coupling of the coupled coils is $k = 0.9$. However, we start by assuming that the coefficient of coupling is 1.0.

In order to achieve a resistance match with the transformer,

$$\frac{1}{n^2} = \frac{R_L}{R_s} = 100 \tag{8.35}$$

Therefore, $n = \frac{1}{10}$. From the results of Section 8.2, the $-3$ dB bandwidth must be $f_o/Q = 50$ kHz. The pole locations must be

$$s_1, s_2 \approx -0.157 \times 10^6 \pm j6.28 \times 10^6 \tag{8.36}$$

where the real part is equal to the radian measure of one-half of the bandwidth and the imaginary part is taken to be $2\pi f_o$, without a correction for the square of the real part, cf., (8.2).

From (8.2), the real part of $s_1, s_2$ is equal to one half of the total conductance divided by the capacitance.

$$\frac{1}{2}\left(\frac{1}{R_p C}\right) = 0.157 \times 10^6 \tag{8.37}$$

where $R_p$ is the parallel combination of $\frac{1}{n^2}R_s$ and $R_L$. This leads to

$$C = 0.636 \text{ nF} \tag{8.38}$$

The required value of the output (secondary) inductance is

$$l_{22} = \frac{1}{(2\pi f_o)^2 C} = 39.8 \ \mu\text{H} \tag{8.39}$$

With the overall turns ratio of $\frac{1}{10}$,

$$l_{11} = n^2 l_{22} = 0.398 \ \mu\text{H} \tag{8.40}$$

The voltage transfer ratio at the center frequency is 5.

A Spice input file is given in Figure 8.7a where $R_1 = R_s$ and $R_2 = R_L$, etc. Note that two circuits are included, one for $k = 1$, the other for $k = 0.9$. The output is requested to be the poles and zeros of the voltage transfer function and the magnitude response about the center frequency. As expected, the correct results are obtained for $k = 1$, as shown in Figures 8.7b and c. The pole locations for $k = 0.9$ are also given in Figure 8.7b. The magnitudes of

```
BANDPASS EXAMPLE
V1 1 0 0 AC 1.0
.AC LIN 21 .95MEG 1.05MEG
.PZ 1 0 3 0 VOL PZ
*.PZ 1 0 6 0 VOL PZ
R1 1 2.1K
L1 2 0 0.398U
L2 3 0 39.8U
K1 L1 L2 1
C2 3 0 0.636N
R2 3 0 10K
R3 1 5 0.1K
L3 5 0 0.398U
L4 6 0 39.8U
K2 L3 L4 0.9
C4 6 0 0.636N
R4 6 0 10K
.WIDTH OUT=80
.END
```

$k = 1,\ n = \frac{1}{10}$

——————————————

$k = 0.9,\ n = \frac{1}{10}$

(a)

```
POLE(1)  = -1.32227E+09,  0.000000E+00
POLE(2)  = -1.42308E+05,  6.284048E+06       k = 0.9
POLE(3)  = -1.42308E+05, -6.28405E+06
----------------------------------------------------
POLE(4)  = -1.57233E+05,  6.283389E+06       k = 1.0
POLE(5)  = -1.57233E+05, -6.28339E+06
```

(b)

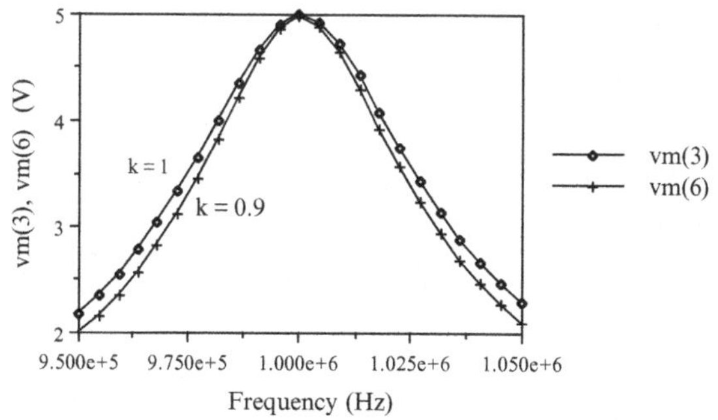

(c)

**Fig. 8.7.** (a) Spice input file. (b) Poles of the circuits. (c) Frequency response for $k$ = 0.9 and 1.0.

the real part of the poles are smaller than with $k = 1$, indicating that the loading reflected from the input is too light. In order to obtain the design specification, the overall turns ratio of the coupled coils must be reduced by $(1/k)$. The value of $l_{11}$ is then 0.491 $\mu$H, and $n$ is changed accordingly to 0.11 as given in the Spice input file of Figure 8.8a. In Figure 8.8b, it is seen that the desired dominant pole locations are closely realized. The magnitude response function as shown in Figure 8.8c is coincident with that for $k = 1$.

```
BANDPASS EXAMPLE
V1 1 0 0 AC 1.0
.AC LIN 21 .95MEG 1.05MEG
.PZ 1 0 3 0 VOL PZ
*.PZ 1 0 6 0 VO1 PZ
R1 1 2 .1K
L1 2 0 0.398U
L2 3 0 39.8U
K1 L1 L2 1
C2 3 0 0.636N
R2 3 0 10K
R3 1 5 0.1K
L3 5 0 0.491U
L4 6 0 39.8U
K2 L3 L4 0.9
C4 6 0 0.636N
R4 6 0 10K
.WIDTH OUT=80
.END
```

$k = 1,\ n = \dfrac{1}{10}$

_____

$k = 0.9,\ n = \dfrac{1}{9}$

(a)

```
POLE(1)  = -1.07177E+09,  0.000000E+00
POLE(2)  = -1.57196E+05,  6.283851E+06        k = 0.9
POLE(3)  = -1.57196E+05, -6.28385E+06
----------------------------------------------------
POLE(4)  = -1.57233E+05,  6.283389E+06        k = 1.0
POLE(5)  = -1.57233E+05, -6.28339E+06
```

(b)

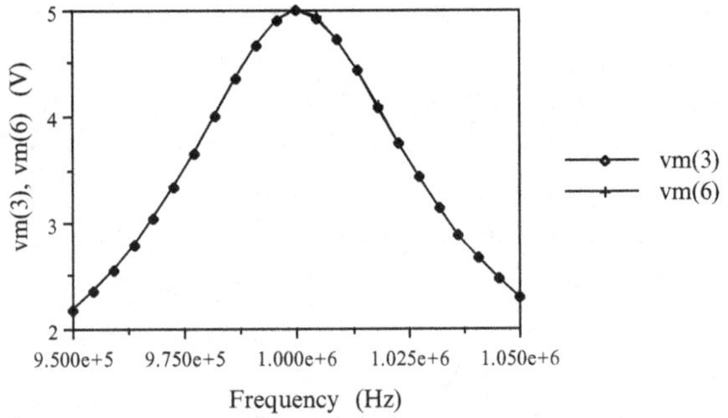

(c)

**Fig. 8.8.** (a) Spice input file. (b) Locations of the poles. (c) Frequency response for $k = 0.9$ and 1.0.

In the development above, the transformer equivalent circuit is chosen so the $L_2$ and $R_S$ are in series. If the capacitor $C$ is located on the left side of the transformer, the primary and secondary sides should be reversed as shown in Figure 8.9. In this figure $l'_{11} = l_{22}$ of Figure 8.6a, $n' = \frac{1}{n}$, etc. (Remember that the primary side of a transformer has the $l_{11}$ inductance parameter.) For a high-Q situation, $(1 - k^2)l'_{11}\omega_o \ll R_L$, and again the simplified configuration of Figure 8.6e is achieved with different element values. If capacitors appear across both sides of the transformer, the simple approximations of this section do not hold well. The results can be used, however, to provide a first estimate, followed by iteration using circuit simulation.

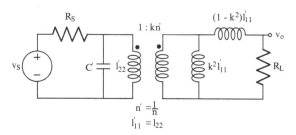

**Fig. 8.9.** Circuit representation for the single-tuned circuit when the capacitor is located on the left side of the transformer.

## 8.6 Double-Tuned Stages

If two single-tuned bandpass stages, as shown in Figure 8.10a, are loosely coupled, a double-tuned situation is produced. The transfer response can provide more gain and/or a better selectivity characteristic. The subject of double-tuned circuits can be extensive in itself. Only a simple case is considered here to illustrate the possibilities.

In the transformer-coupled bandpass stage of Figure 8.10a, in contrast to the example of the last sections, two capacitors are present. Our procedure is to tune (design) the input and output stages for the same center frequency and the same $Q$ (bandwidth). With a small coefficient of coupling of the transformer, the interaction between the two tuned circuits is not great and can be adjusted to achieve a desired overall performance. It is seen that, in spite of a small value of $k$, the overall results are impressive in the selectivity and transfer ratio obtained.

In Figure 8.10b, a pi equivalent for the transformer is used, cf., Figure 7.7d. For a small coupling, $k$ is small and the inductive element $L_y$ is large. It can be shown that

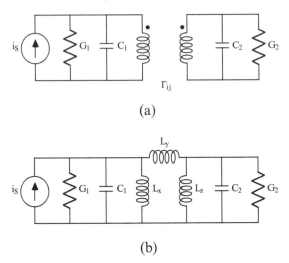

**Fig. 8.10.** (a) A double-tuned circuit obtained by two loosely coupled single-tuned bandpass stages. (b) Circuit with a 'pi'-equivalent circuit model used for the transformer.

$$L_y = \frac{l_{11}(1 - k^2)}{nk} \tag{8.41}$$

A large value of $L_y$ permits a significant simplification in the analysis which follows. From the datum-node circuit equations, the determinant has the form

$$\begin{vmatrix} Y_1(s) & \frac{-\Gamma_{12}}{s} \\ \frac{-\Gamma_{12}}{s} & Y_2(s) \end{vmatrix} \tag{8.42}$$

where $\Gamma_{12}$ is the short-circuit, transfer reciprocal-inductance parameter ($= -1/L_y$, cf. (7.28). $Y_i(s)$ is the input admittance of the input or output single-tuned circuits, including the loading effects of $L_y$.

$$Y_i(s) = sC_i + G_i + \frac{\Gamma_{ii}}{s} \tag{8.43}$$

Expressions for $\Gamma_{ij}$ from (7.28) are $\Gamma_{11} = (1/L_x + 1/L_y)$, $\Gamma_{12} = -1/L_y$, and $\Gamma_{22} = (1/L_y + 1/L_z)$. We now require that the input and output $Y_i$ are equal. (This can be eased later to require the same center frequency and bandwidth for the two tuned circuits.) An alternate form $Y_i$ is

$$Y(s) = \frac{C}{s} \left[ s^2 + 2as + \omega_o^2 \right] \tag{8.44}$$

where the $i$ subscript is removed and two constants are introduced.

$$a = \frac{G}{2C} \qquad (8.45)$$

$$\omega_o^2 = \frac{\Gamma_{11}}{C} = \frac{\Gamma_{22}}{C}$$

$\omega_o$ can be recognized as the resonant frequency of each of the single-tuned circuits, taken alone but including the loading of $L_y$. Notice now that the determinant has a simple form

$$P(s) = \begin{vmatrix} p & q \\ q & p \end{vmatrix} \qquad (8.46)$$

$$= (p+q)(p-q)$$

where $P(s)$ is the characteristic polynomial of the circuit. Using this equation, one obtains

$$P(s) = \left(\frac{C}{s}\right)^2 \left[ s^2 + 2as + \omega_o^2 + \frac{\Gamma_{12}}{C} \right] \left[ s^2 + 2as + \omega_o^2 - \frac{\Gamma_{12}}{C} \right] \qquad (8.47)$$

Remember that the degree of $P(s)$ is four due to the multiplication of the two factors, one involving the $+$, the other the $-$. The right-hand elements of Equation (8.47) can be put into a convenient form using the resonant frequency of the tuned circuits, $\omega_o^2 = \Gamma_{11}/C = \Gamma_{22}/C$.

$$\frac{\Gamma_{12}}{C} = \frac{\Gamma_{12}}{\Gamma_{11}}\omega_o^2 = k\omega_o^2 \qquad (8.48)$$

The second form results because $k = \Gamma_{12}/\sqrt{\Gamma_{11}\Gamma_{22}} = \Gamma_{12}/\Gamma_{11}$ for equal input and output tuned circuits. This leads to

$$P(s) = \left(\frac{C}{s}\right)^2 \left[ s^2 + 2as + \omega_o^2(1+k) \right] \left[ s^2 + 2as + \omega_o^2(1-k) \right] \qquad (8.49)$$

The zeros of $P(s)$ are the natural frequencies of the complete circuit.

$$s_1, s_2, s_3, s_4 = -a \pm j\omega_o \sqrt{1 \pm k - \left(\frac{a}{\omega_o}\right)^2} \qquad (8.50)$$

The $a/\omega_o$ term can be neglected for cases where the bandwidth is small relative to the center frequency. Since $k$ is assumed to be small, the square-root function can be expanded in a power series retaining only the first two terms.

$$s_i = -a \pm j\omega_o \left(1 \pm \frac{k}{2}\right) \qquad (8.51)$$

A locus of the natural frequencies, that are the poles of the voltage transfer function, is given in Figure 8.11a as $k$ is changed. Notice that the real part of the poles is the same as those of the single-tuned circuit, $-a = -\frac{G_1}{2C_1} = -\frac{G_2}{2C_2}$. As $k$ is increased from 0, the poles of the two separate single-tuned circuits, which are coincident initially, move apart parallel to the $j\omega$ axis as shown in the figure; in the third quadrant one moves up, the other down. If $k$ is sufficiently large, one can anticipate a double-humped magnitude response for $|v_o/i_s(j\omega)|$.

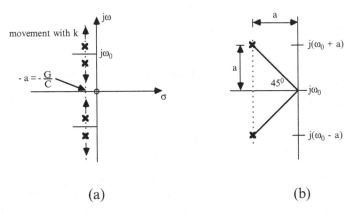

(a)                                         (b)

**Fig. 8.11.** (a) Locus of the natural frequencies as $k$ is changed. (b) Poles with equal real and imaginary parts.

A commonly desired response for the double-tuned circuit is achieved when the real and imaginary parts of the complex poles have equal values when measured from $j\omega_o$. This is illustrated in Figure 8.11b. This leads to a simple requirement for $k$.

$$a = \omega_o \left( 1 + \frac{k}{2} \right) - \omega_o \tag{8.52}$$

$$k = \frac{2a}{\omega_o} \tag{8.53}$$

For this condition, the response is called the *maximally flat magnitude* (MFM) response.

For a numerical example, the input and output circuits are designed to have the same center frequency and bandwidth values used in the example of the last section. Each circuit is tuned to $f_o = 1$ MHz with a $-3$ dB bandwidth of 50 kHz. The Spice input file is given in Figure 8.12a. For convenience in making comparisons, a single-tuned circuit having the same tuning is also included. Because the total shunt resistance for the single-tuned circuit is $R_s \| R_L = 5$ k$\Omega$, the input and output resistors for the double-tuned circuit are lowered to 5 k$\Omega$ to obtain the required bandwidths. A voltage signal source is used for convenience. In effect, a Thevenin equivalent of $i_s$ and $R_1 = \frac{1}{G_1}$ in Figure 8.10 is used. For a MFM response and with $2a/w_o = 1/20$, $k = 0.05$. The values of the poles and zeros of the voltage transfer function are given in Figure 8.12b. It is clear that the design has been realized closely. The difference in the magnitude of the imaginary parts of the poles and $jw_o$ is $1.64 \times 10^5$ whereas the magnitude of the real part is $1.57 \times 10^5$.

The magnitude response for $s = jw$ for both the double-tuned and single-tuned circuits are shown in Figure 8.12c. A flat magnitude response near $f_o$ is obtained for the double-tuned circuit. In addition, the magnitude response outside of the passband is much steeper for the double-tuned circuit. The latter is said to have a steeper skirt selectivity than the single-tuned circuit. Notice that the value of the center frequency 'gain' is 0.5. Thus, in spite of the small value of the coefficient of coupling, maximum power transfer is achieved at $f_o$.

In practice, the expected results of a design are often masked by effects due to internal feedback of the associated gain stages of a complete bandpass amplifier. Therefore, great design precision is often not warranted.

## 8.7 Parallel-to-Series/Series-to-Parallel Transformations

A simple approach for analyzing a $RLC$ circuit that is not a parallel or series resonant circuit is based on the parallel-to-series (or series-to-parallel) transformations. As an example consider the circuit of Figure 8.3b which is reproduced as Figure 8.13a. This circuit cannot be directly analyzed as a series or parallel tuned circuit. However, in a narrowband around the resonance frequency, $w_o$, the parallel $RC$ branch can be transformed into a series $RC$ branch as shown in Figure 8.13b.

With the parallel-to-series transformation, the original circuit is transformed into a series $RLC$ network as shown in Figure 8.14. The advantage of this transformation is that the analysis for a series resonant circuit can now be used. This transformation is also useful in the analysis of impedance transforming or matching networks as described in the next section.

The parallel-to-series transformation can be derived by ensuring that the impedances of the series and parallel branches are equal. Upon equating the impedances of the two branches in Figure 8.13b one obtains,

$$\frac{R_p}{1 + jw_o R_p C_p} = R_s + \frac{1}{jw_o C_s} = \frac{1 + jw_o R_s C_s}{jw_o C_s} \qquad (8.54)$$

or,

$$jw_o R_p C_s = (1 + jw_o R_p C_p)(1 + jw_o R_s C_s) \qquad (8.55)$$
$$= 1 + jw_o(R_p C_p + R_s C_s) - w_o^2 R_p C_p R_s C_s$$

```
DBL-TUNED BANDPASS EXAMPLE
V1 1 0 0 AC 1.0
.AC LIN 41 .90MEG 1.10MEG
.PZ 1 0 3 0 VOL PZ
R1 1 2 5K
C1 2 0 0.636N
L1 2 0 39.8U
L2 3 0 39.8U
K L1 L2 0.05
C2 3 0 0.636N
R2 3 0 5K
R3 1 5 10K
L3 5 0 39.8U
C3 5 0 0.636N
R4 5 0 10K
.WIDTH OUT=80
.END
```

Double-tuned

_____

Single-tuned

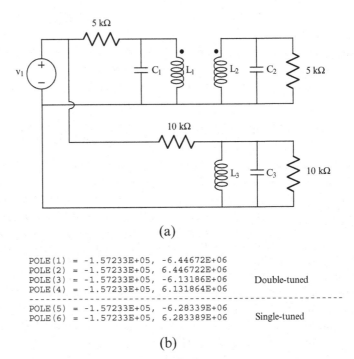

(a)

```
POLE(1) = -1.57233E+05, -6.44672E+06
POLE(2) = -1.57233E+05,  6.446722E+06
POLE(3) = -1.57233E+05, -6.13186E+06    Double-tuned
POLE(4) = -1.57233E+05,  6.131864E+06
-----------------------------------------------------------
POLE(5) = -1.57233E+05, -6.28339E+06
POLE(6) = -1.57233E+05,  6.283389E+06    Single-tuned
```

(b)

**Fig. 8.12.** (a) Spice input file and circuit for double-tuned bandpass example. (b) Poles of the voltage transfer function.

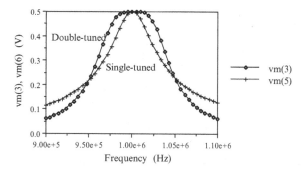

**Fig. 8.12.** (c) Magnitude response for single-tuned and double-tuned circuits.

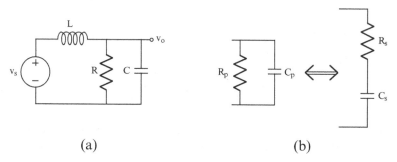

(a)                                    (b)

**Fig. 8.13.** (a) Lowpass circuit. (b) Parallel-to-series transformation applied on the parallel $RC$ branch converts it to a series $RC$ branch.

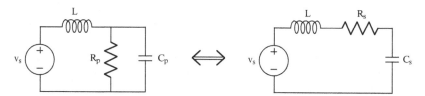

**Fig. 8.14.** Original circuit transformed into a series $RLC$ resonant circuit.

By equating the real and imaginary parts of Equation (8.55), one obtains,

$$0 = 1 - \omega_o^2 R_p C_p R_s C_s \tag{8.56}$$

and,

$$\omega_o R_p C_s = \omega_o R_p C_p + \omega_o R_s C_s \tag{8.57}$$

Equation (8.56) can be rearranged as

$$\frac{1}{\omega_o R_s C_s} = \omega_o R_p C_p = Q \tag{8.58}$$

It is seen from Equation (8.58) that the $Q$s of the series and parallel circuits are the same. Thus, $Q$ is automatically preserved during the transformation.

From Equations (8.57) and (8.58)

$$\omega_o R_p C_s = Q + \frac{1}{Q} = \frac{Q^2 + 1}{Q} \tag{8.59}$$

Noting that $\omega_o C_s = \frac{1}{QR_s}$, Equation (8.59) can be written as

$$\frac{R_p}{QR_s} = \frac{Q^2 + 1}{Q} \tag{8.60}$$

from which the relationship between $R_s$ and $R_p$ is obtained

$$R_s = \frac{R_p}{1 + Q^2} \tag{8.61}$$

The relationship between $C_s$ and $C_p$ is derived by observing that $\omega_o R_p = \frac{Q}{C_p}$ and substituting in (8.59). Then

$$C_s = \frac{Q^2 + 1}{Q^2} C_p = \left(1 + \frac{1}{Q^2}\right) C_p \tag{8.62}$$

A similar analysis can be performed with an inductor as the reactive element. In general, it can be shown that in the series-to-parallel transformation the resistances and reactances are related by,

$$R_p = R_s(1 + Q^2) \approx R_s Q^2 \tag{8.63}$$

and,

$$X_p = \frac{Q^2 + 1}{Q^2} X_s = \left(1 + \frac{1}{Q^2}\right) X_s \approx X_s \tag{8.64}$$

The above equation indicates that the reactances remain approximately the same for circuits with a high $Q$.

## 8.8 Tuned Circuits as Impedance Transformers

Tuned circuits are widely used as matching networks or impedance transformation networks. Here we look at a very simple example of a matching network, that of an L-match network. The two forms of an L-match are shown in Figure 8.15.

In the circuit of Figure 8.15a using a series-to-parallel transformation it can be shown that

$$R_p = R_s(1 + Q^2) \approx R_s Q^2 \tag{8.65}$$

Since $R_p > R_s$, this circuit is referred to as an upward impedance transformer. On the other hand, for the circuit of Figure 8.15b, $R_s = \frac{R_p}{1+Q^2}$, i.e., $R_s < R_p$. The circuit of Figure 8.15b is a downward impedance transformer.

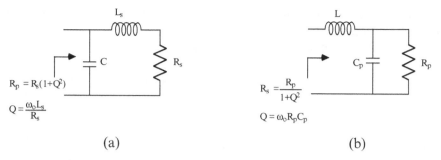

**Fig. 8.15.** (a) Upward impedance transforming L-match circuit. (b) Downward impedance transforming L-match circuit.

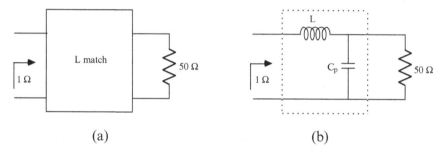

**Fig. 8.16.** (a) Example of downward impedance transformer. (b) Realization of L-match circuit.

As a numerical example, consider a downward transformation of a 50 $\Omega$ resistance to $1\Omega$ at 1 GHz as shown in Figure 8.16a. The realization of the matching network is shown in Figure 8.16b where the values of $L$ and $C$ have to be determined from the given specifications.

Since,

$$R_s = \frac{R_p}{1 + Q^2} \Rightarrow Q = \sqrt{\frac{R_p}{R_s} - 1} \qquad (8.66)$$

From Equation (8.66) the value of $Q$ can be determined.

$$Q = \sqrt{\frac{50}{1} - 1} = 7 \qquad (8.67)$$

With $Q = \omega_o R_p C_p$, $C = \frac{Q}{\omega_o R_p}$ and

$$C_p = \frac{7}{2\pi \times 10^9 \times 50} = 22.28 \text{ pF} \qquad (8.68)$$

Next $L$ is obtained from the resonance frequency of 1 GHz.

$$L = \frac{1}{\omega_o^2 C_p} = \frac{1}{(2\pi \times 10^9)^2 \times 22.28 \times 10^{-12}} = 1.14 \text{ nH} \qquad (8.69)$$

The Spice input file is given in Figure 8.17a and the frequency response around the resonant frequency is shown in Figure 8.17b. It is seen that the input resistance is approximately 1 $\Omega$ as desired.

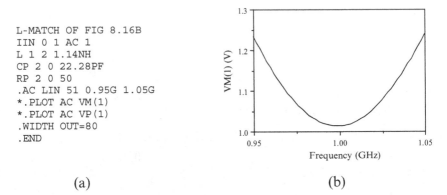

```
L-MATCH OF FIG 8.16B
IIN 0 1 AC 1
L 1 2 1.14NH
CP 2 0 22.28PF
RP 2 0 50
.AC LIN 51 0.95G 1.05G
*.PLOT AC VM(1)
*.PLOT AC VP(1)
.WIDTH OUT=80
.END
```

(a)                                          (b)

**Fig. 8.17.** (a) Spice input file for L-match example. (b) Frequency response of L-match circuit.

For the above example, it is seen that once the impedance transformation ratio and the frequency have been selected, the $Q$ of the circuit is fixed. If the 50 $\Omega$ resistance has to be matched to a 5 $\Omega$ resistance at 1 GHz, the $Q$ would be a low value of 3. If it is desired to meet a given value of $Q$, a $\pi$-match or a T-match network must be used [19]. For additional matching networks the reader is referred to [19].

## Problems

**8.1.** The center frequency of a bandpass tuned circuit is 500 MHz. The Q of this circuit is 50. What is the -3 dB frequency of the circuit?

**8.2.** Derive Equation (8.24).

**8.3.** For the parallel resonant circuit of Figure 8.2a, show that the current in the inductor (or capacitor) at resonance is a factor of $Q$ larger than the input current.

**8.4.** For the series RLC circuit of Figure 8.3a, show that the voltage across the inductor (or capacitor) at resonance is a factor of $Q$ larger than the source voltage.

**8.5.** Design a bandpass stage having the configuration shown in Figure 8.18 to achieve a center frequency of 3 MHz with a -3 dB bandwidth of 50 kHz. Determine the pole locations. Use Spice to find the frequency response for transformer couplings of $k = 1.0$ and $k = 0.8$.

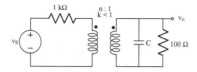

**Fig. 8.18.** Bandpass stage for Problem 8.5.

**8.6.** Derive the equations for the series-to-parallel transformation of the circuit shown in Figure 8.19.

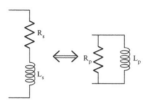

**Fig. 8.19.** Series-to-parallel transformation circuit for Problem 8.6.

**8.7.** Design a L-matching network that transforms a resistive load of 10 $\Omega$ to 50 $\Omega$ at a frequency of 900 MHz. What is the Q of this network. Verify your design with Spice.

**8.8.** Design a L-matching network that transforms a resistive load of 100 $\Omega$ to 10 $\Omega$ at a frequency of 2.4 GHz. What is the Q of this network. Verify your design with Spice.

**8.9.** A BJT circuit is shown in Figure 8.20.
(a) What is the center frequency of the tuned circuit at the output?
(b) Determine the amplitude of the output voltage at the center frequency of the tuned circuit for $I_{in} = 0.25 \cos(2\pi 5 \times 10^6 t)$ mA.
(c) What is the amplitude of the output voltage at a frequency of 1 MHz?
(d) Can you identify the function of the circuit?
(e) Verify your analysis with Spice.

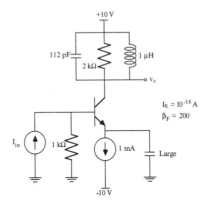

**Fig. 8.20.** Circuit for Problem 8.9.

# 9

## Simple Bandpass Amplifiers

## 9.1 Simple Active RC Bandpass Amplifiers

In addition to the $RLC$ tuned circuits of Chapter 8, highly selective band-pass circuits can be achieved with lowpass amplifiers together with resistor-capacitor combinations in overall negative or positive feedback configurations. These active $RC$ configurations are a most interesting study in themselves. In this section, only two basic types are introduced to illustrate the realization techniques.

In Figure 9.1a is shown the classical 'ring-of-three' configuration consisting of three lowpass amplifiers and two $RC$ lowpass filters. The amplifiers for the moment are assumed to have negative gain, very large input resistances and low output resistances. A series-shunt feedback amplifier provides these characteristics [6]. With an odd number of negative-gain stages, a negative feedback configuration exists. If the feedback loop is opened, the poles and zeros of the open-loop gain function may be as shown in Figure 9.1b. With the loop closed and as the lowpass gain of the system is increased from zero, the natural frequencies of the system move as shown in the root-locus plot. If the natural frequencies, which are the poles of the closed-loop system, move sufficiently far from the original open-loop locations, a bandpass response is produced in the neighborhood of the complex natural frequencies. This is comparable to the situation provided by a lowpass filter configuration properly designed, as brought out in Section 8.3. It is easily shown that the closed-loop voltage gain is:

$$\frac{v_o}{v_s} = \frac{-A\omega_1^2}{s^2 + 3\omega_1 s + (2 + A)\omega_1^2} \tag{9.1}$$

where $A$ is the magnitude of the open-loop gain (the product of the three gains $a_1 a_2 a_3$ in Figure 9.1a) and $\omega_1 = 1/RC$. For $A$ large, the center frequency is approximately $\sqrt{A}\omega_1$ and the -3 dB bandwidth is approximately $3\omega_1$. The effective $Q$ of this system is $\frac{\sqrt{A}}{3}$ and the peak magnitude of the

D.O. Pederson and K. Mayaram, *Analog Integrated Circuits for Communication*, DOI 10.1007/978-0-387-68030-9_9,
© 2008 Springer Science+Business Media, LLC

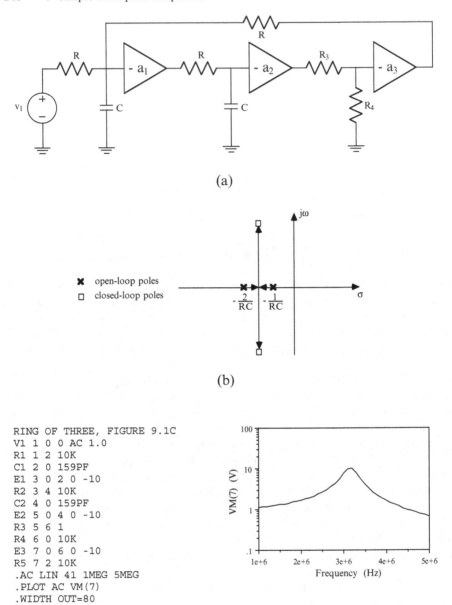

(a)

(b)

(c)

```
RING OF THREE, FIGURE 9.1C
V1 1 0 0 AC 1.0
R1 1 2 10K
C1 2 0 159PF
E1 3 0 2 0 -10
R2 3 4 10K
C2 4 0 159PF
E2 5 0 4 0 -10
R3 5 6 1
R4 6 0 10K
E3 7 0 6 0 -10
R5 7 2 10K
.AC LIN 41 1MEG 5MEG
.PLOT AC VM(7)
.WIDTH OUT=80
.END
```

**Fig. 9.1.** (a) A 'ring-of-three' configuration. (b) Poles and zeros of the open-loop gain function and root-locus plot. (c) Spice input file and magnitude response.

closed-loop voltage gain is approximately $\frac{\sqrt{A}}{3}$. The sensitivity of the effective center frequency of the response is dependent upon the square root of the open-loop gain. However, the open-loop gain can be the value of a cascaded gain of feedback amplifiers. Therefore, the gain value is often the product of ratios of two resistors.

The Spice input file for this configuration is given in Figure 9.1c. Each ideal amplifier is realized as a voltage-controlled voltage source and is chosen to provide a voltage gain of $-10$. The resistors, $R_3 \ll R_4$, provide an established interstage circuit. The parameter $\omega_1 = \frac{1}{RC} = 2\pi \times 100$ kHz. The magnitude response of the output voltage is shown in the same figure. The predicted peak magnitude of the gain is approximately 10. The simulated value is 10.2. The predicted center frequency is 3.16 MHz. The simulated value is 3.2 MHz. Notice that the outband rejection at low frequencies is only 0.1 relative to the peak value.

A modification of the system is shown in Figure 9.1d. The lowpass feedback amplifiers now may be shunt-shunt configurations which together with the feedback capacitors serve as integrators. The third negative-gain amplifier again is necessary to achieve overall negative feedback.

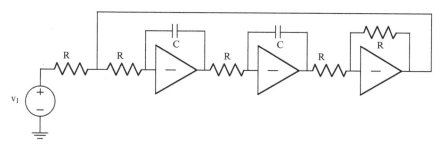

**Fig. 9.1.** (d) A modified 'ring-of-three' configuration.

A positive feedback configuration to obtain a bandpass response is shown in Figure 9.2a. The amplifier must provide a precise small-valued positive gain. An opamp with a large negative feedback can produce the well-defined positive gain. The precision is needed in order to insure that this positive feedback configuration does not become unstable. This aspect is considered below and in Chapters 10 and 11.

A simpler model of the positive gain amplifier is shown in Figure 9.2b. In relation to the opamp of Figure 9.2a, $A \approx 1 + \frac{R_a}{R_b}$ for an opamp voltage gain, $a \gg 1$.

The locii of the natural frequencies of the system of Figure 9.2b as the magnitude of the open-loop gain is changed are shown in Figure 9.2c. Notice that the open-loop response provides a transmission zero at the origin. The locii on the real axis occur to the left of an even number of singularities for a positive-feedback situation, rather than to the left of an odd number for

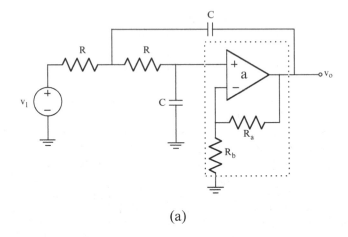

(a)

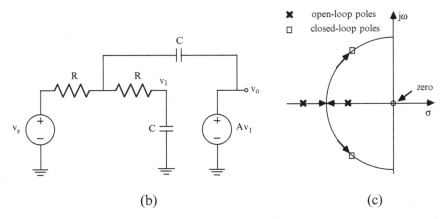

(b)                                              (c)

**Fig. 9.2.** (a) A positive feedback configuration. (b) A simpler model of the positive gain amplifier. (c) Locii of the natural frequencies.

negative feedback. For identical $R_i$ and $C_i$, the locii cross into the right-half plane for $A = 3$, i.e., a loop-gain of 3. This is the condition for instability of the system.

For the configuration of Figure 9.2a, the overall voltage gain is

$$\frac{v_o}{v_s} = \frac{A\omega_1^2}{s^2 + (3 - A)\omega_1 s + \omega_1^2} \tag{9.2}$$

where $\omega_1 = 1/RC$. The center frequency of the response occurs at $\omega_1$ with a -3 dB bandwidth of $(3 - A)\omega_1$. The effective $Q$ of the closed-loop system is approximately $1/(3 - A)$. The value of the peak magnitude is $A/(3 - A)$.

A realization of the positive feedback configuration using a single ECP with an EF output stage is shown in Figure 9.2d. The Spice input file for this circuit is given in Figure 9.2e. Negative feedback is introduced to the base of

$Q_2$. The value of the gain parameter $A$ is changed by adjusting the feedback resistors $R_{B4}$ and/or $R_{F2}$. The input resistance presented to $C_1$ is very large because of the negative feedback. The (logarithmic) magnitude response of the overall voltage gain is shown in Figure 9.2f. For this response, the peak magnitude is 90, corresponding to a value of $A = 2.967$. The value of $Q$ of the response is approximately 30.

Another positive feedback configuration is shown in Figure 9.2g. The feedback circuit is now the series combination of $R_1$ and $C_1$. The $\omega_1^2$ term in the numerator of (9.2) is replaced by $\omega_1(s+\omega_1)$. Approximately the same response can be obtained with this circuit as with that of Figure 9.2b. This circuit is sometimes called a Wien-type circuit and is the basis of the Wien-type oscillator studied in Chapter 10.

The feedback configuration of Figures 9.1a and 9.2a are shunt-shunt combinations. It is an interesting exercise to use the procedures of Section 5.2, including the loading effects of the feedback circuit, to develop the loop gain, the closed-loop gain functions, and the locii of the natural frequencies with loop gain.

Another technique to produce bandpass and other filter responses is based upon the use of MOS transistors as switches together with capacitors and opamps. For an introduction to this topic, see [6].

## 9.2 An ECP Bandpass Amplifier

In this section, simple bandpass amplifiers using emitter-coupled pairs of bipolar transistors and conventional $RLC$ tuned circuits are investigated to illustrate the design to achieve a desired passband. First, a very simple example is designed to bring out the aspects of bandwidth shrinkage. The effects of internal feedback in the gain elements next are studied, making use of circuit simulation. Finally, a multistage amplifier is designed and evaluated with loss and charge storage included in the transistors models.

In the amplifier of Figure 9.3a, notice that one ECP 'stage' is used to provide the gain for the amplifier, as well as isolation between the tuned circuits. There are two tuned circuits: one at the input and one at the output. Therefore, care must be taken in referring to this circuit; it may be called a one-stage amplifier with respect to the gain stage or a two-stage amplifier with respect to the number of tuned circuits.

Rather than analyze a predetermined circuit, a design problem is chosen. The desired value of the center frequency of each tuned circuit is 0.5 MHz with a −3 dB bandwidth of 20 kHz (Q=25). As shown in Figure 9.3a, dc voltage supply levels of ±10 V are given with a common-emitter 'tail' current of 2 mA. The signal-source resistance is 1 k$\Omega$ and the load resistance is 100 $\Omega$. A limit of 10 is specified for the largest value of turns ratio of the transformers.

We design first the input tuned circuit. The coefficient of coupling of both transformers is assumed to be unity. Therefore, the transformer can

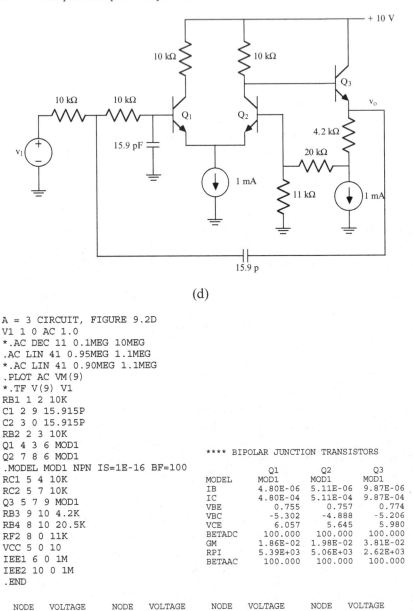

(d)

```
A = 3 CIRCUIT, FIGURE 9.2D
V1 1 0 AC 1.0
*.AC DEC 11 0.1MEG 10MEG
.AC LIN 41 0.95MEG 1.1MEG
*.AC LIN 41 0.90MEG 1.1MEG
.PLOT AC VM(9)
*.TF V(9) V1
RB1 1 2 10K
C1 2 9 15.915P
C2 3 0 15.915P
RB2 2 3 10K
Q1 4 3 6 MOD1
Q2 7 8 6 MOD1
.MODEL MOD1 NPN IS=1E-16 BF=100
RC1 5 4 10K
RC2 5 7 10K
Q3 5 7 9 MOD1
RB3 9 10 4.2K
RB4 8 10 20.5K
RF2 8 0 11K
VCC 5 0 10
IEE1 6 0 1M
IEE2 10 0 1M
.END
```

**** BIPOLAR JUNCTION TRANSISTORS

|  | Q1 | Q2 | Q3 |
|---|---|---|---|
| MODEL | MOD1 | MOD1 | MOD1 |
| IB | 4.80E-06 | 5.11E-06 | 9.87E-06 |
| IC | 4.80E-04 | 5.11E-04 | 9.87E-04 |
| VBE | 0.755 | 0.757 | 0.774 |
| VBC | -5.302 | -4.888 | -5.206 |
| VCE | 6.057 | 5.645 | 5.980 |
| BETADC | 100.000 | 100.000 | 100.000 |
| GM | 1.86E-02 | 1.98E-02 | 3.81E-02 |
| RPI | 5.39E+03 | 5.06E+03 | 2.62E+03 |
| BETAAC | 100.000 | 100.000 | 100.000 |

| NODE | VOLTAGE | NODE | VOLTAGE | NODE | VOLTAGE | NODE | VOLTAGE |
|---|---|---|---|---|---|---|---|
| ( 1) | 0.0000 | ( 2) | -0.0479 | ( 3) | -0.0959 | ( 4) | 5.2062 |
| ( 5) | 10.0000 | ( 6) | -0.8511 | ( 7) | 4.7942 | ( 8) | -0.0942 |
| ( 9) | 4.0203 | ( 10) | -0.1652 | | | | |

(e)

**Fig. 9.2.** (d) A circuit for the positive feedback amplifier. (e) Spice input file, transistor model parameters, and dc operating point.

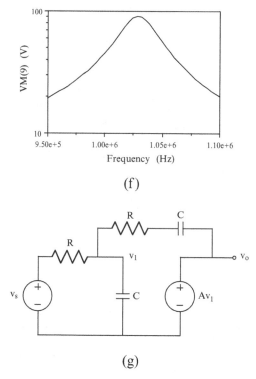

(f)

(g)

**Fig. 9.2.** (f) Magnitude response of the overall voltage gain. (g) Another positive feedback configuration.

be modeled with a simple $L_m$ and an ideal transformer. For this input circuit, the open-circuit inductance at the primary side (right side in this case), is the magnetizing inductance, as shown in Figure 9.3b. This element is labeled $L_{mi} = L_1$. From the definitions of Chapter 8, this is $l_{11i}$ of the coupled coil. We retain the notation that the primary side is that where $n$ is specified. This is also then the side where the magnetizing inductance is specified. The Thevenin equivalent of the input source is also brought across the ideal transformer with the values of $n_i v_s$ and $n_i^2 R_s$, as shown in the figure. The total resistance, $R_1$, across $L_1$ and $C_1$, is the parallel combination of the transferred source resistance and the input resistance of the ECP, $R_i$.

$$R_1 = n_i^2 R_s \parallel R_i \tag{9.3}$$

where

$$R_i = 2r_\pi$$

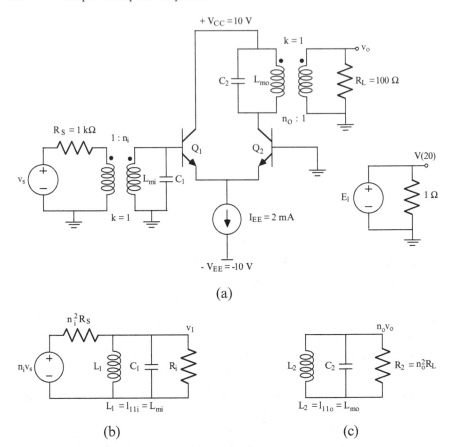

**Fig. 9.3.** (a) An amplifier circuit. (b) Equivalent circuit at the input. (c) Equivalent circuit at the output.

For $I_{EE} = 2$ mA, each collector current is approximately 1 mA (assuming that betas of the transistors are large and that the devices are identical). For $\beta = \beta_{ac} = 100$, the value of $r_\pi$ is 2.585 k$\Omega$, and

$$R_i = 5.17 \text{ k}\Omega \tag{9.4}$$

The input turns ratio, $n_i$, is not specified. One criteria that can be used is to achieve maximum power transfer from the signal source to the pair. This leads to

$$n_i^2 R_s = R_i \tag{9.5}$$

$$n_i = \sqrt{\frac{5.17 \text{ k}\Omega}{1 \text{ k}\Omega}} = 2.27 \tag{9.6}$$

In Figure 9.3b, the total resistance, $R_1$, appearing across $L_1$ and $C_1$ is

$$R_1 = \left(\frac{1}{2}\right) 2r_\pi = r_\pi = 2.59 \text{ k}\Omega \tag{9.7}$$

To obtain the necessary values of $L_1$ and $C_1$, we start with the bandwidth specification. In radian values, $\omega_b = 2\pi \times 20$ kHz.

$$\omega_b = \frac{1}{R_1 C_1}$$

This leads to

$$C_1 = \frac{1}{R_1 \omega_b} \tag{9.8}$$
$$= 3.08 \text{ nF}$$

With this value, $L_1$ is found from the center-frequency specification.

$$\omega_o = \frac{1}{\sqrt{L_1 C_1}}$$

$$L_1 = \frac{1}{\omega_o^2 C_1} = \frac{R_1}{\omega_o Q} \tag{9.9}$$
$$= 32.9 \text{ }\mu\text{H} = L_{mi}$$

Note that the parameter $Q$ can be used in determining $L_1$, since $Q = \omega_o/\omega_b$. $L_1$ is the open-circuit inductance of the primary, $l_{11i}$. The open-circuit inductance of the secondary, $l_{22i}$ is $L_1$ divided by the square of the turns ratio.

$$l_{22i} = \frac{1}{n_i^2} l_{11i} = \frac{1}{n_i^2} L_{mi} = 6.37 \text{ }\mu\text{H} \tag{9.10}$$

In the design of the output tuned circuit, it is convenient to work with the equivalent circuit shown in Figure 9.3c. For the output circuit, the load resistance is transferred to the primary side of the ideal transformer, and the inductance $L_2$ is the open-circuit inductance of the primary, which is also the magnetizing inductance parameter of the transformer, $L_{mo}$. No output conductance for the ECP is present, since the parameter $V_A$ for the transistors is not specified. Similarly, a loss conductance has not been specified for the transformers. For this case, then, we are at liberty to choose the output turns ratio, $n_o$, independently of the maximum-power-transfer criterion. It is a simple matter to show that the largest voltage transfer is obtained with the largest value of $n_o$. However, the specifications include that the turns ratio can be no larger than 10. For $n_o = 10$, the value of resistance presented to the primary side of the transformer, $R_2$, is

$$R_2 = n_o^2 R_L = 10 \text{ k}\Omega \tag{9.11}$$

Using Equations (9.8) and (9.9), we obtain for the same bandwidth and center frequency specifications,

$$C_2 = 0.796 \text{ nF} \tag{9.12}$$

$$L_2 = 127 \ \mu\text{H} = l_{11o} \tag{9.13}$$

$$l_{22o} = \frac{L_2}{n_o^2} = 1.27 \ \mu\text{H}$$

For the complete amplifier of Figure 9.3a, the center frequency of both the input and output tuned circuits has been designed to be equal. At this frequency, the reactances of the inductances and capacitances cancel and the gain calculation is the same as that for a calculation of the small-signal gain at dc for a lowpass amplifier.

$$\frac{v_o}{v_s} \approx \left(\frac{1}{2}\right) n_i \left(\frac{1}{R_i}\right) \beta R_2 \left(\frac{1}{n_o}\right) \tag{9.14}$$

$$= \left(\frac{1}{4}\right)(n_i n_o)(g_m R_L)$$

$$= 21.98$$

The circuit input file for Spice is given in Figure 9.4a. Also included are the simulated values of the dc state and the transistor model parameters at the quiescent operating point. The ac magnitude responses of the ECP input voltage and the output voltage are shown in Figure 9.4b. From the Spice results of Figure 9.4b, $\frac{v_o}{v_s}(j\omega_o) = 21.84$.

The width of the overall passband is not 20 kHz, the $-3$ dB bandwidth of the individual tuned circuits. A shrinkage of the overall bandwidth is produced because of the multiplicative nature of the gain function. This is illustrated by the frequency-response plots obtained from circuit simulation.

Note in Figure 9.4a that a special voltage-controlled, voltage source, $E_1$, is included in the circuit. This permits us to compare the responses of the input circuit and the overall amplifier, both having the same peak value. The relative selectivity of the responses is then easily observed. The scaling parameter of $E_1$ is the ratio of the overall center-frequency transfer responses of the output voltage and the ECP input voltage. From the earlier calculations,

```
BANDPASS AMPLIFIER, FIGURE 9.4A
VS 1 0 0 AC 1
RS 1 2 1K
L22I 2 0 6.37U
L11I 3 0 32.9U
K1 L11I L22I 1
E1 20 0 3 0 19.3
RX1 20 0 1
C1 3 0 3.08N
Q11 6 3 10 MOD1
Q12 5 0 10 MOD1
VCC 6 0 10
IEE 10 13 2MA
VEE 13 0 -10
C2 6 5 0.796N
L11O 5 6 127U
L22O 7 0 1.27U
K2 L11O L22O 1
RL 7 0 100
.MODEL MOD1 NPN BF=100 IS=1E-16
*+RB=100
*+VA=100
*+ TF=0.3N
*+CJC=0.5P CJE=0.5P
.AC LIN 21 480K 520K
.PRINT AC VM(7) VM(20)
.END
```

**** BIPOLAR JUNCTION TRANSISTORS

|        | Q11       | Q12       |
| ------ | --------- | --------- |
| MODEL  | MOD1      | MOD1      |
| IB     | 9.90E-06  | 9.90E-06  |
| IC     | 9.90E-04  | 9.90E-04  |
| VBE    | 0.774     | 0.774     |
| VBC    | -10.000   | -10.000   |
| VCE    | 10.774    | 10.774    |
| BETADC | 100.000   | 100.000   |
| GM     | 3.83E-02  | 3.83E-02  |
| RPI    | 2.61E+03  | 2.61E+03  |
| BETAAC | 100.000   | 100.000   |

| NODE | VOLTAGE  | NODE | VOLTAGE | NODE | VOLTAGE | NODE | VOLTAGE   |
| ---- | -------- | ---- | ------- | ---- | ------- | ---- | --------- |
| ( 1) | 0.0000   | ( 2) | 0.0000  | ( 3) | 0.0000  | ( 5) | 10.0000   |
| ( 6) | 10.0000  | ( 7) | 0.0000  | (10) | -0.7740 | (13) | -10.0000  |
| (20) | 0.0000   |      |         |      |         |      |           |

(a)

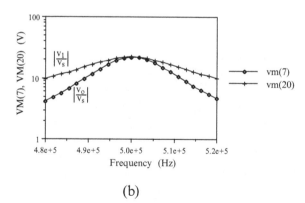

(b)

**Fig. 9.4.** (a) Spice input file, transistor model parameters, and the dc operating point. (b) Ac magnitude response of the EC pair.

$$E_1 = \frac{21.98}{\frac{1}{2}n_i} \tag{9.15}$$

$$= \frac{21.98}{1.14}$$

$$= 19.3$$

The relative input response, $V(20)$, is taken from $E_1$.

From the magnitude frequency responses of Figure 9.4b, it is clear that for frequencies away from the center-frequency, the overall response drops faster than what occurs for the input circuit alone. A bandwidth shrinkage occurs in the overall response relative to the response of either tuned circuit. From the printout of the magnitude responses given in Figure 9.4c and simple interpolations to establish frequencies for 0.707 of peak magnitude, the input circuit response provides a $-3$ dB bandwidth of approximately 20 kHz while the overall response has a $-3$ dB bandwidth of only 13 kHz, about 65% of the bandwidth of the two individual tuned circuits. In the next section, we show that this is the expected result from the cascading of two stages, each tuned the same. In passing, notice that the selectivity factor, $Q$, of the overall amplifier is approximately 38 relative to $Q = 25$ for the individual tuned circuits.

| FREQ | VM(7) | VM(20) |
|---|---|---|
| 4.80000E+05 | 4.122E+00 | 9.672E+00 |
| 4.82000E+05 | 4.878E+00 | 1.053E+01 |
| 4.84000E+05 | 5.831E+00 | 1.152E+01 |
| 4.86000E+05 | 7.039E+00 | 1.267E+01 |
| 4.88000E+05 | 8.572E+00 | 1.399E+01 |
| 4.90000E+05 | 1.050E+01 | 1.550E+01 |
| 4.92000E+05 | 1.287E+01 | 1.716E+01 |
| 4.94000E+05 | 1.562E+01 | 1.888E+01 |
| 4.96000E+05 | 1.845E+01 | 2.047E+01 |
| 4.98000E+05 | 2.078E+01 | 2.163E+01 |
| 5.00000E+05 | 2.184E+01 | 2.206E+01 |
| 5.02000E+05 | 2.122E+01 | 2.161E+01 |
| 5.04000E+05 | 1.919E+01 | 2.046E+01 |
| 5.06000E+05 | 1.648E+01 | 1.890E+01 |
| 5.08000E+05 | 1.374E+01 | 1.722E+01 |
| 5.10000E+05 | 1.131E+01 | 1.561E+01 |
| 5.12000E+05 | 9.300E+00 | 1.416E+01 |
| 5.14000E+05 | 7.688E+00 | 1.288E+01 |
| 5.16000E+05 | 6.410E+00 | 1.176E+01 |
| 5.18000E+05 | 5.397E+00 | 1.080E+01 |
| 5.20000E+05 | 4.589E+00 | 9.966E+00 |

**Fig. 9.4.** (c) Printout of the magnitude response.

## 9.3 Synchronous Tuning, Cascading, and Bandwidth Shrinkage

In the amplifier of the last section, a multiplication (cascading) of the responses of the input and output tuned circuits occurs which leads to bandwidth shrinkage. For an amplifier with several cascaded gain stages, this multiplication effect is enhanced. In this section, the overall response for a particular type of tuning is treated in greater detail.

In *synchronous* tuning of an amplifier, such as the cascade of ECPs shown in Figure 9.5, each tuned circuit is designed (and adjusted) to produce the same center frequency and the same bandwidth. The values of the circuit elements of the tuned circuits may not have identical values. Therefore, one tuned circuit may contribute more to the center-frequency gain than another.

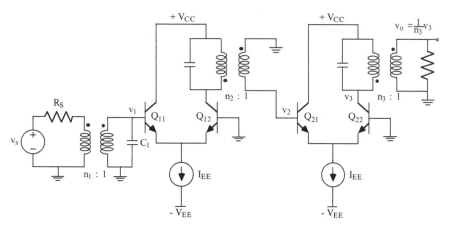

**Fig. 9.5.** A cascade of EC pairs.

Synchronous tuning is not the only method of tuning an overall amplifier, but it is the simplest and leads to easy estimation of performance and simple adjustment or tuning of an actual amplifier circuit.

Another common method of overall tuning is called stagger tuning. For this technique, each tuned circuit is assigned a different, but definite center frequency and an individual $-3$ dB bandwidth according to a specific design criterion. Usually, a flatter magnitude response in the passband can be produced with steeper 'skirts' outside of the passband. Therefore, better selectivity and rejection of signals outside of the passband are obtained relative to the response obtained for synchronous tuning. Because of the inevitable presence of feedback both within the gain stages (devices), as well as within the overall amplifier implementation, precise realization of the desired center frequencies and bandwidths is very difficult, and true stagger tuning is seldom achieved. In practice, even if synchronous tuning is desired, because of the aforementioned feedback effects, an informal stagger tuning is almost always realized in the actual tuning of an amplifier.

We now turn to the estimation of bandwidth shrinkage for a synchronously tuned amplifier. The cascaded amplifier of Figure 9.5 is modeled as in Figure 9.6 to emphasize the tuned circuits as one-port impedance blocks, $Z_i$. For simplicity, only the first two stages are considered for the moment, and each tuned circuit is assumed to be a single-tuned parallel $RLC$ circuit. The overall voltage-gain function can be developed as follows:

$$a_v = \frac{v_2}{v_s} = \left(\frac{n_1}{R_s}\right) Z_1 \left(\frac{g_m}{2}\right) Z_2 \left(\frac{1}{n_2}\right) \tag{9.16}$$

$$= \left(\frac{1}{2}\right) \left(\frac{n_1}{n_2}\right) \left(\frac{g_m}{R_s}\right) Z_1 Z_2$$

where the source voltage and resistance are transferred from left to right across $n_1$ and $Z_1$ includes the transferred loading of the source. $Z_2$ at the collector of $Q_{12}$ includes the transferred loading of $R_L$ (or $R_{in2}$ for the general case). The output voltage $v_2$ includes the division by $n_2$.

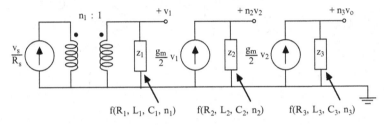

f($R_1$, $L_1$, $C_1$, $n_1$)          f($R_2$, $L_2$, $C_2$, $n_2$)          f($R_3$, $L_3$, $C_3$, $n_3$)

**Fig. 9.6.** A model for the cascaded amplifier of Figure 9.5.

The multiplicative feature of the gain in (9.16) is clear. For parallel $RLC$ circuits, the two impedances have the form

$$Z_i(s) = \frac{\frac{1}{C_i}s}{s^2 + \frac{G_i}{C_i}s + \frac{1}{L_i C_i}} \tag{9.17}$$

$$Z_i(j\omega) = \frac{jR_i \omega_b \omega}{(\omega_o^2 - \omega^2) + j\omega\omega_b} \tag{9.18}$$

$$= \frac{R_i}{1 + jQ(\frac{\omega}{\omega_o} - \frac{\omega_o}{\omega})}$$

$$= \frac{R_i}{1 + jQX}$$

where $R_i$ is the total shunt resistance of the tuned circuit, $C_i$ is the capacitance and $L_i$ is the inductance.

$$\omega_o^2 = \frac{1}{L_i C_i} \tag{9.19}$$

$$\omega_{bi} = \frac{G_i}{C_i}$$

$$Q = \frac{\omega_o}{\omega_{bi}}$$

$$X = \left[ \frac{\omega}{\omega_o} - \frac{\omega_o}{\omega} \right]$$

The last form of Equation (9.18) is convenient in establishing the $-3$ dB bandwidth of the overall response of a cascade of synchronously tuned stages.

For a cascade of $N$ tuned stages, the gain-magnitude function, normalized with respect to the value at the center frequency, has the form

$$\frac{\mid a_v(j\omega) \mid}{\mid a_v(j\omega_o) \mid} = \frac{1}{(1 + Q^2 X^2)^{\frac{N}{2}}} \tag{9.20}$$

Remember that $N$ is the number of 'tuned stages,' not gain stages. At the bandedge frequencies, the magnitude response is down to 0.707 of its value at the center frequency. There are four $-3$ dB frequencies of the magnitude response, since there are upper and lower bandedges for both positive and negative frequencies. At these frequencies,

$$\frac{1}{(1 + Q^2 X_i^2)^{\frac{N}{2}}} = 0.707 = \frac{1}{\sqrt{2}} \tag{9.21}$$

where $X_i$ are the values which satisfy the equation. Both sides of Equation (9.21) are now raised to the power $(\frac{2}{N})$ and the equation is solved for the $X_i$.

$$X_i = \pm \left( \frac{a}{Q} \right) \tag{9.22}$$

where

$$a = \sqrt{2^{\frac{1}{N}} - 1} \tag{9.23}$$

The actual frequency variable, $\omega$, is next reintroduced, and a quadratic equation is solved to obtain the normalized values of the four bandedge frequencies.

$$\frac{\omega_1}{\omega_o}, \frac{\omega_2}{\omega_o}, \frac{\omega_3}{\omega_o}, \frac{\omega_4}{\omega_o} = \pm \frac{a}{2Q} \pm j \sqrt{1 + \frac{1}{4} \left( \frac{a}{Q} \right)^2} \tag{9.24}$$

The bandwidth of interest for positive frequencies is

$$\omega_b = \omega_1 - \omega_2 = a \times \omega_{bi} \tag{9.25}$$

where $\omega_{bi}$ is the $-3$ dB bandwidth for an individual stage. The parameter $a$ is therefore the bandwidth shrinkage factor. In Table 9.1, values of this factor are given for several values of $N$, the number of tuned stages in the cascade.

As expected from the example of the last section, the bandwidth for the simple amplifier with two tuned circuits is 64% of the bandwidths of the individual stages.

| N | a |
|---|------|
| 1 | 1.00 |
| 2 | 0.64 |
| 3 | 0.51 |
| 4 | 0.43 |
| 5 | 0.39 |

**Table 9.1.** Bandwidth shrinkage factor $a$ as a function of the number of tuned stages in the cascade $N$.

## 9.4 Effects of Internal Feedback

Because of the presence of sharply tuned circuits, the effects of internal feedback in a gain stage can drastically modify the performance of a bandpass amplifier. We first illustrate this with a simple common-emitter stage and then return to the emitter-coupled pair as the gain stage. Figure 9.7a shows a variational model of a common-emitter stage at a quiescent operating point. The capacitor $C_{jc}$ is brought out explicitly to emphasize the feedback. The output load is a simple parallel tuned circuit. At frequencies below the resonant frequency of the tuned circuit, the load appears inductive. For simplicity, only $L$ is retained for the moment, as shown in Figure 9.7b to evaluate this circumstance. In Figure 9.7b, a simple circuit model for the transistor is also included. Of particular interest is the effective admittance $Y_x(s)$ presented to the $r_\pi - C_\pi$ combination, due to the feedback effects of $C_{jc}$. Simple 'Miller-effect' reasoning leads to

$$Y_x(s) = \frac{i_x}{v_1} = sC_{jc}\left(1 + g_m sL\right) \tag{9.26}$$

$$Y_x(j\omega) = j\omega C_{jc} - \omega^2 g_m C_{jc} L \tag{9.27}$$

Due to the inductive load, a frequency-dependent negative conductance appears across $r_\pi$. Naturally, when a parallel resistance is included in the load, the negative conductance effect is minimized. But for high $Q$ loads and for frequencies below the center frequency where the tuned circuit appears inductive, a negative conductance is produced at the input of the transistor, and the feedback produced by $C_{jc}$ is positive. Therefore, the tuning of a resonant circuit at the transistor input is affected. In some cases, instability and oscillation can occur. These aspects are pursued below for the ECP and also in Chapter 10.

For the emitter-coupled pair, internal feedback arises from several sources. Almost every element of the circuit model of a bipolar transistor, as illustrated in Figure 9.8, produces feedback, $r_b, r_o, C_{je}, C_{jc}$ and $C_\pi = C_B + C_{je}$ where $C_B = g_m \tau_F$. In Section 5.7, the feedback effects of $r_o = V_A/I_C$ are developed. The input admittance of an ECP is multiplied by $(1 + a_L)$, one plus the loop-gain function. Thus, the tuning of an input circuit definitely is affected by the

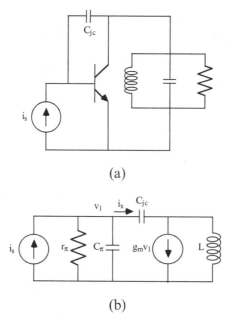

(a)

(b)

**Fig. 9.7.** (a) A variational model for the common-emitter stage at an operating point. (b) Circuit with a simple model for the transistor.

internal feedback. In this section, circuit simulation is used to illustrate the effects of the feedback due to the different parameters of the transistors.

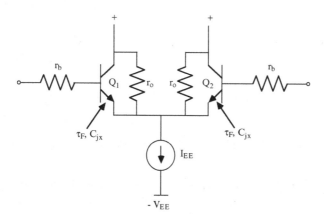

**Fig. 9.8.** Elements of the circuit model of a bipolar transistor that produce internal feedback.

To illustrate the effects of feedback in an ECP, the circuit of Figure 9.9 is simulated with Spice, initially without internal feedback. In the figure, the input circuit file also is included, as are the transistor model parameters at the quiescent operating point with only $BF$ and $IS$ specified. The required output response is the current through the voltage source, $v_s$. Note that its ac value is $-1$. This provides the current out of $v_s$ and into the circuit. This $i(v_s)$ provides the input admittance, $Y_{in}(j\omega) = i(v_s)/(-v_s)$. The real and imaginary parts of $Y_{in}$ indicate the loading and changes of loading with frequency across an input circuit.

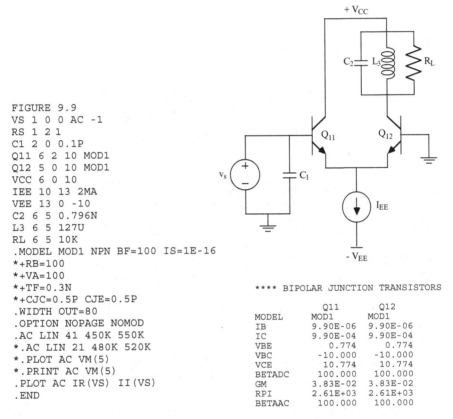

```
FIGURE 9.9
VS 1 0 0 AC -1
RS 1 2 1
C1 2 0 0.1P
Q11 6 2 10 MOD1
Q12 5 0 10 MOD1
VCC 6 0 10
IEE 10 13 2MA
VEE 13 0 -10
C2 6 5 0.796N
L3 6 5 127U
RL 6 5 10K
.MODEL MOD1 NPN BF=100 IS=1E-16
*+RB=100
*+VA=100
*+TF=0.3N
*+CJC=0.5P CJE=0.5P
.WIDTH OUT=80
.OPTION NOPAGE NOMOD
.AC LIN 41 450K 550K
*.AC LIN 21 480K 520K
*.PLOT AC VM(5)
*.PRINT AC VM(5)
.PLOT AC IR(VS) II(VS)
.END
```

**** BIPOLAR JUNCTION TRANSISTORS

|  | Q11 | Q12 |
|---|---|---|
| MODEL | MOD1 | MOD1 |
| IB | 9.90E-06 | 9.90E-06 |
| IC | 9.90E-04 | 9.90E-04 |
| VBE | 0.774 | 0.774 |
| VBC | -10.000 | -10.000 |
| VCE | 10.774 | 10.774 |
| BETADC | 100.000 | 100.000 |
| GM | 3.83E-02 | 3.83E-02 |
| RPI | 2.61E+03 | 2.61E+03 |
| BETAAC | 100.000 | 100.000 |

**Fig. 9.9.** Circuit, Spice input file, and transistor model parameters.

In Figure 9.9, a single-tuned circuit is present at the output. The values of the elements correspond to the output tuned circuit of Figure 9.3a and produce a center frequency of 500 kHz and $bw = 20$ kHz. A small capacitance $C_1$ is added at the ECP input to provide a reference. The real and imaginary parts of the input admittance over the frequency range $450 - 550$ kHz are given in Figure 9.10. As expected for this initial example without feedback, the input conductance is constant and equal to $0.5/r_\pi$. The slope of the input susceptance is constant and equal to the added capacitance 0.1 pF.

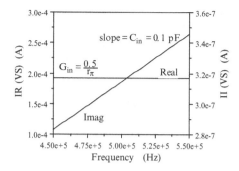

**Fig. 9.10.** Input admittance of the EC pair versus frequency.

In Figure 9.11a, the feedback effects of $r_o = V_A/I_C$ are observed. The sharp dip in $Re(Y_{in})$, because of the feedback due to $r_o$ and the sharply tuned output circuit, appears troublesome. However, if we inspect the values, only a 4% drop occurs at the resonance frequency. From Equation (5.65) in Section 5.7, the expected change due to $r_o$ is approximately $R_L/2r_o = 5\%$ for our values.

The input capacitance of the ECP, which is the slope of the imaginary part of the response, is very much modified by the inclusion of $r_o$. In the vicinity of the resonant frequency of the output circuit, the slope of the imaginary part is approximately 50 pF, but well outside of the passband the slopes are negative.

In Figure 9.11b, the effects due to both $V_A$ and $r_b$ are included. The feedback due to $r_b$ has lowered the input conductance further, but the effects due to the high $Q$ loading still are only 4%. The input capacitance appears worse than for $V_A$ only, but this is due to the scaling of the plots. The input capacitance near the resonant frequency is again approximately 50 pF.

In Figure 9.12, the feedback effects of $\tau_F$ and $C_{j\pi}$ ($C_{je}$ and $C_{jc}$) without $V_A$ and $r_b$ are shown. The input conductance is not affected by either, and the equivalent input capacitances are constant and equal to approximately 37 pF and 4.6 pF, respectively.

In Figure 9.13, the effects of all of these parameters is shown. Again the input conductance varies sharply but by only 4%. The input capacitance is approximately 70 pF over the passband of the output tuned circuit. The input susceptance is almost constant outside of the passband.

In the design of the amplifier of Figure 9.3a, only $2r_\pi$ from the ECP input, is included. To incorporate the effects of the many transistor parameters, equivalent input and output conductances and capacitances need to be established. The values of the $C_i$ and $L_i$ can then be adjusted to incorporate the loading effects. The usual procedure is to start with the conductances at the input and the output of the ECP, together with the values of the source and load resistances. Iterative calculations using circuit simulation may be necessary to obtain the proper values incorporating the feedback effects. The

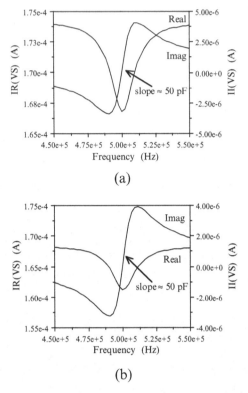

**Fig. 9.11.** Frequency response of $Y_{in}$ with (a) $r_o(V_A)$, and (b) $r_o$ and $r_b$.

necessary turns ratios of the transformers can then be established. The values of the total capacitance at the ECP input and output are next determined to provide the desired bandwidth of the passband. The actual value of $C_i$ to be added as an external element is the total value at a port minus the input or output capacitive loading. Finally, the $L_i$ are calculated, and the transformer parameters are set. Circuit simulation is then needed to validate the design. Usually, some final design changes are necessary to achieve the specifications. Often, for the ECP, the internal feedback effects are not too severe. This is not the case for common-emitter stages unless a neutralization technique is used.

## 9.5 Multistage Bandpass Design Example

To bring together several of the aspects introduced in the earlier sections, a design example is used. The specifications are to realize a three tuned-stage, synchronously tuned bandpass amplifier illustrated in Figure 9.14 with a center frequency of 0.5 MHz and an overall $Q$ of 25. Emitter-coupled pairs

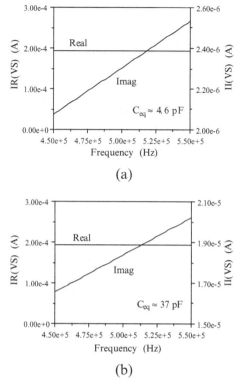

**Fig. 9.12.** Frequency response of $Y_{in}$ with (a) $C_{jx}$, and (b) $\tau_F$.

are to be used for the gain stages as shown in Figure 9.14. The input should include a tuned circuit incorporating a signal-source resistance of 2 k$\Omega$. The output should be a tuned, transformer-coupled circuit with a load resistance of 300 $\Omega$. Thus, three tuned circuits are necessary: the input stage, an interior or interstage, and an output stage. Single-tuned transformers are to be used each with a loss equivalent to 10 k$\Omega$ on the primary side. Maximum power transfer is to be achieved, if possible. The positive dc voltage source has a value of +9 V. The common-emitter currents are to be implemented with 10 k$\Omega$ resistors returned to $V_{EE} = $ -20 V. The available transistor parameters are those listed in the Spice input file given in Figure 9.14.

At the quiescent state, the common-emitter supply currents are

$$I_{EE} \approx \frac{V_{EE} - 0.8}{R_E} \tag{9.28}$$
$$= 1.92 \text{ mA}$$

The dc collectors currents, including $\beta$ effects, are approximately

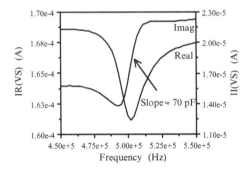

**** BIPOLAR JUNCTION TRANSISTORS

|            | Q11        | Q12        |
|------------|------------|------------|
| MODEL      | MOD1       | MOD1       |
| IB         | 9.01E-06   | 9.01E-06   |
| IC         | 9.91E-04   | 9.91E-04   |
| VBE        | .772       | .772       |
| VBC        | -10.000    | -10.000    |
| VCE        | 10.772     | 10.772     |
| BETADC     | 110.001    | 110.001    |
| GM         | 3.83E-02   | 3.83E-02   |
| RPI        | 2.87E+03   | 2.87E+03   |
| RX         | 1.00E+02   | 1.00E+02   |
| RO         | 1.11E+05   | 1.11E+05   |
| CPI        | 1.23E-11   | 1.23E-11   |
| CMU        | 2.08E-13   | 2.08E-13   |
| CBX        | 0.00E-01   | 0.00E-01   |
| CCS        | 0.00E-01   | 0.00E-01   |
| BETAAC     | 109.975    | 109.975    |
| FT         | 4.86E+08   | 4.86E+08   |

**Fig. 9.13.** Frequency response of $Y_{in}$ with $r_o$, $r_b$, $C_{jx}$, and $\tau_F$, and transistor model parameters at the operating point.

$$I_C = 0.95 \text{ mA} \tag{9.29}$$

The corresponding value of the small-signal parameters are

$$\frac{1}{g_m} = 27.2 \ \Omega \tag{9.30}$$
$$r_\pi = 2720 \ \Omega$$

The simulated values are given in Figure 9.14.

We can start the design at the signal input. As shown in the Spice input file, a 10 k$\Omega$ resistor, $R_{M1}$, is included across the primary of the transformer to simulate the loss of the transformer. (In practice, the loss is usually specified across the 'high' side of the transformer, i.e., the side where $n : 1$ is defined). The total resistance presented to the primary side is $R_s = 2$ k$\Omega$ in parallel with 10 k$\Omega$ or 1.67 k$\Omega$. At the ECP input, the input resistance, without including feedback effects due to $r_o$ and $r_b$, is $2r_\pi$. We neglect for the moment feedback effects and study them using circuit simulation after the initial design

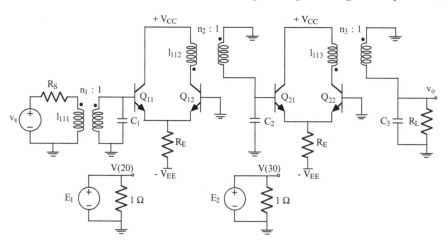

```
BANDPASS AMPLIFIER, FIGURE 9.14
VS 1 0 0 AC 1
RS 1 2 2K
RM1 2 0 10K
L11I 2 0 20.8U
L221 3 0 67.9U                    **** BIPOLAR JUNCTION TRANSISTORS
K1 L11I L221 1                         (WITH ONLY BF AND IS)
E1 20 0 3 0 1076
RX1 20 0 1                                 Q11       Q12       Q21       Q22
C1 3 0 1.49N              MODEL      MOD1      MOD1      MOD1      MOD1
Q11 6 3 10 MOD1          IB        9.52E-06  9.52E-06  9.52E-06  9.52E-06
Q12 5 0 10 MOD1          IC        9.52E-04  9.52E-04  9.52E-04  9.52E-04
RE 10 13 10K            VBE          .773      .773      .773      .773
VEE 13 0 -20            VBC        -9.000    -9.000    -9.000    -9.000
RM2 6 5 10K             VCE         9.773     9.773     9.773     9.773
L112 6 5 126U          BETADC    100.000   100.000   100.000   100.000
L222 7 0 67.9U          GM        3.68E-02  3.68E-02  3.68E-02  3.68E-02
K2 L112 L222 1          RPI       2.72E+03  2.72E+03  2.72E+03  2.72E+03
E2 30 0 7 0 15.9       RX        0.00E-01  0.00E-01  0.00E-01  0.00E-01
RX2 30 0 1             RO        1.00E+12  1.00E+12  1.00E+12  1.00E+12
C2 7 0 1.49N           CPI       0.00E-01  0.00E-01  0.00E-01  0.00E-01
Q21 6 7 11 MOD1         CMU       0.00E-01  0.00E-01  0.00E-01  0.00E-01
Q22 8 0 11 MOD1         CBX       0.00E-01  0.00E-01  0.00E-01  0.00E-01
RE2 11 13 10K           CCS       0.00E-01  0.00E-01  0.00E-01  0.00E-01
RM3 6 8 10K            BETAAC    100.000   100.000   100.000   100.000
L113 6 8 125U          FT        5.86E+17  5.86E+17  5.86E+17  5.86E+17
L223 9 0 3.74U
K L113 L223 1                     (SEE ALSO FIG. 9.17)
C3 9 0 27N
RL 9 0 300
VCC 6 0 9
.MODEL MOD1 NPN BF=100 IS=1E-16
*+RB=100
*+VA=100
*+TF=0.3N
*+CJC=0.5P CJE=0.5P
.AC LIN 21 480K 520K
.PLOT AC VM(9) VM(30) VM(20) (100,1000)
.PRINT AC VM(9) VM(3) VM(7) VM(30) VM(20)
.END
```

**Fig. 9.14.** Cascade of EC pair gain stages, Spice input file, and transistor model parameters at the operating point.

is complete. To achieve maximum power transfer to the ECP input resistance of $2r_\pi = 5.440$ k$\Omega$, the input turns ratio must be

$$n_1 = \sqrt{\frac{1.67}{5.44}} = 0.56 \tag{9.31}$$

The total shunt resistance, $R_1$, at the input of the ECP is then 2.72 k$\Omega$. In the last section, it is determined that for similar frequency specifications on the tuned output circuit, a 4-5% increase of $R_{in}$ is obtained in the passband. For the present case, where transformer loss must be included, the feedback effects will be only one-half of this, since the effective center-frequency resistance of the ECP output will be 5 k$\Omega$ due to resistance matching with the transformers.

The bandwidth of each stage for synchronous tuning with three tuned stages and an overall amplifier bandwidth, $bw_T = f_o/Q = 20$ kHz, is from Table 9.1:

$$bw_i = \frac{bw_T}{0.51} = 39.2 \text{ kHz} \tag{9.32}$$

The required capacitance is

$$C_1 = \frac{1}{2\pi(bw_i)R_1} = 1.49 \text{ nF} \tag{9.33}$$

At this point, one can estimate the capacitive input effects from the ECP. The actual added capacitance at this node can be the value of $C_1$ minus the contribution from the input of the ECP. From the results of the last section, for similar load and transistor conditions, the input capacitances in the passband is of the order of 70 pF. This is much smaller than the required 1490 pF and can be neglected for this initial design.

To obtain a center frequency of 500 kHz,

$$L_1 = \frac{1}{\omega_o^2 C_1} = 67.9 \text{ }\mu\text{H} \tag{9.34}$$

This is also the value of $l_{221}$, the open-circuit inductance at the secondary of the input transformer. The primary side inductance is $n_1^2$ times this value.

$$l_{111} = 20.8 \text{ }\mu\text{H} \tag{9.35}$$

The voltage transfer ratio from $v_s$ to the ECP input is needed for the circuit simulation to obtain (normalized) equal values of the magnitude responses at the center frequency. Note the uses of the voltage-controlled voltage sources, $E_i$, in the schematic diagram and input file of Figure 9.14.

$$\frac{v(3)}{v_s} = \left(\frac{10}{12}\right)\left(\frac{1}{2}\right)\left(\frac{1}{0.56}\right) = 0.744 \tag{9.36}$$

For the interior (interstage) tuned circuit, a resistance match is needed between the transformer resistance and the input resistance of the second ECP, neglecting for the moment the loading of $r_o = \frac{V_A}{I_C}$.

$$n_2 = \sqrt{\frac{10k}{5.44k}} = 1.36 \tag{9.37}$$

The tuned circuit element values at the input to $Q_{21}$ are

$$R_2 = 2.72 \text{ k}\Omega \tag{9.38}$$
$$C_2 = 1.49 \text{ nF}$$
$$L_2 = 67.9 \text{ }\mu\text{H}$$

$L_2$ is the open-circuit inductance of the secondary, $l_{222}$. At the primary,

$$l_{112} = (n_2)^2 L_2 = 126 \text{ }\mu\text{H} \tag{9.39}$$

The voltage transfer ratio from $v_s$ to the input of the second ECP is

$$\frac{v(7)}{v_s} = -0.744 \left(\frac{1}{2}\right) \left(\frac{1}{27.2}\right) 5k \left(\frac{1}{1.36}\right) = -50.7 \tag{9.40}$$

where 0.744 is from Equation (9.36).

For the output tuned circuit, a resistance match is needed between the transformer loss resistance at the primary, again neglecting $r_o$ loading, and the specified load resistance of 300 $\Omega$. With a match, the values of the tuned circuit at the secondary side of the transformer are

$$n_3 = 5.77 \tag{9.41}$$
$$R_3 = 150 \text{ }\Omega$$
$$C_3 = 27.0 \text{ nF}$$
$$L_3 = 3.74 \text{ }\mu\text{H} = l_{223}$$
$$l_{113} = 125 \text{ }\mu\text{H}$$

The overall center-frequency voltage gain of the amplifier is

$$\mid a_v(j\omega_o) \mid = \frac{v(9)}{v_s} = 50.7 \left(\frac{1}{2}\right) \left(\frac{1}{27.2}\right) 5k \left(\frac{1}{5.77}\right) = 807.5 \tag{9.42}$$

The values of the $E_i$, the voltage-controlled voltage sources, to obtain the same center-frequency values of voltage transfer are

$$E_1 = \frac{807.5}{0.75} = 1076 \tag{9.43}$$

$$E_2 = \frac{807.5}{50.7} = 15.9$$

The results from a Spice simulation are shown in Figure 9.15 with only the initial transistor parameters, i.e., BF and IS. The desired center frequency of 500 kHz is achieved, as is the overall −3 dB bandwidth of 20 kHz. The gain magnitude at the center frequency is 828 and at the −3 dB frequencies should be 0.707×828. Linear interpolation can be used to check out the bandedges. It is also seen that approximately equal normalized center frequency values from the $E_i$ are obtained.

| FREQ | VM(9) | VM(3) | VM(7) | VM(30) | VM(20) |
|---|---|---|---|---|---|
| 4.80000E+05 | 2.599E+02 | 5.164E-01 | 2.407E+01 | 3.827E+02 | 5.556E+02 |
| 4.82000E+05 | 3.046E+02 | 5.445E-01 | 2.675E+01 | 4.253E+02 | 5.858E+02 |
| 4.84000E+05 | 3.564E+02 | 5.738E-01 | 2.970E+01 | 4.722E+02 | 6.174E+02 |
| 4.86000E+05 | 4.154E+02 | 6.039E-01 | 3.288E+01 | 5.228E+02 | 6.497E+02 |
| 4.88000E+05 | 4.811E+02 | 6.341E-01 | 3.624E+01 | 5.763E+02 | 6.823E+02 |
| 4.90000E+05 | 5.515E+02 | 6.635E-01 | 3.967E+01 | 6.308E+02 | 7.139E+02 |
| 4.92000E+05 | 6.232E+02 | 6.909E-01 | 4.300E+01 | 6.837E+02 | 7.434E+02 |
| 4.94000E+05 | 6.912E+02 | 7.148E-01 | 4.602E+01 | 7.317E+02 | 7.691E+02 |
| 4.96000E+05 | 7.490E+02 | 7.338E-01 | 4.849E+01 | 7.709E+02 | 7.896E+02 |
| 4.98000E+05 | 7.901E+02 | 7.465E-01 | 5.016E+01 | 7.976E+02 | 8.032E+02 |
| 5.00000E+05 | 8.089E+02 | 7.518E-01 | 5.088E+01 | 8.089E+02 | 8.089E+02 |
| 5.02000E+05 | 8.031E+02 | 7.493E-01 | 5.055E+01 | 8.037E+02 | 8.063E+02 |
| 5.04000E+05 | 7.737E+02 | 7.395E-01 | 4.924E+01 | 7.828E+02 | 7.957E+02 |
| 5.06000E+05 | 7.252E+02 | 7.231E-01 | 4.709E+01 | 7.488E+02 | 7.781E+02 |
| 5.08000E+05 | 6.638E+02 | 7.016E-01 | 4.435E+01 | 7.051E+02 | 7.550E+02 |
| 5.10000E+05 | 5.959E+02 | 6.765E-01 | 4.123E+01 | 6.556E+02 | 7.279E+02 |
| 5.12000E+05 | 5.270E+02 | 6.490E-01 | 3.797E+01 | 6.037E+02 | 6.984E+02 |
| 5.14000E+05 | 4.611E+02 | 6.206E-01 | 3.472E+01 | 5.520E+02 | 6.677E+02 |
| 5.16000E+05 | 4.007E+02 | 5.920E-01 | 3.161E+01 | 5.025E+02 | 6.370E+02 |
| 5.18000E+05 | 3.467E+02 | 5.640E-01 | 2.870E+01 | 4.563E+02 | 6.069E+02 |
| 5.20000E+05 | 2.995E+02 | 5.371E-01 | 2.603E+01 | 4.139E+02 | 5.779E+02 |

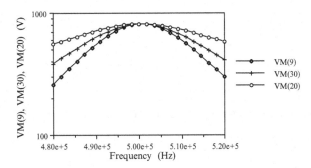

**Fig. 9.15.** Simulated magnitude responses with only $\beta_F$ and $I_S$.

Figures 9.16a, b, and c show plots of the magnitude responses of the three voltage-transfer functions as first, transistor parameters $V_A$ and $r_b$ are added; then, just $\tau_F$ and $C_{je}$ and $C_{jc}$; and finally, with all of these parameters.

A printout of the magnitude responses for the last case is given in Figure 9.17, together with a listing of the transistor parameters at the quiescent operating point. From the listing, the center frequency gain is 828. From a linear interpolation, the final overall $-3$ dB bandwidth is 19.3 kHz, a 3.5% decrease over the initial design. Redesign of the amplifier in this case may not be necessary.

For an ECP amplifier, the combined $r_o$ effects on $R_{in} = 2r_\pi(1 + \frac{R_L}{2r_o})$ and on $R_{out} = 2r_o(1 + \frac{R_S}{2r_\pi})$ usually demand an iterative solution, cf., (5.62) and (5.63).

For higher frequency specifications, the internal feedback effects of the transistors are more severe. In particular, the effects due to charge storage in the transistors are troublesome. For the example above, the charge-storage effects are small for this relatively low-frequency application, $f_o = 0.5$ MHz $\ll$ $f_T = 484$ MHz. In general, a sequential trial-and-error design procedure must be used, since detailed analysis is very cumbersome. In the initial design, a first-order estimate of both resistive and capacitive loading by the ECPs is included. Spice then is used to investigate the response down the cascade. This provides information concerning what adjustments might be made, being careful, of course, to insure that each tuned circuit remains tuned to the center frequency. Otherwise, the tuning procedure becomes very involved. Even with care, an informal stagger tuning, as mentioned in Section 9.3, rather than synchronous tuning, is the usual result.

## 9.6 Cross Modulation

In Chapter 3, it is shown that harmonic distortion is produced due to the nonlinearity of the gain stages of an amplifier. For a broadband lowpass amplifier to which two or more signals are introduced, not only are the harmonics of each signal produced by the nonlinearities, but also 'beats' are produced between the signals and their harmonics. The latter is the intermodulation distortion. For a bandpass amplifier, the frequency response may attenuate undesired signals and their harmonics, unless a lower-frequency signal happens to be an integral divisor of the desired signal. However, the beat terms, such as those developed in intermodulation distortion, may occur within the passband. These distortion components cannot then be eliminated with further filtering.

In this section, we consider the case of two signals, one of which is a pure sinusoid and the other a sinusoidally modulated sinusoidal carrier. A transfer of the modulation of one signal to the unmodulated carrier occurs due to the amplifier nonlinearities and is four times worse than the effects produced by IM3.

Let the input to an amplifier be

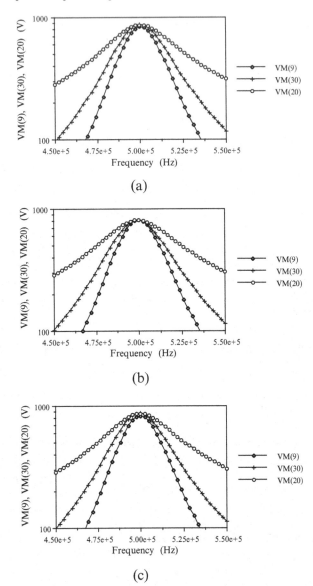

**Fig. 9.16.** Magnitude responses with (a) $V_A$ and $r_b$, (b) $\tau_F$, $C_{je}$, and $C_{jc}$, and (c) $V_A$, $r_b$, $\tau_F$, $C_{je}$, and $C_{jc}$.

```
**** BIPOLAR JUNCTION TRANSISTORS

              Q11        Q12        Q21        Q22
MODEL         MOD1       MOD1       MOD1       MOD1
IB            8.74E-06   8.74E-06   8.74E-06   8.74E-06
IC            9.53E-04   9.53E-04   9.53E-04   9.53E-04
VBE           .772       .772       .772       .772
VBC           -9.000     -9.000     -9.000     -9.000
VCE           9.772      9.772      9.772      9.772
BETADC        109.001    109.001    109.001    109.001
GM            3.68E-02   3.68E-02   3.68E-02   3.68E-02
RPI           2.96E+03   2.96E+03   2.96E+03   2.96E+03
RX            1.00E+02   1.00E+02   1.00E+02   1.00E+02
RO            1.14E+05   1.14E+05   1.14E+05   1.14E+05
CPI           1.19E-11   1.19E-11   1.19E-11   1.19E-11
CMU           2.14E-13   2.14E-13   2.14E-13   2.14E-13
CBX           0.00E-01   0.00E-01   0.00E-01   0.00E-01
CCS           0.00E-01   0.00E-01   0.00E-01   0.00E-01
BETAAC        108.975    108.975    108.975    108.975
FT            4.84E+08   4.84E+08   4.84E+08   4.84E+08


   FREQ           VM(9)        VM(3)        VM(7)        VM(30)       VM(20)

4.50000E+05    3.279E+01    2.659E-01    6.189E+00    9.841E+01    2.861E+02
4.52500E+05    3.783E+01    2.791E-01    6.816E+00    1.084E+02    3.003E+02
4.55000E+05    4.387E+01    2.934E-01    7.532E+00    1.198E+02    3.157E+02
4.57500E+05    5.115E+01    3.091E-01    8.354E+00    1.328E+02    3.326E+02
4.60000E+05    5.998E+01    3.262E-01    9.303E+00    1.479E+02    3.510E+02
4.62500E+05    7.075E+01    3.450E-01    1.040E+01    1.654E+02    3.712E+02
4.65000E+05    8.397E+01    3.657E-01    1.168E+01    1.857E+02    3.935E+02
4.67500E+05    1.003E+02    3.885E-01    1.318E+01    2.095E+02    4.180E+02
4.70000E+05    1.205E+02    4.136E-01    1.493E+01    2.373E+02    4.451E+02
4.72500E+05    1.458E+02    4.414E-01    1.698E+01    2.700E+02    4.749E+02
4.75000E+05    1.772E+02    4.720E-01    1.940E+01    3.084E+02    5.079E+02
4.77500E+05    2.165E+02    5.056E-01    2.223E+01    3.535E+02    5.441E+02
4.80000E+05    2.654E+02    5.424E-01    2.554E+01    4.060E+02    5.836E+02
4.82500E+05    3.254E+02    5.820E-01    2.935E+01    4.667E+02    6.263E+02
4.85000E+05    3.977E+02    6.241E-01    3.366E+01    5.352E+02    6.715E+02
4.87500E+05    4.820E+02    6.674E-01    3.839E+01    6.103E+02    7.181E+02
4.90000E+05    5.750E+02    7.099E-01    4.329E+01    6.884E+02    7.639E+02
4.92500E+05    6.692E+02    7.487E-01    4.799E+01    7.630E+02    8.056E+02
4.95000E+05    7.525E+02    7.799E-01    5.190E+01    8.252E+02    8.391E+02
4.97500E+05    8.096E+02    7.994E-01    5.439E+01    8.648E+02    8.601E+02
5.00000E+05    8.280E+02    8.043E-01    5.497E+01    8.741E+02    8.654E+02
5.02500E+05    8.036E+02    7.939E-01    5.355E+01    8.514E+02    8.542E+02
5.05000E+05    7.429E+02    7.701E-01    5.042E+01    8.017E+02    8.286E+02
5.07500E+05    6.595E+02    7.365E-01    4.619E+01    7.345E+02    7.925E+02
5.10000E+05    5.678E+02    6.972E-01    4.148E+01    6.596E+02    7.502E+02
5.12500E+05    4.789E+02    6.558E-01    3.678E+01    5.847E+02    7.056E+02
5.15000E+05    3.990E+02    6.146E-01    3.237E+01    5.147E+02    6.613E+02
5.17500E+05    3.304E+02    5.753E-01    2.842E+01    4.518E+02    6.190E+02
5.20000E+05    2.733E+02    5.386E-01    2.495E+01    3.967E+02    5.795E+02
5.22500E+05    2.265E+02    5.048E-01    2.195E+01    3.490E+02    5.432E+02
5.25000E+05    1.885E+02    4.740E-01    1.938E+01    3.081E+02    5.100E+02
5.27500E+05    1.577E+02    4.460E-01    1.718E+01    2.732E+02    4.799E+02
5.30000E+05    1.327E+02    4.206E-01    1.530E+01    2.432E+02    4.526E+02
5.32500E+05    1.124E+02    3.976E-01    1.368E+01    2.175E+02    4.278E+02
5.35000E+05    9.583E+01    3.767E-01    1.229E+01    1.954E+02    4.053E+02
5.37500E+05    8.220E+01    3.577E-01    1.109E+01    1.764E+02    3.849E+02
5.40000E+05    7.095E+01    3.404E-01    1.005E+01    1.598E+02    3.663E+02
5.42500E+05    6.159E+01    3.246E-01    9.145E+00    1.454E+02    3.493E+02
5.45000E+05    5.377E+01    3.101E-01    8.353E+00    1.328E+02    3.337E+02
5.47500E+05    4.719E+01    2.969E-01    7.657E+00    1.217E+02    3.194E+02
5.50000E+05    4.163E+01    2.847E-01    7.043E+00    1.120E+02    3.063E+02
```

**Fig. 9.17.** Transistor model parameters at the operating point and printout of the magnitude responses.

$$v_i = v_1 + v_2 = V_{1A} \cos \omega_1 t + V_{2A}(1 + m \cos \omega_m t) \cos \omega_2 t \qquad (9.44)$$

where $m$ is the modulation index. Following the same procedure as used in Chapter 3, we characterize the incremental output response of an amplifier in terms of a power series.

$$v_o = a_1 v_i + a_2 v_i^2 + a_3 v_i^3 + \dots \qquad (9.45)$$

Consider the quadratic term of $v_o$ after the input has been included.

$$
\begin{aligned}
a_2 v_i^2 = {} & a_2 V_{1A}^2 \cos^2 \omega_1 t \\
& + V_{1A} V_{2A}(1 + m \cos \omega t) \cos(\omega_1 \pm \omega_2)t \\
& + V_{2A}^2(1 + m^2 \cos^2 \omega_m t + 2m \cos \omega_m t) \cos^2 \omega_2 t
\end{aligned}
\qquad (9.46)
$$

Notice that the unmodulated carrier is not modulated by the $\omega_m$ term. That is, the modulation has not been transferred. From the cubic term of $v_o$, we are interested particularly in the component due to the multiplication of $v_1$ and the square of $v_2$.

$$
\begin{aligned}
a_3 v_i^3 = {} & \dots \qquad (9.47) \\
& + 3a_3 \left[ V_{1A} \cos \omega_1 t \right] \left[ V_{2A}^2(1 + m \cos \omega_m t)^2 \cos^2 \omega_2 t \right] \\
= {} & \dots + 3a_3 V_{1A} V_{2A}^2 \cos \omega_1 t \cdot \\
& \left[ (1 + 2m \cos \omega_m t + m^2 \cos^2 \omega_m t) \left( \frac{1}{2} \right) (1 + \cos 2\omega_2 t) \right] + \dots \\
= {} & \dots 3a_3 V_{1A} V_{2A}^2 m \cos \omega_m t \cos \omega_1 t + \dots
\end{aligned}
$$

The modulation of the $\omega_2$ carrier has been transferred to the $\omega_1$ carrier by the cubic term of the nonlinearity.

The cross-modulation index, $CM$, is defined as the ratio of the transferred modulation index to $m$, the original modulation index. The newly modulated $\omega_1$ component is

$$a_1 V_{1A} \left( 1 + \frac{3a_3 V_{2A}^2 m}{a_1} \cos \omega_m t \right) \cos \omega_1 t \qquad (9.48)$$

The CM index is

$$CM = \frac{3a_3 V_{2A}^2}{a_1} \qquad (9.49)$$

From the results of Chapter 3, it can be seen that CM defined in terms of equal inputs is four times larger than $IM_3$ and is 12 times greater than $HD_3$, all due to the same cubic term of the nonlinearity.

CM is usually specified in terms of the maximum allowable second signal for a given level of CM, often 1%.

For a numerical example, assume that the gain stage is a simple common-emitter stage (BJT). The desired signal has a center frequency of 560 kHz while the interfering modulated signal has a center frequency of 610 kHz. The $Q$ of the tuned circuit is 25. Therefore, the second signal is rejected by the tuned circuit at the stage input by 4.37. From Equation (3.13) in Section 3.2, the coefficients of the power series are

$$a_1 = I_{CA} \left( \frac{1}{V_T} \right) \tag{9.50}$$

$$a_3 = \frac{1}{6} I_{CA} \left( \frac{1}{V_T} \right)^3$$

where $I_{CA}$ is the quiescent value of the collector current. For an allowable CM of 1%,

$$0.01 = 3 \frac{\left( \frac{V_{2A}}{V_T} \right)^2}{6} \tag{9.51}$$

$$V_2 = 3.66 \text{ mV}$$

for $V_t = 25.85$ mV. This is the allowed signal level at the nonlinearity. Because of the rejection of the tuned circuit, the maximum allowable input signal at 610 kHz is

$$v_{in}(610 \text{ kHz}) = (4.37)(3.66 \text{ mV}) = 16 \text{ mV} \tag{9.52}$$

Any signal larger than this will introduce unacceptable modulation transfer to the desired signal. Again this modulation cannot be eliminated by further filtering. This example illustrates why design attention in amplifiers is spent on reducing the third-order nonlinearities.

## Problems

**9.1.** Derive Equations (9.1) and (9.2).

**9.2.** A bandpass amplifier based upon an active $RC$ circuit is shown in Figure 9.18. Design the circuit to provide an approximate single-tuned bandpass response with a center frequency of 10 MHz with a Q of 30. Confirm your design values with Spice simulations.

**9.3.** A simple bandpass amplifier using a SCP is shown in Figure 9.19.
(a) Design the amplifier for a center frequency of 0.5 MHz with a Q of 20. Provide maximum power transfer at the output.
(b) What is the value of the center frequency voltage gain?

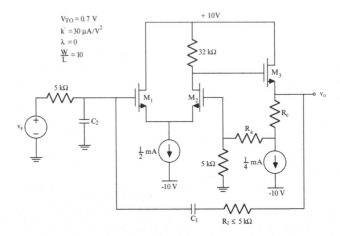

**Fig. 9.18.** Active $RC$ bandpass amplifier circuit for Problem 9.2.

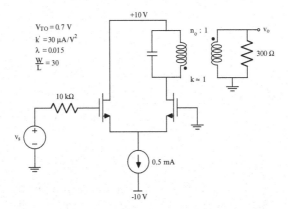

**Fig. 9.19.** SCP bandpass amplifier circuit for Problem 9.3.

(c) By how much is a signal at 600 kHz rejected by the amplifier relative to a signal at the center frequency?

(d) Verify your analysis results with Spice.

**9.4.** Derive Equation (9.16).

**9.5.** A bandpass amplifier is shown in Figure 9.20a.

(a) Design the amplifier to produce a center frequency of 1 MHz with a -3 dB bandwidth of 50 kHz. Provide maximum power transfer at the output.

(b) What is the value of the center frequency voltage gain?

(c) What is the rejection of a signal at 1.5 times the center frequency?

(d) Redo Part (a) including a tuned circuit at the EC pair input as illustrated in Figure 9.20b.

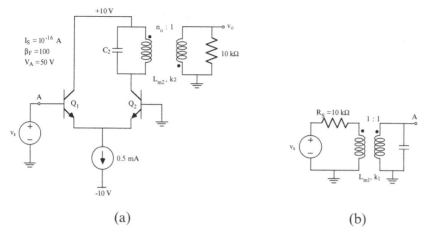

(a)                                    (b)

**Fig. 9.20.** Bandpass amplifier circuit for Problem 9.5.

**9.6.** A simple bandpass amplifier is shown in Figure 9.21.
(a) Design the input tuned circuit to produce a center frequency of 10 MHz with a -3 dB bandwidth of 0.5 MHz.
(b) What is the value of the center frequency voltage gain of the amplifier ($L_2$ is not to be included here)?
(c) If $L_2$ is added across the 50 k$\Omega$ resistor to achieve resonance at the center frequency, how is the value of the midband gain affected? How is the value of the overall bandwidth affected?
(d) Verify your analysis results with Spice.

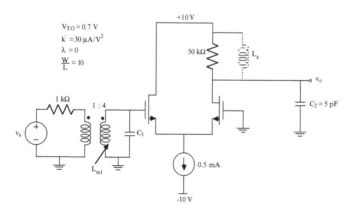

**Fig. 9.21.** Bandpass amplifier circuit for Problem 9.6.

**9.7.** A bandpass amplifier is shown in Figure 9.22. Design the circuit to obtain a synchronously tuned response with a center frequency of 10 MHz and a -3 dB bandwidth of 300 kHz. For the transformers $n \leq 8$ and $k = 0.9$. Achieve maximum power transfer if possible.

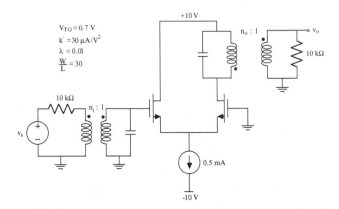

**Fig. 9.22.** Bandpass amplifier circuit for Problem 9.7.

**9.8.** Design the MOS bandpass amplifier in Figure 9.23 to provide a center frequency of 10 MHz with an overall Q of 30. The turns ratios of the transformers are limited to $1/10 \leq n \leq 10$ with a coefficient of coupling approximately 1. Confirm your design values of center frequency and bandwidth with Spice simulation.

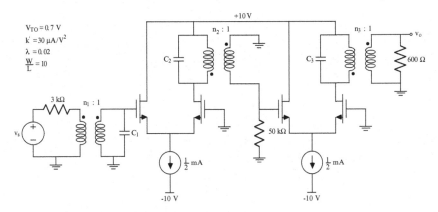

**Fig. 9.23.** MOS bandpass amplifier circuit for Problem 9.8.

# 10

# Basic Electronic Oscillators

## 10.1 Instabilities, Oscillations and Oscillators

The possibility of instabilities in amplifiers has been introduced earlier. Because of internal feedback in the active devices, external feedback paths, or connections in the amplifier, a feedback signal proportional to the output is produced. If this feedback signal is in phase with the input, a regenerative situation exists, and if the magnitude of the feedback is large enough, an unstable circuit is obtained. That is, if any input, including noise, is applied to or is present in a circuit which is initially at rest, growing transients occur. After a period of time, these growing transients are sufficiently large to produce a nonlinear response of the circuit elements, and these nonlinearities finally stop the growth of the signals. Eventually, then, steady-state oscillations can occur, and the circuit becomes an (unwanted) oscillator.

The problem in amplifiers is to avoid the possibilities of oscillation by proper design. Ideally, we desire an absolutely stable circuit; however, we may have to settle for a potentially unstable situation. In the latter, cf., Section 9.1, oscillations are possible for some tuning or operational conditions which, of course, are to be avoided for our amplifier.

To achieve the oscillator circuit function, in contrast to the amplifier situation, we must insure an unstable situation. However, we cannot be content with the mere fact that a circuit will oscillate. Our task also includes the development of the oscillatory power at a desired frequency, with a given, adequate amplitude, and with excellent constancy of envelope amplitude and frequency. Therefore, even though an oscillation of some kind is relatively easy to produce, the task of realizing a true oscillator can be even more difficult than for an amplifier. This is true at least for one important reason: nonlinearities are a basic necessity in the oscillator, as indicated above; thus, the governing equations of an oscillator are nonlinear, differential equations. On the other hand, for the amplifier, the basic description can be linear, at least initially, and nonlinearities can often be introduced as perturbations. As a consequence, oscillator analysis and design cannot be as advanced as that for linear circuits.

D.O. Pederson and K. Mayaram, *Analog Integrated Circuits for Communication*, DOI 10.1007/978-0-387-68030-9_10,
© 2008 Springer Science+Business Media, LLC

By and large, typical oscillator analysis involves reasonably simple approximate analyses of linearized or piece-wise-linear-circuit models of the oscillator together with perturbation and power-series techniques. However, there are a few oscillator circuits for which the nonlinear equations can be solved, at least approximately. These are the emphasis of the next two chapters. The results from these special cases provide guides, checks, and insight into the operation and design of all oscillators.

The introduction above implies that the feedback approach is the key issue in oscillators. Feedback, however, is not the entire story. Devices such as the tunnel diode exist which produce a negative-conductance characteristic. These devices, associated with resonant circuits or even $RC$ circuits, can produce oscillations and oscillators. In addition, circuits such as potentially unstable feedback amplifiers can be viewed on the basis of the negative conductance that appears at a port. The situation is, in some respects, like the familiar 'chicken-and-egg' controversy. We have two points of view, both of which can be used to achieve and to study oscillators. By suitable manipulation, one can always move from one basis to the other for a given circuit. Nonetheless, each approach has its advantages, and it is helpful to have an appreciation for both. To this end, we consider, in the next section, the negative-conductance (negative-resistance) approach to oscillators. The feedback approach is used with other examples later in the chapter.

For the active devices in this chapter, the bipolar junction transistor (BJT) and the MOS field effect transistor (MOSFET) are used, as well as one negative-conductance device, the tunnel diode. The analysis techniques in this chapter include linear, piece-wise linear and nonlinear analyses. In this chapter, two basic oscillators are considered that, in a sense, operate in an almost Class-A manner. The excursion into the 'off' regions of the device behavior is not great. In the last example in this chapter and in the next chapter, operation corresponding to Class-C behavior is introduced. Deep penetration into the off region of the devices usually is present. From the study of both classes of oscillator, a general oscillator situation usually can be explored leading to adequate design and operational information.

## 10.2 The Ideal Electronic Oscillator

An electrical model of an ideal, harmonic oscillator is shown in Figure 10.1a and is a lossless $LC$ circuit. Since the circuit is lossless, energy is conserved once the circuit is excited and alternates between electrical and magnetic forms. The voltage and current are pure sinusoids.

The electrical-circuit equations for this circuit lead to the differential equation

$$\frac{d^2v}{dt^2} + \omega_o^2 v = 0 \qquad (10.1)$$

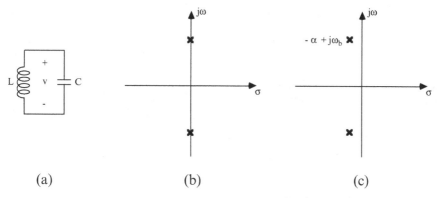

**Fig. 10.1.** (a) An electronic model of an ideal harmonic oscillator. (b) Natural frequencies of the ideal oscillator. (c) Natural frequencies of an oscillator with losses.

where
$$\omega_o^2 = \frac{1}{LC} \tag{10.2}$$

The solutions of this equation have the form

$$v = V_1 \exp\left(j\omega_o t\right) + V_2 \exp\left(-j\omega_o t\right) \tag{10.3}$$
$$= V \sin \omega_o t$$

where for the latter expression, the time origin is chosen to produce a zero phase angle.

In the frequency domain, the characteristic equation of the circuit is

$$s^2 + \omega_o^2 = 0 \tag{10.4}$$

The natural frequencies of this linear situation are the roots of this equation

$$s_1, s_2 = \pm j\omega_o \tag{10.5}$$

These lie on the imaginary axis of the complex frequency plane as shown in Figure 10.1b.

An actual resonator includes losses, and ideal behavior cannot be achieved. After an initial excitation a damped sinusoid is the time-domain response of the circuit. In the frequency domain, the characteristic equation includes a linear term, for the simplest case,

$$s^2 + 2\alpha s + \omega_o^2 = 0 \tag{10.6}$$

The natural frequencies lie in the left-half plane as shown in Figure 10.1c. For $\omega_o > \alpha$,

$$s_1, s_2 = -\alpha \pm j\sqrt{\omega_o^2 - \alpha^2} \approx -\alpha \pm j\omega_o \qquad (10.7)$$

If oscillations are to be maintained, energy must continuously be supplied to the circuit on a time average, i.e., average real power must flow into the circuit to compensate for the circuit losses. This power usually is supplied by dc bias sources to the electronic devices that convert the bias power into signal power in the form of either a negative, nonlinear conductance presented to the lossy resonator or as regenerative feedback. As mentioned above, both a negative conductance approach or a feedback approach may be used in analyzing, evaluating, and designing electronic oscillators.

## 10.3 A Tunnel-Diode Oscillator

For a first example of the analysis of a simple, nonlinear oscillator, the tunnel-diode oscillator of Figure 10.2a is used. The I-V characteristic of a tunnel diode is typically that of Figure 10.2b. If the diode is biased at $V_{DD}$ as shown, the incremental input conductance presented to the passive resonant circuit is negative and can compensate for the positive losses of the inductance, capacitance, leads, etc. Such losses are modeled for this example with a single conductance, $G = 1/R$.

It is convenient for the analysis to introduce a new variable which shifts the origin of the diode I-V characteristic to the quiescent operating point, $I_o$, $V_{DD}$. The shift in the characteristic is shown in Figures 10.2c and 10.2d. The new voltage variable is

$$v' = V - V_{DD} \qquad (10.8)$$

The original diode I-V characteristic is described functionally as

$$I = f_1(V) \qquad (10.9)$$

and is illustrated in Figure 10.2b. The translated diode-current variable is

$$i' = I - I_o = I - f_1(V_{DD}) \qquad (10.10)$$

The functional description of the shifted diode I-V characteristic is

$$i' = f_2(v') = f_1(V) - f_1(V_{DD}) \qquad (10.11)$$

The circuit equation in the original variables is

$$\frac{1}{L} \int (V_{DD} - V)\, dt = GV + C\frac{dV}{dt} + f_1(V) \qquad (10.12)$$

In terms of the new variables,

$$\frac{1}{L} \int v'\, dt + Gv' + C\frac{dv'}{dt} + f_2(v') + GV_{DD} + I_o = 0 \qquad (10.13)$$

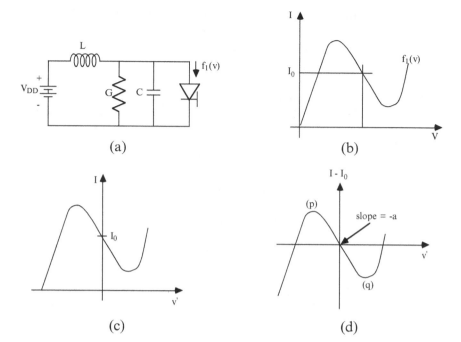

**Fig. 10.2.** (a) A tunnel diode oscillator. (b) I-V characteristics of the tunnel diode. (c) (d) I-V characteristics with a shifted origin.

where the following are used:

$$\frac{dv'}{dt} = \frac{dV}{dt} \tag{10.14}$$
$$I_o = f_1(V_{DD})$$

Both sides of Equation (10.13) are next differentiated and multiplied by $L$ to obtain

$$LC\frac{d^2v'}{dt^2} + L\frac{d}{dt}\left[Gv' + f_2(v')\right] + v' = 0 \tag{10.15}$$

Equation (10.15) is the present, desired form of the differential characteristic equation of this oscillator. Note in particular the combined first-order term of the equation. It is shown below that this combined entity can be viewed as the *net* nonlinearity of the oscillator. This form of the characteristic equation appears in many of the oscillators to be studied in this and the next chapter. (It is to be noted that $V_{DD}$ and $I_o = f_1(V_{DD})$ are assumed in (10.13) to be constant. As brought out in Chapter 11, if $v'(t)$ in the steady-state has a dc harmonic component, a shift in the effective bias point must occur, i.e., $I_o$ must change since $V_{DD}$ is fixed.)

We now investigate the *starting conditions* for the oscillator. That is, we conduct an analysis of the quiescent state of the circuit and determine whether oscillations can build up for an initial excitation.

From another point of view, we need to determine whether the equilibrium point of the circuit is unstable. At the quiescent bias point, which is the equilibrium point, an incremental analysis is made. For the diode, the slope of the translated I-V characteristic at $v' = 0$, as illustrated in Figure 10.2d, is designated $-a$

$$\frac{df_2}{dv'}\,|_{v'=0} = -a \tag{10.16}$$

The characteristic equation at the bias point and in the frequency domain is

$$LCs^2 + L(G - a)s + 1 = 0 \tag{10.17}$$

The natural frequencies of the linearized circuit about the bias point are

$$s_1, s_2 = -\frac{G - a}{2C} \pm j\sqrt{\frac{1}{LC} - \left(\frac{G - a}{2C}\right)^2} \tag{10.18}$$
$$= -\alpha \pm j\beta$$

where it is assumed that $s_1, s_2$ are complex.

If the magnitude of the slope of the diode characteristic at the bias point, $a$, is greater than $G$, the loss conductance of the oscillator, the natural frequencies lie in the right-half plane. Therefore, given an excitation, the oscillatory response grows exponentially. If one assumes that the natural frequencies are complex, as expressed in Equation (10.18), the response is an exponentially growing sinusoid. The voltage variable has the form:

$$v'(t) \approx Ae^{+|\alpha|t} \cos \beta t$$

This growth cannot continue indefinitely. As the diode voltage excursion gets large, the end points, $p$ and $q$, of the negative slope region of the diode I-V characteristic, as illustrated in Figure 10.2d, are encountered, and the diode introduces more loss into the system. The voltage response can be illustrated as in Figure 10.3. For very small excursions of the bias point, an exponential growth is produced. As the oscillatory response becomes larger, the total losses introduce a limiting condition, and a steady-state oscillation is produced. The output cannot be a pure sinusoid. The tips of the voltage output must be compressed, and the output waveform must contain harmonics. For the tunnel-diode oscillator, the output voltage will be quite nonsinusoidal in the best of cases, since the nonlinear I-V characteristic is not antisymmetrical about the bias point in the negative-conductance region. This aspect is brought out in greater detail later in this chapter.

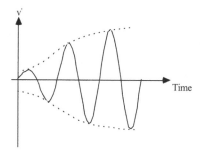

**Fig. 10.3.** Voltage response of the oscillator.

One might ask: why not adjust the value of $G$ to just equal $a$, the magnitude of the negative slope of the diode? From Equations (10.4) and (10.17), it is seen that this circumstance produces an ideal resonator. The answer is that the I-V characteristic is not linear at the bias point, or anywhere. Thus, for any excitation, the equality condition is not satisfied over an excursion of the variable. However, even if the characteristic is linear at the bias point, exact cancellation of the losses is not possible, since the characteristic is undoubtedly temperature and age sensitive, and the cancellation is only temporary.

The question now is: oscillations may build up, but what is the frequency and the magnitude of the steady-state behavior? Three different approximation techniques to answer this question are introduced in this chapter. Another estimation procedure is developed in the next chapter for a different class of electronic oscillators. A numerical example of the approximate analysis of a tunnel-diode oscillator is given in Section 10.5.

## 10.4 The van der Pol Approximation

B. van der Pol, in the late 1920s, evaluated electronic oscillators using vacuum tubes that provided a negative conductance characteristic [25] [26]. He proposed to model the essential features of the I-V characteristic in the simplest possible manner. About the bias point of the total I-V characteristic, he proposed a cubic polynomial approximation as illustrated in Figure 10.4. The slope at the bias point is equal in magnitude to the total conductance. The cubic term of the approximation produces the essential limiting action.

For the tunnel-diode oscillator of the last section, the nonlinear differential equation is repeated for easy reference.

$$LC\frac{d^2v}{dt^2} + L\frac{d}{dt}\left[Gv + f(v)\right] + v = 0 \tag{10.19}$$

where for convenience the voltage variable is labeled $v$, not $v'$, and the translated I-V characteristic is labeled simply $f(v)$. Again notice that the first-order term combines the effects of the passive conductance and of the diode's I-V

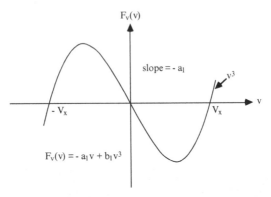

**Fig. 10.4.** I-V characteristics represented by a cubic polynomial.

characteristic. It is convenient next to introduce a time scaling (normalization). The dimensionless independent variable is

$$T = \frac{t}{\sqrt{LC}} \tag{10.20}$$

With this variable, the differential equation becomes

$$\frac{d^2v}{dT^2} + \sqrt{\frac{L}{C}} \frac{d}{dT}[F(v)] + v = 0 \tag{10.21}$$

where the total device and conductance function is defined

$$F(v) = Gv + f(v) \tag{10.22}$$

A plot of the total function is the addition of $f(v)$, as shown in Figure 10.2d, and a linear curve, $Gv$, drawn through the bias point of the I-V plane. Examples of this composite curve are given shortly. (Remember that in setting up the analysis, a change to incremental voltage and current variables has been made with respect to the bias point.) The slope of this total function at the origin (the bias point of the oscillator) is $-(a-G)$ where $-a$ is the slope of the diode characteristic alone at the bias point. The van der Pol approximation for $F(v)$ is labeled $F_v(v)$.

$$F_v(v) = -a_1v + b_1v^3 \tag{10.23}$$

For $v = 0$, the slope of the approximation is set equal to the slope of the actual total characteristic.

$$-a_1 = -(a - G) \tag{10.24}$$

To determine an appropriate value for $b_1$, note that the crossover points on the voltage axis, $\pm V_x$, for the van der Pol approximation are

$$V_x = \pm\sqrt{\frac{a_1}{b_1}} \tag{10.25}$$

This leads to

$$b_1 = \frac{a_1}{V_x^2} \tag{10.26}$$

Van der Pol continued his oscillator analysis by transforming the oscillator equation containing the cubic approximation into a standard form and then showed that an approximate closed-form solution for this equation is possible for the situation where the output is nearly sinusoidal, e.g., has low harmonic content. To obtain the *van der Pol Equation*, introduce first the parameter $\epsilon$, which is called the *van der Pol parameter*. This is the negative of the slope of the normalized total nonlinear function at the origin, multiplied by $\sqrt{\frac{L}{C}}$.

$$\epsilon = \sqrt{\frac{L}{C}}a_1 = \sqrt{\frac{L}{C}}(a - G) \tag{10.27}$$

Next, a scaling of the voltage variable is introduced.

$$v = hu \tag{10.28}$$

where

$$h^2 = \frac{\epsilon}{3b_1\sqrt{\frac{L}{C}}} \tag{10.29}$$

Using these parameters, we obtain the standard form of the *van der Pol Equation* [25].

$$\frac{d^2u}{dT^2} - \epsilon\left(1 - u^2\right)\frac{du}{dT} + u = 0 \tag{10.30}$$

Notice that about the equilibrium (bias) point, $u = 0$, the differential equation of the system becomes the following characteristic equation

$$p^2 - \epsilon p + 1 = 0 \tag{10.31}$$

where $p$ is the normalized complex frequency variable, $p = \sqrt{LC}s$. The natural frequencies of the linearized, normalized system are

$$p_1, p_2 = \frac{\epsilon}{2} \pm j\sqrt{1 - \left(\frac{\epsilon}{2}\right)^2} \tag{10.32}$$

For positive $\epsilon$, the natural frequencies are in the right-half plane (RHP) and are complex if $\epsilon < 2$. For the case $\epsilon > 0$ but very small, the buildup of oscillation has the form

$$u(T) = A\exp\left(\frac{\epsilon T}{2}\right)\cos T \tag{10.33}$$

We expect that after a steady state has been reached, the form of the response should be

$$u(T) = const \times \cos T \tag{10.34}$$

Van der Pol used this type of reasoning to propose a form of the solution that is then substituted back into the nonlinear differential equation. After making several reasonable assumptions and simplifications, which are valid for near-sinusoidal oscillations, the solution in terms of the original variables becomes

$$v(t) = \sqrt{\frac{4}{3}\frac{a_1}{b_1}} \frac{1}{\sqrt{1 + \exp\left\{\frac{-(t-t_o)\epsilon}{\sqrt{LC}}\right\}}} \cos\left[\frac{1}{\sqrt{LC}}t + \phi_o\right] \tag{10.35}$$

In this equation, a phase angle $\phi_o$ is introduced to provide the proper phase for the cosine term in relation to the choice of the constant $t_o$. For small values of time, the envelope of the oscillation grows exponentially, as expected from (10.33). After a long period of time, the zero-to-peak amplitude reaches a maximum value.

$$V_{max} = \sqrt{\frac{4}{3}\frac{a_1}{b_1}} \tag{10.36}$$

From (10.26), the maximum amplitude is related simply to the value of the nearest crossover voltage.

$$V_x = \sqrt{\frac{a_1}{b_1}} = \sqrt{\frac{a - G}{b_1}} \tag{10.37}$$

$$V_{max} = \sqrt{\frac{4}{3}V_x} = 1.15V_x \tag{10.38}$$

Since the value of the voltage at the negative peak of the total characteristic is

$$V^- = \sqrt{\frac{a_1}{3b_1}} \tag{10.39}$$

$$V_{max} = 2V^- \tag{10.40}$$

The maximum amplitude is twice the value of the voltage at which the negative maximum occurs. These relations are illustrated in Figure 10.5.

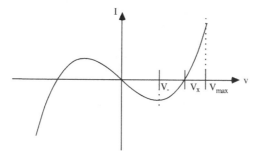

**Fig. 10.5.** Definitions of $V^-$, $V_x$, and $V_{max}$.

## 10.5 Tunnel-Diode Oscillator Example

For a numerical example of a tunnel-diode oscillator and the use of the van der Pol analysis, consider the diode I-V characteristic sketched in Figure 10.6a. The quiescent bias point is chosen to be $V_{DD} = 0.45$ V. At this bias point, the slope of the characteristic is estimated from the curve to be approximately

$$-a = -2.8 \times 10^{-3} \tag{10.41}$$

The tuned circuit is chosen to have an inductor of $L = 1$ $\mu$H and a capacitor of $C = 30$ pF. The loss in the tank circuit is assumed to be modeled with a shunt resistor, $R = 1$ k$\Omega$. For this circuit the *van der Pol parameter* $\epsilon$ is

$$\epsilon = \sqrt{\frac{L}{C}}\left(a - \frac{1}{R}\right) = 183(2.8 - 1) \times 10^{-3} = 0.33 \tag{10.42}$$

The value of $\epsilon$ is small and positive. Therefore, we can expect that oscillations will build up and be sinusoidal in nature, with a frequency $f_o = 1/(2\pi\sqrt{LC}) = 29$ MHz.

To estimate the steady-state oscillation voltage magnitude, we construct the 'total' I-V characteristic, i.e., we plot $F(v)$ from Equation (10.15) or (10.22) in Figure 10.6a by first plotting the $\frac{1}{R}v' = Gv'$ line through the bias point and then adding the two characteristics. The average crossover voltage of the total characteristic, with the origin translated to the bias point, is $V_x$. Therefore, the estimated voltage amplitude of the steady-state oscillation is 1.15 $V_x$. Since the Spice programs do not contain a device model for the tunnel-diode, we cannot verify an estimate with circuit simulation. We can, however, use the polynomial characteristics available for controlled sources in the Spice programs to produce a 'van der Pol' negative-conductance characteristic and a negative-conductance oscillator. The schematic diagram of such an oscillator is the same as that of Figure 10.2a, where the tunnel diode is replaced with a dependent current source. The controlling node pair of this current source is

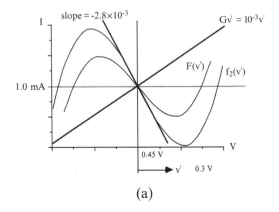

(a)

```
FIG 10.6
* DEVICE CHARACTERISTICS
V1 1 0 0 PULSE 0.3 0 0 0 0 01
*.TRAN 2N 900N 800N
*.PLOT TRAN V(2)
*.FOUR 29.4MEG V(2)
L1 1 2 1U
R1 2 0 1K
C1 2 0 30P
VD 2 3 0
G11 3 0 POLY(1) 2 0 0 -2.8M 0 20M
*G12 3 0 POLY(1) 2 0 0 -2.8M 2M 20M
*THE FOLLOWING ELEMENT IS NEEDED FOR SPICE3
*IN PLACE OF G11 IN SPICE2
*B12 3 0 I = -2.8E-3*(V(3)-V(0)) + 20E-3*((V(3)-V(0))^3)
.DC V1 -.4 .4 .02
.PLOT DC I(VD) I(VD1)
*TOTAL CHARACTERISTIC
L11 1 21 1U
R21 31 0 1K
C11 21 0 30P
VD1 21 31 0
G11T 31 0 POLY(1) 2 0 0 -2.8M 0 20M
G12T 31 0 POLY(1) 2 0 0 -2.8M 2M 20M
.WIDTH OUT=80
.OPTIONS ITL5=0 LIMPTS=2001
.OPTIONS RELTOL=1E-5
.END
```

(b)

**Fig. 10.6.** (a) Diode I-V and total characteristics. (b) Spice input file for the oscillator.

made the node pair of the element; therefore, a nonlinear conductance characteristic can be produced. The Spice input file for the oscillator is given in Figure 10.6b. Note that the new element has the description:

$$G11 \ 3 \ 0 \ \text{poly}(1) \ 3 \ 0 \ 0 \ -a \ 0 \ b_1 \tag{10.43}$$

The current through this element flows from node 3 to node 0 and is described by a polynomial of one dimension. The polynomial is chosen to have the form of the van der Pol approximation.

$$I(G11) = 0 - a \ v(3) + 0 + b_1(v(3))^3 \tag{10.44}$$

The coefficients $a$ and $b_1$ are found from a cubic approximation of one of the given nonlinear characteristics in Figure 10.6a. The van der Pol cubic approximations for both the device characteristic, $f_2(v')$, and the total characteristic, $F(v')$, are

$$f_{2v}(v') = -av' + b_1(v')^3 \tag{10.45}$$
$$F_v(v') = -a_1v' + b_1(v')^3$$

where $a_1 = (a - G)$. Note that the $b_1$ coefficient is the same for both approximations. From the slope at the bias point of Figure 10.6a, $a = 2.8 \times 10^{-3}$. From the voltage crossover of the total characteristic of 0.3V, we obtain $b_1 = 1.8 \times 10^{-3}/(0.3)^2 = 20 \times 10^{-3}$. The corresponding voltage crossover of the device characteristic alone should be $V_x = \sqrt{a/b_1} = 0.37$V.

A plot of the approximation can be obtained by monitoring the current through the sensing voltage source $V_D$ as $V_1$ is varied and is shown in Figure 10.6c. The slope of the approximation at the bias point is $-2.8 \times 10^{-3}$ and the voltage crossovers are $\pm 0.374$V. A plot of the approximation of the total characteristic can be obtained by moving the resistor, $R$, from nodes 2 and 0 to nodes 3 and zero, and again monitoring the current through $V_D$ as $V_1$ is varied. The plot is also shown in Figure 10.6c. The slope of the curve at $V = 0$ is $-1.8 \times 10^{-3}$ and the voltage crossovers are $\pm 0.3$V.

The other elements of the oscillator are those used in the example above, except that $V_{DD} = 0.45$V is not needed since the nonlinear characteristic has been chosen to have a bias of $V(3) = 0$V. To initiate the oscillation, a step (pulse) voltage is applied at the $V_{DD}$ node. The waveform of the oscillator voltage and its Fourier components are given in Figure 10.6d. The amplitude of the fundamental is 0.34 V, which is the estimated value $= 1.15 \times 0.30$ V.

The results for additional Spice runs are given in Figures 10.6e through h. In Figure 10.6e, the I-V characteristic is given when a quadratic term is added to the polynomial. The characteristic is no longer antisymmetrical about the origin, $V(3) = 0$. The oscillator output voltage waveform and its Fourier components are given in Figure 10.6f. Notice that the amplitude of the fundamental is approximately unchanged. However, a dc shift of the bias point has

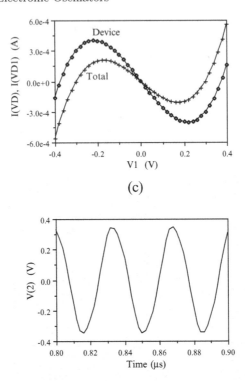

(c)

(d)

**Fig. 10.6.** (c) Dc transfer characteristics. (b) Oscillator voltage waveform and its Fourier components.

occurred and that the harmonic content of the output voltage has increased significantly. In Figures 10.6g and h, the corresponding plots are given when an input offset voltage is applied at $V_{DD}$, with the initial nonlinearity (without the quadratic term). For this case the amplitude of the fundamental has decreased considerably, the dc shift of the bias point is large and the total harmonic content of the output voltage is much larger. The results from these last two examples can be used to illustrate the nature and effects of bias-point shift in the next chapter.

```
FIG 10.6E
* DEVICE CHARACTERISTICS WITH QUADRATIC
V1 1 0 0 PULSE 0.3 0 0 0 01
*.TRAN 2N 900N 800N
*.PLOT TRAN V(2)
*.FOUR 29.4MEG V(2)
L1 1 2 1U
R1 2 0 1K
C1 2 0 30P
VD 2 3 0
*G11 3 0 POLY(1) 2 0 0 -2.8M 0 20M
G12 3 0 POLY(1) 2 0 0 -2.8M 2M 20M
*THE FOLLOWING ELEMENT IS NEEDED FOR SPICE3
*IN PLACE OF G11 IN SPICE2
*B12 3 0 I = -2.8E-3*(V(3)-V(0)) + 20E-3*((V(3)-V(0))^3)
.DC V1 -.4 .4 .02
.PLOT DC I(VD)
.OPTIONS ITL5=0 LIMPTS=2001
.OPTIONS RELTOL=1E-5
.END
```

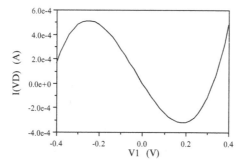

**Fig. 10.6.** (e) Spice input file and dc characteristics for device with quadratic term.

## 10.6 Wien-Type Oscillators

A classic oscillator is the Wien-bridge configuration shown in Figure 10.7a. In this oscillator, a positive-gain, differential input, differential output amplifier is used to provide positive feedback of the proper magnitude for regeneration at a single frequency. The frequency selectivity is produced by a $RC$ bridge circuit in the feedback path. From an impedance-bridge viewpoint, the bridge is in balance at a single frequency. For the case where $R_1 = R_2$, $C_1 = C_2$, and $R_a = 2R_b$ the balance occurs at $w_o = \frac{1}{RC}$. The voltage, $V_{in}$ is then in phase with $V_{out}$, i.e., the feedback to the amplifier input is positive. If the resistive arms of the bridge satisfy the condition $R_a \geq 2R_b$, the voltage gain around the loop at $w_o$ is equal to or greater than one, and instability results.

To evaluate the performance of the oscillator, it is convenient to redraw the bridge configuration to that shown in Figure 10.7b. In the new circuit, the high-gain differential amplifier can be identified as an operational amplifier. The resistors $R_a$ and $R_b$ provide negative feedback. The subcircuit shown

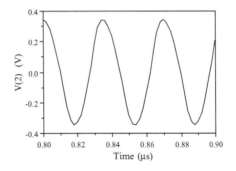

```
FOURIER COMPONENTS OF TRANSIENT RESPONSE V(2)
DC COMPONENT =   -4.726D-03
HARMONIC    FREQUENCY      FOURIER      NORMALIZED      PHASE       NORMALIZED
   NO         (HZ)        COMPONENT     COMPONENT       (DEG)      PHASE (DEG)

    1       2.940D+07     3.467D-01     1.000000       36.194        0.000
    2       5.880D+07     1.842D-02     0.053118       67.262       31.067
    3       8.820D+07     1.843D-02     0.053144       27.969       -8.226
    4       1.176D+08     5.418D-03     0.015627       40.836        4.641

    TOTAL HARMONIC DISTORTION =          7.878759     PERCENT
```

**Fig. 10.6.** (f) Voltage waveform and its Fourier components.

in dashed lines in Figure 10.7b, can be identified as a precise, positive gain, voltage amplifier. For $R_a \approx 2R_b$ and with a high differential amplifier gain, the negative feedback (the loop gain) is very large. The closed-loop amplifier therefore has a very precisely set positive voltage gain in the active region of the amplifier, $a = 1 + \frac{R_a}{R_b}$. This is the slope of the $v_2$ - $v_1$ characteristic of Figure 10.7c at the origin. The slope of the characteristic is approximately zero outside of the active region of the differential amplifier. The negative feedback amplifier in the active region also has a high input impedance and a very low output resistance. The amplifier can then be modeled as an ideal voltage-controlled piece-wise-linear voltage source as shown in Figure 10.7d, where the controlled voltage-source parameter, $a_v(v_1) = f(v_1)$, is described by the characteristic shown in Figure 10.7c.

The configuration of Figure 10.7d is referred to as a *Wien-type* oscillator. The circuit equations which describe the Wien-type configuration are

$$\frac{v_1}{R_2} + C_2 \frac{dv_1}{dt} = i_1 \tag{10.46}$$

$$i_1 R_1 + \frac{1}{C_1} \int i_1 dt = v_2 - v_1 = f(v_1) - v_1$$

where the output voltage of the gain block, $v_2$, can also be designated as the function $f(v_1)$, a plot of which has the shape of Figure 10.7c. It is assumed that $v_1$ and $v_2$ are incremental variables about the quiescent bias state. For a good op amp with little or no offset, the incremental and total input or

```
FIG 10.6G
* DEVICE CHARACTERISTICS WITH 0.1V OFFSET
V1 1 0 0 PULSE 0.3 0 0 0 0 01
*.TRAN 2N 900N 800N
*.PLOT TRAN V(2)
*.FOUR 29.4MEG V(2)
L1 1 2 1U
R1 2 0 1K
C1 2 0 30P
VD 2 3 0
G11 3 4 POLY(1) 2 0 0 -2.8M 0 20M
VDD 4 0 -0.1
*G12 3 0 POLY(1) 2 0 0 -2.8M 2M 20M
*THE FOLLOWING ELEMENT IS NEEDED FOR SPICE3
*IN PLACE OF G11 IN SPICE2
*B12 3 0 I = -2.8E-3*(V(3)-V(0)) + 20E-3*((V(3)-V(0))^3)
.DC V1 -.4 .4 .02
.PLOT DC I(VD)
.WIDTH OUT=80
.OPTIONS ITL5=0 LIMPTS=2001
.OPTIONS RELTOL=1E-5
.END
```

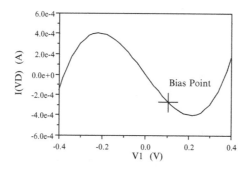

**Fig. 10.6.** (g) Spice input file and dc transfer characteristics.

output variables are the same. The equations of (10.46) can be manipulated to obtain the nonlinear differential equation of the oscillator.

$$\frac{d^2v_1}{dt^2} + \frac{d}{dt}\left[\left(\frac{1}{R_1C_1} + \frac{1}{R_2C_2} + \frac{1}{R_1C_2}\right)v_1 - \frac{1}{R_1C_2}f(v_1)\right] + \frac{v_1}{R_1R_2C_1C_2} = 0$$
(10.47)

For convenience, we let

$$R_1 = R_2 = R \qquad\qquad (10.48)$$
$$C_1 = C_2 = C$$
$$T = \frac{t}{RC}$$

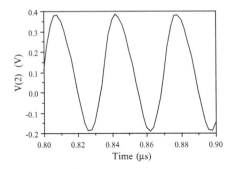

FOURIER COMPONENTS OF TRANSIENT RESPONSE V(2)
DC COMPONENT =    1.078D-01

| HARMONIC NO | FREQUENCY (HZ) | FOURIER COMPONENT | NORMALIZED COMPONENT | PHASE (DEG) | NORMALIZED PHASE (DEG) |
|---|---|---|---|---|---|
| 1 | 2.940D+07 | 2.817D-01 | 1.000000 | -46.247 | 0.000 |
| 2 | 5.880D+07 | 3.427D-02 | 0.121657 | -70.229 | -23.982 |
| 3 | 8.820D+07 | 2.092D-03 | 0.007426 | 171.483 | 217.730 |
| 4 | 1.176D+08 | 2.417D-03 | 0.008582 | 37.960 | 84.207 |

TOTAL HARMONIC DISTORTION =    12.294816  PERCENT

**Fig. 10.6.** (h) Output voltage waveform and its Fourier components.

In the last expression, a time normalization is introduced, similar to what is done in the van der Pol oscillator analysis. The oscillator differential equation becomes

$$\frac{d^2 v_1}{dT^2} + \frac{d}{dT}[3v_1 - f(v_1)] + v_1 = 0 \tag{10.49}$$

Notice the nonlinearity of the gain block's transfer characteristic appears only in the first-order term. A total, composite, nonlinear characteristic is defined as

$$F(v_1) = 3v_1 - f(v_1) \tag{10.50}$$

and can be developed and drawn as shown in Figures 10.8a and 10.8b. The voltage gain of the gain block at the bias point of the amplifier is the slope of the gain characteristic at $v_1 = 0$ and is labeled $a$.

$$\left. \frac{df(v_1)}{dv_1} \right|_{v_1=0} = +a \tag{10.51}$$

The slope of the total characteristic at $v_1 = 0$ is, for the choices of $R_1, R_2, C_1$ and $C_2$ in Equation (10.48),

$$\left. \frac{dF(v_1)}{dv_1} \right|_{v_1=0} = -(a - 3) \tag{10.52}$$

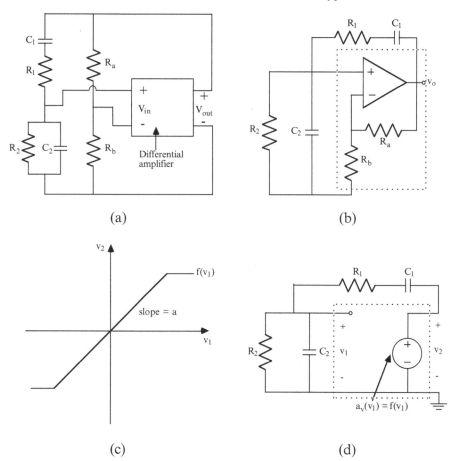

**Fig. 10.7.** (a) Wien-bridge configuration. (b) A convenient representation of the Wien-bridge oscillator. (c) Dc voltage transfer characteristics of the amplifier. (d) Amplifier modeled by an ideal voltage-controlled piece-wise-linear voltage source.

From an inspection of the final differential equation, (10.49), it is clear that it has exactly the same form as that obtained for the tunnel-diode oscillator of Section 10.3. We have a mathematically dual situation. Of course, the total nonlinearities are not identical. However, by proper design of the circuit elements and device operation and biasing, both oscillators can provide a total characteristic which has a shallow 'n-type' shape with a negative slope at the origin and positive-slope regions at large positive and negative excursions. Given an excitation, we can expect that oscillations will build up, and that a steady-state oscillation will be obtained, provided that the equilibrium point of the oscillators is properly unstable.

Following the notation of the van der Pol analysis, we correlate the magnitude of the slope of the total characteristic $F(v_1)$ at the equilibrium point

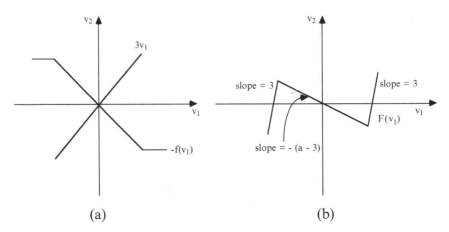

**Fig. 10.8.** (a) Dc characteristics of the amplifier and resistors. (b) Total composite nonlinear characteristics.

to the parameter $\epsilon$. Since for the present case, $F(v_1)$ in (10.46) has a unit multiplier, the slope at $v_1 = 0$ is $\epsilon$.

$$\epsilon = a - 3 \tag{10.53}$$

The normalized characteristic equation at the equilibrium is then

$$p^2 - \epsilon p + 1 = 0 \tag{10.54}$$

The normalized natural frequencies at this point are

$$p_1, p_2 = \frac{\epsilon}{2} \pm j\sqrt{1 - \left(\frac{\epsilon}{2}\right)^2} \tag{10.55}$$

For $\epsilon$ positive and less than 2, the natural frequencies are complex and lie in the RHP. For $\epsilon$ small, they lie close to the $j\omega$ axis. If the oscillator at the equilibrium point receives a small excitation, we expect an oscillatory buildup of circuit variables.

The van der Pol approximation of the total characteristic is not an appropriate approximation of the virtual piece-wise-linear characteristic of the Wien-type oscillator as shown in Figure 10.8b. Studies of the oscillator performance, such as those described below using circuit simulation, show that severe top and bottom flattening or 'squashing' of the voltage waveforms is obtained if $\epsilon$ is greater than approximately 0.1. As brought out in the next chapter, the frequency selectivity of the $RC$ feedback circuit is not sufficient to damp out the harmonics which are introduced by the flattening due to the piece-wise-linear amplifier characteristic. Therefore, the design procedure must be to achieve a small $\epsilon$. The steady-state amplitude of the amplifier input voltage is then approximately equal to the 'peak'-point values of $F(v_1)$, the total nonlinear characteristic.

For a numerical example, the block diagram of a Wien-type oscillator using a 741 opamp is the same as that shown in Figure 10.7b. The Spice input circuit file is given in Figure 10.9a where the Wien-type oscillator circuit elements are identified. For the values $R_1 = R_2 = 1$ k$\Omega$ and $C_1 = C_2 = 0.159$ $\mu$F, the

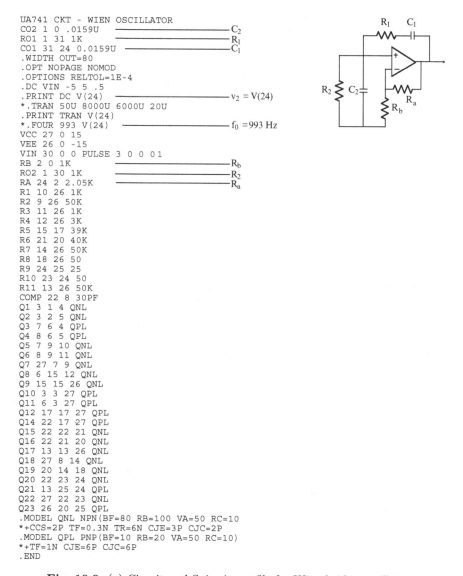

```
UA741 CKT - WIEN OSCILLATOR
CO2 1 0 .0159U    ──────────────C₂
RO1 1 31 1K       ──────────────R₁
CO1 31 24 0.0159U ──────────────C₁
.WIDTH OUT=80
.OPT NOPAGE NOMOD
.OPTIONS RELTOL=1E-4
.DC VIN -5 5 .5
.PRINT DC V(24)   ──────────────v₂ = V(24)
*.TRAN 50U 8000U 6000U 20U
.PRINT TRAN V(24)
*.FOUR 993 V(24)  ──────────────f₀ =993 Hz
VCC 27 0 15
VEE 26 0 -15
VIN 30 0 0 PULSE 3 0 0 01
RB 2 0 1K         ──────────────Rᵦ
RO2 1 30 1K       ──────────────R₂
RA 24 2 2.05K     ──────────────Rₐ
R1 10 26 1K
R2 9 26 50K
R3 11 26 1K
R4 12 26 3K
R5 15 17 39K
R6 21 20 40K
R7 14 26 50K
R8 18 26 50
R9 24 25 25
R10 23 24 50
R11 13 26 50K
COMP 22 8 30PF
Q1 3 1 4 QNL
Q2 3 2 5 QNL
Q3 7 6 4 QPL
Q4 8 6 5 QPL
Q5 7 9 10 QNL
Q6 8 9 11 QNL
Q7 27 7 9 QNL
Q8 6 15 12 QNL
Q9 15 15 26 QNL
Q10 3 3 27 QPL
Q11 6 3 27 QPL
Q12 17 17 27 QPL
Q14 22 17 27 QPL
Q15 22 22 21 QNL
Q16 22 21 20 QNL
Q17 13 13 26 QNL
Q18 27 8 14 QNL
Q19 20 14 18 QNL
Q20 22 23 24 QNL
Q21 13 25 24 QPL
Q22 27 22 23 QNL
Q23 26 20 25 QPL
.MODEL QNL NPN(BF=80 RB=100 VA=50 RC=10
*+CCS=2P TF=0.3N TR=6N CJE=3P CJC=2P
.MODEL QPL PNP(BF=10 RB=20 VA=50 RC=10)
*+TF=1N CJE=6P CJC=6P
.END
```

**Fig. 10.9.** (a) Circuit and Spice input file for Wien-bridge oscillator.

frequency of oscillation should be 1 kHz. The negative feedback resistors are chosen to be $R_b = 1$ k$\Omega$ and $R_a = 2.05$ k$\Omega$ and should provide $\epsilon = 0.05$. The

closed-loop voltage-transfer characteristic of the 741 with negative feedback is
obtained from a Spice simulation and is plotted in Figure 10.9b. As expected,

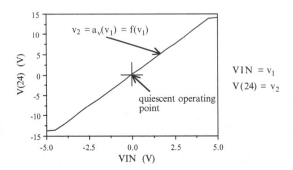

**Fig. 10.9.** (b) Closed-loop voltage transfer characteristics.

the slope at the bias point is + 3.05, since the ratio of the negative feedback
resistors is 2.05. The value of $\epsilon$ is therefore 0.05. The expected period of the
oscillation is

$$\frac{1}{f_o} = 2\pi RC = 1 \text{ ms} \tag{10.56}$$

and the frequency of oscillation is 1 kHz. The amplitude of the steady-state
output voltage should be approximately equal to the break points of the transfer characteristic which in turn should be the magnitude of the voltage supplies for a nominal opamp. From the plot of Figure 10.9b, the extremes of the
dc output voltage obtained from the Spice simulation of the voltage transfer
characteristic of the closed-loop amplifier are -13.4 and +11.7. The amplitude
of the output oscillation voltage should be approximately the average of these
magnitudes, 12.5 V.

In the input file for a transient Spice simulation of the oscillator, note that
a voltage input pulse is used to set up an initial excitation of the oscillator;
otherwise, the slow buildup of oscillations leads to excessive simulation time
and cost. The steady-state output voltage waveform is shown in Figure 10.10a.
It is seen that the peak-to-peak magnitude is 27.44 V, or a zero-to-peak value
of 13.72 V. From the output voltage waveform, the period of oscillation is 1.007
ms, which is very close to the predicted value of 1 ms. It is essential to use
a close estimate of the oscillation frequency for the Fourier series analysis in
Spice. A ±0.1% difference produces a noticeable change in harmonic content.
For a frequency of 993 Hz, the Fourier components of the output voltage, are
given in Figure 10.10b. The fundamental has a magnitude of 14.34V, and the
output waveform has a total harmonic distortion of 2.88% with $HD_2 = 0.03\%$
and $HD_3 = 2.2\%$.

If $C$ is decreased by a factor of 10 to produce a design frequency of 10
kHz, the value of the simulated period is 108 $\mu$s, not 100 $\mu$s, for an oscillation

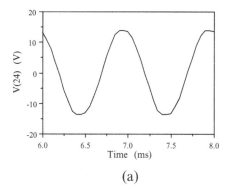

(a)

```
FOURIER COMPONENTS OF TRANSIENT RESPONSE V(24)
DC COMPONENT =    5.581D-03
HARMONIC    FREQUENCY     FOURIER      NORMALIZED      PHASE       NORMALIZED
   NO         (HZ)       COMPONENT     COMPONENT      (DEG)       PHASE (DEG)

    1       9.930D+02    1.435D+01     1.000000      108.085          .000
    2       1.986D+03    5.007D-03      .000349     -132.161      -240.246
    3       2.979D+03    3.202D-01      .022319      -66.199      -174.284
    4       3.972D+03    6.258D-03      .000436      -53.848      -161.933

       TOTAL HARMONIC DISTORTION =      2.893581   PERCENT
```

(b)

**Fig. 10.10.** (a) Steady-state output voltage waveform. (b) Fourier components of the output voltage.

frequency of 925.9 Hz. The drop in frequency is due to the small lagging phase shift of the opamp, even with its large negative feedback. The negative phase shift must be compensated by a leading phase shift from the $RC$ feedback circuit. Because of the low $Q$ of this $RC$ circuit, an appreciable shift in frequency is necessary to obtain the needed leading phase shift. The impact of the broad resonance of the $RC$ feedback circuit is commented upon further in Section 11.2.

To study the effects of the choice of $\epsilon$, it is convenient to use a macromodel for the 741 opamp [27]. A series of simulations of the complete 741-oscillator circuit can be quite costly. The gain block of the 741 with negative feedback can be modeled with a simple voltage-controlled, voltage source as shown in Figure 10.11a which includes the Spice input file. At the output of the model are clamping diodes and voltage sources to provide the desired piece-wise-linear, voltage-transfer characteristic of Figure 10.7c. These clamp diodes are driven very hard into the forward region since no diode resistance is used, but this is not a problem for the simulator. We can expect that the extremes of the output voltage are approximately equal to the values of the clamp voltage sources, ±14V, plus or minus the forward diode drop. In Figure 10.11b, the voltage transfer characteristic of the gain block is given. The slope at the bias point is equal to 3.05, as expected, and the extremes have values slightly larger than ±15V.

```
MACROMODEL OF WIEN-TYPE OSC
V1 1 0 0 PULSE 4 0 0 0 0 1000U
*.DC V1 -10 101
*.PRINT DC V(7)
.TRAN 5U 800U 600U 2U
.PLOT TRAN V(7)
.FOUR 10K V(7)
R1 1 2 1K
C1 2 0 0.0159U
R2 2 6 1K
C2 6 7 0.0159U
EO 5 0 2 0 3.2
RO 5 7 1
DCL1 7 8 MOD1
VBCL1 8 0 14
DCL2 0 9 MOD1
VCL2 9 7 14
.MODEL MOD1 D IS=1E-16
.OPTIONS RELTOL=1E-4
.OPTIONS NOPAGE NOMOD
.WIDTH OUT=80
.END
```

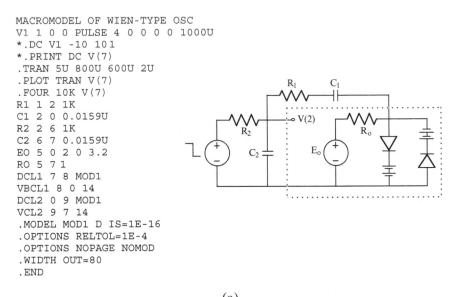

(a)

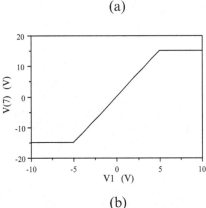

(b)

**Fig. 10.11.** (a) Circuit and Spice input file for a simple gain block model. (b) Dc voltage transfer characteristics of the gain block.

The Spice input file in Figure 10.11a includes the oscillator feedback elements and the oscillator excitation voltage source. The output voltage waveform and the harmonic components for $\epsilon = 0.05$ are shown in Figure 10.11c. The oscillation period for the macromodel is approximately 100 $\mu$s and the amplitude of the fundamental is 15.6 V. The harmonic output components for $\epsilon = 0.1$ and 0.2 are given in Figures 10.11d and e. As $\epsilon$ is increased, the clipping of the output waveform increases, and the harmonic distortion increases. It is clear that $\epsilon$ must be small to obtain a nearly sinusoidal output. Note in Figure 10.11d that for $\epsilon = 0.1$, a smaller THD is obtained for a Fourier frequency of 10.1 kHz. A similar situation occurs for $\epsilon = 0.2$.

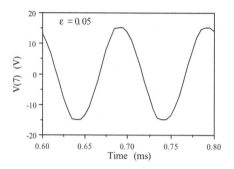

```
FOURIER COMPONENTS OF TRANSIENT RESPONSE V(7)
DC COMPONENT =   -4.623D-02
HARMONIC    FREQUENCY      FOURIER     NORMALIZED      PHASE     NORMALIZED
  NO          (HZ)        COMPONENT    COMPONENT       (DEG)    PHASE (DEG)

   1       1.000D+04     1.560D+01     1.000000      118.047      0.000
   2       2.000D+04     3.854D-02     0.002470      140.119     22.072
   3       3.000D+04     3.262D-01     0.020906      -39.821   -157.868
   4       4.000D+04     9.419D-03     0.000604     -161.375   -279.422

      TOTAL HARMONIC DISTORTION =      2.677468   PERCENT
```

**Fig. 10.11.** (c) Output voltage waveform and Fourier components for $\epsilon = 0.05$.

An alternate form of the Wien-type configuration is that of Figure 9.2b which is used there as the basis of a bandpass amplifier. In the latter, the closed-loop gain at the equilibrium point is restricted to be less than 3. If $A > 3$, ($a = A$ in the analysis of Section 9.1), a simple effective oscillator is achieved.

## 10.7 Transformer-Coupled ECP Oscillators

Another example of a simple electronic oscillator is a transformer-coupled configuration which uses an emitter-coupled pair (ECP) as the gain stage. The circuit is shown in Figure 10.12a. For simplicity, an emitter-bias current source of 4 mA is used, and only a very small resistor is included with the collector of Q1. Once more, the nonlinear differential equation which describes this oscillator has the same form as obtained for the two previous oscillators. Again, the van der Pol approximation cannot be used to predict the amplitude of the steady-state output.

From our earlier developments for the ECP, the input voltage-to-output current transfer characteristic is that shown in Figure 10.12b. The slope of this

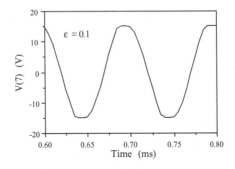

```
FOURIER COMPONENTS OF TRANSIENT RESPONSE V(7)
DC COMPONENT =  -1.066D-01
HARMONIC   FREQUENCY    FOURIER      NORMALIZED    PHASE      NORMALIZED
   NO        (HZ)       COMPONENT    COMPONENT     (DEG)      PHASE (DEG)

    1      1.000D+04    1.598D+01     1.000000    107.857        .000
    2      2.000D+04    9.705D-02      .006075    122.596      14.739
    3      3.000D+04    6.376D-01      .039910    -65.581     -173.438
    4      4.000D+04    2.336D-02      .001462   -177.986     -285.843

       TOTAL HARMONIC DISTORTION =       4.772156  PERCENT


FOURIER COMPONENTS OF TRANSIENT RESPONSE V(7)
DC COMPONENT =  -2.560D-01
HARMONIC   FREQUENCY    FOURIER      NORMALIZED    PHASE      NORMALIZED
   NO        (HZ)       COMPONENT    COMPONENT     (DEG)      PHASE (DEG)

    1      1.010D+04    1.591D+01     1.000000    109.831        .000
    2      2.020D+04    2.330D-01      .014645    126.734      16.902
    3      3.030D+04    5.921D-01      .037211    -62.608     -172.439
    4      4.040D+04    5.832D-02      .003666   -167.804     -277.635

       TOTAL HARMONIC DISTORTION =       4.693735  PERCENT
```

**Fig. 10.11.** (d) Output voltage waveform and Fourier components for $\epsilon = 0.1$.

characteristic at a zero-volt bias point is $-g_m/2$. The extremes of the current excursion are $+I_{EE}$ and 0 that are attained at input voltages of approximately $\pm 4V_t \approx \pm 0.1V$. In Figure 10.12c, a translation is made of $I_{c2}$ and $V_1$ to incremental variables, $i_{c2}$ and $v_1$, about the quiescent values, $I_{c2} = I_{EE}/2$ and $V_1 = 0$. This translated characteristic is labeled as the function $f(v_1)$.

$$i_{c2} = f(v_1) \tag{10.57}$$

(The case where an offset, input bias is present is considered in Section 11.1. For the present, the translated characteristic can be assumed to be anti-symmetrical about the quiescent operating point.)

A circuit equation for the oscillator in the transform domain is obtained from KCL applied at the collector of $Q_2$, as illustrated in Figure 10.12d.

$$- i_{c2} = G(nv_1) + Cs(nv_1) + \frac{1}{sL}(nv_1) + \frac{1}{n}i_{b1} \tag{10.58}$$

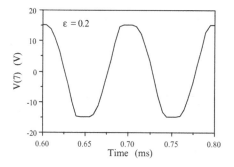

FOURIER COMPONENTS OF TRANSIENT RESPONSE V(7)
DC COMPONENT =   -2.762D-01

| HARMONIC<br>NO | FREQUENCY<br>(HZ) | FOURIER<br>COMPONENT | NORMALIZED<br>COMPONENT | PHASE<br>(DEG) | NORMALIZED<br>PHASE (DEG) |
|---|---|---|---|---|---|
| 1 | 1.000D+04 | 1.646D+01 | 1.000000 | 79.840 | 0.000 |
| 2 | 2.000D+04 | 2.943D-01 | 0.017877 | 65.942 | -13.899 |
| 3 | 3.000D+04 | 1.173D+00 | 0.071232 | -142.973 | -222.813 |
| 4 | 4.000D+04 | 6.151D-02 | 0.003736 | 105.342 | 25.502 |

TOTAL HARMONIC DISTORTION =    8.087821  PERCENT

**Fig. 10.11.** (e) Output voltage waveform and Fourier components for $\epsilon = 0.2$.

where $(1/n)i_{b1}$ is the input current transferred across the transformer and $L$ is the magnetizing inductance, $l_{11}$, of the transformer. The ECP variational input current, $i_{b1}$, if transistor operation is restricted to the normal active and off regions, and for large beta, can be easily related to $i_{c2}$ (assuming $\beta = \beta_{ac}$).

$$i_{b1} \approx -\frac{1}{\beta} i_{c2} \qquad (10.59)$$

These equations can be rearranged to obtain the usual oscillator equation,

$$LCs^2 v_1 + sL \left[ Gv_1 + \frac{1}{n}\left(1 - \frac{1}{n\beta}\right) f(v_1) \right] + v_1 = 0 \qquad (10.60)$$

A time normalization is introduced with $T = t/\sqrt{LC}$. In the normalized time domain the oscillator equation is

$$\frac{d^2 v_1}{dT^2} + \sqrt{\frac{L}{C}} \frac{d}{dT}\left[F(v_1)\right] + v_1 = 0 \qquad (10.61)$$

where

$$F(v_1) = Gv_1 + \frac{1}{n}\left(1 - \frac{1}{n\beta}\right) f(v_1) \qquad (10.62)$$

$$\approx Gv_1 + \frac{1}{n} f(v_1)$$

where the second version is valid for $n\beta \gg 1$. The oscillator equation in (10.61) has the same form as that obtained for the other oscillators of this chapter.

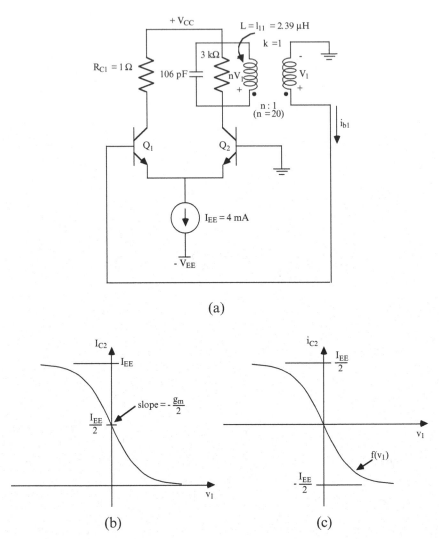

**Fig. 10.12.** (a) Transformer-coupled EC pair oscillator circuit. (b) Dc transfer characteristics of the EC pair. (c) Dc transfer characteristics with a translation of the origin.

Only the first-order term contains the nonlinearity. We can use the results obtained for the earlier oscillators of this chapter to study the performance of the present circuit.

Typical plots of the two components of $F(v_1)$ are shown in Figure 10.12e, as well as a plot of the combined total nonlinearity. Again, a shallow N-shaped characteristic is produced with peak points occurring at approximately $v_1 = \pm 0.1$ V.

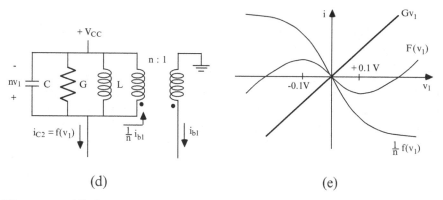

(d)                                    (e)

**Fig. 10.12.** (d) Circuit at the collector of $Q_2$. (e) Plots of $F(v_1)$ and its two components.

The starting condition of the oscillator is determined by the value of the *van der Pol parameter*, $\epsilon$, which is based on the normalized characteristic equation.

$$p^2 - \epsilon p + 1 = 0 \tag{10.63}$$

From (10.61) and (10.62),

$$\epsilon = \sqrt{\frac{L}{C}} \left[ \frac{1}{n} \left( 1 - \frac{1}{n\beta} \right) \frac{g_m}{2} - G \right] \tag{10.64}$$

$$\approx \sqrt{\frac{L}{C}} \left( \frac{g_m}{2n} - G \right)$$

$$\approx \frac{1}{Q} \left( \frac{g_m R}{2n} - 1 \right)$$

where $-g_m/2$ is the slope of $f(v_1)$ at the bias point, $R = 1/G$ and $Q = R/\sqrt{\frac{L}{C}}$. The last two forms of (10.64) are valid for large $n\beta$. In the last expression of (10.64), the $Q$ of the tuned circuit, independent of the loading of the oscillator, is introduced. If $\epsilon$ is positive and small, an exponentially increasing sinusoid results, following a small excitation. (It is to be noted that the same expression for $\epsilon$ is obtained from an incremental circuit analysis about the bias point including $G_m = \frac{g_m}{2}$, $R_i = 2r_\pi$, $L, C, R$ and the transformer turns ratio, $n$.)

From the smooth shape of $F(v_1)$, it might seem reasonable to assume that the van der Pol approximation can be applied to this ECP oscillator. On the basis of the tunnel-diode example of Section 10.5, an estimate of the amplitude of the steady-state oscillation for a small value of $\epsilon$ is somewhat larger than the value of the input at the peaks of $F(v_1)$ or the $V_x$ crossover voltages of

$F(v_1)$ ($\approx 0.1V$ for the present example).[1] However, this estimate is poor. The magnitude of the input voltage in steady-state is much larger. The problem can be appreciated by comparing the plot of $F(v_1)$ with its normalized van der Pol approximation, $F_v(v_1) = (-a_1v_1 + b_1v_1^3)$, as in Figure 10.13. Although the slopes at the equilibrium point are the same, and the crossover points are also equal, the two curves are significantly different. For the van der Pol approximation, the advent of limiting for large voltage excursions is gradual, but increasingly strong as the cubic term predominates. For the actual ECP circuit, the limiting action of the positive slope regions of the approximation, due to the linear I-V characteristic of $G$, is not as severe as that of the cubic term. We can expect, then, a larger voltage excursion for the actual circuit in order to produce the necessary loss per cycle in the steady state.

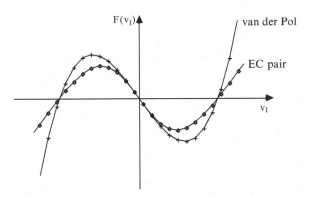

**Fig. 10.13.** Comparison of nonlinear characteristics of the EC pair and van der Pol's equation.

To investigate the steady-state amplitude, we first look at the transfer characteristic of the ECP alone, Figure 10.12b. For a small input voltage excursion, the collector current waveform is as shown in Figure 10.14a. Since both transistors are always on, we can label this Class-A operation. For a large input voltage amplitude, the output collector current is definitely clipped and can approach a square wave, as shown in Figure 10.14b. Class-AB operation is the result when one device or the other is off during a portion of the input excursion. Because of the almost linear nature of $F(v_1)$ in the active region of the devices, an increasingly growing response can be expected during the excursion across the active region. (The design of the ECP must be such that the transistors are not driven into saturation.)

As a simplification to achieve an estimate of the steady-state behavior, the waveform of the collector current in the steady state is assumed to be a square wave. The Fourier representation of a square wave is

---

[1] The output voltage at the collector of $Q_2$ is $n$ times this value.

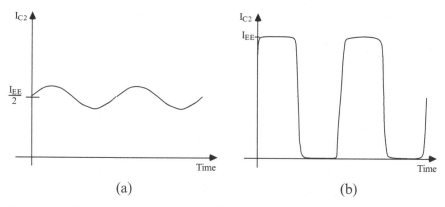

**Fig. 10.14.** Collector current for (a) small input voltage, and (b) large input voltage amplitude.

$$i_{c2}(t) = I_{EE} \left[ \frac{1}{2} + \frac{2}{\pi} \left( \cos \omega_o t + \frac{1}{3} \cos 3\omega_o t + \dots \right) \right] \tag{10.65}$$

The fundamental component has a magnitude of

$$i_{c2}(t) \mid_{\omega_o} = \frac{2}{\pi} I_{EE} = 0.64 I_{EE} \tag{10.66}$$

The estimate of the steady-state amplitude of the output voltage is therefore

$$V_o \mid_{\omega_o} \approx 0.64 I_{EE} R \tag{10.67}$$

where it is assumed that the high-Q tuned circuit, resonant at the oscillation frequency, rejects all upper harmonics. (This estimate of the fundamental can be justified from results of an analysis of the ECP with a sinusoidal input. This is brought out at the end of this section.) From the circuit values of Figure 10.12a, $I_{EE} = 4$ mA, and $R = 3$ k$\Omega$. The predicted amplitude of the output voltage is then approximately 7.7 V. The frequency of oscillation should be 10 MHz.

The Spice input file for the circuit is given in Figure 10.15a. Notice that $I_{EE}$ has a step, oscillator-excitation function. The waveforms of the output voltage and the collector current of $Q_2$ are shown in Figures 10.15b, c and d. In the first, the buildup of oscillation is shown. In Figure 10.15c, it can be assumed that a steady-state condition exists, and the output voltage amplitude is seen to be approximately 7.5 V. The waveform of the collector current in Figure 10.15d shows definite clipping although it is not a perfect square wave. From the Fourier components of the collector current, given in Figure 10.15d, the ratio of the fundamental current to $I_{EE}$ is 0.63. Also listed in Figure 10.15c are the harmonics of the output voltage for two different estimates of the period of oscillation (oscillation frequency) which differ by only

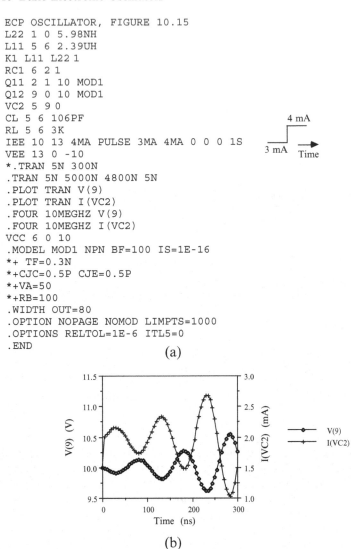

```
ECP OSCILLATOR, FIGURE 10.15
L22 1 0 5.98NH
L11 5 6 2.39UH
K1 L11 L22 1
RC1 6 2 1
Q11 2 1 10 MOD1
Q12 9 0 10 MOD1
VC2 5 9 0
CL 5 6 106PF
RL 5 6 3K
IEE 10 13 4MA PULSE 3MA 4MA 0 0 0 1S
VEE 13 0 -10
*.TRAN 5N 300N
.TRAN 5N 5000N 4800N 5N
.PLOT TRAN V(9)
.PLOT TRAN I(VC2)
.FOUR 10MEGHZ V(9)
.FOUR 10MEGHZ I(VC2)
VCC 6 0 10
.MODEL MOD1 NPN BF=100 IS=1E-16
*+ TF=0.3N
*+CJC=0.5P CJE=0.5P
*+VA=50
*+RB=100
.WIDTH OUT=80
.OPTION NOPAGE NOMOD LIMPTS=1000
.OPTIONS RELTOL=1E-6 ITL5=0
.END
```

(a)

(b)

**Fig. 10.15.** (a) Spice input file for EC pair oscillator. (b) Waveforms of the output voltage and collector current.

1%. For 10 MHz, THD $\approx$ 0.59%. Thus, the input ECP voltage is also very sinusoidal. The turns ratio of the transformer is 20. (The ratio of $L_{11}$ and $L_{22}$ is 400.) The corresponding amplitude of the input voltage of the ECP is 7.5 $rmV/20 \approx$ 0.38 V which is sufficiently large to provide a large overdrive and an approximate square wave for the output collector current.

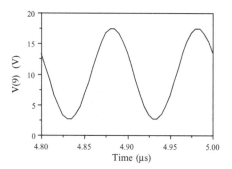

```
FOURIER COMPONENTS OF TRANSIENT RESPONSE V(9)
DC COMPONENT =    9.998D+00
HARMONIC    FREQUENCY      FOURIER      NORMALIZED      PHASE      NORMALIZED
  NO          (HZ)        COMPONENT     COMPONENT       (DEG)      PHASE (DEG)

   1       1.000D+07     7.494D+00      1.000000      152.947       0.000
   2       2.000D+07     5.042D-03      0.000673      174.092      21.145
   3       3.000D+07     4.113D-02      0.005488        8.506    -144.442
   4       4.000D+07     2.002D-03      0.000267     -169.423    -322.371

    TOTAL HARMONIC DISTORTION =           0.585557   PERCENT


FOURIER COMPONENTS OF TRANSIENT RESPONSE V(9)
DC COMPONENT =    1.000D+01
HARMONIC    FREQUENCY      FOURIER      NORMALIZED      PHASE      NORMALIZED
  NO          (HZ)        COMPONENT     COMPONENT       (DEG)      PHASE (DEG)

   1       9.990D+07     7.492D+00      1.000000      152.739       0.000
   2       1.998D+07     4.047D-03      0.000540      -12.285    -165.024
   3       2.997D+07     4.614D-02      0.006158        7.225    -145.514
   4       3.996D+07     1.354D-03      0.000181        4.307    -148.432

    TOTAL HARMONIC DISTORTION =           0.649306   PERCENT
```

**Fig. 10.15.** (c) Steady-state output voltage waveform and its Fourier components.

The design of this oscillator can proceed as follows: assume that $V_{CC}$ and $I_{EE}$ are chosen; to avoid transistor saturation, the output voltage amplitude should be less than $V_{CC}$; from Equation (10.67),

$$R \leq 1.5 \frac{V_{CC}}{I_{EE}} \qquad (10.68)$$

In the example above, $R$ should be less than 3.75 k$\Omega$. A choice of 3 k$\Omega$ was made. For circuit simulation, one does not want too large a $Q$, since excessive

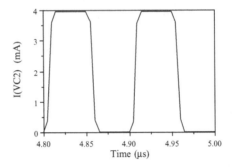

```
FOURIER COMPONENTS OF TRANSIENT RESPONSE I(VC2)
DC COMPONENT =    1.981D-03
0HARMONIC    FREQUENCY      FOURIER      NORMALIZED      PHASE      NORMALIZED
   NO          (HZ)       COMPONENT     COMPONENT       (DEG)     PHASE (DEG)

    1        1.000D+07     2.501D-03     1.000000      -27.463       0.000
    2        2.000D+07     1.021D-06     0.000408       31.527      58.990
    3        3.000D+07     7.816D-04     0.312565      -82.424     -54.961
    4        4.000D+07     8.301D-07     0.000332      -27.229       0.234

    TOTAL HARMONIC DISTORTION =          37.281453   PERCENT
```

**Fig. 10.15.** (d) Steady-state collector current waveform and its Fourier components.

CPU time can be consumed reaching the steady state. Also in practice, large $Q$s are not readily available. This limits the value of $\sqrt{L/C}$.

$$Q = \frac{R}{\sqrt{\frac{L}{C}}} \tag{10.69}$$

$$\sqrt{\frac{L}{C}} = \frac{R}{Q}$$

For the example, a $Q$ of 20 was chosen which leads to $\sqrt{L/C} = 150$. From the desired oscillation frequency,

$$f_o = \frac{1}{2\pi\sqrt{LC}} \tag{10.70}$$

$$C = \frac{1}{2\pi f_o \sqrt{\frac{L}{C}}} = 106 \text{ pF}$$

$$L_{11} = \left(\sqrt{\frac{L}{C}}\right)^2 C = 2.39 \ \mu\text{H}$$

The turns ratio of the transformer follows from a specification of $\epsilon$, the *van der Pol parameter*.

$$\epsilon = \sqrt{\frac{L}{C}} \left( \frac{I_{EE}}{4V_t n} - \frac{1}{R} \right) \tag{10.71}$$

$$n = \frac{\frac{I_{EE}}{4V_t}}{\frac{\epsilon}{\sqrt{L/C}} + \frac{1}{R}}$$

For an $\epsilon$ of approximately 0.3, the required value of $n$ is 16.6; a value of 20 is chosen and $\epsilon = 0.24$. This leads to a value $L_{22} = 2.39 \ \mu H/400 = 5.98$ nH as used in the example.

In Figure 10.16, a family of curves is plotted, a repeat of Figure 2.16, giving the values of the Fourier harmonics of the output collector current of an ECP with varying zero-to-peak sinusoidal input voltages normalized with respect to $V_t$. Notice that for large values of $V_{1A}/V_t$, the magnitude of the fundamental approaches $0.64 I_{EE}$. Values from these curves can be used in an iterative fashion to obtain a closer estimate of the output voltage of the oscillator than provided by (10.67). From an assumed output voltage, the input voltage is obtained by dividing by the turns ratio, $n$. The corresponding value of the fundamental of $I_{c2}$ is obtained from the curve of Figure 10.16. This leads to new value for the output voltage, etc. For the example above, the input drive to the ECP is sufficiently large to use the approximate 0.64 multiplier.

Referring to the ECP oscillator configuration of Figure 10.12a, we can propose another configuration in which the transformer and ECP inputs are changed as in Figure 10.17. The ECP transfer characteristic and the transformer polarity are both changed to maintain the required positive feedback for the oscillator.

On the basis of our earlier investigations of ECPs and the corresponding source-coupled pairs, it is clear that transformer-coupled SCP oscillators can also be readily designed. Although the voltage-to-current transfer characteristic of the SCP does not limit as sharply as for the ECP, the square wave approximation to the drain current can be used to obtain an estimate of the steady-state amplitude of the SCP oscillator. Additional details on the analysis of transformer-coupled SCP oscillators can be found in [29].

## 10.8 Transformerless ECP Oscillators

As brought out in Chapters 7 and 9, transformers have both a lower and an upper frequency limit. The latter is due to the inherent resonant frequency of the magnetizing inductance and the capacitances of the coil windings. A single winding coil usually has a higher resonant frequency. In this section two ECP oscillator configurations are introduced which eliminate the need for the transformer feedback connection. In the first, a capacitor divider is used to achieve the feedback connection. In the other a direct dc connection is used.

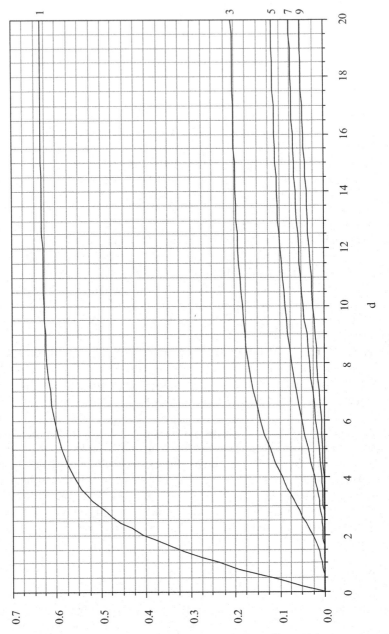

**Fig. 10.16.** Normalized harmonics of the collector current.

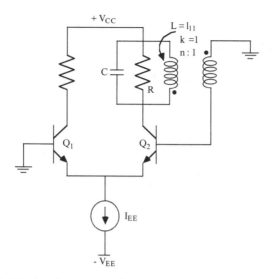

**Fig. 10.17.** An alternate configuration for the EC pair oscillator.

In Figure 10.18a, an ECP oscillator configuration is shown in which the capacitance of the tuned circuit is the series combination of two capacitors, $C_1$ and $C_2$. The interior connection is returned to the base of transistor $Q_1$. In Figure 10.18b, the capacitor subcircuit is shown. Assume for the moment that the input voltage $v_a$ is a pure sinusoid. If no load element is present across $C_2$, the amplitude of the voltage $v_b$ is

$$v_b = \frac{C_1}{C_1 + C_2} v_a \tag{10.72}$$

$C_1$ and $C_2$ in the sinusoidal steady state produce a voltage divider. This divider can be modeled with a step-down transformer as illustrated in Figure 10.18c. The capacitors do not transfer dc voltages or currents, but do provide an ac connection as well as resistance transformation. It is readily shown that if a resistance $R_2$ is placed across $C_2$, the input resistance seen by $v_1$ is

$$R_{in} = n_e^2 R_2 \tag{10.73}$$

where

$$n_e = \frac{C_1 + C_2}{C_1} \tag{10.74}$$

This relation can also be proven by circuit analysis and illustrated by circuit simulation, where the latter is useful to substantiate the transformation for more complicated load situations.

In the ECP oscillator, a bias return resistance, $R_{B1}$, must be used at the base of $Q_1$. To achieve a balanced dc state, an equal resistance, $R_{B2}$, may be

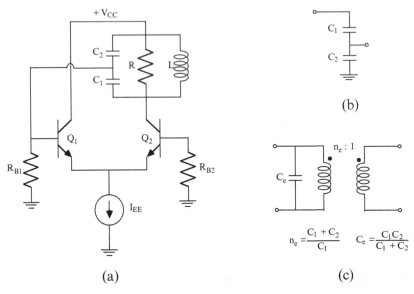

$$n_e = \frac{C_1 + C_2}{C_1} \qquad C_e = \frac{C_1 C_2}{C_1 + C_2}$$

(a)                              (c)

```
ECP OSCILLATOR WITH C DIVIDER, FIGURE 10.18
LL 6 5 2.39UH
RC1 6 2 1
RB1 1 0 1K
Q11 2 1 10 MOD1
Q12 5 7 10 MOD1
RB2 7 0 1
VCC 6 0 10
CL2 6 1 2120P
CL1 1 5 111.58P
RL 6 5 3K
IEE 10 13 4MA PULSE 3MA 4MA 0 0 0 1S
VEE 13 0 -10
*.TRAN 5N 300N
.TRAN 5N 5000N 4800N 5N
.PLOT TRAN V(9)
.PLOT TRAN I(VC2)
.FOUR 10MEGHZ V(5)
.MODEL MOD1 NPN BF=100 IS=1E-16
*+ TF=0.3N
*+CJC=0.5P CJE=0.5P
*+VA=50
*+RB=100
.WIDTH OUT=80
.OPTION NOPAGE NOMOD LIMPTS=1000
.OPTIONS RELTOL=1E-4 ITL5=0
.END
```

(d)

**Fig. 10.18.** (a) An EC pair oscillator. (b) The capacitor divider subcircuit. (c) An equivalent circuit model for the capacitor divider. (d) Spice input file for EC pair oscillator.

placed from the base of $Q_2$ to ground. In order not to affect the nonlinear transfer function of the ECP, it is best to bypass $R_{B2}$ with a capacitor. As brought out in Section 11.6, the value of this bypass capacitor cannot be too large. The loading of resistance $R_{B1}$ on the tuned circuit of the ECP oscillator can be included in the oscillator equation by replacing $R_L$ with $R_L \| n_e^2 R_{B1}$. All of the results of Section 10.7 can be now used directly.

In Figure 10.18d, the Spice input file for the modification of the ECP oscillator of the last section is shown. The transformer is deleted and the capacitor voltage divider is used. For bias return resistors, $R_{B1} = R_{B2} = 1\,k\Omega$, and the latter is not bypassed, for convenience. The collector voltage waveform and its harmonic content are given in Figure 10.18e. Approximately the same performance is achieved as for the ECP oscillator of the last section.

In Section 11.1, it is shown that the bias resistor $R_{B2}$ is not needed. Although the dc bias state is not the same as for the balanced case, a dynamic shift of the bias state occurs and often approximately the same steady-state behavior of the oscillator is obtained. (See also the results with the van der Pol oscillators of Section 10.5.) For the circuit of Figure 10.18a, the output voltage waveform and its harmonic content are given in Figure 10.18f for the case where $R_{B2} = 1\,\Omega$. Little change from the earlier results is noted.

In Figure 10.19a, a configuration of an interesting ECP oscillator is shown. IC designers from the Sony Corporation first published the use of this configuration in 1982 [28]. Note that an ECP is the basis of the oscillator and that dc coupling exists from each collector to the base of the other transistor. In effect, the basic configuration without the tuned circuit is a dc bistable configuration related, but not identical, to the famous Schmitt circuit.

To evaluate and design this 'Sony Oscillator', we establish the output I-V characteristic of the circuit looking back into the circuit from the tuned circuit. In effect, we establish the output conductance characteristic into this node pair. If the base currents are neglected, the circuit equations are:

$$V_{BE1} = V_o - V_e \tag{10.75}$$
$$V_{BE2} = V_{CC} - V_e$$
$$V_{BE1} - V_{BE2} = V_o - V_{CC} = V_o'$$
$$I_{C1} + I_{C2} = I_{EE}$$

where $V_o'$ is the voltage across the tuned circuit. Solution of these equations provides:

$$I_o = I_{C2} = f(V_o) = \frac{I_{EE}}{2}\left[1 - \tanh\left(\frac{V_o'}{2V_t}\right)\right] \tag{10.76}$$

$I_o - V_o'$ is the desired output I-V characteristic and has the same form as the transfer characteristic of the ECP in Chapter 2. A plot of this characteristic is drawn in Figure 10.19b. Of critical importance again is the region

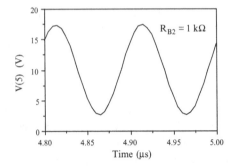

```
FOURIER COMPONENTS OF TRANSIENT RESPONSE V(5)
DC COMPONENT =    9.971D+00
HARMONIC   FREQUENCY    FOURIER     NORMALIZED     PHASE      NORMALIZED
   NO        (HZ)      COMPONENT    COMPONENT      (DEG)     PHASE (DEG)

    1      1.000D+07   7.344D+00    1.000000      36.627        0.000
    2      2.000D+07   6.488D-02    0.008834      28.310       -8.317
    3      3.000D+07   6.608D-02    0.008997      12.997      -23.630
    4      4.000D+07   2.483D-02    0.003381      26.112      -10.515

    TOTAL HARMONIC DISTORTION =           1.388551   PERCENT
```

(e)

```
FOURIER COMPONENTS OF TRANSIENT RESPONSE V(5)
DC COMPONENT =    9.971D+00
HARMONIC   FREQUENCY    FOURIER     NORMALIZED     PHASE      NORMALIZED
   NO        (HZ)      COMPONENT    COMPONENT      (DEG)     PHASE (DEG)

    1      1.000D+07   7.371D+00    1.000000      36.518        0.000
    2      2.000D+07   5.692D-02    0.007722      16.329      -20.188
    3      3.000D+07   7.308D-02    0.009914      11.025      -25.493
    4      4.000D+07   2.771D-02    0.003759      21.587      -14.931

    TOTAL HARMONIC DISTORTION =           1.413080   PERCENT
```

(f)

**Fig. 10.18.** (e) Collector voltage waveform and its Fourier components for (e) $R_{B2}$ = 1 k$\Omega$, and (f) $R_{B2}$ = 1 $\Omega$.

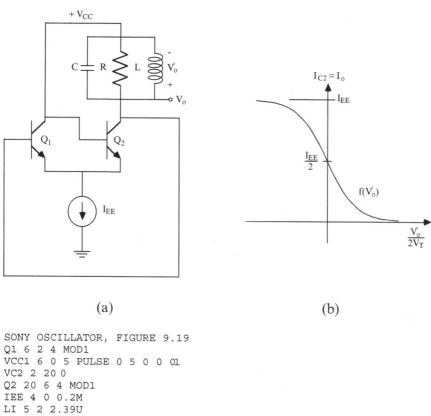

(a)                                    (b)

```
SONY OSCILLATOR, FIGURE 9.19
Q1 6 2 4 MOD1
VCC1 6 0 5 PULSE 0 5 0 0 01
VC2 2 20 0
Q2 20 6 4 MOD1
IEE 4 0 0.2M
LI 5 2 2.39U
CI 5 2 106P
RI 5 2 5K
.MODEL MOD1 NPN BF=100 IS=1E-16
+RB=50
*+VA=50
VCC 5 0 5
.TRAN 5N 1U 0.8U
.FOUR 10MEG V(2)
.PLOT TRAN V(2,4)
.PLOT TRAN I(VC2)
.WIDTH OUT=80
.END
```

(c)

**Fig. 10.19.** (a) Another EC pair oscillator circuit. (b) Dc characteristics of the circuit. (c) Spice input file.

of the characteristic with a negative slope. In effect, the input conductance looking into the ECP from the output node pair is negative in this region. It is possible to develop an oscillator using the negative-conductance concepts of the tunnel-diode oscillator, as brought out in Section 10.3.

Although the oscillator of Figure 10.19a can be now described as a negative-conductance oscillator, the basic oscillator differential equation is the same as those of the other oscillators of this chapter, cf., (10.15) and (10.61). Using the latter, we can design an oscillator based upon the choices of the ECP oscillator of the last section for the van der Pol parameter and the steady-state frequency.

The Spice input file for the circuit of Figure 10.19a is given in Figure 10.19c. The expression for $\epsilon$ is $\epsilon = \sqrt{L/C}[I_{EE}/(4V_t) - 1/R_L]$. The values of $L$ and $C$ are the same as those used in the ECP oscillator of the last section. For $R_L = 0.8$ k$\Omega$, $\epsilon = 0.10$, the waveform of the output voltage is shown in Figure 10.19d. Clearly, a very sinusoidal output is produced with an output-voltage amplitude across the tuned circuit of approximately 0.092V. Note that $THD = 1.2\%$. The waveform for the collector current of $Q_2$ is shown in Figure 10.19e.

If $R_L$ is increased to 5 k$\Omega$, $\epsilon$ is still small (0.26), but the output voltage amplitude has increased to 0.65 V. The waveform is still very sinusoidal as shown in Figure 10.19f. The waveform of the collector current, $I_{C2}$, is shown in Figure 10.19g. This waveform is almost a rectangular pulse train. An overdrive situation exists as is encountered with the ECP oscillators of Section 10.7. Again, the high-Q tuned circuit provides the sinusoidal output voltage with $THD = 0.61\%$.

Returning to the original configuration of the Sony circuit, note that the dc voltage supply can be translated to produce a ground at the top of the tuned circuit. This is a very important practical consideration since parasitic elements can be eliminated. Oscillators with a tuned circuit which has one node grounded are called 'one-pin' oscillators.

In Chapter 12, it is shown that a relaxation (nonsinusoidal) oscillator can be realized starting from the configuration of a basic dc bistable circuit. With appropriate biasing changes and resistive loading, a sinusoidal oscillation can also be achieved. This aspect is illustrated with a variant of the Sony circuit, as shown in Figure 10.20a. As brought out in Chapter 12, a loop-coupled, dc bistable circuit is the same basic circuit as that of an emitter-coupled dc bistable circuit with a change of the ground point of the circuit and with minor changes in the biasing elements.

In the circuit of Figure 10.20a, the basic configuration is the loop-coupled dc bistable circuit. The 'output' node pair of this circuit is chosen to be the collector-collector node pair of the circuit. Again the input I-V characteristic of this node pair can be obtained from simple circuit analysis. The circuit equations and the final result, neglecting the base currents, are:

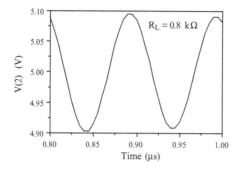

FOURIER COMPONENTS OF TRANSIENT RESPONSE V(2)
DC COMPONENT =    4.999D+00

| HARMONIC NO | FREQUENCY (HZ) | FOURIER COMPONENT | NORMALIZED COMPONENT | PHASE (DEG) | NORMALIZED PHASE (DEG) |
|---|---|---|---|---|---|
| 1 | 1.000D+07 | 9.226D-02 | 1.000000 | 115.304 | 0.000 |
| 2 | 2.000D+07 | 8.850D-04 | 0.009592 | 49.135 | -66.169 |
| 3 | 3.000D+07 | 5.184D-04 | 0.005619 | -74.647 | -189.951 |
| 4 | 4.000D+07 | 2.778D-04 | 0.003011 | 39.009 | -76.295 |

TOTAL HARMONIC DISTORTION =    1.225898  PERCENT

(d)

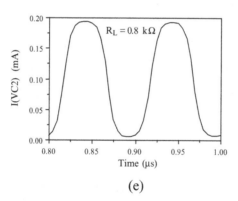

(e)

**Fig. 10.19.** (d) Output voltage waveform and its Fourier components. (e) Collector current waveform.

$$I_o = I_{C1} - I_1 = -I_{C2} + I_1 \tag{10.77}$$
$$V_o = V_{BE1} - V_{BE2}$$

$$I_o = f(V_o) = -I_1 \tanh\left(\frac{V_o}{2V_t}\right) \tag{10.78}$$

The same type of negative conductance characteristic is obtained, differing from the original Sony circuit only in a dc shift. The Spice input file of the circuit of Figure 10.20a is given in Figure 10.20b. Note that the dc collector

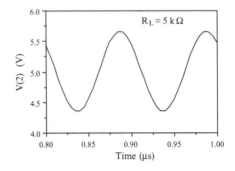

```
FOURIER COMPONENTS OF TRANSIENT RESPONSE V(2)
DC COMPONENT =    4.997D+00
HARMONIC    FREQUENCY      FOURIER      NORMALIZED      PHASE      NORMALIZED
  NO          (HZ)       COMPONENT     COMPONENT       (DEG)     PHASE  (DEG)

   1       1.000D+07     6.513D-01     1.000000      134.576       0.000
   2       2.000D+07     3.278D-03     0.005033      142.392       7.816
   3       3.000D+07     1.109D-03     0.001703      -85.537    -220.113
   4       4.000D+07     8.833D-04     0.001356      173.600      39.024

     TOTAL HARMONIC DISTORTION =              0.607797   PERCENT
```

(f)

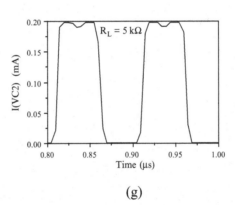

(g)

**Fig. 10.19.** (f) Output voltage waveform and its Fourier components for $R_L = 5$ k$\Omega$. (g) Collector current waveform for $R_L = 5$ k$\Omega$.

current sources are realized with $V_{CC} - R_C$ combinations. The total load resistance for the tuned circuit is $R_L \| 2R_C$. The output voltage waveform for $R_L = 1$ k$\Omega$ providing $\epsilon = 0.14$ is shown in Figure 10.20c and is comparable to that produced by the original Sony circuit.

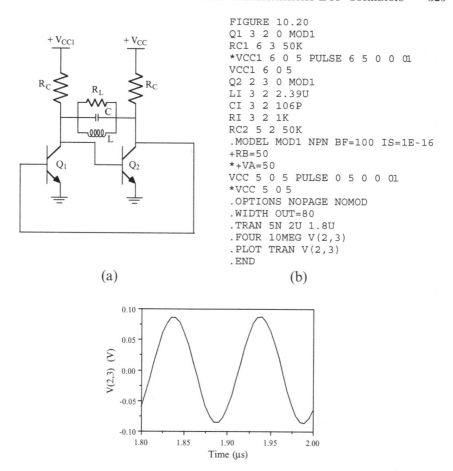

FIGURE 10.20
```
Q1 3 2 0 MOD1
RC1 6 3 50K
*VCC1 6 0 5 PULSE 6 5 0 0 01
VCC1 6 0 5
Q2 2 3 0 MOD1
LI 3 2 2.39U
CI 3 2 106P
RI 3 2 1K
RC2 5 2 50K
.MODEL MOD1 NPN BF=100 IS=1E-16
+RB=50
*+VA=50
VCC 5 0 5 PULSE 0 5 0 0 01
*VCC 5 0 5
.OPTIONS NOPAGE NOMOD
.WIDTH OUT=80
.TRAN 5N 2U 1.8U
.FOUR 10MEG V(2,3)
.PLOT TRAN V(2,3)
.END
```

(a)                                    (b)

FOURIER COMPONENTS OF TRANSIENT RESPONSE V(2,3)
DC COMPONENT =   3.284D-04

| HARMONIC NO | FREQUENCY (HZ) | FOURIER COMPONENT | NORMALIZED COMPONENT | PHASE (DEG) | NORMALIZED PHASE (DEG) |
|---|---|---|---|---|---|
| 1 | 1.000D+07 | 8.704D-02 | 1.000000 | -47.725 | 0.000 |
| 2 | 2.000D+07 | 5.317D-04 | 0.006109 | -16.557 | 31.169 |
| 3 | 3.000D+07 | 4.863D-04 | 0.005588 | 101.574 | 149.299 |
| 4 | 4.000D+07 | 1.917D-04 | 0.002202 | 6.240 | 53.965 |

TOTAL HARMONIC DISTORTION =     0.925347 PERCENT

(c)

**Fig. 10.20.** (a) Another oscillator circuit. (b) Spice input file. (c) Output voltage waveform and its Fourier components.

## Problems

**10.1.** A *RC* oscillator is shown in Figure 10.21.
(a) Find $R_x$ to achieve a van der Pol parameter $\epsilon$ of 0.1.
(b) Estimate the frequency of the buildup of oscillation. Estimate the frequency of steady-state operation.
(c) Estimate and justify the amplitude of the output voltage for steady-state operation.
(d) What is the amplitude of $V_{in}$ in steady-state operation?

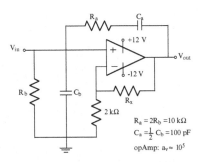

**Fig. 10.21.** *RC* oscillator circuit for Problem 10.1

**10.2.** An EC pair oscillator with transformer feedback is shown in Figure 10.22.
(a) Choose the polarity of the transformer to achieve positive feedback.
(b) Determine whether oscillations will build up from the quiescent state and, if so, with what van der Pol parameter.
(c) Assuming oscillations build up to steady state, estimate the value of the steady-state amplitude of the output voltage.
(d) Verify your analysis results with Spice.

**10.3.** An EC pair oscillator is shown in Figure 10.23a.
(a) Design the oscillator to achieve a steady-state output with a frequency of 10 MHz. The maximum turns ratio for the transformer is 10. Estimate and confirm the value of the steady-state output voltage.
(b) Use Spice simulation to investigate the effects of the BJT parameters $V_A = 100$ V, $R_B = 100\Omega$, and $C_{JC} = 1$ pF.
(c) Can oscillation be achieved if one winding of the transformer is reversed and feedback connection is made to the base of $Q_2$ as shown in Figure 10.23b. If so, how do your results compare with those of Part (a).
(d) Replace the BJTs of Figure 10.23a with MOS devices, having the size and parameter values of Figure 10.23c. Adjust the circuit parameters to achieve oscillations. Compare your results with those of Part (a).

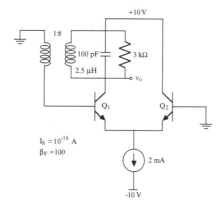

**Fig. 10.22.** EC pair oscillator circuit for Problem 10.2

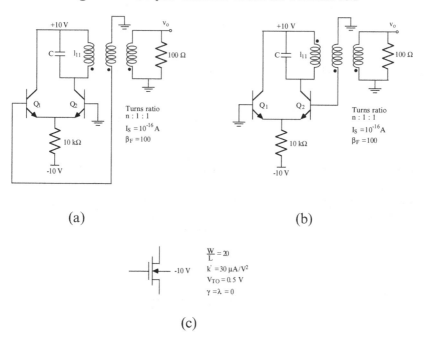

(a)                                              (b)

(c)

**Fig. 10.23.** EC pair oscillator circuit for Problem 10.3

**10.4.** An SC pair oscillator with transformer feedback is shown in Figure 10.24.

(a) Design this oscillator for a frequency of 10 MHz. The $Q$ of the tuned circuit is 20 and the desired value of the van der Pol parameter is 0.25.

(b) Estimate the value of the steady-state amplitude of the fundamental of the output voltage.

(c) Verify your analysis results with Spice.

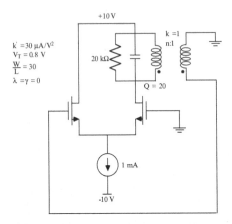

**Fig. 10.24.** SC pair oscillator circuit for Problem 10.4

**10.5.** An emitter-coupled pair oscillator is shown in Figure 10.25.
(a) Determine whether oscillations will build up from the quiescent state and, if so, with what van der Pol parameter.
(b) What is the frequency of steady-state oscillations?
(c) Estimate the value of the steady-state amplitude of the fundamental of the output voltage.
(d) Verify your analysis results with Spice.
(e) Using Spice, estimate the effects of charge storage in the BJT on the build up and steady-state performance of the oscillator. Take $CJC = CJE = 0.02$ pF and $TF = 0.1$ ns.
(f) What changes would be required in your design to keep the performance with charge storage the same as that without charge storage.

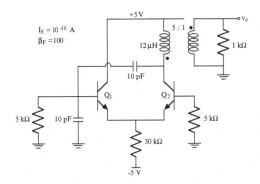

**Fig. 10.25.** EC pair oscillator circuit for Problem 10.5

**10.6.** A source-coupled pair oscillator is shown in Figure 10.26.
(a) Design the oscillator to achieve a steady-state output with a frequency
of 10 MHz. The $Q$ of the tuned circuit including the load resistance cannot
exceed 20. Justify your choice of the ratio of $C_a$ and $C_b$ and of the values of
$R_{G1}$ and $R_{G2}$.
(b) Can $R_{G2}$ be reduced to zero?
(c) Estimate and confirm the value of the steady-state output voltage.
(d) Use Spice simulation to investigate the effects of MOS parameters $\lambda =$
0.02, $C_{gd0} = C_{gs0} = 50$ fF.

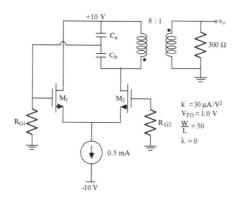

**Fig. 10.26.** Source-coupled pair oscillator circuit for Problem 10.6.

**10.7.** An oscillator circuit is shown in Figure 10.27.
(a) What is the minimum value of the current $I_{SS}$ for oscillations to build up
from the quiescent state?
(b) What is the frequency of steady-state oscillations?
(c) What is the $Q$ of the tuned circuit at the drain of the MOSFETs.
(d) Will oscillations build up from the quiescent state for $I_{SS} = 6$ mA? If so,
estimate the value of the steady-state amplitude of the fundamental of the
output voltage.
(e) Verify your analysis results with Spice.
(f) What is the minimum value of the current $I_{SS}$ for oscillations to build
up from the quiescent state, when the MOSFETs are replaced by BJTs with
$\beta = 100$ and $I_S = 10^{-16}$A?

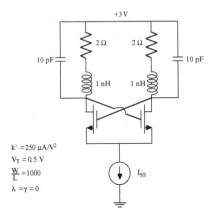

**Fig. 10.27.** Oscillator circuit for Problem 10.7.

# Electronic Oscillators with Bias-Shift Limiting

## 11.1 Bias Shift during Oscillator Buildup in an ECP Oscillator

In this chapter a major, yet simple oscillator prototype is studied, together with several derivative configurations. Of particular interest is the fact that although the basic differential equation which describes the oscillator has the same form as that encountered in the last chapter, there is now a new limiting phenomenon. The essential limiting of the growth of oscillatory buildup is due to the shift of the virtual or dynamic operating point of the configuration. That is, the effective bias point of the oscillator shifts from the initial values at the quiescent state in the active region of the device(s) to equivalent bias values corresponding to a near-off state of the devices. Steady-state operation of the oscillator corresponds to Class-C operation wherein the devices are off during more than one-half of the period of the oscillation.

In the next section, an alternate oscillator equation is established which is helpful in studying the new prototype and its derivatives. First, in this section, an ECP oscillator similar to that of Section 10.7 is examined when an offset bias voltage is applied. A definite and favorable shift of the virtual operating point of the oscillator is produced as oscillations build up to a steady state. An understanding of this shift aids us in the study of the new oscillators of this chapter.

In Figure 11.1a, a transformer-coupled oscillator is shown including a bias voltage source, $V_{B2}$, at the base of transistor $Q_2$. This voltage source provides an offset voltage for the balanced pair. The operating point of the input-voltage-to-output-current transfer characteristic moves to the left with the offset, as illustrated in Figure 11.1b. (Alternately, the characteristic moves down and to the right with the offset.) At the new operating point with an offset voltage of $V_{B2}$, the collector current of $Q_2$ and the transconductance of the ECP are

D.O. Pederson and K. Mayaram, *Analog Integrated Circuits for Communication*, DOI 10.1007/978-0-387-68030-9_11,
© 2008 Springer Science+Business Media, LLC

$$I_{C2} = \frac{I_{EE}}{1 + \exp(-d)} \tag{11.1}$$

$$G_m = \frac{dI_{C2}}{dV_1} = -\frac{I_{EE}}{V_T}\left(\frac{\exp(-d)}{[1 + \exp(-d)]^2}\right)$$

where $d$ is the normalized offset voltage $= \frac{V_{B2}}{V_t}$. For a bias offset of 60 mV, a thermal voltage of 25.85 mV, and an emitter source current of $I_{EE} = 4$ mA, $d = 2.32, I_{C2} = 3.64$ mA and $G_m = 0.0124$ ℧ assuming that $\beta \gg 1$. The values at the original bias point without offset are $d = 0, I_{C2} = 2$ mA and $G_m = g_m/2 = 0.0387$ ℧.

For the total nonlinear characteristic, $F(v_1)$, for the oscillator a corresponding shift of the quiescent bias point is produced as sketched in Figure 11.1c. Clearly, the quiescent bias point is very close to one of the off regions of the characteristic, corresponding to $Q_1$ off and $Q_2$ full on, and the value of $I_{C2}$ is close to $I_{EE}$. Because the slope of the characteristic is smaller in magnitude at the shifted bias point, the value of the *van der Pol parameter* $\epsilon$ is smaller unless the oscillator is redesigned. In terms of the equations of Section 10.7, cf., (10.61),

$$\epsilon = \frac{1}{Q}\left(\frac{G_m R_L}{n} - 1\right)$$

For the original balanced oscillator, $\epsilon = 0.24$. At the new quiescent bias point, $\epsilon = 0.043$. Since $\epsilon > 0$ oscillations will build up. After some buildup, note that the positive excursions of $v_1$ remain mainly in the negative slope region while the negative excursions soon encounter the positive slope region on the left. It can be expected, then, that the negative portions of $i_{C2}$ will continue to grow while the positive portions will be clamped to $I_{EE}$. As in the previous large-signal studies of amplifiers, a shift of the effective dc or average value of the collector current is produced because of the growing negative portions of the current waveform. This constitutes a shift of the bias state. Simulations of the oscillator illustrate these aspects easily.

In Figure 11.1d, the Spice input file is given. This file pertains to the same circuit as studied in Section 10.7 except for the offset voltage $V_{B2}$. Several new control lines are also included. The dc transfer characteristic is obtained by introducing a voltage source $V_{B1}$ while 'removing' the inductance $l_{22}$ and is shown in Figure 11.1e. The values for a .TF run are also shown in the figure. As estimated above, $G_m$ at the quiescent operating point is 0.012 ℧.

With $l_{11}$ reintroduced, the growth of the oscillation for $I_{C2}$ is shown in Figure 11.1f. It is clear that the positive excursions of $I_{C2}$ clamp to $I_{EE}$. From the harmonic components of Figure 11.1g, the dc component of $I_{C2}$ has moved toward $\frac{I_{EE}}{2}$. In Figure 11.1h, the growth of the ECP input voltage is seen to remain a sinusoidal growth, while the waveform of $V_{BE1} = V(1, 10)$ becomes very distorted. Waveforms of $V_o = V(9)$ and $I_{C2} = I(VC2)$ are shown in Figures 11.1i and j for a near steady-state condition. The harmonic

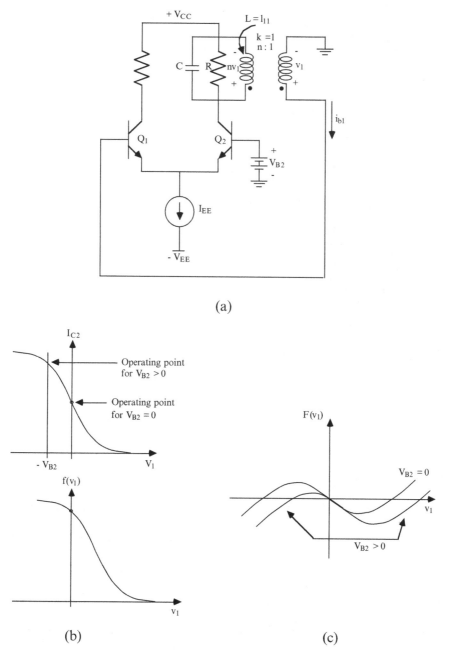

(a)

(b)                                    (c)

**Fig. 11.1.** (a) A transformer-coupled oscillator. (b) Movement of the operating point with offset. (c) The total nonlinear characteristic.

```
ECP OSCILLATOR WITH BIAS OFFSET
*L22 1 0 5.98NH
L11 5 6 2.39UH
*K1 L11 L22 1
VB1 1 0 0
RC1 6 2 1
Q11 2 1 10 MOD1
Q12 9 20 10 MOD1        .
VB2 20 0 0.06
VC2 5 9 0
CL 5 6 106PF
RL 5 6 3K
IEE 10 13 4MA
.TF I(VC2) VB1
.DC VB1 -0.2 0.2 .01
.PLOT DC I(VC2)
*IEE 10 13 4MA PULSE 3MA 4MA 0 0 0 1S
*VEE 13 0 -10
*.TRAN 10N 3000N 0 5N
*.PLOT TRAN V(1,10) V(9) I(VC2)
*.TRAN 5N 5000N 4800N 5N
*.FOUR 10MEGHZ V(9) V(1,10) I(VC2)
VCC 6 0 10
.MODEL MOD1 NPN BF=100 IS=1E-16
*+ TF=0.3N
*+CJC=0.5P CJE=0.5P
*+VA=50
*+RB=100
.OPTIONS RELTOL=1E-4 ITL5=0 LIMPTS=1000
.END
```

(d)

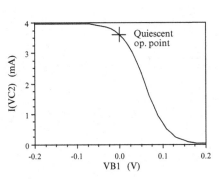

```
SMALL-SIGNAL CHARACTERISTICS

   I(VC2)/VB1                              = -1.248D-02
   INPUT RESISTANCE AT VB1                 =  8.014D+03
   OUTPUT RESISTANCE AT I(VC2)             =  9.103D+11
```

(e)

**Fig. 11.1.** (d) Spice input file. (e) Dc transfer and small-signal characteristics.

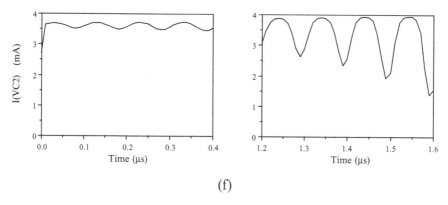

(f)

```
FOURIER COMPONENTS OF TRANSIENT RESPONSE I(VC2)
DC COMPONENT =    2.237D-03
HARMONIC   FREQUENCY    FOURIER      NORMALIZED     PHASE     NORMALIZED
  NO         (HZ)      COMPONENT     COMPONENT      (DEG)     PHASE (DEG)

   1      1.000D+07    2.423D-03     1.000000      -98.568       0.000
   2      2.000D+07    4.599D-04     0.189826     -106.685      -8.117
   3      3.000D+07    5.842D-04     0.241149       63.655     162.223
   4      4.000D+07    3.337D-04     0.137735       55.957     154.525
   5      5.000D+07    1.768D-04     0.072993     -137.352     -38.785
   6      6.000D+07    1.988D-04     0.082057     -143.630     -45.062
   7      7.000D+07    3.969D-05     0.016383        6.194     104.762
   8      8.000D+07    1.005D-04     0.041502       10.042     108.610
   9      9.000D+07    9.800D-06     0.004045       93.317     191.884

   TOTAL HARMONIC DISTORTION =     35.668725  PERCENT
```

(g)

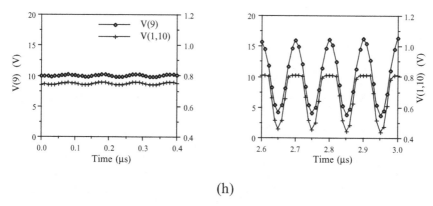

(h)

**Fig. 11.1.** (f) Growth of oscillation for $I_{C2}$. (g) Fourier components of $I_{C2}$. (h) Growth of the input voltages.

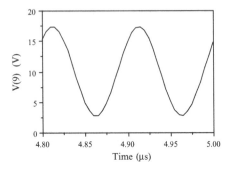

```
FOURIER COMPONENTS OF TRANSIENT RESPONSE V(9)
DC COMPONENT =    9.968D+00
HARMONIC    FREQUENCY      FOURIER     NORMALIZED      PHASE       NORMALIZED
  NO          (HZ)        COMPONENT    COMPONENT      (DEG)       PHASE (DEG)

   1       1.000D+07      7.302D+00    1.000000       41.153        0.000
   2       2.000D+07      4.402D-02    0.006029      -19.006      -60.159
   3       3.000D+07      6.390D-02    0.008751       23.776      -17.377
   4       4.000D+07      2.878D-02    0.003941       12.966      -28.186

      TOTAL HARMONIC DISTORTION =        1.208141   PERCENT
```

(i)

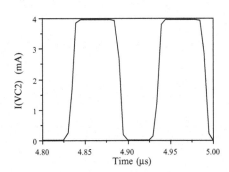

```
FOURIER COMPONENTS OF TRANSIENT RESPONSE I(VC2)
DC COMPONENT =    2.200D-03
HARMONIC    FREQUENCY      FOURIER     NORMALIZED      PHASE       NORMALIZED
  NO          (HZ)        COMPONENT    COMPONENT      (DEG)       PHASE (DEG)

   1       1.000D+07      2.448D-03    1.000000     -139.746        0.000
   2       2.000D+07      4.050D-04    0.165453      169.998      309.744
   3       3.000D+07      6.427D-04    0.262562      -59.409       80.338
   4       4.000D+07      3.205D-04    0.130933     -110.751       28.995

      TOTAL HARMONIC DISTORTION =       36.738048   PERCENT
```

(j)

**Fig. 11.1.** Steady-state waveform and Fourier components of (i) output voltage, and (j) collector current.

components of the waveforms are also given. The ratio of the amplitude of the fundamental of $I_{C2}$ in the steady state to the value of $I_{EE} = 4$ mA is 0.61. This is close to the ratio of 0.64 from the simple square-wave estimate of Section 10.7 for a balanced ECP and to the ratio 0.63 for the actual simulated balanced oscillator.

Because of the constraints imposed by the circuit configuration and its bias supplies including the fact that $F(v_1)$, the net or total nonlinearity of the oscillator, is not antisymmetrical about the quiescent operating point, a bias shift occurs which provides very effective steady-state operation. In developing the differential equation of this oscillator in Section 10.7, cf., (10.57) through (10.62), a translation is made of the transfer characteristic of the ECP. This involves the quiescent bias state of the circuit. As mentioned for the tunnel-diode oscillator in Section 10.3, and as noted above, the effective bias point of the oscillator must shift as the oscillation grows. Thus, a new term or consideration must be added. This is developed further in the next section.

## 11.2 The Basic Oscillator Equation

In the last chapter, three seemingly different oscillator configurations are analyzed and evaluated: a negative-conductance (tunnel-diode) circuit, an $RC$ feedback, positive gain block (Wien-type) configuration, and a transformer-feedback, emitter-coupled pair arrangement. With simple device models, all three are described by the same nonlinear differential equation and produce the same linearized characteristic equation at the equilibrium point (the quiescent bias point). However, the nature of the total nonlinearity of the oscillator for each is different, leading to different methods to estimate the steady-state oscillation magnitude. In addition, the selectivity of the tuned circuits is not the same for the three oscillators, even though each can be designed to provide the same natural frequencies of the linearized circuit at the equilibrium point. This aspect is developed further in this chapter.

By a slight manipulation of the differential equation describing the oscillators, another basic form can be established that is helpful in studying other configurations. The tunnel-diode, negative-conductance configuration is used to illustrate the procedure.

For the tunnel-diode oscillator, used as an example, from Section 10.3 and Figure 10.2a, the circuit equation is now used without translated variables.

$$\frac{1}{L} \int (V - V_{DD})\, dt + GV + C\frac{dV}{dt} + f(V) = 0 \tag{11.2}$$

Notice that the bias voltage source is included. Multiply through by $L$ and take the derivative of this equation with respect to time.

$$V - V_{DD} + GL\frac{dV}{dt} + LC\frac{d^2V}{dt^2} + L\frac{d}{dt}f(V) = 0 \tag{11.3}$$

In operator form, this equation, after some manipulation, becomes

$$(LCs^2 + LGs + 1)V + Lsf(V) = V_{DD} \tag{11.4}$$

Alternatively, the following can be obtained.

$$V + N(s)f(V) = N_o(s)V_{DD} \tag{11.5}$$

where

$$N(s) = \frac{Ls}{LCs^2 + LGs + 1} \tag{11.6}$$

$$N_o(s) = \frac{1}{LCs^2 + LGs + 1}$$

The oscillator equation in the form of Equation (11.5) is defined as the 'total basic oscillator equation.' It pertains to the class of oscillators studied in the last chapter, where one nonlinear function is encountered in the first-order term of the differential equation. In particular, take note in Equation (11.5) that the nonlinear function appears only once and is operated on by a linear time-invariant network function.

Both $N(s)$ and $N_o(s)$ pertain to passive networks. The right-hand side of Equation (11.5) represents the transient response of the circuit when the power supply $V_{DD}$ is applied. Considered separately, this function in the transform domain represents the response of a passive system (poles in the left-half plane) and for large elapsed time (the steady-state, i.e., $s = 0$) becomes simply $V_{DD}$, since $N_o(0) = 0$.

The 'basic oscillator equation for the steady-state,' or just the 'basic oscillator equation' then has the form

$$x + N(s)f(x) = A_o \tag{11.7}$$

where $f(x)$ is a real nonlinear function of the variable $x$, $N(s)$ is the network function of a linear, time-invariant network, and $A_o$ is a 'constant' and represents the dc power source to the oscillator in the steady state. (In the case of an effective bias point shift, $A_o$ becomes a function of the dc component of a steady-state oscillator variable. This is taken up in the next section.)

The 'basic oscillator equation' can be represented by the block diagram of Figure 11.2a. The oscillator feedback configuration contains a nonlinearity, $f(x)$, the linear, time-invariant circuit characterized by the 'transfer function,' $N(s)$, and the power source, $A_o$. For near-harmonic oscillations, $N(s)$ needs to provide a magnitude function $|N(j\omega)|$ with a peaked response at $j\omega_o$ as illustrated in Figure 11.2b. The loop gain of the system at this frequency must have a magnitude greater than one with a zero-degree phase. The magnitude response of the loop gain, due to $|N(j\omega)|$, should fall off rapidly away from $j\omega_o$, so that other frequencies are not provided with a condition for oscillatory buildup. In addition, the loop gain should not support appreciable harmonics. In this regard, the $RC$ Wien-type feedback circuit, studied in Section 10.6, does provide a bandpass response with transmission zeros at the origin and at infinity. The basic oscillator equation for this Wien-type oscillator is

$$V_1 + \frac{-s/RC}{s^2 + 3s/RC + (1/RC)^2} f(V_1) = 0 \tag{11.8}$$

The poles of its transfer function are on the real axis as shown in Figure 11.2b, and the passband is very broad as sketched in the figure. It is clear that little rejection of the harmonics produced by the nonlinearity of the amplifier occurs. Further, the phase function of the bandpass response of the $RC$ feedback circuit is very flat, as illustrated in Figure 11.2b. Therefore, a significant shift in oscillation frequency is necessary to compensate for phase shift in the transfer function of the amplifier.

The basic oscillator equation can also be developed for the other oscillator circuits presented in Chapter 10. Throughout this chapter, it is used as other oscillators are investigated. The basic oscillator equation permits us to see the critical similarities of the several seemingly different oscillator configurations.

## 11.3 Single-Device, Transformer-Coupled Oscillators

Figure 11.3a shows an oscillator configuration containing a single bipolar transistor (BJT) with transformer coupling from the collector to the emitter. Although an important oscillator configuration in its own right, the circuit provides the basic foundation for an entire class of practical oscillators. Our procedure is to establish first the basic oscillator equation for the configuration and to investigate the starting conditions. This is followed by a study of the means by which limiting and a steady-state situation are achieved.

It is convenient to redraw the transformer-coupled circuit as in Figure 11.3b. The coupling capacitor is assumed for the moment to be large and can be modeled as a dc voltage source, $V_{EX}$. The $R_E - V_{EE}$ combination is modeled as a dc current source, $I_{EE}$. The transistor is operated in common-base and can be modeled as shown in Figure 11.3c under the restriction that operation is limited to the off and normal active regions. For a npn transistor, the emitter current is modeled as

$$I_E = -\frac{I_C}{\alpha} = -\frac{I_S}{\alpha} \exp\left(\frac{V_{BE}}{V_t}\right) = \frac{-I_S}{\alpha} \exp\left(\frac{-V_{EB}}{V_t}\right) \tag{11.9}$$
$$= f(V_{EB})$$

Notice that the nonlinear I-V characteristic $f(V_{EB})$ is defined in terms of the emitter-base voltage, $V_{EB}$, $(V_{EB} = -V_{BE})$. At the collector, the circuit and device equations are developed.

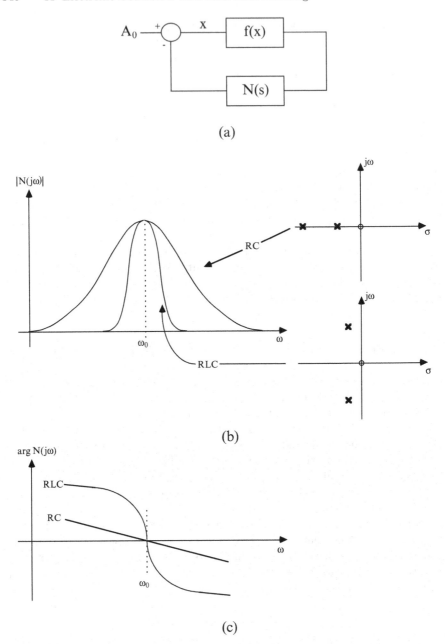

(a)

(b)

(c)

**Fig. 11.2.** (a) Block diagram of the 'basic oscillator equation'. (b) Magnitude responses for $N(s)$ and poles of the transfer functions. (v) Phase responses for $N(s)$.

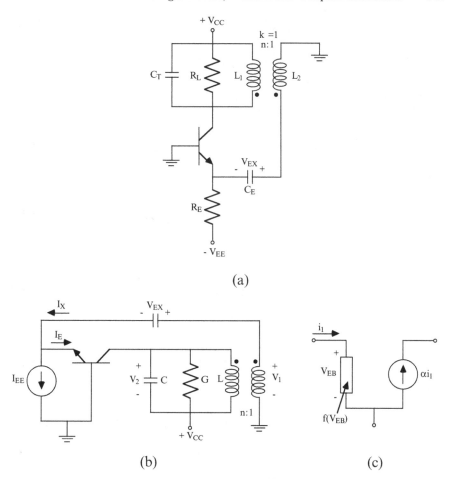

**Fig. 11.3.** (a) A single BJT transformer-coupled oscillator. (b) A convenient representation of the oscillator. (c) Model for the transistor.

$$I_C = -\alpha I_E = -\alpha f(V_{EB}) \tag{11.10}$$

$$V_1 = V_{EB} + V_{EX}$$

$$V_2 = nV_1 = n(V_{EB} + V_{EX})$$

where $V_{EX}$ is the voltage source due to the charged capacitor, $I_X = I_{EE} + I_E$ is the current through it, and $V_2$ is the voltage drop across the parallel $RLC$ circuit. For the transformer, a perfectly coupled set of coils is assumed, i.e., $k \approx 1$. Using transform notation for convenience, we obtain

$$\left(\alpha - \frac{1}{n}\right)I_E = G s(nV_{EB} + nV_{EX}) + C s(nV_{EB} + nV_{EX}) \quad (11.11)$$

$$+ \frac{1}{Ls}(nV_{EB} + nV_{EX}) + \frac{1}{n}I_{EE}$$

After some manipulation, the result is

$$V_{EB} + N(s)f(V_{EB}) = -V_{EX} - \frac{1}{n^2}\frac{Ls}{LCs^2 + LGs + 1}I_{EE} \quad (11.12)$$

where

$$N(s) = -\frac{1}{n}\left(\alpha - \frac{1}{n}\right)\frac{Ls}{LCs^2 + LGs + 1} \quad (11.13)$$

Equation (11.12) is in the form of the 'total basic oscillator equation.' In the steady state, the right-hand side becomes $-V_{EX}$.

$$V_{EB} + N(s)f(V_{EB}) = A_o = -V_{EX} \quad (11.14)$$

For the stability analysis of this circuit, incremental variables are introduced about the quiescent operating point.

$$v_e = dV_{EB}, \quad i_e = dI_E \quad (11.15)$$

and the differential of the nonlinearity is taken.

$$dI_E = i_e = df(V_{EB}) = \frac{df}{dV_{EB}}dV_{EB} = \frac{df}{dV_{EB}}v_e \quad (11.16)$$

where from Equation (11.9),

$$\frac{df}{dV_{EB}} = \frac{1}{\alpha}\frac{I_C}{V_t} = \frac{g_m}{\alpha} \quad (11.17)$$

The basic oscillator equation in terms of the incremental variables is

$$v_e + N(s)\frac{g_m}{\alpha}v_e = 0 \quad (11.18)$$

The characteristic equation is obtained by multiplying through by the denominator of $N(s)$ and rearranging.

$$LCs^2 + L\left[G - \frac{1}{n}\left(\alpha - \frac{1}{n}\right)\frac{g_m}{\alpha}\right]s + 1 = 0 \quad (11.19)$$

The frequency variable is next normalized with respect to $\sqrt{LC}$, $p = s\sqrt{LC}$. (This is the same as introducing the time normalization $T = t/\sqrt{LC}$.) The characteristic equation becomes

$$p^2 - \epsilon p + 1 = 0 \quad (11.20)$$

where the *van der Pol parameter* $\epsilon$ is

$$\epsilon = \sqrt{\frac{L}{C}} \left[ \frac{1}{n} \left( \alpha - \frac{1}{n} \right) \frac{g_m}{\alpha} - G \right] \qquad (11.21)$$

$$= \frac{\sqrt{\frac{L}{C}}}{R_L} \left[ \frac{1}{n} \left( \alpha - \frac{1}{n} \right) \frac{g_m R_L}{\alpha} - 1 \right]$$

$$= \frac{1}{Q} \left[ \frac{1}{n} \left( \alpha - \frac{1}{n} \right) \frac{g_m R_L}{\alpha} - 1 \right]$$

where $Q = \frac{R_L}{\sqrt{\frac{L}{C}}}$. This is the selectivity factor for the tuned circuit alone without loading from the transistor. Notice that an equivalent total conductance can be identified in the brackets in the first expression.

Thus far in this development, the basic oscillator equation has been obtained, and, as established in the last chapter, for $\epsilon$ positive and small, sinusoidal oscillations will buildup from a small excitation. However, for this oscillator which has a dc current source for bias, the effective bias point of the oscillator shifts as the amplitude of oscillation grows. This bias point shift involves a change of $V_{EX} = -A_o$; the dc voltage across the coupling capacitor, $C_E$. It is this shift which limits the growth of the output amplitude. This is comparable to the bias shift observed for the ECP oscillator of Section 11.1. It is to be remembered that the transformer-coupled, single-device oscillator has three independent energy storage elements, $L$, $C$ and $C_E$. In this section, we have modeled the oscillator with a second-order differential equation involving only $L$ and $C$. The effects of $C_E$ are slow changing effects, but must be included in determining the effective bias point. The bias shift for the present oscillator is developed further in the next section and in Section 11.6.

## 11.4 Bias Shift and Harmonic Balance

To understand the bias shift during oscillation buildup in the single-device, transformer-coupled configuration, it is helpful to sketch the total nonlinear characteristic of the first-order term of the oscillator differential equation. From Equations (11.13) and (11.14) and changing to the normalized time domain,

$$\frac{d^2 V_{EB}}{dT^2} + \sqrt{\frac{L}{C}} \frac{d}{dT} F(V_{EB}) + V_{EB} = A_o \qquad (11.22)$$

where

$$F(V_{EB}) = G V_{EB} - \frac{1}{n} \left( \alpha - \frac{1}{n} \right) f(V_{EB}) \qquad (11.23)$$

Equation (11.22) has the same form as the oscillator equations of Chapter 10.

Representative plots of $f(V_{EB})$ for a npn transistor and $F(V_{EB})$ are shown in Figures 11.3d and e. Note that a N-type shape is not present for $F(V_{EB})$ as observed for the oscillators of the last chapter. The dashed, vertical portion of the curve on the left in Figure 11.3e, which does complete a N-type plot, is a representation of what occurs if the device is driven into saturation. If saturation is encountered, the simple nonlinear circuit model for the BJT of Figure 11.3c is not valid. A second independent nonlinearity is present, and the differential equation describing the oscillator has a different form than that of Equation (11.22).

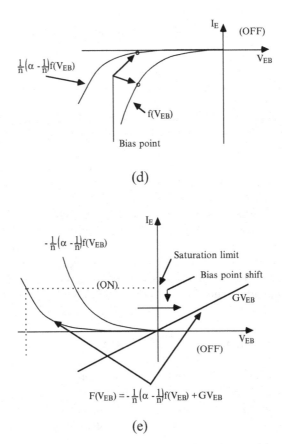

**Fig. 11.3.** Representative plots of (d) $f(V_{EB})$, and (e) $F(V_{EB})$.

Also note that for the simple nonlinearity in Figure 11.3e, the upward portion of the characteristic in the second quadrant has an increasing negative slope with a decrease of the input voltage (increasing in the negative direction). Thus, incrementally, the effective 'negative conductance' of the oscillator increases in magnitude as one moves out to the left from the equilibrium point. It is clear intuitively that during the oscillatory buildup, the positive peaks of $-I_E \approx I_C$ have larger magnitudes than the negative peaks, as illustrated in Figure 11.4a.

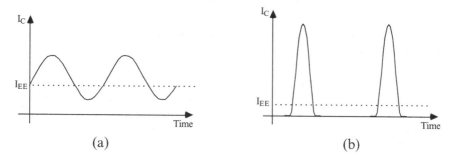

(a)                                    (b)

**Fig. 11.4.** (a) Collector current waveform during buildup of oscillations. (b) Steady-state collector current waveform.

During buildup, the distorted waveform of $I_C$ develops significant even harmonics. Of particular interest, a zero-order harmonic is produced; that is, a dc component. If the bias arrangement is for a constant $V_{BE}$, the average dc collector current increases, and the growth in oscillations continues until the transistor saturates during the positive peaks of $I_C = -I_E$. The introduction of this new nonlinearity limits the growth, and a distorted waveform of the output voltage results in the steady state. (If the tuned circuit has a high $Q$, the output voltage is a distorted sinusoid with a frequency comparable to the resonant frequency of the tuned circuit even though the saturated transistor provides a short across the tuned circuit for a small portion of the cycle.)

In the present case, the dc collector and emitter currents are fixed by the bias circuit, i.e., by $I_{EE}$. There is still the 'tendency' for the average collector current to increase. Since it cannot increase, the effective, average value of the base-emitter voltage must decrease and $V_{EX}$, the average value of the voltage of the coupling capacitor, $C_E$, must decrease. In effect, the bias current can be viewed as having two components: one from the bias state of the circuit, the other from the harmonic generation. The sum of the two is constant. Since the harmonic component increases from zero as the oscillation builds up, the other component, due to the bias state, must decrease. The effective bias state moves toward an off state, as $V_{BE}$ becomes less positive and $V_{EX}$ changes accordingly. From the standpoint of the I-V plot of Figure 11.3d, the effective bias point moves to the right toward the off region of transistor operation.

The aspects of steady-state operation are brought out in the analysis which follows. For the moment, assume that the buildup of oscillations continues until a steady-state equilibrium is reached, where the bias state shifts toward the off region and the subsequent 'gain and loss' compensates on a cycle-by-cycle basis for the growth of the oscillation during the positive peaks of the collector current. As seen below, this can occur before the transistor saturates. The waveform for the collector current takes the form of a pulse train as shown in Figure 11.4b. The average value of the pulse train, assuming that the transistor beta is large, is the value $I_{EE}$, the bias current source. In effect, we have a Class-C operation where the transistor is off for over 180 degrees of the fundamental cycle. This type of operation is very efficient in the conversion of dc power to fundamental ac power. Examples using Spice simulation which illustrate these aspects are given shortly.

The following analysis of the steady-state operation is based on a major assumption. The *GLC* resonant circuit of the oscillator at the collector of the BJT is assumed to be highly resonant, i.e., to have a very high $Q$. Therefore, the voltage across this resonant circuit contains primarily the fundamental component, even though the driving collector current is a pulse train, as illustrated in Figure 11.4b. The higher harmonics of the voltage are inhibited. (This is the same circumstance as occurs for the ECP oscillators of Sections 10.7 and 11.1). Of course, the currents in the $L$ and $C$ are still rich in harmonics. The voltage transformed by the transformer to the emitter-base is also then almost a pure sinusoid.

With an ideal exponential I-V characteristic for the emitter-base, a known mathematical relation can be used to obtain the Fourier coefficients of the emitter-current waveform. This is done for the CE stage in Section 3.3. We include here portions of the same development. We start the analysis with an assumed form of the emitter-base voltage, based on the high-$Q$ assumption.

$$V_{EB} = V_a - V_b \cos \omega_o t \qquad (11.24)$$

where $\omega_o^2 = 1/(LC)$, $V_a$ is the average value of $V_{EB}$ in the steady state, and $V_b$ is the amplitude of the fundamental. (The '-' sign is used for later convenience and ties this development closely to that of Section 3.3). This expression is used in the basic oscillator equation for this circuit.

$$V_a - V_b \cos \omega_o t + N(s)f(V_a - V_b \cos \omega_o t) = -V_{EX} \qquad (11.25)$$

The nonlinear function from (11.9) is

$$f(V_a - V_b \cos \omega_o t) = -\frac{1}{\alpha} I_S \exp \left[ -\frac{(V_a - V_b \cos \omega_o t)}{V_t} \right] \qquad (11.26)$$

$$= -\frac{I_S}{\alpha} \exp \left( \frac{-V_a}{V_t} \right) \exp \left( \frac{V_b \cos \omega_o t}{V_t} \right)$$

The exponential containing the sinusoid can be rewritten as

$$\exp \left( \frac{V_b \cos \omega_o t}{V_t} \right) = \exp(d \cos \omega_o t) \qquad (11.27)$$

where $d = V_b/V_t$. The right-hand side has a known expansion in terms of a Fourier series of cosines.

$$\exp(d\cos\omega_o t) = I_o(d) + 2I_1(d)\cos\omega_o t + \ldots \tag{11.28}$$
$$\ldots + 2I_n(d)\cos n\omega_o t + \ldots$$

where the $I_n(d)$ are modified Bessel functions of order $n$. Using this expansion in (11.26), we obtain

$$f = I_{dc} + I_{dc}\left[\frac{2I_1(d)}{I_o(d)}\cos\omega_o t + \frac{2I_2(d)}{I_o(d)}\cos 2\omega_o t + \ldots\right] \tag{11.29}$$

where

$$I_{dc} = \frac{-I_S}{\alpha}\exp\left(-\frac{V_a}{V_t}\right)I_o(d) \tag{11.30}$$

This expression for $f(V_{EB})$ is now inserted into the basic oscillator equation, and harmonic balance is used. That is, the equation is solved for like frequency components, in increasing harmonic order. This is similar to the procedure in Section 4.2, although in the latter, an approximation concerning higher-order terms had to be made. For dc, $N(0)$ is equal to zero.

$$V_a + (0 \cdot I_{dc}) = -V_{EX} \tag{11.31}$$

This is the capacitor bias voltage, as expected. For the fundamental terms,

$$-V_b\cos\omega_o t + \frac{\frac{-1}{n}\left(\alpha - \frac{1}{n}\right)j\omega_o L}{-1 + j\omega_o LG + 1}I_{dc}\frac{2I_1(d)}{I_o(d)}\cos\omega_o t = 0 \tag{11.32}$$

This leads to (solving for the magnitude)

$$V_b = \frac{1}{n}\left(\alpha - \frac{1}{n}\right)\frac{|I_{dc}|}{G}\frac{2I_1(d)}{I_o(d)} \tag{11.33}$$

$$d = \frac{V_b}{V_t} = \frac{1}{n}\left(\alpha - \frac{1}{n}\right)R_L\frac{|I_{dc}|}{V_t}\frac{2I_1(d)}{I_o(d)}$$

where for later convenience the load resistance, $R_L = 1/G$, is introduced. The last equation can be solved by iteration to obtain the value of $d$ in the steady state. Values of the ratios of these special Bessel functions are plotted in Figure 3.6 which is repeated here as Figure 11.5. One starts with an assumed value of $d$. The ratio $I_1(d)/I_o(d)$ is found from the plots and used in Equation (11.33). From the new value of $d$ from the right-hand side, the procedure is continued until a solution is reached. For large magnitudes of the fundamental ($d > 6$), the value of $I_1(d)/I_o(d) \approx 0.95$ and the value of the amplitude of the emitter-base voltage is approximately

$$V_b \approx \frac{1.9}{n}\left(\alpha - \frac{1}{n}\right) R_L I_{EE} \tag{11.34}$$

where $I_{EE} = -I_{dc} = |I_{dc}|$. The variational output voltage amplitude, $V_{oA}$, is

$$V_{oA} = nV_b \approx 1.9\left(\alpha - \frac{1}{n}\right) R_L I_{EE} \tag{11.35}$$

Since $\alpha \approx 1$ and $n > 1$, a first approximation for $V_{oA}$ is

$$V_{oA} \approx 1.9 R_L I_{EE} \tag{11.36}$$

The upper harmonics of $V_o(t)$ are assumed to be inhibited by $N(s)$, i.e., the resonant circuit at the output.

For a numerical example, let $\alpha = 0.99$ and the dc current source $I_{EE} = 2$ mA. Also choose $L = 5$ $\mu$H, $C = 450$ pF, $R_L = 750$ $\Omega$, and $n = 10$. The oscillation frequency, both for the buildup and for the steady state is approximately the resonant frequency of the tuned circuit.

$$f_o = \frac{1}{2\pi\sqrt{LC}} = 3.36 \text{ MHz} \tag{11.37}$$

and the time period is $T = 0.3$ $\mu$s. The value of $\epsilon$ for the startup condition from Equation (11.21) is

$$\epsilon = \sqrt{\frac{L}{C}}\left[\frac{1}{n}\left(\alpha - \frac{1}{n}\right)\frac{g_m}{\alpha} - \frac{1}{R_L}\right] \tag{11.38}$$
$$= 0.6$$

In the steady state, the simple estimate for the value of the output voltage using Equation (11.36) is $V_{oA} = 2.85$ V. If actual values for $\alpha$ and $n$ are used, Equation (11.35) yields $V_{oA} = 2.54$ V. For this value, $d = 2.5/(10V_t) = 9.67$. Clearly, from Figure 11.5, the approximation is good.

The input file for a Spice simulation is given in Figure 11.6a, and the output waveforms of $I_C$ and $V_o = V(2)$ are given in Figure 11.6b for the steady state. The period of oscillation is observed to be approximately 295ns. As brought out in Chapter 10, care must be used in specifying the frequency of oscillation for a Fourier analysis. For this example, $f_o = 1/295$ $\mu$s $= 3.39$ MHz. The harmonics of the steady-state output voltage are given in Figure 11.6c. (If the frequency in the .FOUR specification is decreased by $1-2\%$, the value of the second harmonic drops while the third harmonic increases.) The amplitude of the output voltage is 2.55 V.

Notice from the waveform of the collector current in Figure 11.6b, that the peak current is approximately 15 mA, and that the transistor is 'on' for only 100ns, while it is 'off' for over 190ns. Class-C operation is obtained. The conversion efficiency is, for an ideal current source $I_{EE}$,

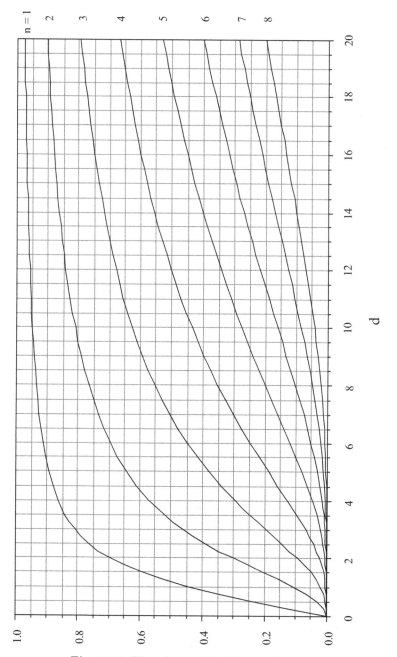

**Fig. 11.5.** Plot of normalized Bessel functions.

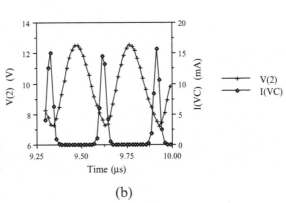

```
XSFMR-COUPLED OSCILLATOR, FIGURE 11.
.TRAN 15N 10U 9.3U 15N
.PLOT TRAN V(2) I(VC)
.FOUR 3.39MEGHZ V(2)
*.TRAN 30N 12U 10U 15N
*.PLOT TRAN V(3,5)
R1 1 0 1
Q1 7 1 3 MOD1
VC 2 7 0
VCC 4 0 10
RL 4 2 750
*RL 4 2 2.6K
CT 4 2 450PF
L1 4 2 5UH
L2 0 5 0.05UH
K1 L1 L2 1
CE 5 3 5.0NF
RE 3 6 4.65K
VEE 6 0 -10 PULSE -15 -10 0 0 0 1S
.MODEL MOD1 NPN IS=1E-16 BF=100 RC=01
.OPTIONS NOPAGE NOMOD ITL5=0
.OPTIONS LIMPTS=5000 RELTOL=1E-6
.WIDTH OUT=80
.END
```

(a)

(b)

```
FOURIER COMPONENTS OF TRANSIENT RESPONSE V(2)
DC COMPONENT =    9.999D+00
```

| HARMONIC NO | FREQUENCY (HZ) | FOURIER COMPONENT | NORMALIZED COMPONENT | PHASE (DEG) | NORMALIZED PHASE (DEG) |
|---|---|---|---|---|---|
| 1 | 3.390D+06 | 2.550D+00 | 1.000000 | 6.159 | 0.000 |
| 2 | 6.780D+06 | 2.128D-01 | 0.083452 | 28.243 | 22.084 |
| 3 | 1.017D+07 | 8.597D-02 | 0.033711 | 124.262 | 118.103 |
| 4 | 1.356D+07 | 4.004D-02 | 0.015701 | -128.261 | -134.420 |

```
   TOTAL HARMONIC DISTORTION =      9.203121 PERCENT
```

(c)

**Fig. 11.6.** (a) Circuit and Spice input file. (b) Steady-state output voltage and collector current waveforms. (c) Fourier components of the steady-state output voltage.

$$\eta = \frac{V_{orms}^2 / R_L}{I_{EE}(V_{CC} + V_{BE})} \qquad (11.39)$$

$$= \frac{\frac{(0.707 \times 2.53)^2}{750}}{2(10 + 0.8)}$$

$$= 20\%$$

If the current source $I_{EE}$ is produced with a $R_E - V_{EE}$ combination, $\eta$ drops to 11%. Another simulation run of the circuit choosing $V_{BE}$ as the output shows that the average bias shifts from approximately 0.79 V to 0.59 V over the buildup of oscillations.

The maximum amplitude of the output voltage is approximately $V_{CC}$ if transistor saturation is to be avoided. From (11.35), the maximum value for $R_L$ in this example is 2.96 k$\Omega$. However, simulation shows that this value is too large. The output voltage waveform for $R_L = 2.6$ k$\Omega$ is shown in Figure 11.7. The output voltage waveform varies from zero to twice $V_{CC}$. Clipping is not observed.

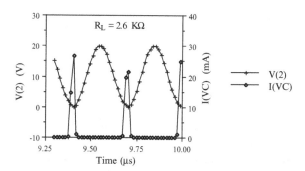

```
FOURIER COMPONENTS OF TRANSIENT RESPONSE V(2)
DC COMPONENT =   1.010D+01
HARMONIC   FREQUENCY     FOURIER      NORMALIZED    PHASE     NORMALIZED
   NO        (HZ)       COMPONENT     COMPONENT     (DEG)    PHASE (DEG)

   1       3.390D+06    9.832D+00     1.000000     -84.416      0.000
   2       6.780D+06    2.649D-01     0.026942    -150.904    -66.487
   3       1.017D+07    1.408D-01     0.014318    -149.119    -64.703
   4       1.356D+07    9.245D-02     0.009402    -145.223    -60.807

   TOTAL HARMONIC DISTORTION =         3.342086  PERCENT
```

**Fig. 11.7.** Output voltage and collector current waveforms and Fourier components of the output voltage.

Another transformer-coupled oscillator can be produced by coupling back from the collector to the base as illustrated in Figure 11.8. For the resistor-voltage source-bypass capacitor arrangement as shown, a fixed bias current is established, and bias-shift limiting can be achieved. An analysis similar to that above can be used to establish the starting conditions and the estimate of the steady-state output. The result is approximately that of Equation (11.36). If a voltage source in the base circuit is used for biasing, or if diode current-mirror biasing is used in the base circuit, no voltage bias shift can occur, and transistor saturation performs the limiting of the oscillation voltage.

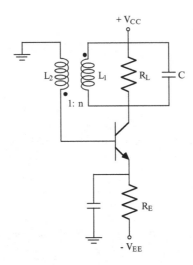

**Fig. 11.8.** Another transformer-coupled oscillator.

As a final comment in this section, it must be stressed that a shift in the quiescent bias point occurs for any oscillator for which the total $F(v)$ characteristic is not anti-symmetrical about the bias point. In such cases, the increasing positive and negative peaks are not equal in magnitude as the oscillation builds up and produce a dc component which shifts the virtual bias point. As mentioned earlier, the tunnel-diode oscillator of Section 10.3 displays this effect. Except for special circuits, such as studied in this chapter, the estimation of the steady-state oscillation output is in general either difficult or very crude.

## 11.5 Transformer-Coupled MOS Oscillators

A MOSFET version of the BJT oscillator of the last section can be simply realized by direct substitution of the same polarity, same mode device and by adjusting the bias levels. A transformer-coupled configuration using a n-channel MOS device is shown in Figure 11.9. The Spice input file of Figure 11.10 provides the circuit and device parameters for the circuit. The output voltage waveform is given in Figure 11.11.

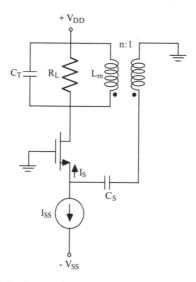

**Fig. 11.9.** A transformer-coupled NMOS oscillator.

```
MOS XSFRMR-COUPLED OSCILLATOR, FIG 11.10
.OPTION ITL5=0 LIMPTS=60000 RELTOL=1E-6
VDD 1 0 10
LH 9 1 43.76UH
VL 9 2 0
LL 8 0 2.735UH
KA LH LL 1
CL 1 7 5.78845885PF
VC 7 2 0
RL 1 6 27.5K
VR 6 2 0
VM 2 5 0
M1 5 0 3 3 MODN W=100U L=10U
ISS 3 4 .25M PULSE .5M .25M 1NS 0 0 1M
VSS 4 0 -10
CC 3 8 30PF
.MODEL MODN NMOS LEVEL=1 KP=30U VTO=.45
.TRAN 5N 6U 5.8U UIC
.PLOT TRAN V(2)
.FOUR 10MEG V(2)
.END
```

**Fig. 11.10.** Spice input file for NMOS oscillator.

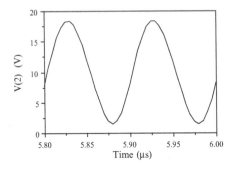

```
FOURIER COMPONENTS OF TRANSIENT RESPONSE V(2)
DC COMPONENT =    9.988D+00
HARMONIC    FREQUENCY       FOURIER     NORMALIZED      PHASE      NORMALIZED
  NO          (HZ)         COMPONENT    COMPONENT      (DEG)      PHASE (DEG)

   1        1.000D+07      8.409D+00     1.000000     -13.123       0.000
   2        2.000D+07      2.442D-01     0.029036     -14.028      -0.905
   3        3.000D+07      3.670D-02     0.004364     113.468      126.591
   4        4.000D+07      1.740D-02     0.002069    -157.303     -144.180

   TOTAL HARMONIC DISTORTION =         2.961371  PERCENT
```

**Fig. 11.11.** Output voltage waveform and its Fourier components.

The estimation of the steady-state oscillation amplitude is not as simple as for the BJT configuration. In contrast to the Bessel-function series for the BJT with a sinusoidal drive, a single expression for the MOS device is not available to model the transfer characteristic in both the off and normal active regions of operation. A piece-wise solution is needed.

A solution for the steady-state oscillation amplitude has been developed [29]. The result is a set of curves in terms of new variables $f_1(\theta_c)$ and $f_2(\theta_c)$. The variables are related to the circuit parameters of the oscillator as follows:

$$f_1(\theta_c) = \frac{d}{nV_T}\left(1 - \frac{1}{n}\right)R_L I_M \qquad (11.40)$$

where $I_M = \frac{k'}{2}\frac{W}{L}V_T^2$ and $V_T$ is the threshold voltage of the device. The variable $d = V_{GA}/V_T$ is the normalized value of the assumed sinusoidal amplitude at the gate of the MOS device. The amplitude of the output voltage is $V_{out} = ndV_T$.

$$f_2(\theta_c) = \frac{I_{SS}}{\dfrac{1}{I_M}\left[\dfrac{nV_T}{\left(1-\frac{1}{n}\right)R_L}\right]^2}$$

A study of these curves has lead to simple linear approximations [29].

$$f_1(\theta_c) = 1.83 f_2(\theta_c) - 0.186, \ f_2(\theta_c) \le 2.0 \tag{11.41}$$
$$f_1(\theta_c) = 1.89 f_2(\theta_c) - 0.3, \ 2 \le f_2(\theta_c) \le 7.0$$

The independent variable, $\theta_c$, can be described as a cutoff angle. The device is off for $\theta_c < \theta < (2\pi - \theta_c)$ where $\theta = \omega_o t$.

For a numerical example, choose the values given in Figure 11.10: $I_{SS} = 0.25$ mA, $R_L = 27.5$ k$\Omega$, $n = 4$, $V_T = 0.45$ V, $W/L = 10$, and $V_{DD} = 10$ V. For the circuit and device values, the calculated value of the *van der Pol parameter* is $\epsilon = 0.1$ and the oscillation frequency should be 10 MHz. The value of $I_M = 30.38$ $\mu$A leads to a value $f_2(\theta_c) = 0.997$. Using the first approximation of (11.41), we obtain $f_1(\theta_c) = 1.64$. From (11.40), $d = 4.71$ and $V_{out} = ndV_T = 8.48$ V. From the waveform of Figure 11.11, the amplitude of the output obtained from the simulation is 8.41 V.

## 11.6 Squegging

The transformer-coupled oscillators described and analyzed in the last two sections may exhibit an output such as shown in Figure 11.12. After a buildup in oscillations, a collapse is suffered which is due to a large change in the charge on the coupling capacitor, $C_E$. This sequence of events means that we have two oscillatory phenomena present at the same time. One is the near-sinusoidal oscillation studied in the last two sections; the other is a relaxation-mode oscillation produced by the charging and discharging of the bias coupling capacitor. The combination of the two modes of oscillations is called 'squegging' [20]. In this section, we analyze the situation and come up with a criterion to avoid the squegging.

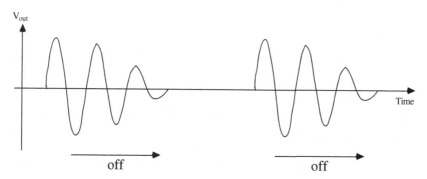

**Fig. 11.12.** Output from a single-device transformer-coupled oscillator that exhibits squegging.

The circuit diagram of a BJT transformer-coupled oscillator is repeated in Figure 11.13. We concentrate our attention now on the charging and discharging of $C_E$. There is a change of charge of $C_E$ due to the pulsed current flowing through it. The current flowing out of the emitter, $-I_E = I_C$, has larger and larger peaks as the oscillation builds up. This current must flow into $C_E$ from left to right, since $I_{EE}$ is a constant bias source, and discharge $C_E$. At the quiescent bias state, $C_E$ is charged as illustrated in the figure. ($V_{EX} = V_{BE}$ across $C_E$ is positive from right to left at the quiescent bias point.) After a period of time while the oscillation is building up to a steady-state level, the average voltage across $C_E$ becomes sufficiently small to turn the transistor off. The near-sinusoidal oscillation dies down as illustrated in Figure 11.12. The capacitor voltage then charges because of the current source $I_{EE}$. When the average voltage, $V_{EX}$, has built up sufficiently, the transistor reenters the active region and near-sinusoidal oscillations again grow. With time, there is the simultaneous existence of two modes of oscillation.

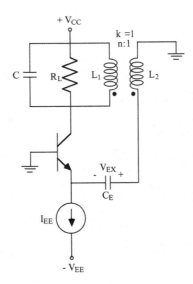

**Fig. 11.13.** A BJT transformer-coupled oscillator.

To avoid squegging, the value of $C_E$ must be small. From the above argument, as $C_E$ is made larger and larger, the change of charge will be slower, but more and more certain. Little charge will decay during each cycle of the near-sinusoidal buildup. We cannot expect that the charge of $C_E$ should decay completely during a cycle of the near-sinusoidal oscillation. This would defeat the need for ac coupling. But we can achieve a balance in the buildup with the 'free' (transient) response of the circuit. In essence, we make the time constant for the free response of the oscillator approximately equal to the time constant for the buildup of charge on $C_E$. Therefore, the charge on $C_E$ during the buildup of oscillations can gradually approach the value needed for bias-shift limiting without turning the oscillations off.

The basic oscillator equation of the circuit of Figure 11.13 is from Equation (11.12):

$$V_{EB} + N(s)f(V_{EB}) = -V_{EX} - N_1(s)I_{EE} \tag{11.42}$$

where $N_1(s)$ describes the free response of the circuit, after the application of the bias sources.

$$N_1(s) = \frac{1}{n^2} \frac{Ls}{LCs^2 + LGs + 1} \tag{11.43}$$

The characteristic equation for the free response alone is

$$s^2 + \frac{G}{C}s + \frac{1}{LC} = 0 \tag{11.44}$$

The natural modes of the free response are

$$\exp(s_1 t), \exp(s_2 t) \tag{11.45}$$

and the natural frequencies of the free response are the roots of Equation (11.44),

$$s_1, s_2 = -\frac{G}{2C} \pm j\sqrt{\frac{1}{LC} - \left(\frac{G}{2C}\right)^2} \tag{11.46}$$

The time constant, $T_f$, of the exponential decay envelope of the free response is the reciprocal of the real part of the $s_i$.

$$T_f = \frac{2C}{G} = 2R_L C \tag{11.47}$$

Inspecting the charging of $C_E$, one can propose that the time constant for the buildup of charge during initial growth of the oscillation is

$$T_g = \frac{1}{g_m} C_E \tag{11.48}$$

where the input resistance of the transistor, looking into the emitter, is approximately $\frac{1}{g_m}$ and $g_m$ is the small-signal transconductance of the transistor at the quiescent bias point, i.e., the value during the startup of oscillations. When the steady state has been attained or at least when the steady state is approached, a 'large-signal' value of the device transconductance must be used to investigate the buildup of charge on $C_E$.

$$T_s = \frac{1}{G_m} C_E > T_g \tag{11.49}$$

where $G_m$ is the ratio of the amplitude of the fundamental component of the collector current, $I_{cA}$, and the fundamental component of the base-emitter voltage [20]. From the developments of Section 11.3, the base-emitter voltage in the steady state is $V_{BE} = -V_{EB} = -V_a + V_b \cos \omega_o t$. $G_m$ is

$$G_m = \frac{I_{cA}}{V_b} \tag{11.50}$$

We next introduce the fundamental of the output voltage, $V_{oA}$.

$$V_{oA} = \left(\alpha - \frac{1}{n}\right) R_L I_{cA} = nV_b \tag{11.51}$$

Therefore,

$$G_m = \frac{n}{\left(\alpha - \frac{1}{n}\right) R_L} \tag{11.52}$$

To avoid squegging, $T_s \approx T_f$. Since squegging occurs during the buildup of oscillation, (11.49) is too large an estimate and an intermediate value between (11.48) and (11.49) is more appropriate, leading to

$$C_E < \frac{2nC}{\left(\alpha - \frac{1}{n}\right)} \tag{11.53}$$

For the circuit and device values of the example of Figure 11.6a, $C_E < 10$ nF. The results of Figures 11.6b and c are obtained with $C_E = 5$ nF. For $C_E = 6$ nF, squegging occurs as illustrated in Figure 11.14. Two periods of oscillation are observed. After a cessation of near-sinusoidal oscillation, the collector current builds up rapidly at the expected oscillation period $\approx 300$ ns. The output voltage grows quickly. However, the transistor quickly turns off and the collector current is negligible while the voltage across the tuned circuit decays passively with a period of approximately 300 ns. After 3.2 $\mu$s, the relaxation is over, and the collector current is again observed producing a new burst of output voltage. For the circuit values of Figure 11.6, $C_E$ must lie in the range $3 - 5$ nF.

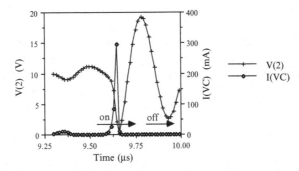

**Fig. 11.14.** Output voltage and collector current waveforms.

For a high-$Q$ resonant circuit, the buildup of oscillations is very slow. In these cases, squegging may be missed during a simulation run unless a large transient simulation time is used.

From an analysis of MOS oscillators of the same transformer-coupled configuration, the identical criterion to avoid squegging is obtained. In practice, however, the squegging phenomenon is less severe because the square-law nonlinearity of the MOS device in the active region produces less-strong harmonics in the drain and source currents. Therefore, the pulse-type waveform is not as effective in charging the coupling capacitor, and squegging is less of a problem. Nonetheless, the coupling capacitor definitely cannot be too large, and its value should satisfy Equation (11.53) with $\alpha = 1$.

Some oscillators which are equivalent to or can be modeled by transformer-coupled, single-device configurations are not plagued with squegging. An example is the Colpitts oscillator which is studied in a later section.

## 11.7 Phase-Shift Oscillators

A useful technique to develop, study, and categorize oscillator configurations is to concentrate on the stability condition at the equilibrium point. In particular, the positive-feedback arrangement for the linearized circuit about a quiescent operating point is investigated. As an example, for the Wien-type configuration of Section 10.6, the amplifier (gain block) at the quiescent bias point provides positive gain with zero phase shift. The $RC$ feedback circuit provides a zero phase shift at the 'resonant' frequency. At the frequency of interest, if the loop gain magnitude is greater than one with a phase shift of zero, or a multiple of 360 degrees, the conditions for the build up of oscillations exist. The natural frequencies of the linearized oscillator are then in the right-half plane. If they are also complex, a sinusoidal buildup of oscillations occurs. Of course, whether a near-sinusoidal steady state is achieved depends upon the limiting situation as described in the last sections.

A zero-phase loop gain is also provided by the transformer-coupled ECP oscillator of Section 10.7. In the transformer-coupled, single-device circuit of this chapter, the common-base transistor in Figure 11.3b again provides a zero-phase current gain, and the transformer coupling also provides zero phase shift. If the circuit is modified to provide collector-to-base feedback, as in Figure 11.8, the common-emitter stage provides a net phase reversal and the polarity of the transformer must be such as to provide another phase reversal to achieve a phase for the loop gain of $180° + 180° = 360°$.

Phase-shift oscillators are those that provide a positive loop gain condition at the quiescent-bias state, but do not use transformers directly. The Wien-type oscillator is then a phase-shift oscillator. Another example is shown in Figure 11.15a, where biasing elements are not shown for simplicity. The common-emitter transistor provides a net phase reversal and a lowpass, one $L$, two $C$ filter can produce at infinite frequency a total of $-270$ degrees phase shift. The magnitude and phase characteristics for the sinusoidal steady state are sketched in Figure 11.15b. At a finite frequency, $\omega_o$, the phase shift is -180 degrees, and the phase condition for the buildup of oscillations is present. If the loop-gain magnitude exceeds one, build up occurs. This configuration is one form of the Colpitts oscillator and is studied in the next section.

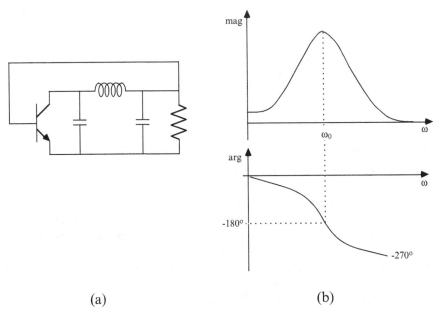

(a)                                    (b)

**Fig. 11.15.** (a) A phase-shift oscillator. (b) Magnitude and phase responses for the sinusoidal steady state.

In Figure 11.16a, the phase-shift network is a high-pass filter, with two $L$ and one $C$. As shown in Figure 11.16b, a total of +270 degree of phase shift is available near dc, and at one frequency the phase shift is 180°. This circuit is one form of the configuration of the Hartley Oscillator.

In Figure 11.17a, a bandpass filter is present. This circuit can be arranged to obtain the configuration shown in Figure 11.17b. This circuit is called the tuned-input, tuned-output oscillator. Notice that the series capacitor of the filter becomes the feedback capacitor from output to input.

Of course, MOS devices can be used in place of the BJT to obtain other phase-shift oscillators.

As a final example of phase-shift oscillators, $RC$ filters can be used to provide the necessary phase shift as in the Wien-type oscillator. A lowpass version is shown in Figure 11.18a. At least three capacitors are needed to provide a phase shift greater than 180 degrees as illustrated in Figure 11.18b. Because of the loss due to the $RC$ filter, the gain element must provide more

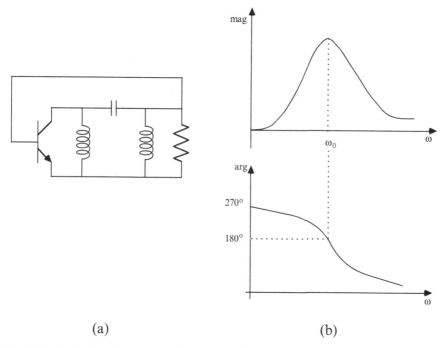

(a)                                             (b)

**Fig. 11.16.** (a) Oscillator with a high-pass filter phase network. (b) Magnitude and phase responses for the sinusoidal steady state.

gain magnitude at the 180 degree-phase frequency than is usually needed for an $LC$ filter. A distributed $RC$ delay line can also be used in place of the lumped filter, as illustrated in Figure 11.18c. Often, this is the configuration usually referred to when the term phase-shift oscillator is used without other qualifiers.

Continuing the above line of reasoning, one can also propose to use an $LC$ distributed structure, i.e., an $LC$ delay line, to produce the desired phase shift at a given frequency.

It should now be clear that innumerable oscillator configurations can be proposed using the phase-shift approach as the vehicle. This is particularly helpful when ac coupling elements are present or desired. For dc coupled circuits, and including transformer-coupled circuits in which the transformers have close coupling, circuit manipulations can be used to provide a negative conductance I-V characteristic to the resonant circuit. Alternately, a differential equation of the type developed in the last chapter can be obtained. For either case, it is often possible to utilize the evaluation and design techniques of the tunnel-diode, the Wien-type or the transformer-coupled ECP oscillators of the last chapter. Alternately, a form of the basic oscillator equation can

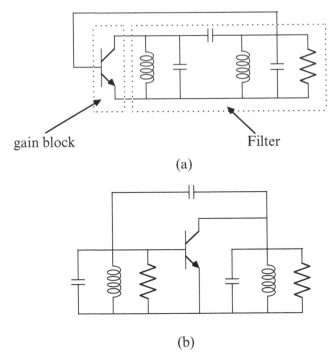

gain block                                    Filter

(a)

(b)

**Fig. 11.17.** (a) Oscillator with a band-pass filter phase network. (b) An alternate representation.

be developed, often using approximations to obtain simple desired network functions together with multiplying modifiers, as used in Section 8.3.

In the next section, the nonlinear analysis technique of the present chapter is used for one of the phase-shift circuits. In the last section of this chapter, phase-shift arguments are used to study crystal-controlled oscillators, oscillators which provide a very precise steady-state output frequency.

## 11.8 The Colpitts Oscillator

In the last section, the Colpitts configuration is introduced from the standpoint of a phase-shift oscillator using a lowpass filter to provide $-180°$ phase shift at the desired oscillation frequency. In this section, we relate this configuration to the single-device, transformer-coupled circuit in order to use its steady-state results.

The circuit presented in the last section for the Colpitts oscillator is repeated in Figure 11.19a, where the loss element (the resistor) has been relocated across the inductance. Again, biasing elements are not included. One rearrangement of the circuit is that of Figure 11.19b. In this figure, bias arrangements are also included. The current source, $I_{EE}$, is usually realized

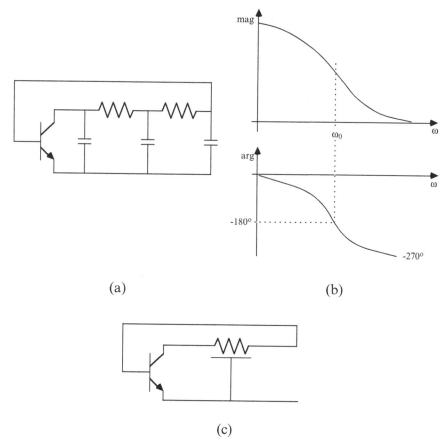

(a)                                                    (b)

(c)

**Fig. 11.18.** (a) A $RC$ lowpass phase network. (b) Magnitude and phase responses. (c) A distributed $RC$ delay line filter.

with a resistor, negative-voltage-supply combination. Another circuit rearrangement is shown in Figure 11.19c. The transistor is now shown in the common-base connection, and the feedback to the emitter lead is via a capacitor voltage divider. This use of a capacitive voltage divider is comparable to that of the transformer-less ECP oscillator of Section 10.8. If the circuit of Figure 11.19c is compared with the configuration of Figure 11.3b, it is seen that for the present circuit, the capacitive voltage divider takes the place of the single capacitor and the (ideal) transformer. As in Section 10.8, we model the two capacitors with this combination as shown in Figure 11.19d and Figure 11.20. The equivalent turns ratio is determined by the voltage-divider ratio.

$$n_e = \frac{C_1 + C_2}{C_1} \tag{11.54}$$

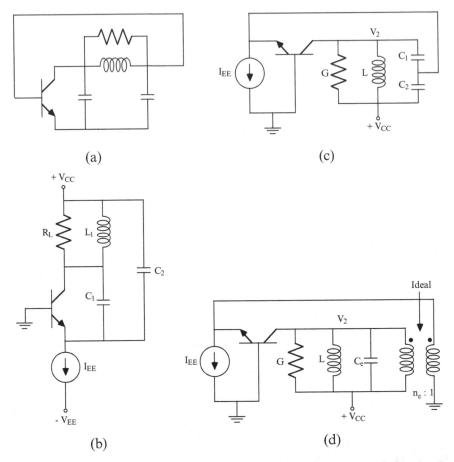

**Fig. 11.19.** (a) Colpitts oscillator circuit. (b) One rearrangement of the circuit. (c) Another circuit rearrangement. (d) Equivalent circuit with the capacitor divider replaced by a transformer.

The equivalent tuned-circuit capacitor is the input capacitance of the voltage divider.

$$C_e = \frac{C_1 C_2}{C_1 + C_2} = \frac{C_2}{n_e} \tag{11.55}$$

The modeling is confirmed with the simulations of Section 10.8 and is further confirmed through the use of the basic oscillator equation with the exact circuit of Figure 11.19c. This is done later in this section.

With the circuit of Figure 11.19d, we have returned to the oscillator configuration of Section 11.3 and can use the results of that section directly in evaluation and design.

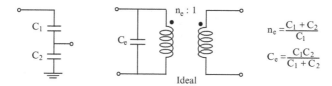

**Fig. 11.20.** Equivalent circuit model for the capacitor divider.

For a numerical example, consider the circuit in Figure 11.21a. The circuit and device values are given in the Spice input file of Figure 11.21b. The quiescent bias state is

$$I_C \approx I_{EE} = \frac{(10 - .8)}{4.65k} = 1.98 \text{ mA} \tag{11.56}$$
$$V_{CE} = 10 - (-.8) = 10.8 \text{ V}$$

The equivalent turns ratio and the tuned-circuit capacitance are

$$n_e = 10 \tag{11.57}$$
$$C_e = 450 \text{ pF}$$

For the starting condition of oscillation, we calculate $\epsilon$ from (11.21).

$$\epsilon = \sqrt{\frac{L}{C_e}} \left[ \left( \alpha - \frac{1}{n} \right) \frac{g_m}{\alpha n} - \frac{1}{R_L} \right] \tag{11.58}$$

For the numerical values, $1/g_m$ has the value of 13.2, and $\alpha = 0.99$.

$$\epsilon = \sqrt{\frac{5 \times 10^{-8}}{4.5 \times 10^{-12}}} \left( \frac{1}{147} - \frac{1}{750} \right) = 0.58$$

Therefore, we expect an oscillatory buildup with the rate $\epsilon/2 = 0.29$. The oscillation frequency should be approximately

$$f_o = \frac{1}{2\pi\sqrt{LC_e}} = 3.36 \text{ MHz} \tag{11.59}$$

The period of the oscillator waveform should be $1/f_o = 298$ ns.

In the steady state, we assume initially that the emitter-base drive is large, and that the parameter $d > 6$. The output voltage magnitude is

$$V_{oA} = 1.9 \left( \alpha - \frac{1}{n} \right) R_L I_{EE} = 2.5 \text{ V} \tag{11.60}$$

The ac power output developed in $R_L$ is

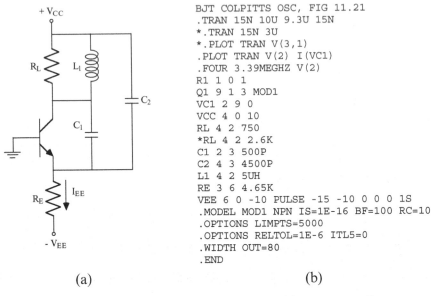

```
BJT COLPITTS OSC, FIG 11.21
.TRAN 15N 10U 9.3U 15N
*.TRAN 15N 3U
*.PLOT TRAN V(3,1)
.PLOT TRAN V(2) I(VC1)
.FOUR 3.39MEGHZ V(2)
R1 1 0 1
Q1 9 1 3 MOD1
VC1 2 9 0
VCC 4 0 10
RL 4 2 750
*RL 4 2 2.6K
C1 2 3 500P
C2 4 3 4500P
L1 4 2 5UH
RE 3 6 4.65K
VEE 6 0 -10 PULSE -15 -10 0 0 0 1S
.MODEL MOD1 NPN IS=1E-16 BF=100 RC=10
.OPTIONS LIMPTS=5000
.OPTIONS RELTOL=1E-6 ITL5=0
.WIDTH OUT=80
.END
```

(a)                                    (b)

**Fig. 11.21.** (a) Colpitts oscillator circuit. (b) Spice input file.

$$P_{ac} = 4.2 \text{ mW} \tag{11.61}$$

The dc power input to this circuit depends on how the current source is realized. For the $V_{EE} = -10$ V and $R_{EE} = 4.65$ k$\Omega$ combination,

$$P_{dc} = (V_{CC} + V_{EE})I_{EE} = 39.6 \text{ mW}. \tag{11.62}$$

The estimated conversion efficiency is then

$$\eta = 11\% \tag{11.63}$$

For the Spice simulation of this circuit, the collector voltage and collector current waveforms are as shown in Figure 11.22a. In the steady state, the observed period of the oscillation is approximately 295 ns. From the harmonic components of the output voltage in Figure 11.22b, $V_{oA} = 2.55$ V, $HD_2 = 8.4\%$ and $THD = 9.2\%$. Another simulation run shows that the average bias shifts from 0.79 V to 0.59 V over the buildup of oscillations. The collector current waveform is also shown in Figure 11.22a. It is clear that the operation is definitely Class C with a high peak current.

It is shown easily that squegging is not a problem for the Colpitts oscillator in spite of the ac (capacitor) coupling from the tuned circuit to the gain element input (emitter). Using Equation (11.53), we note that the coupling capacitor in this case can be considered to be $C_1$ while the tuned-circuit capacitor is $C_e = C_1 C_2/(C_1 + C_2)$. Therefore from (11.53), $C_1$ must be less

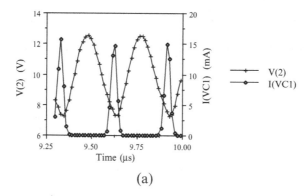

(a)

```
FOURIER COMPONENTS OF TRANSIENT RESPONSE V(2)
DC COMPONENT =   1.000D+01
HARMONIC    FREQUENCY     FOURIER      NORMALIZED      PHASE      NORMALIZED
  NO          (HZ)       COMPONENT     COMPONENT       (DEG)     PHASE (DEG)

   1       3.390D+06     2.551D+00      1.000000       2.504       0.000
   2       6.780D+06     2.139D-01      0.083874      21.122      18.618
   3       1.017D+07     8.662D-02      0.033959     113.031     110.527
   4       1.356D+07     3.886D-02      0.015234    -144.288    -146.792

     TOTAL HARMONIC DISTORTION =       9.241260   PERCENT
```

(b)

**Fig. 11.22.** (a) Steady-state collector voltage and current waveforms. (b) Fourier components of the output voltage.

than $2n_eC_e = 2C_2$. This condition is usually true, and squegging does not occur.

To check on the voltage-divider, transformer approximation, the basic oscillator equation for the circuit of Figure 11.19b can be developed. From the circuit equations, the result has the form

$$v_1 + N(s)N_a(s)f(V_{EB}) = N_o(s)A_o \tag{11.64}$$

where $A_o$ represents the bias sources, and

$$N(s) = \frac{-1}{n_e} \frac{s/C_e}{\left(s^2 + \frac{G}{C_e} + \frac{1}{LC_e}\right)} \tag{11.65}$$

$$N_a(s) = \alpha - 1 - \frac{1}{s^2 L C_1} - \frac{G}{C_1 s}$$

$C_e = (C_1 C_2)/(C_1 + C_2)$ is the pertinent capacitance of the passive, resonant circuit consisting of $L$, $C_e$, and $G$, and the parameter $n_e = (C_1 + C_2)/C_1$ can be identified as the turns ratio of the equivalent transformer introduced in the earlier development. $N(s)$ is the function developed on the basis of an equivalent transformer except for the absence of the constant multiplier $(\alpha - 1/n_e)$.

The function $N_a(s)$ for the ideal transformer-coupled oscillator is

$$N_a(s)|_{\text{ideal}} = \left(\alpha - \frac{1}{n}\right) \tag{11.66}$$

For the present circuit, we investigate $N_a(j\omega_o)$, the response of $N_a(s)$ at the resonant frequency of the tuned circuit, $\omega_o^2 = 1/(LC_e)$. The result is

$$N_a(j\omega_o) = \left(\alpha - \frac{1}{n_e}\right) + j\left(\frac{n_e - 1}{n_e}\right)\frac{1}{Q} \tag{11.67}$$

where $Q$ is the quality factor of the passive tuned circuit

$$Q = \frac{R_L}{\sqrt{\frac{L}{C_e}}} \tag{11.68}$$

For large $Q$ and for $n_e > 1$, the reactive term in Equation (11.67) is negligible, and the basic oscillator equation for the transformer-coupled approximation is valid.

The Hartley oscillator configuration also can be modeled with a transformer, and the results of Section 11.3 can be used to evaluate and design the Hartley oscillator. For this circuit, the rearrangement of the circuit leads to that of Figure 11.23a. The inductive combination $L_1$ and $L_2$ constitutes a voltage divider that can be replaced with an equivalent inductance $L_e = L_1 + L_2$, and an ideal transformer of turns ratio, $n_e = (L_1 + L_2)/L_2$, as shown in Figure 11.23b. The evaluation and design proceed as above for the Colpitts.

MOS transistors can be used in place of the BJT to obtain the MOSFET versions of the Colpitts and Hartley oscillators. A solution for the steady-state oscillation amplitude for the MOS Colpitts oscillator that includes the cutoff, triode, and saturation regions of operation is available in [30].

## 11.9 Crystal-Controlled Oscillators

The piezoelectric crystal is an electro-mechanical device which exhibits a sharp frequency selective response. Equivalent $Q$ values in excess of 10,000 are available. When used in bandpass amplifiers and oscillators, the very sharp resonances produce excellent frequency rejection for amplifiers and very sinusoidal outputs for oscillators with little harmonic content.

The piezoelectric crystal (often abbreviated xtal) is a material such as quartz. The xtal is precisely ground and polished to achieve exact dimensions. Two opposite sides of the xtal are plated and thus act as the plates of a capacitor. When excited by a voltage, the xtal displays a mechanical resonance. This resonance is coupled back into the electric circuit and can be

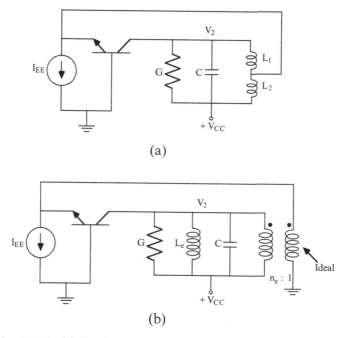

(a)

(b)

**Fig. 11.23.** (a) Hartley oscillator circuit. (b) An equivalent circuit.

modeled as an electrical resonant circuit. The schematic diagram of a xtal and the electric circuit model of a xtal and its 'holder' or plates are shown in Figures 11.24a and b. The series resonant circuits model the mechanical resonances, one for each harmonic of the structure. The value of the series-loss resistors can be as low as 10 $\Omega$. The shunt capacitor, $C_o$, represents the capacitance of the 'holder.'

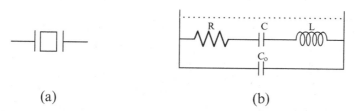

(a)                                (b)

**Fig. 11.24.** (a) Schematic diagram and (b) circuit model of a crystal.

For a typical xtal which is series resonant at its fundamental of 1.59 MHz, the values of the circuit elements of the model are

$$R = 125 \ \Omega, L = 250 \text{ mH}, C_1 = 0.04 \text{ pF}, C_o = 4 \text{ pF} \qquad (11.69)$$

The $Q$ of the series resonance is

$$Q = \frac{1}{R}\sqrt{\frac{L}{C_1}} = 20,000 \qquad (11.70)$$

For another example, a xtal resonant at 20 MHz has a value of $R$ of 16.3 $\Omega$, $C_1 = 0.009$ pF with a $Q$ of 54,600. The value of $C_o$ is 2.3 pF.

The Pierce oscillator is a common configuration for a xtal-controlled oscillator. To evolve a MOS version, we start with a phase-shift version of the Colpitts circuit shown in Figure 11.25a. Next, the inductance of the filter is replaced with a xtal as illustrated in Figure 11.25b. A bias and load arrangement for the inverter is shown in Figure 11.25c. Note that a depletion-mode MOS device is used as a load resistor. Bias resistors $R_1$ and $R_2$ can also be realized with small MOS devices. The xtal, for a frequency very near the series resonance of the xtal and just above resonance, will provide the necessary inductance for the feedback filter. The circuit can be redrawn as in Figure 11.25d to obtain the usual Pierce configuration for a MOS oscillator [31], [32]. An alternative configuration of this type of oscillator is shown in Figure 11.25e. One side of the xtal is grounded for this circuit. In this circuit, the gain transistor is $M_1$ while $M_2$ provides a current source. $M_3$ is a high-resistance coupling element connecting the gate of $M_1$ to the voltage divider, $R_1$ and $R_2$. By a change of ground point, the relation to the Pierce oscillator can be established.

A xtal-controlled oscillator for a bipolar circuit is shown in Figure 11.25f. The amplifier element of the oscillator is an emitter-coupled pair. As used in the circuit of Figure 10.18a, a capacitive voltage divider is included in the resonant circuit. Feedback to the ECP input is provided by the xtal. Because of the very sharp resonance property of the xtal, feedback is only effective very near the series resonance of the xtal. Thus, although the 'tank' circuit is tuned close to the desired oscillation frequency, the actual oscillation frequency is set by the xtal. If two balanced bias resistors are used with the ECP, the one on the right is bypassed with a capacitor which cannot be made too large, or squegging can occur.

From these examples, it is clear that xtal-controlled oscillators can be achieved starting from a large variety of oscillator configurations, using either the series resonance or the shunt resonance of the xtal to provide the necessary reactance or to inhibit or permit oscillation at the xtal frequency. Usually the design of an oscillator for steady-state performance starts without the xtal. The xtal is then introduced, and the design modified as needed [32], [33].

In commercially available ICs, the equivalent circuit within the IC contains many more devices than shown in the circuit diagrams of the last two chapters. Usually, very precise bias circuits are included so that the oscillator is not supply-voltage or temperature dependent. Terminals are made available to connect a desired resonant circuit, e.g., a $LC$ tuned circuit or a xtal.

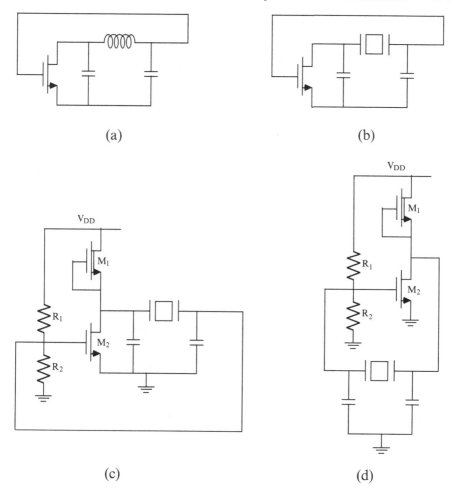

**Fig. 11.25.** (a) Phase-shift version of the Colpitts oscillator. (b) Inductance replaced by a crystal. (c) Bias and load arrangement for the inverter. (d) A rearrangement of the circuit.

## Problems

**11.1.** An oscillator configuration is shown in Figure 11.26.
(a) Design the circuit to achieve an oscillation frequency of 100 MHz with a van der Pol parameter of 0.3.
(b) Establish the effects on the performance of the circuit if the transistor parameters include $V_A = 100$ V, $R_B = 100$ $\Omega$, $C_{je0} = C_{jc0} = 0.1$ pF and $\tau_F = 0.1$ ns.

**11.2.** A transformer-coupled oscillator configuration is shown in Figure 11.27.
(a) Determine whether oscillations will build up from the quiescent state and,

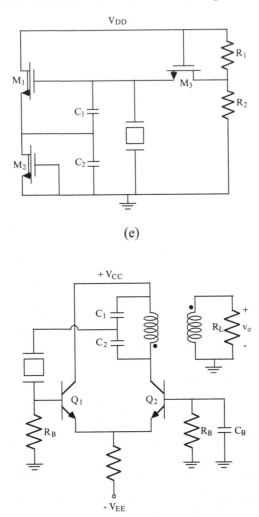

**Fig. 11.25.** (e) An alternate configuration for the Pierce oscillator. (f) A crystal-controlled bipolar oscillator circuit.

if so, with what van der Pol parameter.

(b) What is the frequency of steady-state oscillations?

(c) Estimate the value of the steady-state amplitude of the fundamental of the output voltage.

(d) Verify your analysis results with Spice.

**11.3.** An oscillator configuration is shown in Figure 11.28. The quiescent collector current is 2 mA.

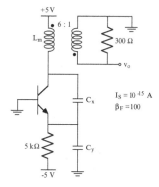

**Fig. 11.26.** Oscillator circuit for Problem 11.1.

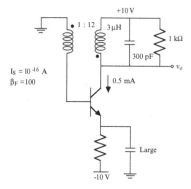

**Fig. 11.27.** Oscillator circuit for Problem 11.2.

(a) Design the circuit to achieve an oscillation frequency of 100 MHz with a van der Pol parameter of 0.5.

(b) Estimate the value of the steady-state amplitude of the fundamental output voltage.

(c) Determine the ac power dissipated in the 10 k$\Omega$ resistor.

(d) Verify your analysis results with Spice.

**11.4.** Design a BJT Colpitts oscillator for an oscillation frequency of 30 MHz. The power supply is $\pm 10$ V and the average power dissipation cannot exceed 20 mW. The BJT parameters are $\beta = 100$, $I_S = 10^{-16}$ A.

(a) Estimate the value of the steady-state amplitude of the output voltage.

(b) Redesign the oscillator with the BJT replaced by a MOSFET with $k' = 30$ $\mu$A/V$^2$, $V_T = 0.5$ V.

(c) Verify your analysis results with Spice.

**11.5.** An oscillator configuration is shown in Figure 11.29.

(a) Find the minimum value of $I_{SS}$ for which oscillations will build up from the quiescent state.

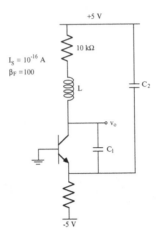

**Fig. 11.28.** Oscillator circuit for Problem 11.3.

(b) What is the frequency of steady-state oscillations?
(c) Estimate the value of the van der Pol parameter and the steady-state amplitude of the fundamental of the output voltage for $I_{SS} = 1$ mA.
(d) Verify your analysis results with Spice.

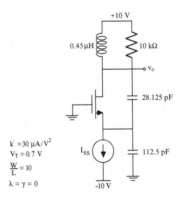

**Fig. 11.29.** Oscillator circuit for Problem 11.5.

**11.6.** An oscillator circuit is shown in Figure 11.30.
(a) Derive an expression for the start-up condition of this oscillator.
(b) What is the frequency of steady-state oscillations?
(c) What is the Q of the tuned circuit?
(d) For the given bias conditions will oscillations build up from the quiescent state?

(e) Calculate the amplitude of the voltage at the gate. Assume that the fundamental of the drain current is $1.9I_{SS}$.
(f) Verify your analysis results with Spice.

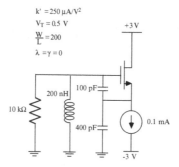

**Fig. 11.30.** Oscillator circuit for Problem 11.6.

**11.7.** An oscillator circuit is shown in Figure 11.31.
(a) Derive an expression for the start-up condition of this oscillator.
(b) What is the frequency of steady-state oscillations?
(c) What is the minimum value of the resistance $R$ for this circuit to oscillate?
(e) Calculate the amplitude of the voltage at the gate. Assume that the fundamental of the drain current is $1.9I_{SS}$ and $R = 10$ k$\Omega$.
(f) Verify your analysis results with Spice.

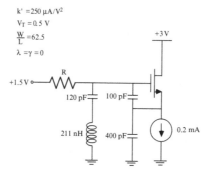

**Fig. 11.31.** Oscillator circuit for Problem 11.7.

**11.8.** An oscillator circuit is shown in Figure 11.32a. Assume the inductor is on chip and has a series resistance, $R_S$.
(a) Show that the equivalent impedance between the drain and gate of the MOSFET, $Z_{DG}(s)$ is given by the equivalent circuit representation of Figure

11.32b.
(a) Derive an expression for the start-up condition of this oscillator.
(b) What is the frequency of steady-state oscillations?
(c) What is the minimum value of the inductor $Q$ for this circuit to oscillate?
(e) Calculate the amplitude of the voltage at the drain of the MOS transistor. Assume that the fundamental of the drain current at the oscillation frequency is 1 mA and the series resistance of the inductor is 2 $\Omega$.
(f) Using the result of (d) calculate the amplitude of the gate voltage.
(g) Verify your analysis results with Spice.

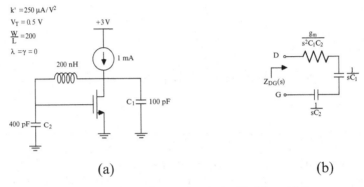

(a)                                    (b)

**Fig. 11.32.** Oscillator circuit for Problem 11.8.

# 12

## Relaxation and Voltage-Controlled Oscillators

## 12.1 Relaxation-Mode Oscillations

In the last two chapters, oscillators that were studied can be described or modeled in the steady state with the basic oscillator equation,

$$x + N(s)f(x) = A_o \tag{12.1}$$

where $f(x)$ is a device and/or circuit nonlinearity and $A_o$ represents a constant voltage or current source.

The nonlinear, time normalized, differential equation describing these oscillators has the form

$$\frac{d^2x}{dT^2} + \alpha\frac{d}{dT}F(x) + x = A'_o \tag{12.2}$$

where $F(x)$ is the combined, total nonlinearity and $\alpha$ and $A'_o$ are constants.

In Chapter 10, oscillators are studied for which the total effective nonlinearity, $F(x)$, is antisymmetric about the quiescent operating point. For these oscillators, near-harmonic oscillations are obtained in the steady state particularly if the *van der Pol parameter* $\epsilon$ of the characteristic equation at the equilibrium point is positive and small.

The normalized characteristic equation is

$$p^2 - \epsilon p + 1 = 0 \tag{12.3}$$

$$\epsilon = -\alpha\frac{dF(x)}{dx}\bigg|_{x=0} \tag{12.4}$$

For a small value of $\epsilon$, the natural frequencies $(p_1, p_2)$ of the linearized system about the equilibrium point are complex and lie in the right-hand plane near the $j\omega$ axis, as illustrated in Figure 12.1a. For a small excitation,

D.O. Pederson and K. Mayaram, *Analog Integrated Circuits for Communication*, DOI 10.1007/978-0-387-68030-9_12,
© 2008 Springer Science+Business Media, LLC

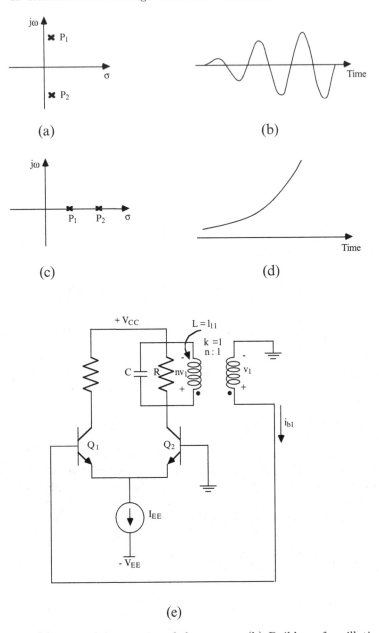

**Fig. 12.1.** (a) Natural frequencies of the system. (b) Buildup of oscillations. (c) Natural frequencies for $\epsilon > 2$. (d) Exponential growth of the oscillation. (e) An EC pair oscillator circuit.

the buildup of oscillations is a sinusoidal growth, as sketched in Figure 12.1b and as illustrated by the examples of the last two chapters.

If the $\epsilon$ parameter is positive and greater than 2, the natural frequencies of the linearized system at the equilibrium point are real, as illustrated in Figure 12.1c, and the growth of the oscillation is exponential as shown in Figure 12.1d. The ever-increasing exponential grows until a loss portion of the total nonlinearity is reached. The growth of the circuit variable then stops. If the positive-slope regions of $F(x)$ are not steep, the passive circuit can provide an almost half cycle, slightly damped sinusoidal response. The complete cycle of steady-state operation is a distorted sinusoid. This type of response is explored more fully in the next section but can also be illustrated with circuit simulation using an ECP oscillator, such as that given in Figure 10.12a and repeated in Figure 12.1e. In the Spice input circuit file of Figure 12.1f, the inductance and capacitor values are such as to provide $\epsilon = 3.84$. The output voltage response is shown in Figure 12.1g, together with the harmonic components of the output voltage waveform. As noted on the waveform, exponential growth segments can be identified, followed by a resonant half-cycle response. The oscillation period is 105 $\mu s$, somewhat longer than $2\pi\sqrt{L_2C} = 100$ $\mu s$. The even harmonics of $V_o(t)$ are very small and $HD_3 = 9.1\%$ is the major harmonic distortion factor. The behavior of this oscillator is described further in the next section.

If during the exponential growth of the voltage or current variable, new nonlinear regions of device operation are encountered, e.g., BJT saturation, the initial nonlinear differential equation, (12.2), is no longer valid. However, the oscillation frequency can remain close to the resonant frequency of the tuned circuit.

Usually, tuned circuits per se are not included in a relaxation oscillator although energy-storage elements are essential and govern the recovery periods. In a relaxation oscillator, the active devices of the oscillator act more like on-off switches rather than piece-wise-linear amplifiers. The major time segments of the oscillation period occur when the devices are off or full on rather than in the active regions.

Initially, in this chapter, a graphical analysis of a basic oscillator is used to develop further the ideas presented above. Relaxation oscillator examples are then brought in for both BJT and MOS devices. The concepts which are gained from the graphical analysis are very helpful in understanding the fundamentals of the operation of these examples. However, the steady-state operation of the oscillator can be established directly from the understanding of the circuit operation, and the recovery periods can be evaluated independently. The exponential growth segments of the oscillation period are treated separately or are assumed to be very short and are neglected.

The chapter concludes with an examination of oscillators with periods that can be easily controlled by voltage or current sources.

```
ECP OSCILLATOR, FIG 12.1
L22 1 0 95.68N
L11 5 6 38.24U
K1 L11 L22 1
RC1 6 2 1
Q11 2 1 10 MOD1
Q12 9 7 10 MOD1
VB2 7 0 0.0
VC2 5 9 0
CL 5 6 6.625P
RL 5 6 3K
IEE 10 13 PULSE 3MA 4MA 0 0 0 1S
VEE 13 0 -10
.TRAN 5N 4U 3.8U 5N
.FOUR 9.524MEGHZ V(9)
.PLOT TRAN V(9)
VCC 6 0 10
.MODEL MOD1 NPN BF=100 IS=1E-16
.OPTIONS RELTOL=1E-4 ITL5=0 LIMPTS=1000
.END
```

(f)

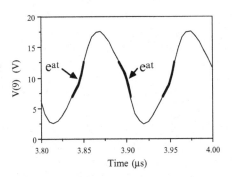

```
FOURIER COMPONENTS OF TRANSIENT RESPONSE V(9)
DC COMPONENT =    9.996D+00
```

| HARMONIC NO | FREQUENCY (HZ) | FOURIER COMPONENT | NORMALIZED COMPONENT | PHASE (DEG) | NORMALIZED PHASE (DEG) |
|---|---|---|---|---|---|
| 1 | 9.524D+06 | 7.469D+00 | 1.000000 | 172.923 | 0.000 |
| 2 | 1.905D+07 | 3.336D-02 | 0.004466 | -175.348 | -348.271 |
| 3 | 2.857D+07 | 6.808D-01 | 0.091148 | 67.138 | -105.785 |
| 4 | 3.810D+07 | 2.110D-03 | 0.000282 | 121.612 | -51.311 |
| 5 | 4.762D+07 | 1.996D-01 | 0.026721 | 28.293 | -144.630 |
| 6 | 5.714D+07 | 8.066D-03 | 0.001080 | -159.719 | -332.642 |
| 7 | 6.667D+07 | 8.335D-02 | 0.011158 | -6.956 | -179.879 |
| 8 | 7.619D+07 | 7.018D-03 | 0.000940 | 141.051 | -31.871 |
| 9 | 8.572D+07 | 4.643D-02 | 0.006216 | -44.142 | -217.064 |

```
    TOTAL HARMONIC DISTORTION =       9.595365  PERCENT
```

(g)

**Fig. 12.1.** (f) Spice input file. (g) Output voltage waveform and its Fourier components.

## 12.2 Oscillator Graphical Analysis

To obtain a graphical representation of the growth and steady-state nature of the oscillation, we start with the nonlinear, time normalized, differential equation of the oscillator (12.2) and introduce state variables and plots on the state plane. The state-variables are labeled $x$ and $y$ where

$$y = \frac{dx}{dT} \tag{12.5}$$

From (12.2) and (12.5), using incremental variables whereby $A'_o$ can be neglected, we obtain

$$\frac{dy}{dT} + \alpha \frac{dF(x)}{dT} + x = 0 \tag{12.6}$$

Alternately,

$$\frac{dy}{dT} + \alpha \frac{dF(x)}{dx} \frac{dx}{dT} + x = 0$$

Equations (12.5) and (12.6) can be rearranged to obtain the usual state-equation form

$$\frac{dx}{dT} = y \tag{12.7}$$

$$\frac{dy}{dT} = -\left[ \alpha \frac{dF(x)}{dx} y + x \right] \tag{12.8}$$

The state plane is the $x - y$ plane. A plot of the locus of the oscillation in the state plane provides a graphical portrait of the response.

At a point in the state plane, $(x_1, y_1)$, the ratio of the two state equations gives the slope of the oscillatory trajectory of the response when it is at that point.

$$\text{slope} \,|_{(x_1,y_1)} = \frac{dy}{dx}\bigg|_{(x_1,y_1)} = \frac{\frac{dy}{dT}}{\frac{dx}{dT}} \tag{12.9}$$

$$= -\frac{\alpha F'(x_1)y_1 + x_1}{y_1}$$

where

$$F'(x_1) = \frac{dF(x)}{dx}\bigg|_{x_1} \tag{12.10}$$

The slope, (12.9), can be plotted as shown in Figure 12.2a. The locus of the response of the oscillator, following an initial state or excitation, can be estimated by choosing a sequence of points, choosing each new point a short distance along the slope line of the previous point. This can be a tedious procedure since a new calculation and slope plot are needed at each point. But the principal value of the technique lies in the insight the procedure provides. In particular, notice that for a steady-state oscillation, the trajectory of the response (the locus) must be a closed path as illustrated in Figure 12.2b.

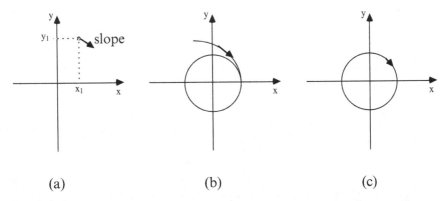

(a)                        (b)                        (c)

**Fig. 12.2.** (a) Plot of the slope. (b) The trajectory of the response for a steady-state oscillation. (c) Locus of the state at steady state.

It is necessary to stress that no bias-point shift is permitted during the build up of oscillations, such as encountered with the oscillators of Chapter 11. With a virtual bias-point shift, $A'_o$ changes and the incremental variables have a new reference, modifying the plot.

For an ideal resonator, it is a simple matter to show that the state-plane locus is a circle. This fact can be used to evaluate the performance and characteristics of other oscillators. For the ideal resonator, the total nonlinear function in (12.2) is absent, $F(x) = 0$. The oscillator differential equation and the corresponding state equations are

$$\frac{d^2x}{dT^2} + x = 0 \tag{12.11}$$

$$\frac{dx}{dT} = y \tag{12.12}$$

$$\frac{dy}{dT} = -x \tag{12.13}$$

The solution of these equations is

$$x(T) = C \cos T \tag{12.14}$$

$$y(T) = -C \sin T \qquad (12.15)$$

Notice that

$$x^2 + y^2 = C^2 \left[\cos^2 T + \sin^2 T\right] = C^2 \qquad (12.16)$$

Therefore, the locus of the state with time in the $x - y$ plane is circular for any initial excitation. This is illustrated in Figure 12.2c. In general, it is clear from the locus and from the corresponding time responses for $x(T)$ and $y(T)$, if the closed locus for an oscillator is almost circular, near-harmonic oscillation is realized.

It is convenient to plot $\alpha F(x)$ on the state plane as shown in Figure 12.3. The slope of $\alpha F(x)$ at the equilibrium point, $x = 0$, is equal to $-\epsilon$. We know from the results of the last two chapters that if the parameter $\epsilon$ is small, the plot of $\alpha F(x)$ is very shallow as shown by Curve A in the figure. It follows that if $\epsilon$ is small, the steady-state oscillation is near sinusoidal, and that the locus of the state-plane trajectory is almost circular.

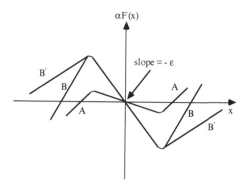

**Fig. 12.3.** Plot of $\alpha F(x)$ on the state plane.

For $\epsilon > 2$, we expect that the plot of $\alpha F(x)$ is not shallow. In some cases a sharp N-type characteristic is obtained as illustrated by Curve B in Figure 12.3. However, the positive slope regions may be quite flat, as in Curve $B'$. The latter is the case for the oscillator of Figure 12.1e. If no bias point shift is to be produced during the buildup of oscillations, the plot of $\alpha F(x)$ should be antisymmetric about the quiescent bias point, i.e., the origin of the state plane with incremental variables.

Although we can continue to deal with the regular state plane, it is more convenient for $\epsilon > 2$ cases to use a 'modified state-plane.' An auxiliary variable, rather than the usual state variable, is chosen.

$$y = \frac{dx}{dT} + \alpha F(x) \qquad (12.17)$$

The modified state equations from the original equation, (12.2), are

$$\frac{dx}{dT} = y - \alpha F(x) \tag{12.18}$$

$$\frac{dy}{dT} = -x \tag{12.19}$$

where incremental variables are used and $A'_o$ can again be neglected. The slope of the response trajectory in the new plane at a point $x_1$, $y_1$ is

$$\left.\frac{dy}{dx}\right|_{(x_1,y_1)} = -\frac{x_1}{y_1 - \alpha F(x_1)} \tag{12.20}$$

Lienard [34] proposed a very simple graphical technique to obtain the slope of the trajectory in this new plane. It is shown easily that the slope in (12.20) is the negative normal of the straight line drawn from the point $(x_1, y_1)$ to the point $(x_a = 0, y_a = \alpha F(x_1))$. This is illustrated in Figure 12.4a. As a consequence, the desired slope of the trajectory is the tangent of the arc of a circle drawn through $(x_1, y_1)$ with a center at $(x_a, y_a)$.

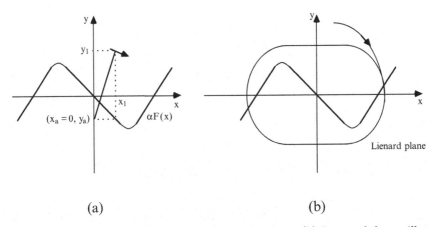

(a)                                    (b)

**Fig. 12.4.** (a) Graphical technique to obtain the slope. (b) Locus of the oscillator response in the Lienard plane.

In the modified state plane, usually referred to as the Lienard plane, the construction of the locus of the oscillator response starts with the plot of the total nonlinearity multiplied by the constant, $\alpha$, as shown in Figure 12.4b. An arbitrary point can be chosen, the points $x_a$ and $y_a$ are determined from the plot, a circular arc is drawn through $x_1, y_1$ and a new point is chosen near the first along the arc and a new arc center is established. Ultimately, we expect to achieve a closed locus for the steady-state situation.

The above construction is of course only possible and valid if the bias point does not shift due to even-ordered harmonic generation. The plot must remain fixed with respect to the equilibrium point. Nonetheless, the portrait of the oscillator in the state plane or the Lienard plane is valuable in establishing the nature of the oscillator behavior.

For typical relaxation oscillators, the parameter $\epsilon$ is greater than 2 and the positive slope regions are steep as shown in Figure 12.5a. A typical Lienard-plane locus is also shown in Figure 12.5a. It can be shown from an investigation of the instantaneous angular velocity of a point on the closed locus that the locus is very close to the steep positive slope regions of $\alpha F(x)$. In the positive slope segments, the active devices are passive. If the slopes of these segments are steep, the circuit is in a slow recovery segment of the oscillation. The locus is almost horizontal at the top and the bottom of the locus, moving from one passive region to the other. These last segments of the locus are due to the fast regenerative switching intervals. For relaxation oscillators, the locus for the steady-state response is almost a parallelogram in the Lienard plane.

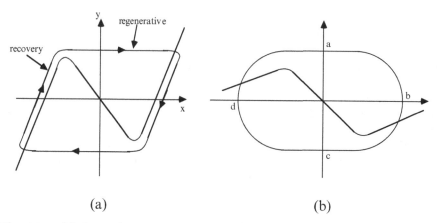

(a)                                             (b)

**Fig. 12.5.** (a) A typical Lienard-plane locus for a relaxation oscillator. (b) A different shape for the Lienard-plane locus.

If $\epsilon > 2$, but the positive regions are shallow, as for Curve B' in Figure 12.3, the Lienard-plane locus can have a shape as sketched in Figure 12.5b. As brought out in the Spice simulations leading to Figure 12.1g, a fast exponential growth occurs during Segments a and c of the locus in Figure 12.5b. During Segments b and d, the damping of the tuned circuit is small and a 'fly wheel'-type of response is produced for approximately one-half cycle for each segment. The locus is somewhat circular.

As a final point, it is to be noticed from (12.5) and (12.18) that the state plane is a distorted Lienard plane with respect to the $x$ axis, and vice versa. In the Lienard plane, the $x$ axis for the state plane is the $\alpha F(x)$ curve. The closed locus in the state plane for a relaxation oscillator can be deduced from this fact. The state-plane locus corresponding to that of Figure 12.5a is shown

in Figure 12.6. Clearly in the state plane, the geometric form of the locus of
a relaxation oscillation is not a simple parallelogram.

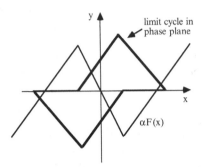

**Fig. 12.6.** State-plane locus corresponding to Figure 12.5 (a).

## 12.3 Regenerative Switching in a Relaxation Oscillator

As pointed out in the earlier sections, in a relaxation oscillator, exponential
growth is produced rather than a growing sinusoidal response when the circuit
is in the 'active' region of operation, i.e., the negative conductance region, or
when the devices are in the active region. This type of response is referred to
as the 'regenerative-switching' mode. In this section, this type of switching is
examined for the typical loop-coupled (which can also be redrawn as cross-
coupled) circuit shown in Figure 12.7a. For this circuit, it is assumed that
the bias resistors, $R_B$ and $R_C$, are such that without capacitor coupling, both
devices are in the active region of operation. With coupling and for reasonably
large values of capacitances, the capacitors can be considered to be dc voltage
sources (generalized short circuits). Clearly, a positive feedback connection
exists.

We now 'follow through' the response of the circuit where it is assumed
that both transistors are in the normal operating region. A small voltage or
current perturbation at the base of $Q_1$ is amplified by $Q_1$ with a voltage
phase reversal. This is coupled directly to the base of $Q_2$ where again an
amplification occurs with another phase change leading to positive feedback at
the base of $Q_1$. If the loop gain is greater than one, regenerative buildup occurs
until one or the other transistor leaves its active region and enters saturation
or cutoff. We show below that the regenerative switching is very fast, and
therefore the voltages of the coupling capacitors do not change significantly
during regeneration.

After regeneration, the bias arrangement of the circuit attempts to return
the circuit to the bias state of the circuit without capacitors. However, when

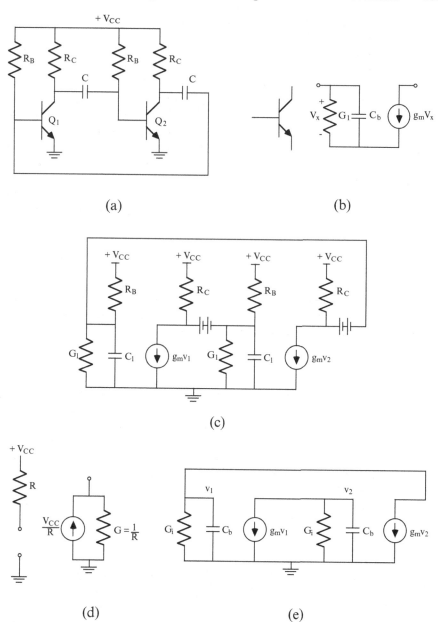

**Fig. 12.7.** (a) A typical loop-coupled circuit. (b) Circuit model for the BJT. (c) Coupling capacitors modeled as dc voltage sources. (d) Norton equivalent circuit. (e) Circuit model of the oscillator for the variational response.

the off or saturated transistors re-enter the active region, the loop gain is again greater than unity, and regenerative switching again occurs, but this time in the opposite direction until the transistors once more enter their off or saturated condition. Another recovery takes place, and the cycle starts over again. A steady-state oscillation exists.

From the follow through of the operation of the circuit, it is seen that the circuit has no stable states but does have two 'quasi-stable' states, i.e., the two recovery periods. This is the definition of an *astable* circuit.

The recovery segments of the steady-state oscillation are analyzed in the next section. In the present section, a simplified analysis of the regenerative switching is made. For the analysis, assume that both transistors are in their active regions and can be modeled with linear circuit elements as in Figure 12.7b. The capacitor $C_b$ models the charge storage in the base region and the emitter-base junction of the transistor.

$$C_b = g_m T_t \tag{12.21}$$

where $T_t$ is the equivalent transit time of the base region and includes effects of charge storage in the base and emitter-base depletion regions of the transistors. Charge storage in the collector-base depletion regions is neglected for the moment. The conductance $G_1$ models recombination effects $(G_1 \approx 1/r_\pi)$.

$$G_1 = \frac{g_m}{\beta} \tag{12.22}$$

The ohmic base resistance and the collector resistance are neglected for simplicity.

In Figure 12.7c, the coupling capacitors are modeled as dc voltage sources. The bias voltage source $V_{CC}$ and resistor $R_C$ or $R_B$ combinations can be replaced with Norton equivalents. The latter are shown in Figure 12.7d. With these models, the circuit model of the oscillator for the variational response during regenerative switching is that of Figure 12.7e. In this circuit, all dc sources are reduced to zero. Voltage sources are short circuits and current sources are open circuits. The conductances are combined:

$$G_i = G_B + G_1 + G_C \tag{12.23}$$

where $G_B = 1/R_B$ and $G_C = 1/R_C$. The nodal circuit equations for this reduced situation are

$$0 = (G_i + sC_b)\, v_1 + g_m v_2 \tag{12.24}$$
$$0 = g_m v_1 + (G_i + sC_b)\, v_2$$

The zeros on the left-hand side are the result of omitting the dc sources and the initial voltages on the capacitors. The characteristic polynomial for this situation is

$$P(s) = \det \begin{bmatrix} (G_i + sC_b) & g_m \\ g_m & (G_i + sC_b) \end{bmatrix} \tag{12.25}$$

$$= (s - s1)(s - s2)$$

where $s_1$ and $s_2$ are the natural frequencies of the linearized system about the assumed equilibrium point. Note that $P(s)$ has the form

$$P(s) = a^2 - b^2 = (a - b)(a + b) \tag{12.26}$$

Therefore, the $P(s)$ is simply factored.

$$P(s) = (sC_b - g_m + G_i)(sC_b + g_m + G_i) \tag{12.27}$$

The natural frequencies are

$$s_1 = \frac{g_m - G_i}{C_b} = \frac{g_m}{C_b}\left(1 - \frac{G_i}{g_m}\right) \tag{12.28}$$

$$s_2 = -\frac{g_m + G_i}{C_b} = -\frac{g_m}{C_b}\left(1 + \frac{G_i}{g_m}\right) \tag{12.29}$$

From an inspection of the circuit of Figure 12.7e, the loop gain of the linearized circuit at the equilibrium point is

$$\text{loop gain} = \left(\frac{g_m}{G_i}\right)^2 \tag{12.30}$$

For a loop gain greater than one, the ratio $\frac{G_i}{g_m}$ is less than one and a representative plot of $s_1$ and $s_2$ is given in Figure 12.8a. One natural frequency lies on the positive real axis and leads to a growing exponential as shown in Figure 12.8b. The other natural frequency lies on the negative real axis and its natural mode is a decaying exponential as shown in Figure 12.8b. The 'time constant' of the regenerative mode is

$$T_1 = \frac{1}{s_1} = \frac{C_b}{g_m - G_i} \tag{12.31}$$

$$= T_t \frac{1}{1 - \frac{G_i}{g_m}}$$

The time constant of regeneration is approximately equal to the simplified, effective transit time of the devices. For present-day transistors, $T_t$ is less than 0.1ns. Since a 100-to-1 increase in the value of the regenerative mode takes only 4.6 time constants, regenerative switching can be very fast.

If capacitors to model charge-storage effects in the base-collector junction are included, a two-node circuit model is still present, and the simple factoring

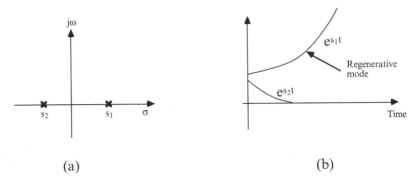

**Fig. 12.8.** (a) A representative plot of the natural frequencies. (b) Time-domain response.

of the circuit determinant can also be obtained. Four natural frequencies are produced, two in the right-half plane and two in the left-half plane. The largest natural frequency on the positive-real axis, $s_1$, establishes the regenerative switching speed. Although the magnitude of $s_1$ is smaller by the addition of $C_{jc}$ to $C_b$, regeneration is still very fast.

## 12.4 Recovery Analysis in a BJT Relaxation Feedback Oscillator

In this section, simple analyses of the recovery segments of the steady-state oscillation of a relaxation oscillator are made. In contrast to the analysis of the last section, we must now include the charging and discharging of the coupling capacitors, and the transistors are assumed to be either off or on and saturated. For simplicity, the base-bias circuits ($V_{CC}$ and $R_B$) in the loop-coupled, relaxation oscillator of Figure 12.7a are replaced in Figure 12.9a with current sources, $I_{B1}$ and $I_{B2}$. In contrast to the active-region biasing of the earlier description, these current sources are assumed to be large enough to produce saturation, $I_B > \frac{1}{\beta}\left(\frac{V_{CC}}{R_C}\right)$.

It is helpful to once more follow through the operation of the circuit. We propose a steady-state oscillation sequence. Let $Q_1$ be off and $Q_2$ be on and saturated. The biasing current $I_{B2}$ tends to hold $Q_2$ in the assumed on state. $I_{B1}$, however, can bring $Q_1$ into the active region. This involves the change of charge of $C_2$. When this occurs, regenerative switching occurs leaving $Q_1$ on and saturated and $Q_2$ off. $I_{B2}$ now changes the charge of the coupling capacitor $C_1$ and brings $Q_2$ back into the active region, leading to another regenerative switching.

During the recovery segments of operation, which are often called the quasi-stable states, significant changes of charge occur for the coupling capacitors. The time durations for these charge changes establishes the lengths

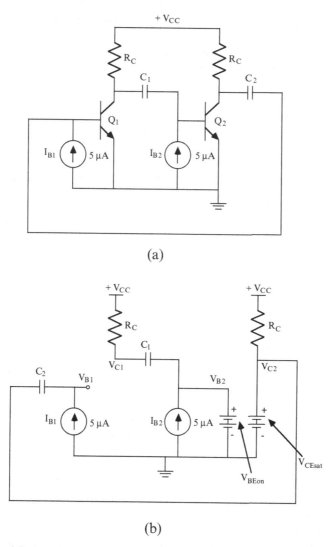

**Fig. 12.9.** (a) Another loop-coupled oscillator circuit. (b) Circuit model for the oscillator when $Q_1$ is off and $Q_2$ is on and in saturation.

of the recovery segments. For the initial recovery analysis, we assume that, after the end of a regenerative switching, transistor $Q_1$ is off and transistor $Q_2$ is on and saturated. The circuit model of the oscillator at this point of time is that of Figure 12.9b. Notice that the on transistor is modeled with two voltage sources, $V_{BEon}$ and $V_{CEsat}$. The off transistor is modeled as a three-node open circuit.

Just before the last regenerative switching, the coupling capacitor $C_1$ had an initial voltage from the collector node to the base node of $(V_{CEsat} - V_{BEon})$, since $Q_2$ was just returning from an off state into conduction, and $Q_1$ had been on with its collector at $V_{CEsat}$. We check for consistency and the validity of such assumptions after the analysis. The initial voltage of coupling capacitor $C_2$ after regenerative switching is $(V_{CC} - V_{BEon})$ for the present analysis since before regenerative switching $Q_2$ was off with a collector voltage of $V_{CC}$ and $Q_1$ was on with a base voltage of $V_{BEon}$. Looking at the two circuit segments, we see that a large base current flows into the base of $Q_2$, as shown in Figure 12.10a, because the initial voltage across $C_1$ is small. The initial current is almost $\frac{V_{CC}}{R_C}$. With time, the flow of current through $C_1$ charges it to $(V_{CC} - V_{BEon})$ and ceases to flow when $C_1$ is fully charged. The base-supply current $I_{B2}$ maintains $Q_2$ on. The waveform of $V_{C1}(t)$ starts at $V_{CEsat}$ and recovers to $V_{CC}$. It is sketched in Figure 12.10a.

For the other circuit segment, a drop of $V_{C2}(t)$ from $V_{CC}$ to $V_{CEsat}$ as shown in Figure 12.10b occurs after regenerative switching as $Q_2$ moves from off to on. This change of voltage is coupled across $C_2$ to $V_{B1}$ with little change. Thus, $V_{B1}$ drops by $(V_{CC} - V_{CEsat})$ from $V_{BEon}$ to

$$V_{B1}(0) = V_{BEon} - (V_{CC} - V_{CEsat}) \tag{12.32}$$

For a numerical example, let $V_{CC} = 5$ V, $V_{BEon} \approx 0.8$ V and $V_{CEsat} \approx 0$ V. Other circuit elements are assumed to be $R_C = 10$ k$\Omega$, $C_1 = C_2 = 100$ pF and $I_{B1} = I_{B2} = 5$ $\mu$A. For these values,

$$V_{B1}(0) = -4.2 \text{ V} \tag{12.33}$$

This negative value of $V_{BE1}$ holds $Q_1$ off. However, the base-bias source $I_{B1}$ flows through $C_2$ to the saturated collector of $Q_2$ since $Q_1$ is off. This discharges $C_2$, and $V_{BE1}$ rises as shown in Figure 12.10b. The left-hand side of $C_2$ is held at $V_{CEsat}$ by the on transistor $Q_2$. For the simple circuit of Figure 12.9b, $V_{B1}(t)$ will attempt to increase without limit because of the current drive $I_{B1}$. However, when $V_{B1}(t) = V_{BEon} \approx 0.8$ V, $Q_1$ returns to the active region, the circuit model of Figure 12.9b is no longer valid, and regenerative switching occurs.

The time for $V_{B1}(t)$ to reach $V_{BEon}$ can be found from the simple current-charge relation for $C_2$.

$$i = C_2 \frac{dv}{dt} \tag{12.34}$$

$$dt = C_2 \frac{dv}{i} \tag{12.35}$$

The charging current is $I_{B1}$, and increments of time and voltage can be used instead of the differentials, leading to

$$\Delta t = C_1 \frac{\Delta V}{I_{B1}} \tag{12.36}$$

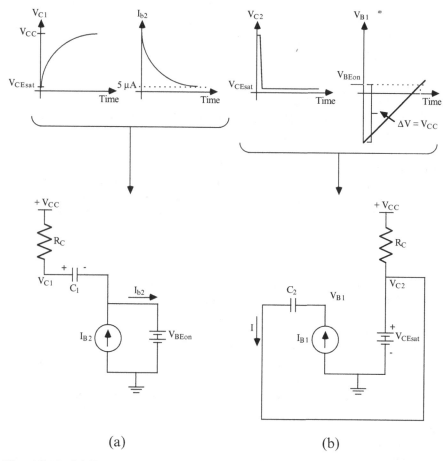

**Fig. 12.10.** (a) First circuit segment and associated waveforms. (b) Second circuit segment and waveforms.

The necessary change of voltage is equal to the change of $V_{B1}$. From (12.33) and if $V_{CEsat}$ is neglected, this change is from $-4.2$ V to $0.8$ V $= 5$ V $= V_{CC}$. The time duration of this charging is, using the circuit value above,

$$\Delta t = C_2 \frac{V_{CC}}{I_{B1}} = 100 \text{ pF} \frac{5 \text{ V}}{5 \text{ }\mu\text{A}} = 100 \text{ }\mu\text{s} \tag{12.37}$$

$\Delta t$ is the governing length of the recovery, since as described above, the other portion of the circuit remains in a state keeping $Q_2$ on.

Note from Figure 12.10a that $V_{C1}(t)$ returns to $V_{CC}$ long before the next regenerative switching occurs. It takes about 5 time constants for a capacitor to charge fully. Therefore, $V_{C1}(t)$ returns to $V_{CC}$ in approximately

$$5T_{RC} = 5R_C C_1 = 5 \ \mu s. \tag{12.38}$$

This is much less than the recovery period.

At the end of the recovery period, $C_1$ is charged to $(V_{CC} - V_{BEon})$ and $C_2$ is discharged to $-(V_{BEon} - V_{CEsat})$, again with positive node as the collector side.

We have obtained a consistent pattern of circuit behavior, starting with assumed initial capacitor voltages at the beginning of a recovery period (and the end of a regenerative switching) ending with the same capacitor voltages for the opposite set at the beginning of the next recovery segment of operation.

For a complete cycle of oscillation, the second recovery time must also be calculated. For equal capacitors and bias elements, the second recovery is the same as the first. Therefore, the period of oscillation for the values in (12.37) is

$$T_p = 2 \times 100 \ \mu s = 200 \ \mu s \tag{12.39}$$

The input file for a Spice simulation of this circuit is given in Figure 12.11a. The circuit values are the same as used above. The small node-to-ground capacitors are included to insure convergence of the simulation, since no charge storage is included in the transistors. A pulse input is used with one base bias current source to initiate action properly. The waveforms of $V_{B1} = V(1)$ and $V_{C2} = V(6)$ variables are shown in Figure 12.11b. The waveforms across the coupling capacitors are given in Figure 12.11c. The waveforms substantiate our evaluation above; however, the simulated period is only 184 $\mu s$. Actual values of $V_{BEon}$ and $V_{CEsat}$ differ from those assumed and the change of capacitor voltage is seen from Figure 12.11c to be approximately 4.5 V, not 5 V. In addition, the transistors turn on for $V_{BE} < 0.8$ V.

The base-current sources can be replaced with the $V_{CC}$, $R_B$ combinations of Figure 12.7a. These elements are included in Figure 12.11a as commented lines. It is a simple matter to show that the recovery period is now

$$\Delta t = \frac{T_p}{2} = R_B C_1 \ln \left[ \frac{2V_{CC} - V_{CEsat} - V_{BEon}}{V_{CC} - V_{BEon}} \right] \tag{12.40}$$

In order to achieve the 5 $\mu A$ base current, $R_B$ must be 840 k$\Omega$. For values of $V_{CEsat} = 0.1$ V and $V_{BEon} = 0.8$ V, the estimated period is

$$T_p \approx 130 \ \mu s \tag{12.41}$$

The observed period from a Spice simulation is 116 $\mu s$.

## 12.5 Other Astable Oscillators

A common method to produce an astable relaxation oscillator is to start with a dc-coupled bistable circuit. A general example is shown in Figure 12.12a.

```
FIGURE 12.11
.PLOT TRAN V(1) V(6) (-5,5)
.PLOT TRAN V(3,4) V(6,1) (-5,5)
.OPTION RELTOL=1E-6 ITL5=0
I1 0 1 5U
*RB1 5 1 840K
CB1 1 0 1P
Q1 3 1 0 MOD1
RC1 5 3 10K
C1 3 4 100P
I2 0 4 5U PULSE 0 5U 0 0 0 500U
*RB2 7 4 840K
*VCC2 7 0 5 PULSE 0 5 0 0 0 500U
CB2 4 0 1P
Q2 6 4 0 MOD1
RC2 5 6 10K
C2 6 1 100P
VCC 5 0 5
.MODEL MOD1 NPN IS=1E-16
.OPTION NOPAGE NOMOD
.WIDTH OUT=80
.TRAN 4U 380U 180U 2U
.END
```

(a)

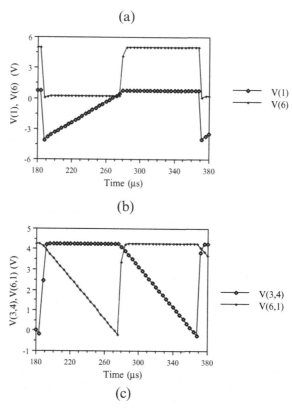

(b)

(c)

**Fig. 12.11.** (a) Spice input file for relaxation oscillator. (b) Waveforms of voltages $V_{B1}$ and $V_{C2}$. (c) Voltage waveforms across the coupling capacitors.

The two inverters connected in a loop produce a bistable circuit if the input on voltage of an inverter is greater than the output on voltage. To achieve an astable circuit, ac coupling, usually with capacitor coupling, is introduced. New bias elements are also introduced to provide each inverter with a definite bias state in the absence of coupling. The arrangement shown in Figure 12.12b is that used in the previous astable BJT oscillator examples.

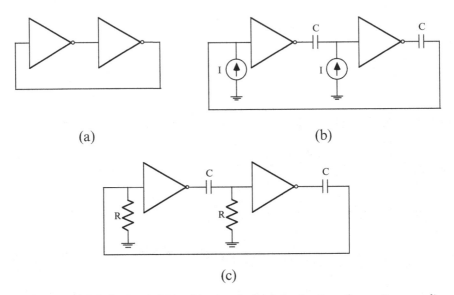

(a)                                (b)

(c)

**Fig. 12.12.** (a) A dc-coupled bistable circuit. (b) Introduction of capacitor coupling and bias elements. (c) Another biasing alternative.

Another bias possibility is to return the bias resistors to ground, as illustrated in Figure 12.12c. This corresponds to an off bias for the transistors. However, with an adequate excitation, often obtained by just turning on the circuit, steady-state relaxation oscillations are produced. To determine the length of the recovery for the configuration of Figure 12.7a, but returning the $R_B$ to ground, one must analyze the time for the base current to drop from the initial heavy overdrive into the on transistor to a value corresponding to the base current at the edge of the active region. This analysis is left for a problem.

Since the early days of electronics, there have been devices which have a V-I or I-V characteristics which display a negative-resistance (conductance) shape, i.e., have a negative-slope region bordered by two positive-slope regions. In addition to the tunnel diode introduced in Section 10.2, examples include the point-contact transistor, the unijunction transistor, the four-layer switch and the silicon-controlled rectifier (SCR). Using a properly chosen resistive load together with a voltage or current source, one can obtain bistable operation from these devices. As brought out above, by suitably modifying the bias arrangement and by introducing the correct energy-storage element, monostable or astable operation can be achieved. There is extensive literature on these devices and their use. For low-power applications and in particular for integrated circuit realization, these negative resistance (conductance) devices have not proven to be the best basis to achieve these relaxation-circuit functions. Because the characteristics are usually temperature and aging sensitive and because of the problems of triggering and output loading, the basic device must be augmented with many other elements. Consequently, the preferred approach has been to start with a two inverter configuration, such as shown in Figure 12.12a. Additional transistors can be added with little penalty in an IC design, and a designable, insensitive configuration can be obtained. An example of this is introduced with the voltage-controlled oscillator of Section 12.10.

## 12.6 A CMOS Relaxation Oscillator

In the circuit of Figure 12.13a, only one capacitor is needed to achieve the relaxation oscillation. For the example to be analyzed here, CMOS inverters are used, and the complete circuit is that of Figure 12.13b. The bulk connections are returned to the lowest voltage for the n-channel devices and to the highest voltage for the p-channel devices. It is important to note that the inverters in this oscillator can either supply both current to charge the capacitor and current to the bias circuitry of the other inverter, or can also 'sink' current from the other inverter, i.e., can accept current and discharge the coupling capacitor. For the BJT circuit of Section 12.4, the saturated transistor sinks the current during one segment of the oscillation while the collector resistor 'sources' (supplies) current while its transistor is off. For a CMOS inverter, the NMOS transistor can sink the discharging current while the PMOS transistor sources the charging current.

As usual, it is necessary to think through the steady-state operation of the relaxation oscillation before setting up the analysis. Assume that the input voltage to the left-hand inverter, A, is low. For Inverter A, its NMOS device is off, the PMOS device is on, and the output voltage is high. Because of this, the other inverter, B, has its PMOS device off and its NMOS device on and can supply current to a load. The reduced circuit is shown in Figure 12.14a. For Inverter A, only if there is an open-circuit load will the PMOS unit have no current. This is the usual condition for dc coupling as in logic circuits. In our case, the P device can supply current through the resistor R to charge the capacitor C with the current being sunk by the (on) N device of Inverter B. Of course, this charging of C increases the input voltage of Inverter A with time as shown in Figure 12.14b for $t$ approaching $t_1$. Ultimately, Inverter A reaches its threshold voltage at $t = t_1$ and becomes active leading to a regenerative switching which leaves Inverter B in the off state with its input voltage low.

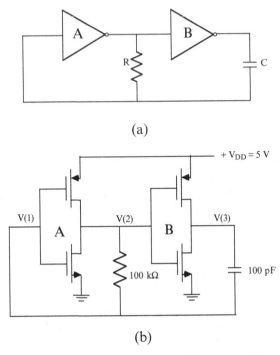

(a)

(b)

**Fig. 12.13.** (a) Schematic of CMOS relaxation oscillator. (b) Complete circuit of the relaxation oscillator.

Now Inverter B through its P device can source current through C and R and into the output of Inverter A. This is illustrated in Figure 12.14c. Inverter A's N device now sinks the current, and C discharges until $V(1)$ reaches the threshold value and a second regenerative switching occurs.

The waveform of $V(1)$ can be expected to be that of Figure 12.14b. At $t_1$, regenerative switching occurs, turning Inverter A on. Since the capacitor voltage cannot change instantly, the input voltage of A jumps from the threshold voltage by the same amount of the change of the output voltage of B. Before switching, $V(3)$ is clamped to almost zero voltage, and after switching $V(3)$ is almost $V_{DD}$. Therefore, $V(1)$ jumps by $V_{DD}$.

Similarly, $V(1)$, at the end of the next regenerative switching, drops from the threshold voltage by an amount $V_{DD}$ after the next switching, since $V(3)$ drops from $V_{DD}$ to zero.

To estimate the half-period of the oscillation, we need the values of the turn-on voltage and the turn-off voltage of the inverters. These are approximately $V_{TN}$ and $V_{DD} - V_{TP}$. However, it is simpler to assume that the critical input voltage for the inverter, $V_x$, for appreciable current conduction is the midpoint voltage of the voltage transfer characteristic of the CMOS inverter.

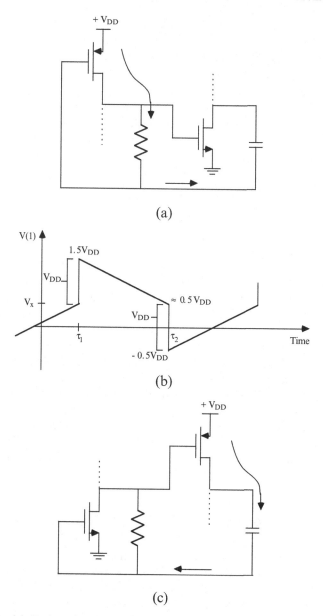

**Fig. 12.14.** (a) Reduced circuit with inverter A's NMOS device off. (b) Voltage waveforms across the capacitor. (c) Inverter B sourcing current through C and R.

$$V_x = \frac{V_{DD} + 0}{2} \tag{12.42}$$

The time for $V(1)$ to change from $V_{DD} + V_x$ to $V_x$ is estimated to be

$$\Delta t = RC \ln \left[ \frac{V_{DD} + V_x}{V_x} \right] = RC \ln 3 = 1.1 RC \tag{12.43}$$

The same result is obtained for the recovery of $V(1)$ to $V_x$ during the next recovery segment of operation. The period of oscillation is then

$$T_p = 2\Delta t = 2.2 RC \tag{12.44}$$

For $R = 100$ k$\Omega$ and $C = 100$ pF,

$$T_p = 22 \ \mu s. \tag{12.45}$$

The Spice input file for this circuit is given in Figure 12.15a. The waveforms of $V(1), V(2), V(3)$ and $V(3, 1)$ are given in Figures 12.15b and c. Note that the threshold voltage at switching is approximately 1.77 V when Inverter A turns on and 2.1 V when turning off. The estimated value of $V_x$ is 2.5 V. The simulated period of oscillation is 22.5 $\mu s$. $V(3)$ has a very rectangular waveform.

This single-capacitor relaxation oscillator can also be used with other MOS inverters. In Figure 12.15d, the circuit and the Spice input file are given for an oscillator in which the inverters are simple NMOS enhancement-mode devices with a resistive load. The waveforms of the capacitor and node voltages are given in Figures 12.15e and f.

Bipolar versions of this circuit are possible but care must be taken to incorporate the transistor base currents. If BJTs are substituted for the MOS devices in the circuit of Figure 12.15d, the right-hand device can be a Darlington connection of a double transistor.

## 12.7 Voltage- and Current-Controlled Oscillators

In many applications, the oscillation frequency of the relaxation oscillator needs to be varied with a control source. An example is the phase-locked loop circuit studied in Chapter 15. For the BJT example of the Section 12.4, the period of oscillation is directly proportional to the value of the base-bias current source. Usually, current sources are not available directly, but are approximated by transistor circuits which are voltage-source driven. An example for the BJT oscillator is that shown in Figure 12.16. Positive base bias first is produced by the resistors $R_B$ and $V_{CC}$. The control currents reduce this level. The control voltage is $V_{BS}$, but could be a combination of $V_{BS}$ and $V_{EE}$. In Section 12.10, a similar bias configuration is used in a single-capacitor, voltage-controlled oscillator (VCO).

```
CMOS REL OSC
.TRAN 0.5U 50U 20U 0.5U
.PLOT TRAN V(3,1) V(3) (-2,6)
.PLOT TRAN V(1) V(2) V(3) (-3,7)
M1 2 1 4 4 MOD1 W=80U L=8U
M2 2 1 0 0 MOD2 W=80U L=8U
M3 3 2 4 4 MOD1 W=80U L=8U
M4 3 2 0 0 MOD2 W=80U L=8U
C1 1 0 1P
I1 0 1 PULSE 10U 0 0 0 0 100U
C2 2 0 1P
R1 2 1 100K
CC 3 1 100P
VCC 4 0 5
.OPTIONS RELTOL=1E-6 ITL5=0
.MODEL MOD1 PMOS VTO=-0.5 KP=10U
.MODEL MOD2 NMOS VTO=0.5 KP=30U
.END
```

(a)

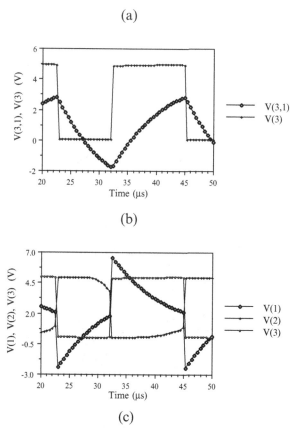

(b)

(c)

**Fig. 12.15.** (a) Spice input file. (b) Waveforms of V(3,1) and V(3). (c) Waveforms of V(1), V(2), and V(3).

```
NMOS REL OSC
.TRAN 0.5U 50U 20U 0.5U
.PLOT TRAN V(3,1) V(3) (-5,5)
.PLOT TRAN V(1) V(2) V(3) (-5,5)
RD1 4 2 10K
RD2 4 3 10K
M2 2 1 0 5 MOD2 W=80U L=8U
M4 3 2 0 5 MOD2 W=80U L=8U
VSS 5 0 -0
C1 1 0 1P
I1 0 1 PULSE 10U 0 0 0 0 100U
C2 2 0 1P
R1 2 1 100K
CC 3 1 100P
VCC 4 0 5
.OPTIONS RELTOL=1E-6 ITL5=0
.MODEL MOD1 PMOS VTO=-0.5 KP=10U
.MODEL MOD2 NMOS VTO=0.5 KP=30U
.END
```

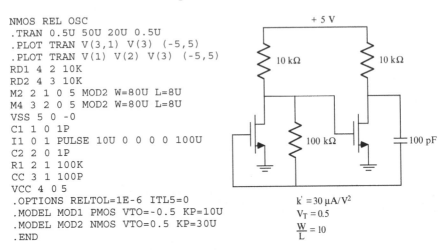

$k' = 30\,\mu A/V^2$

$V_T = 0.5$

$\dfrac{W}{L} = 10$

(d)

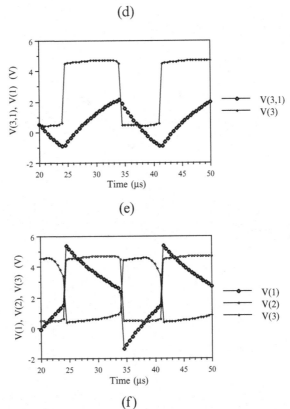

(e)

(f)

**Fig. 12.15.** (d) Circuit and Spice input file. (e) Waveforms of V(3,1) and V(3). (f) Waveforms of V(1), V(2), and V(3).

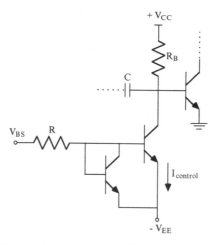

**Fig. 12.16.** A voltage-controlled bias arrangement.

For the CMOS oscillator of the last section, the period of oscillation is dependent on the value of the supply voltage, $V_{DD}$. However, one does not usually make the control variable the major system supply voltage. It is possible to achieve current (voltage) control by introducing a bias offset current at the input, $V(1)$, of Inverter A of Figure 12.13b.

## 12.8 An Astable Schmitt Circuit

A basic voltage-controlled oscillator can be produced by a modification of the emitter-coupled bistable circuit of Figure 12.17a. The predecessor of this circuit is the cathode-coupled Schmitt trigger circuit using vacuum tubes and was first described in 1938 [35]. (The basic loop-coupled bistable circuit was first described by Eccles and Jordan in 1919 [36].) The Schmitt circuit, using BJT or MOS devices, has many variants and has many applications as a bistable, monostable, or astable circuit. In this section, we briefly review the basic bistable circuit using BJTs and modify it for astable operation. In Section 12.10, a common VCO version is analyzed.

A practical Schmitt circuit is illustrated in Figure 12.17a and contains several resistive bias elements. A simple emitter resistor source is used to realize a current source $I_{EE}$. The circuit of Figure 12.17b contains only the essential elements for bistable behavior (and even $R_{C2}$ can be omitted if a separate voltage supply is used). As with the loop-coupled bistable circuit, bistability for the basic circuit depends upon the fact that $V_{CEsat}$ is less than $V_{BEon}$.

$$V_{BEon} > V_{CEsat} \qquad (12.46)$$

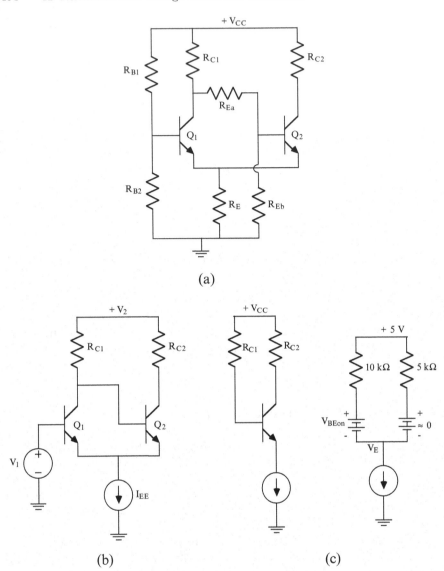

(a)

(b)                                    (c)

**Fig. 12.17.** (a) An emitter-coupled bistable circuit. (b) Circuit with only the essential elements for bistable behavior. (c) Simple circuit with $Q_1$ off and $Q_2$ on.

The bistability of the circuit is easiest to establish with a numerical example. Let $V_{BEon} = 0.8$ V, $R_{C1} = 10$ k$\Omega$, $R_{C2} = 5$ k$\Omega$, $V_{CC} = 5$ V, and $I_{EE} = 1$ mA. For $Q_1$ off and $Q_2$ on, the simple circuit of Figure 12.17c can be used to determine that $V_E = 1.4$ V. $V_1$ must then be less than 2.2 V for bistability. Choose $V_1 = 2.0$ V which leads to $V_E = 1.2$ V when $Q_1$ is on, (and saturated). The collector voltage of $Q_1$ is then also 1.2 V. Since $Q_2$ is off with its $V_{BE} < 0.8$ V, i.e., its base voltage $< 2.0$ V, the circuit is bistable.

To produce an astable circuit, the emitters are ac coupled with a capacitor as shown in Figure 12.18a. The emitter-supply current has been separated into two. Note that the bias arrangements of both transistors are for on operation. Once again, we 'walk through' the astable operation. If $Q_1$ is off, $V_{E1}$ must be greater than $V_1 - V_{BEon} = 1.2$ V, as illustrated in Figure 12.18b. $V_{E2}$ is fixed by the bias arrangement and is equal to 1.4 V. Note that, in spite of the presence of C, the emitter current of $Q_2$ is $I_{EE}$, since $Q_1$ is off, and $\frac{I_{EE}}{2}$ must flow through C. With time, the flow of $\frac{I_{EE}}{2}$ through C changes its charge and decreases $V_{E1}$ until $Q_1$ enters the active region at $V_{E1} = 1.2$ V, as shown in Figure 12.18b. Regenerative switching then occurs, leaving $Q_1$ on and $Q_2$ off. In this new quasi-stable state, $V_{E1}$ is fixed by the bias arrangement at $V_{E1} = 1.2$ V, and $V_{E2}$ must equal 1.2 V + 0.2 V = 1.4 V, since the capacitor had 0.2 V across it when switching occurred. $V_{E2}$ now decreases with time due to the flow of $\frac{I_{EE}}{2}$ through the capacitor, until $V_{E2} = 0.4$ V when $V_{BE}$ of $Q_2$ is equal to 0.8 V. After the second regenerative switching, $V_{E1}$ jumps to 2.2 V, since the capacitor voltage has changed to 1 V. The voltage change for both quasi-stable states (recovery periods) is 1 V. For $C = 3$ nF and $I_{EE} = 1$ mA, the period of the steady-state oscillation is

$$T_p = 2 \times C \frac{V}{\frac{I_{EE}}{2}} = 12 \ \mu s \tag{12.47}$$

The collector voltages have approximate rectangular waveforms with an amplitude of $R_C \frac{I_{EE}}{2}$, as shown in Figure 12.18b.

The Spice input file for this circuit is given in Figure 12.19a, where a step function for the voltage source, $V_1$, is used to excite the oscillation. The waveforms for the emitter voltages are shown in Figure 12.19b and for the collector waveforms in Figure 12.19c. The voltage levels as estimated are only approximately realized, and it is seen that the period of oscillation is 9 $\mu s$, which is less than the predicted value. However, the predicted shapes of the waveforms for $V_{E1}$ and $V_{E2}$ are obtained. From a sequence of simulation runs, it is found that proper operation of this circuit depends upon critical values of the sources and excitation. For this reason, this circuit is not used in practice. Instead, a more complicated version is used which employs both loop and emitter coupling together with voltage limiting diodes. This circuit is examined in the last section of this chapter.

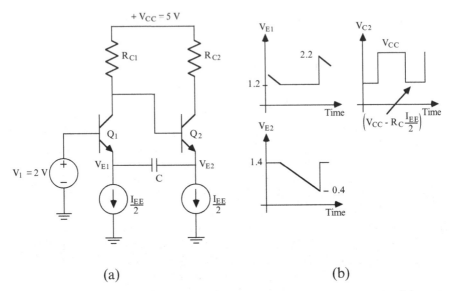

(a)                                          (b)

**Fig. 12.18.** (a) Circuit with capacitor coupling between the emitters. (b) Node voltage waveforms during astable operation.

## 12.9 Equivalence of the Schmitt and Loop-Coupled Bistable Circuits

It is of interest to illustrate that the basic emitter-coupled and collector-base (loop) coupled circuits are equivalent in circuit configuration as well as functionally as bistable circuits. We start with the Schmitt circuit of Figure 12.17b, which is repeated here as Figure 12.20a.

If only the essential circuit elements are retained, we produce the circuit of Figure 12.20b. $I_{EE}$ is relabeled $I_2$. $R_{C1}$ and $V_{CC}$ are replaced with a pure current source $I_1$. $R_{C2}$ can be omitted if a separate voltage supply $V_2$ is used. Bistability is assured with $I_1 \leq I_2$ and $V_1 \leq V_2$. The circuit is next redrawn, and the ground point is moved, as in Figure 12.20c. If $I_2$ is moved across $V_2$, only the power dissipation is changed. Bistability is maintained for $V_1 = V_2$ and for $I_2 = I_1$. We thus have the basic prototype of the loop coupled circuit. If the currents sources are approximated with $V_{CC}$, $R_C$ elements, the familiar loop-coupled circuit of Figure 12.20e is produced.

## 12.10 A BJT VCO

A popular voltage-controlled oscillator in integrated-circuit realizations is shown in Figure 12.21a. The essential circuit is shown in Figure 12.21b. Two emitter current sources, $I_{1a}$ and $I_{1b}$, are used, which are voltage controlled in

```
EC-COUPLED OSC, FIG 12.20
.OPTIONS RELTOL=1E-6
VCC 5 0 5
V1 1 0 2 PULSE 2.3 2.1 0 0 0 200U
Q1 2 1 3 MOD1
.MODEL MOD1 NPN IS=1E-16
RC1 5 2 10K
IE1 3 0 0.5M
Q2 4 2 6 MOD1
RC2 5 4 5K
IE2 6 0 0.5M
CC 3 6 3N
CC1 1 0 1P
CC2 2 0 1P
.TRAN .4U 40U 20U 0.2U
.PLOT TRAN V(3) V(6) (0,3)
.PLOT TRAN V(2) V(4) (0,5)
.END
```

(a)

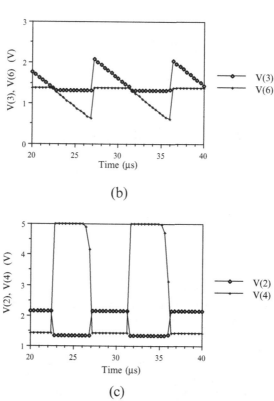

(b)

(c)

**Fig. 12.19.** (a) Spice input file. Voltage waveforms at the (b) emitters, and (c) collectors.

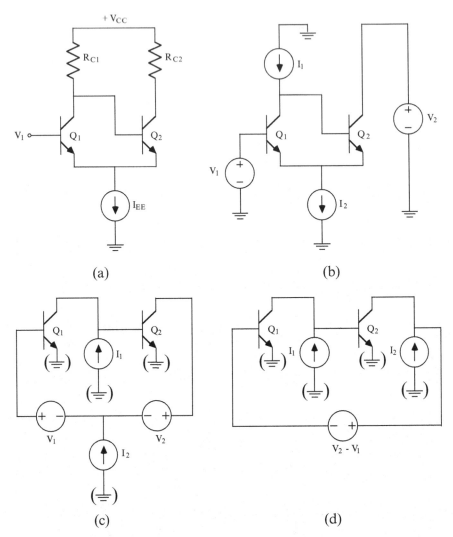

**Fig. 12.20.** (a) Schmitt circuit. (b) Circuit with only the essential elements retained. (c) Circuit redrawn with a new ground point. (d) Circuit with voltage and current sources rearranged.

the practical circuit. Notice that cross coupling (which can be drawn as loop-coupling) of $Q_1$ and $Q_2$ is used, as well as emitter coupling via the capacitor $C$. Buffering emitter followers $Q_3$ and $Q_4$ are used as coupling elements from the transistor collectors to transistor bases in the loop. Two additional current sources, $I_B$, are used at the bases of $Q_1$ and $Q_2$ to provide bias for the emitter followers, $Q_3$ and $Q_4$. The diodes $Q_5$ and $Q_6$ (usually diode-connected transistors) act as voltage clamps and limit the value of the voltage drop across

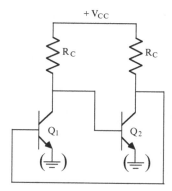

**Fig. 12.20.** (e) The familiar loop-coupled circuit.

the resistors, $R_1$ and $R_2$. Recall that there is virtually no penalty in an IC realization to use a large number of small transistors to achieve a well-defined design.

The operation of the oscillator is first established by following through a cycle of operation using the waveforms of Figure 12.21c. A numerical example and Spice simulations are then introduced. We assume that the regenerative switching is very fast and that the voltage across $C$ does not change appreciably during regeneration. We also assume that when current flows through resistors $R_1$ or $R_2$, the voltage drop is sufficient to turn on the diodes $Q_5$ or $Q_6$. Base currents are assumed to be negligible with respect to collector currents, i.e., the values of $\beta$ are large.

We start by setting $Q_1$ off and $Q_2$ on. Since $Q_2$ is on, $Q_6$ conducts with a voltage drop, across $R_2$, of $V_{BEon}$. The base voltage of $Q_4$ is

$$V_{B4} = V_{CC} - V_{BEon} \tag{12.48}$$

The base voltage of $Q_1$, as shown in the idealized waveforms of Figure 12.21c, is

$$V_{B1} = V_{B4} - V_{BEon} = V_{CC} - 2V_{BEon} \tag{12.49}$$

Since base currents are neglected, the base voltage of the emitter follower $Q_3$ is

$$V_{B3} = V_{CC} \tag{12.50}$$

and, as shown in the waveforms,

$$V_{B2} = V_{CC} - V_{BEon} \tag{12.51}$$

$$V_{E2} = V_{CC} - 2V_{BEon} \tag{12.52}$$

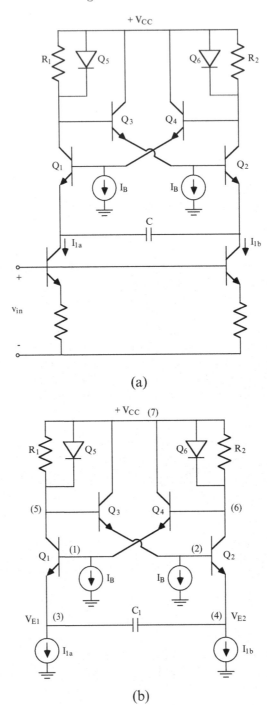

(a)

(b)

**Fig. 12.21.** (a) A popular voltage-controlled oscillator in integrated realizations. (b) The essential circuit of the voltage-controlled oscillator.

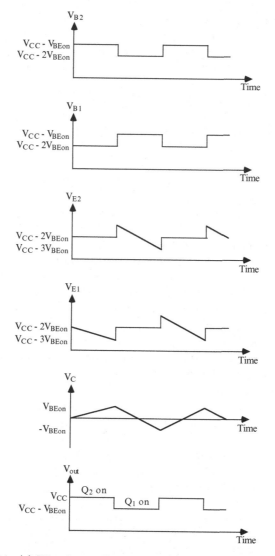

**Fig. 12.21.** (c) Waveforms illustrating the operation of the oscillator.

Let the initial voltage of $C$ be zero (uncharged). This leads to

$$V_{E1} = V_{E2} = V_{CC} - 2V_{BEon} \tag{12.53}$$

As shown in the waveforms, notice that the above state confirms that $Q_1$ is off. The current source $I_{1a}$ on the left charges $C_1$, and $V_{E1}$ is pulled lower with time. Ultimately, $Q_1$ turns on when $V_{E1} = V_{CC} - 3V_{BEon}$, and regenerative switching occurs to the other quasistable state with $Q_1$ on and $Q_2$ off. At the end of regenerative switching, $V_{E2} = V_{CC} - 2V_{BEon}$ and with time is pulled lower by the right-hand current source, $I_{1b}$, ending with another regenerative switching and the cycle is complete.

During a half cycle, i.e., during one quasi-stable state, the capacitor voltage changes by $2V_{BEon}$. The time for this to occur is

$$\frac{T_p}{2} = C_1 \frac{2V_{BEon}}{I_1} \qquad (12.54)$$

where $T_p$ is the period of the oscillation and $I_1 = I_{1a} = I_{1b}$. The frequency of oscillation, $f_o$, is

$$f_o = \frac{1}{T_p} = \frac{I_1}{4C_1 V_{BEon}} \qquad (12.55)$$

Since $I_1$ can be linearly proportional to $V_{in}$, a linear voltage controlled oscillator is produced. Since $f_o$ depends upon $V_{BEon}$ which is temperature dependent, the frequency of oscillation depends upon temperature. However, the current sources $I_1$ can be compensated to cancel this effect.

For a numerical example, the device and circuit parameters are chosen as shown in the Spice input file of Figure 12.22a. Note that the base bias currents are 0.5 mA and $I_1 = 1$ mA. A pulse start is used to initiate operation. Assume that $V_{BEon} = 0.8$ V. For $V_{CC} = 10$ V, $V_{B1}$ and $V_{B2}$ should switch between 8.4 V and 9.2 V. $V_{E1}$ and $V_{E2}$ should switch between 7.6 V, 8.4 V, and 9.2 V. The midportion occurs when its transistor is conducting. The output voltage should switch between 9.2 V and 10 V. Finally, the period of the oscillation should be

$$T_p = 4 \times 0.1 \ \mu F \times \left( \frac{0.8 \ V}{1 \ mA} \right) = 320 \ \mu s \qquad (12.56)$$

From the Spice simulation results of Figures 12.22b and c, the switching points of the $V_E$ are 7.83 V, 8.45 V and, 9.07 V, and the period of oscillation is 255 $\mu s$. The small voltage differences in the value of $V_{BEon}$ make a large change in $T$.

Another version of this type of relaxation oscillator is shown in Figure 12.23. Notice that fixed current supplies are used that include bias setting diodes $D_1$ and $D_2$. The clamp diodes across $R_1$ and $R_2$ are not used. If an analysis similar to that above is followed, it can be shown that the time duration of one quasistable state is $4R_1C_1$. The frequency of oscillation is then

$$f_o = \frac{1}{8R_1C_1} \qquad (12.57)$$

This relaxation oscillator is not voltage controlled and is not even supply-voltage dependent.

```
VOLTAGE CONTROLLED OSCILLATOR
.WIDTH OUT=80
RC1 7 5 1K
RC2 7 6 1K
Q5 7 7 5 MODN
Q6 7 7 6 MODN
Q3 7 5 2 MODN
Q4 7 6 1 MODN
IB1 2 0 .5MA
IB2 1 0 .5MA
CB1 2 0 1PF
CB2 1 0 1PF
Q1 5 1 3 MODN
Q2 6 2 4 MODN
C1 3 4 .1UF
IS1 3 0 PULSE 5MA 1MA 1NS 1NS 1NS 50MS
IS2 4 0 1MA
VCC 7 0 10
.MODEL MODN NPN IS=1E-16 BF=100 RB=10 RC=10 VA=100
.OPTION ITL5=0 LIMPTS=5000
.OPTIONS NOPAGE NOMOD
.TRAN 5US 500US 200U 5US
.PLOT TRAN V(5) V(1) V(3) (7,10)
.PLOT TRAN V(3) V(4) (7,10)
*.PLOT TRAN V(4,3) (-2, 2)
.END
```

(a)

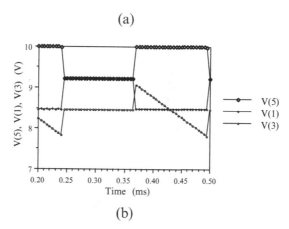

(b)

**Fig. 12.22.** (a) Spice input file. (b) Waveforms of voltages V(5), V(1), and (V3).

## Problems

**12.1.** A voltage-controlled oscillator is shown in Figure 12.24. Note the transistor current sources.

(a) For $V_{in} = 0$ V, determine the frequency of oscillation and sketch the waveforms at the base and collector of $Q_2$.

(b) Determine the range of voltage control and the corresponding range of

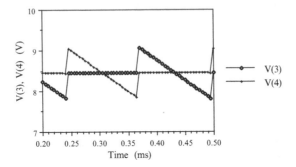

**Fig. 12.22.** (c) Waveforms of voltages V(3) and V(4).

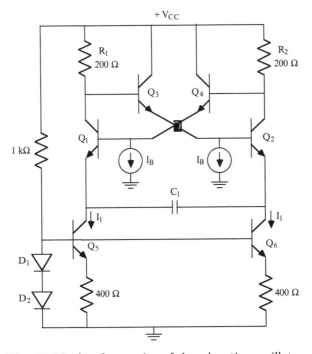

**Fig. 12.23.** Another version of the relaxation oscillator.

frequency of oscillation.
(c) Verify your analysis results with Spice.

**12.2.** A relaxation oscillator is shown in Figure 12.25.
(a) For $V_{BB} = 0$ V determine the frequency of steady-state oscillation. Note the initial current source excitation.
(b) What is the change in oscillation frequency, if any, for $V_{BB} = 1$ V?
(c) What is the change in oscillation frequency, if any, for $V_{BB} = -1$ V?
(d) Verify your analysis results with Spice.

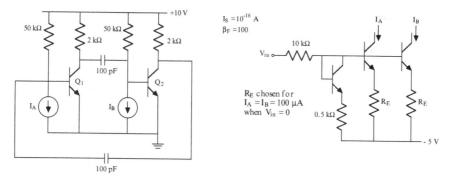

**Fig. 12.24.** Circuit for the voltage-controlled oscillator of Problem 12.1.

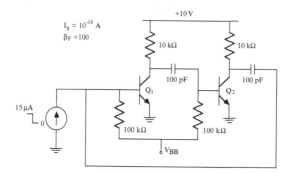

**Fig. 12.25.** Circuit for relaxation oscillator of Problem 12.2.

**12.3.** A relaxation oscillator using NMOS inverters is shown in Figure 12.26.
(a) Sketch all voltage waveforms and label the appropriate voltage levels for steady-state operation.
(b) What is the frequency of oscillation?
(c) Verify your analysis results with Spice.

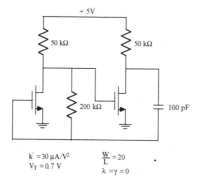

**Fig. 12.26.** Circuit for the NMOS relaxation oscillator of Problem 12.3.

**12.4.** A MOS relaxation oscillator is shown in Figure 12.26.
(a) Sketch all voltage waveforms and label the appropriate voltage levels for steady-state operation.
(b) What is the frequency of oscillation?
(c) Verify your analysis results with Spice.

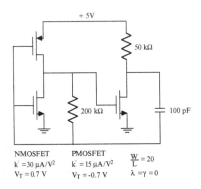

+ 5V

50 kΩ

200 kΩ

100 pF

| NMOSFET | PMOSFET | |
|---|---|---|
| $k' = 30\ \mu A/V^2$ | $k' = 15\ \mu A/V^2$ | $\dfrac{W}{L} = 20$ |
| $V_T = 0.7\ V$ | $V_T = -0.7\ V$ | $\lambda = \gamma = 0$ |

**Fig. 12.27.** Circuit for relaxation oscillator of Problem 12.4.

**12.5.** A bipolar relaxation oscillator is shown in Figure 12.27.
(a) Determine the steady-state operation of the circuit and the frequency of oscillation. Sketch all voltage waveforms.
(b) Estimate the change in performance if devices $Q_3$ and $Q_4$ are replaced with short circuits from base to emitter.

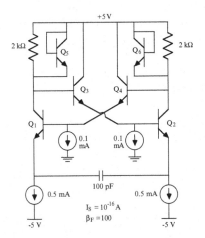

+5 V

2 kΩ          2 kΩ

$Q_5$          $Q_6$

$Q_3$     $Q_4$

$Q_1$                    $Q_2$

0.1 mA    0.1 mA

100 pF

0.5 mA              0.5 mA

$I_S = 10^{-16}\ A$
$\beta_F = 100$

-5 V                          -5 V

**Fig. 12.28.** Circuit for the bipolar relaxation oscillator of Problem 12.5.

# 13

---

# Analog Multipliers, Mixers and Modulators

## 13.1 The Emitter-Coupled Pair as a Simple Analog Multiplier

The multiplication of two analog, real-time signals is an important, required circuit function. A particular application concerns the translating of a frequency spectrum from one passband to another. This is often referred to by the term, mixing. In the early days of electronics, before the availability of reliable analog multipliers, two signals were combined (mixed) in a nonlinear device and the mixing (beating) of the two signals produced sum and difference components of the signals and their harmonics comparable to intermodulation distortion presented in Chapter 3. For $IM_2$, the critical factor is the presence of the square-law term of the device transfer characteristic which leads to the (cross) product of the two signals. With a more complicated circuit, easily realizable in IC form, the multiplication can be achieved directly.

The simple emitter-coupled pair of Figure 13.1a provides a very simple example of an elementary multiplier. From Chapter 2, we note that the collector currents and the input voltages are related by

$$I_{C1} = \frac{I_{EE}}{2} \left[ 1 + \tanh\left(\frac{d}{2}\right) \right] \tag{13.1}$$

$$I_{C2} = \frac{I_{EE}}{2} \left[ 1 - \tanh\left(\frac{d}{2}\right) \right] \tag{13.2}$$

where $d = \frac{v_{i1}}{V_t}$ and $v_{i1} = V_{i1} - V_{i2}$ is the differential-mode input signal. As usual, base currents are neglected. A common-mode (bias) input, $V_{B1} = \frac{V_{i1}+V_{i2}}{2}$, may affect the value of $I_{EE}$, the dc value of the common-emitter current source. For simplicity in the following developments, we assume that $V_{B1} = 0$. The differential output voltage is

D.O. Pederson and K. Mayaram, *Analog Integrated Circuits for Communication*, DOI 10.1007/978-0-387-68030-9_13,
© 2008 Springer Science+Business Media, LLC

$$v_o = V_{o1} - V_{o2} = -R_C(I_{C1} - I_{C2}) = -I_{EE}R_C \tanh\left(\frac{d}{2}\right) \tag{13.3}$$

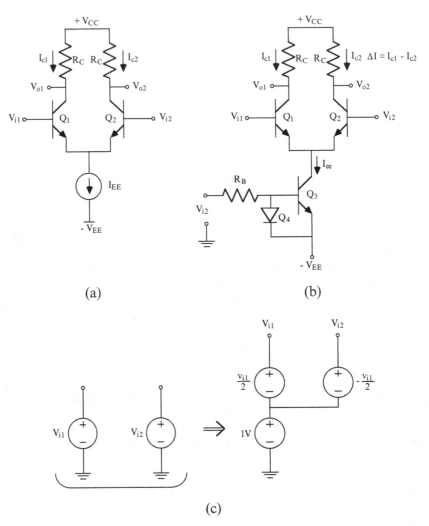

(a)                                    (b)

(c)

**Fig. 13.1.** (a) A simple emitter-coupled pair. (b) A second input introduced through $I_{ee}$. (c) Common-mode and differential-mode input signals.

As brought out in Chapter 2, for $d \ll 1$, the tanh function can be approximated well by the first term of its power-series expansion

$$\tanh\left(\frac{v_{i1}}{2V_t}\right) \approx \frac{v_{i1}}{2V_t} \tag{13.4}$$

$$v_o \approx -I_{EE}R_C \frac{v_{i1}}{2V_t} \tag{13.5}$$

A second input signal can be introduced into a transistor realization of $I_{EE}$ as shown in Figure 13.1b.

$$I_{ee} \approx \frac{V_{i2} - V_{BEon} - (-V_{EE})}{R_B} \tag{13.6}$$

This second input in terms of its bias and variational components is

$$V_{i2} = v_{i2} + V_{B2} \tag{13.7}$$

The total common-emitter current source is

$$I_{ee} = i_{ee} + I_{EE} = \frac{v_{i2}}{R_B} + \frac{V_{B2} - V_{BEon} + V_{EE}}{R_B} \tag{13.8}$$

$$= \frac{v_{i2}}{R_B} + I_{EE}$$

where $i_{ee}$ is the incremental value dependent upon $v_{i2}$ and $I_{EE}$ is the dc value. (Remember that $I_{EE}$ may also depend upon $V_{B1}$). The differential output voltage of the ECP becomes

$$v_o = -\frac{R_C}{R_B}\frac{v_{i1}v_{i2}}{2V_t} - \frac{R_C I_{EE} v_{i1}}{2V_t} \tag{13.9}$$

The output voltage can then be written

$$v_o = v_{om} + f(I_{EE}, v_{i1}) \tag{13.10}$$

The $v_{om}$ term above is the one of immediate interest and can be considered the 'ideal' multiplier output.

$$v_{om} = -K v_{i1} v_{i2} \tag{13.11}$$

where the multiplier coefficient is

$$K = \frac{R_C}{R_B}\frac{1}{2V_t} \tag{13.12}$$

(The multiplier constant $K$ is used throughout this chapter as a generic term. Care must be taken to insure the correct usage.)

For a numerical example, let $R_C = 1\ k\Omega$ and $R_B = 5\ k\Omega$. The upper input voltages are assumed to have a common-mode dc bias component of 1 V and a sinusoidal differential-mode signal component with an amplitude of 0.02 V and a radial frequency of $10^6$ r/s, as shown in Figure 13.1c.

$$V_{B1} = 1\ V \tag{13.13}$$

$$v_{i1} = 0.02 \cos 10^6 t \qquad (13.14)$$

Because a transistor-diode current source is used in Figure 13.1b, the dc value of $I_{EE}$ does not depend upon $V_{B1} = 1$ V. (However, the values of the $V_{CE}$ of the transistors do depend upon $V_{B1}$.) The lower signal source is chosen to have a dc component of zero and a sinusoidal input with an amplitude of 1 V and a radial frequency of $10^7$ r/s.

$$V_{i2} = 0 + 1 \cos 10^7 t \qquad (13.15)$$

The dc state of the circuit for $V_{BEon} = 0.8$ V, $V_{CC} = 10$ V, $-V_{EE} = -10$ V, $R_C = 1$ k$\Omega$ and $R_B = 5$ k$\Omega$ is

$$I_{EE} = \frac{0 - V_{BEon} + V_{EE}}{R_B} \qquad (13.16)$$
$$= 1.84 \text{ mA}$$

$$V_{01} = V_{02} = V_{CC} - \frac{I_{EE}}{2} R_C = 9.08 \text{ V} \qquad (13.17)$$

These values do not enter into the signal multiplication in $v_{om}$, provided that normal active region operation of the devices is maintained. The multiplier coefficient from (13.12) is 3.85 V$^{-1}$. Therefore, the 'signal multiplication' output is

$$v_{om} = -3.85(1.0 \cos 10^7 t)(0.02 \cos 10^6 t) \text{ V} \qquad (13.18)$$
$$= -38.5 \text{ mV} \cos[(1 - .1)10^7 t] - 38.5 \text{ mV} \cos[(1 + .1)10^7 t]$$

As expected, the sum and difference of the two sinusoidal signals (the $IM_2$ terms) are produced in the output.

From (13.5), the upper-input signal $v_{i1}$ is also amplified by the ECP and appears in the output. Its amplitude is

$$v_{oA} = \frac{(1.84 \text{ mA})(1 \text{ k}\Omega)(20 \text{ mV})}{2(25.85 \text{ mV})} = 712 \text{ mV} \qquad (13.19)$$

This signal is approximately 20 times that of the $v_{om}$ terms above. If the difference-frequency output component is of main interest, substantial frequency filtering is needed to reject the $v_{i1}$ component.

To continue with the example, it is also of interest to determine the range of the input, $v_{i1}$, over which linear multiplication is obtained, i.e., the range over which the transfer characteristic slope deviates from linearity by a given percent. We start with (13.3) and take the derivative with respect to $v_{i1}$.

$$\frac{dv_o}{dv_{i1}} = \frac{R_C I_{EE}}{2V_t} \frac{1}{\cosh^2\left(\frac{d}{2}\right)} \tag{13.20}$$

For $v_{i1} = 0$, the value of the slope is

$$\frac{dv_o}{dv_{i1}} = \frac{R_C I_{EE}}{2V_t} \tag{13.21}$$

To find the value of $v_{i1}$ for which the slope is, say, 1% different than a linear curve, we solve for the value of $v_{i1}$ where

$$\frac{dv_o}{dv_{i1}} = 0.99\frac{R_C I_{EE}}{2V_t} \tag{13.22}$$

This leads to

$$\cosh\left(\frac{d}{2}\right) = \sqrt{\frac{1}{.99}} \tag{13.23}$$

$$v_{i1} = \pm 5.2 \text{ mV} \tag{13.24}$$

Therefore, the upper-signal amplitude must be less than 5.2 mV to maintain 99% linearity. The linearity restriction for the lower-signal source is much less severe because of the linearity introduced by the bias source resistance, $R_B$.

This simple type of analog multiplier has many uses but is restricted to the particular 'single-ended input' for the lower signal source. Often, it is desired to use a differential input signal for both inputs. In addition, the voltage excursion for the upper signal is limited for accurate multiplication. In the next sections, new circuitry is introduced to remove these limitations.

The lower-signal source can be introduced in several ways into the bias current source. Another example is shown in Figure 13.1d. In this circuit, the lower-signal source is ac coupled into the emitter of the bias transistor, $Q_3$. For this signal, this transistor operates in a common-base mode. It is easily shown that the multiplier gain constant is

$$K \approx \frac{R_C I_{EE}}{2V_t^2} \tag{13.25}$$

For $R_C = 1$ k$\Omega$, $R_E = 5$ k$\Omega$, $V_{EE} = 10$ V, one obtains $I_{EE} = 1.84$ mA and $K = 1.4 \times 10^3$ V$^{-1}$. The linearity constraint for the amplitude of the lower signal is now much more severe because of the direct input to transistor $Q_3$.

## 13.2 A Subtraction Improvement

The simple ECP analog multiplier suffers from the small input signal restriction and the offset voltage due to $v_{i1}$ alone. In terms of the original equations,

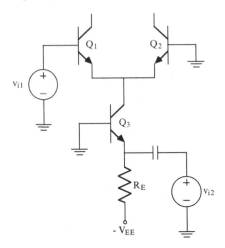

**Fig. 13.1.** (d) Another example of introducing the lower-signal source.

$$v_o = V_{o1} - V_{o2} = -\frac{R_C}{R_B} \tanh\left(\frac{v_{i1}}{2V_t}\right) v_{i2} - R_C I_{EE} \tanh\left(\frac{v_{i1}}{2V_t}\right) \quad (13.26)$$

The second term can be eliminated by an addition (subtraction) configuration involving an additional ECP. In the configuration of Figure 13.2a, a
second pair, $Q_3 - Q_4$, is in parallel with the first ECP, $Q_1 - Q_2$, except that
the input voltage drive to the second pair is inverted, i.e., has the opposite
phase as of the first pair. The differential output voltage is

$$v_o = -\left[(I_1 - I_2) + (I_3 - I_4)\right] R_C \quad (13.27)$$

$$= -\left[I_{EE1} \tanh\left(\frac{v_{i1}}{2V_t}\right) + I_{EE2} \tanh\left(\frac{-v_{i1}}{2V_t}\right)\right] R_C$$

$$= -(I_{EE1} - I_{EE2}) R_C \tanh\left(\frac{v_{i1}}{2V_t}\right)$$

Of course, if the two pairs have equal common-emitter current sources,
the differential output voltage cancels and is zero. This cancellation aspect
at first does not seem promising unless we also consider the case where the
two common-emitter currents sources have incremental components which are
also out of phase and are proportional of $v_{i2}$. Let

$$I_{ee1} = I_{EE} + i_{ee} \quad (13.28)$$
$$I_{ee2} = I_{EE} - i_{ee}$$

Using these in (13.27), we obtain

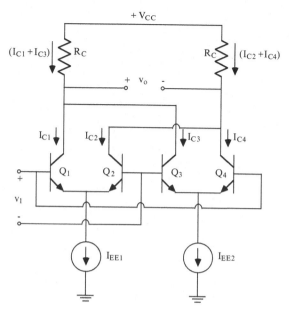

**Fig. 13.2.** (a) Circuit to eliminate the second term of Equation (13.26).

$$v_o = 2i_{ee}R_C \tanh\left(\frac{v_{i1}}{2V_t}\right) \tag{13.29}$$

The dc offset in $v_o$ is eliminated provided that we can produce the currents in (13.28). If one starts with the configuration of either Figure 13.1b or Figure 13.1c, a new inverting stage is necessary to obtain $-v_{i2}$ and thus $-i_{ee}$. A simpler scheme uses a third ECP, $Q_5 - Q_6$, as shown in Figure 13.2b. The difference of the 'source' currents $I_{C5}$ and $I_{C6}$ is

$$I_{C5} - I_{C6} = I_{EE} \tanh\left(\frac{v_{i2}}{2V_t}\right) \tag{13.30}$$

Starting with (13.27), we obtain

$$
\begin{aligned}
v_o &= -R_C[(I_{C1} + I_{C3}) - (I_{C2} + I_{C4})] \tag{13.31}\\
&= -R_C[(I_{C1} - I_{C2}) + (I_{C3} - I_{C4})]\\
&= -R_C\left[I_{C5}\tanh\left(\frac{v_{i1}}{2V_t}\right) + I_{C6}\tanh\left(\frac{-v_{i1}}{2V_t}\right)\right]\\
&= -R_C\left[(I_{C5} - I_{C6})\tanh\left(\frac{v_{i1}}{2V_t}\right)\right]\\
&= -R_C I_{EE}\tanh\left(\frac{v_{i2}}{2V_t}\right)\tanh\left(\frac{v_{i1}}{2V_t}\right)
\end{aligned}
$$

For small inputs,

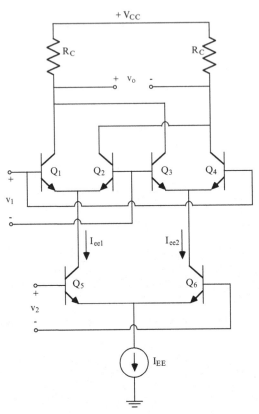

**Fig. 13.2.** (b) A fully balanced four-quadrant multiplier circuit.

$$v_o = v_{om} = -\frac{R_C I_{EE}}{4V_t^2}v_{i1}v_{i2} \qquad (13.32)$$

This is one-half of the corresponding small-signal output for a single ECP multiplier. The magnitude restriction on the input signals for linearity remains the same now for both $v_{i1}$ and $v_{i2}$.

The final circuit of this section, Figure 13.2b, is often referred to as a fully balanced multiplier and as a four-quadrant multiplier. The latter is the result of the output of the multiplier having the same relative behavior in all of the four quadrants of the $v_{i1} - v_{i2}$ plane, dependent only on the sign of the input signals.

## 13.3 Predistortion and Linearity Improvement in the ECP

B. Gilbert first proposed a scheme to predistort the input signal to an emitter-coupled pair, at the expense of gain, which greatly extends the linearity range of the circuit [37]. This is of particular importance for analog multipliers, but also has application for other large-signal uses of the ECP. Although Gilbert first published the predistortion technique in conjunction with a fully balanced multiplier, as described in the last section, the technique, for simplicity, is first introduced with the ECP alone.

In the Gilbert predistortion scheme, diodes are introduced at the input of an ECP to compress the input signal logarithmically. This compressed signal is expanded by the transfer characteristic of the ECP to obtain an almost linear transfer characteristic of the total circuit. In Figure 13.3a, the input signals are chosen for the moment to be the currents $I_a$ and $I_b$. These two currents flow down through the diodes (diode-connected transistors) developing the voltages $V_a$ and $V_b$.

$$V_a = V_B - V_t \ln \left( \frac{I_a}{I_S} \right) \tag{13.33}$$

$$V_b = V_B - V_t \ln \left( \frac{I_b}{I_S} \right)$$

where $I_S$ is the saturation current of the bipolar device. The difference voltage, $V_a - V_b$, is taken as the differential input voltage to an ECP.

$$v_{i1} = V_a - V_b = -V_t \ln \left( \frac{I_a}{I_b} \right) = V_t \ln \left( \frac{I_b}{I_a} \right) \tag{13.34}$$

For the pair shown in Figure 13.3b, the collector currents, from (13.1) and (13.2) and from Chapter 2, are

$$I_{C1} = \frac{I_{EE}}{1 + \exp(-d)} = \frac{I_{EE}}{2} \left[ 1 + \tanh \left( \frac{d}{2} \right) \right] \tag{13.35}$$

$$I_{C2} = \frac{I_{EE}}{1 + \exp(d)} = \frac{I_{EE}}{2} \left[ 1 - \tanh \left( \frac{d}{2} \right) \right]$$

where $d = v_{i1}/V_t$. The difference of the two currents and the ratio of the two are

$$I_{C1} - I_{C2} = I_{EE} \tanh \left( \frac{d}{2} \right) \tag{13.36}$$

$$\frac{I_{C1}}{I_{C2}} = \exp(d) \tag{13.37}$$

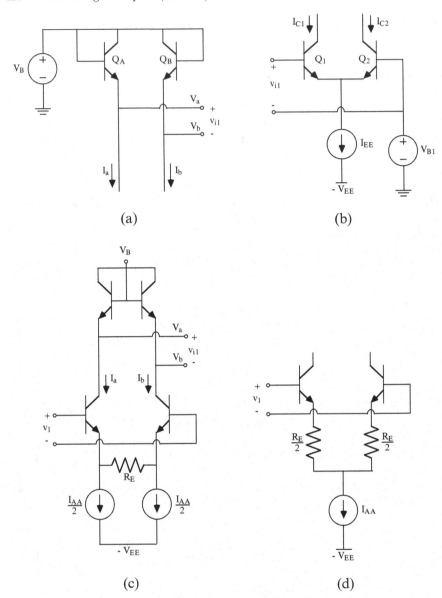

**Fig. 13.3.** (a) Circuit with currents as inputs. (b) An EC pair circuit. (c) An input voltage-to-current transducer. (d) An alternate arrangement to linearize the transfer characteristics.

If (13.34) is solved for $\exp\left(\frac{v_{i1}}{V_t}\right) = \exp(d)$ and used in (13.37),

$$\frac{I_{C1}}{I_{C2}} = \frac{I_b}{I_a} \tag{13.38}$$

Next recognize that $I_{EE} = I_{C1} + I_{C2}$ (assuming that the base currents can be neglected). In addition, (13.38) is introduced into (13.36) to obtain

$$I_b - I_a = (I_a + I_b)\tanh\left(\frac{d}{2}\right) \tag{13.39}$$

This is solved for the tanh function which is used in (13.36).

$$I_{C1} - I_{C2} = \frac{I_{EE}}{I_a + I_b}(I_b - I_a) \tag{13.40}$$

We therefore have a linear relationship between the differential input currents, $(I_b - I_a)$ and the differential output currents, $(I_{C1} - I_{C2})$, of the pair.

An input voltage-to-current 'transducer' is next introduced as shown in Figure 13.3c. In this circuit, a large common-emitter resistor is used to linearize the transfer characteristic of an ECP. An alternate arrangement, as shown in Figure 13.3d, is to use a resistance $R_E/2$ in each emitter lead and a single current source supply, $I_{AA}$. For the circuit of Figure 13.3c and for $R_E \gg \frac{V_t}{I_a}$ and $\gg \frac{V_t}{I_b}$,

$$I_a = \frac{\frac{v_1}{2}}{\frac{R_E}{2}} + \frac{I_{AA}}{2} \tag{13.41}$$

$$I_b = \frac{-\frac{v_1}{2}}{\frac{R_E}{2}} + \frac{I_{AA}}{2}$$

where $v_1$ is the differential-input voltage to the transducer pair. The difference-output current of the transducer is

$$I_a - I_b = \frac{2v_1}{R_E} \tag{13.42}$$

The sum of the two collector currents is a constant

$$I_a + I_b = I_{AA} \tag{13.43}$$

The complete predistorted pair is shown in Figure 13.4a. The differential-output voltage is

$$v_o \approx -R_C(I_{C1} - I_{C2}) = +2\left(\frac{R_C}{R_E}\right)\left(\frac{I_{EE}}{I_{AA}}\right)v_i \tag{13.44}$$

The linearity of the actual transfer characteristic can be found from a Spice simulation. The input file is given in Figure 13.4b, where the local feedback and bias arrangement of Figure 13.3d is used. Several plots of the

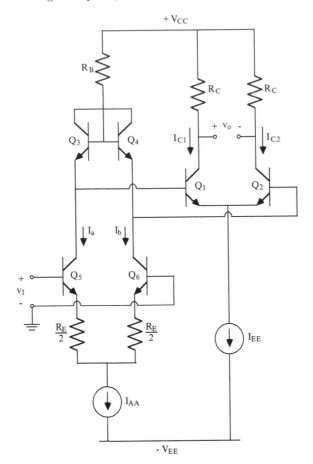

**Fig. 13.4.** (a) The complete predistorted pair.

voltage transfer characteristic are shown in Figure 13.4c. The improvement in linearity with $R_E$ is clear, although the voltage gain suffers. For modest values of $R_E/2$ relative to the output resistance at the emitter of $Q_5$ or $Q_6$, $1/g_{m5}$, the error in (13.44) can be significant. For $R_E = 2$ k$\Omega$, (13.44) estimates a differential voltage gain of 50. The observed value is 32.8.

This predistortion scheme can be used with the simple analog multiplier of the last section by also introducing a second input signal into the current source, $I_{EE}$. However, from the results of Section 13.1, the output still contains both the desired multiplication term as well as the offset term involving $v_{i1}$ and the dc component, $I_{EE}$. The fully differential circuit of the last section eliminates this second term.

```
FIGURE 13.4B
Q1 2 8 3 MOD1
.MODEL MOD1 NPN IS=1E-16 BF=100
RC1 5 2 5K
VCC 5 0 10
RC2 5 4 5K
Q2 4 9 3 MOD2
.MODEL MOD2 NPN IS=1E-16 BF=100
IEE1 3 20 2M
VEE 20 0 -10
RBB 5 7 37K
Q3 7 7 8 MOD1
Q4 7 7 9 MOD1
Q5 8 10 11 MOD1
Q6 9 0 12 MOD1
RE1 11 13 1K
RE3 12 13 1K
IAA 13 20 0.2MA
V1 10 0 0
.DC V1 -0.5 0.5 0.05
.PLOT DC V(2,4)
.OPTIONS RELTOL=1E-6
.WIDTH OUT=80
.END
```

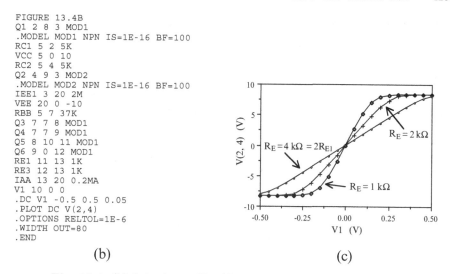

(b)

(c)

**Fig. 13.4.** (b) Spice input file. (c) Voltage transfer characteristics.

## 13.4 The Gilbert Cell

As mentioned in Section 13.2, a so-called four-quadrant multiplier is obtained when two simple analog multipliers of a single pair each are arranged in parallel with a push-pull input connection as shown in Figure 13.5a. In this arrangement, the common-emitter current sources for the upper pairs are also an emitter-coupled pair with emitter degeneration as used in the last section for the input voltage-to-current converter. Repeating some of the analysis of the last two sections, we obtain

$$I_{C3} - I_{C4} = I_{C1} \tanh\left(\frac{v_1}{2V_t}\right) \tag{13.45}$$

$$I_{C5} - I_{C6} = I_{C2} \tanh\left(\frac{-v_1}{2V_t}\right) = -I_{C2} \tanh\left(\frac{v_1}{2V_t}\right) \tag{13.46}$$

$$i_o = (I_{C3} + I_{C5}) - (I_{C4} + I_{C6}) \tag{13.47}$$

$$= (I_{C1} - I_{C2}) \tanh\left(\frac{v_1}{2V_t}\right)$$

For large values of the emitter feedback resistors,

$$I_{C1} - I_{C2} \approx \frac{2v_2}{R_E} \tag{13.48}$$

The differential-output voltage is

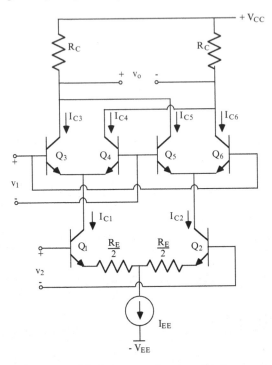

**Fig. 13.5.** (a) A four-quadrant multiplier core.

$$v_o = [V_{CC} - R_C(I_{C3} + I_{C5})] - [V_{CC} - R_C(I_{C4} + I_{C6})] \qquad (13.49)$$
$$= -R_C i_o = -kv_2 \tanh\left(\frac{v_1}{2V_t}\right)$$

where

$$k = \frac{2R_C}{R_E} \qquad (13.50)$$

The low-level input restriction on $v_1$ can be lifted by using predistortion as in the last section. The complete circuit becomes that of Figure 13.5b. Note that both series and coupling resistors are used in the emitters of the transducer pairs. The total, effective emitter resistors, $R_{E1}$ and $R_{E2}$, comparable to that used in the last section are twice one series resistor plus the coupling resistor. An example is given below. From the earlier results, the input-output relation is

$$v_o = -Kv_1v_2 \qquad (13.51)$$

where

$$K = \frac{4R_C}{I_{AA}R_{E1}R_{E2}} \tag{13.52}$$

Often, the parameters are chosen to achieve $K = 0.1$ V$^{-1}$ which can be obtained in a numerical example, with $R_{E1} = R_{E2} = 15$ k$\Omega$, $R_C = 11$ k$\Omega$, and $I_{AA} = 2$ mA.

The Spice input file for the multiplier of Figure 13.5b is shown in Figure 13.6a. The two input voltages are labeled $v_x$ and $v_y$ and correspond to $v_1$ and $v_2$ above. The appropriate values of the transducer resistors are $R_{E1} = R_{E2} = 85$ k$\Omega$, as given in a comment line in the listing. The source current $I_{AA}$ for these transducers is approximately 0.3 mA. Therefore, from (13.52), $K = 0.0092$ V$^{-1}$. In Figure 13.6b, the dc node voltages are given with $v_x = v(1) = 1$ V and $v_y = v(3) = 1$ V. The differential output voltage is $v_0 = v(23) - v(24) = -0.0089$ V, which is close to the estimate from (13.51) and (13.52). Three voltage transfer characteristics are shown in Figure 13.6c. The values of $v_0$ with $v_x$ for $v_y = 2$ V, with $v_y$ for $v_x = 2$ V and with $v_x$ for $v_y = 1$ V. The zero crossings are at zero volts and the linearity is very good over inputs of $\pm 6$ V. The breaks in the characteristics at the input of 9 V are due to transistor saturation.

The circuit shown in Figure 13.5b is the original circuit developed by Gilbert. This basic circuit is often referred to as the "Gilbert Cell." It has been the prototype of many integrated circuit versions produced commercially by many manufacturers. The four-quadrant multiplier is, with the operational amplifier, a workhorse for analog signal processing.

## 13.5 MOS Analog Multipliers

MOS devices can also be used to produce an analog multiplier. As seen repeatedly in previous chapters for other circuit function realizations, MOS enhancement-mode devices can be substituted for the original bipolar, enhancement mode devices, biases adjusted, and a new circuit achieved. Of course, the different gain properties of the new devices, and the differing non-linearities of the input and transfer characteristics may lead to modified performance.

For analog multipliers, the easiest way to proceed is to start directly with source-coupled pairs as illustrated in Figure 13.7. In this circuit three source-coupled pairs are used in the stacked arrangement also used for bipolar multipliers. For the BJT circuit, a straight-forward multiplication of the two input signals is produced due to the simple tanh function provided by the BJT exponential nonlinearity. For MOS SCPs, a closed-form transfer function can be obtained only for operation in the MOS saturation region [38]. In addition, the expressions for SCPs are not convenient to work with. It can be shown that linear multiplication is obtained for typical input voltage ranges.

Input predistortion circuitry and linear voltage-to-current converter sources can also be added to the MOS analog multiplier of Figure 13.7.

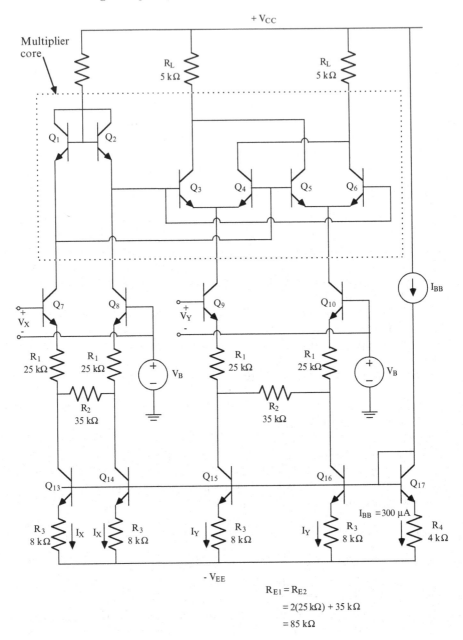

**Fig. 13.5.** (b) Complete circuit of the multiplier.

```
GILBERT CELL, FIG. 13.6A
.OPTIONS NOMOD NOPAGE RELTOL=1E-6
* VX IS BETWEEN 1 AND 2 AND VY BETWEEN 3 AND 4
* SET INPUTS TO APPROPRIATE VALUES
VX 1 2 1
VY 3 4 1
* USE NODE 2 AND NODE 4 AS THE INPUT BIAS POINTS
VX1 2 0 0
VY1 4 0 0
* SUPPLY VOLTAGES
VCC 100 0 10
VEE 101 0 -10
* LOAD RESISTORS, OUTPUT NODES ARE 23 AND 24
RL1 100 23 5K
RL2 100 24 5K
* MULTIPLIER CORE TRANSISTORS
Q1  100 100 20  MODN
Q2  10     100 21  MODN
Q3  23  21  22  MODN
Q4  24  20  22  MODN
Q5  23  20  25  MODN
Q6  24  21  25  MODN
* INPUT STAGE TRANSISTORS + RESISTORS
Q7  20  2   5   MODN
Q8  21  1   6   MODN
Q9  22  4   12  MODN
Q10 25  3   13  MODN
R11 5   7   25K
R12 6   8   25K
R13 12  14  25K
R14 13  15  25K
R21 7   8   35K
R22 14  15  35K
* NOTE: RE1 AND RE2 = 85K
* CURRENT SOURCE TRANSISTORS + RESISTORS
Q13 7   9   10  MODN
Q14 8   9   11  MODN
Q15 14  9   16  MODN
Q16 15  9   17  MODN
Q17 9   9   18  MODN
R31 10  101 8K
R32 11  101 8K
R33 16  101 8K
R34 17  101 8K
R35 18  101 4K
* CURRENT SOURCE AS BIAS GENERATOR
IBB 100 9 300U
* MODEL DEFINITION
.MODEL MODN NPN IS=1E-16 BF=100
*ANALYSIS TYPES
.DC VX -10 10 .5
*.DC VY -10 10 .5
```

**Fig. 13.6.** (a) Spice input file for Gilbert cell.

| NODE | VOLTAGE | NODE | VOLTAGE | NODE | VOLTAGE | NODE | VOLTAGE |
|------|---------|------|---------|------|---------|------|---------|
| ( 1) | 1.0000 | ( 2) | 0.0000 | ( 3) | 1.0000 | ( 4) | 0.0000 |
| ( 5) | -0.7224 | ( 6) | 0.2735 | ( 7) | -4.1235 | ( 8) | -3.7134 |
| ( 9) | -8.0813 | ( 10) | -8.8061 | ( 11) | -8.8061 | ( 12) | -0.7224 |
| ( 13) | 0.2735 | ( 14) | -4.1235 | ( 15) | -3.7134 | ( 16) | -8.8061 |
| ( 17) | -8.8061 | ( 18) | -8.8236 | ( 20) | 9.2776 | ( 21) | 9.2736 |
| ( 22) | 8.5715 | ( 23) | 9.2713 | ( 24) | 9.2802 | ( 25) | 8.5674 |
| (100) | 10.0000 | (101) | -10.0000 | | | | |

(b)

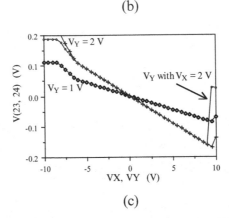

(c)

**Fig. 13.6.** (b) Dc node voltages for input voltages of 1 V. (c) Three voltage transfer characteristics.

## 13.6 Mixing, Modulation and Frequency Translation

The most common form of radio receiver is the superheterodyne configuration which is once again shown in Figure 13.8a. In operation, the mixer must achieve an analog multiplication. With multiplication, sum and difference frequency components at $(\omega_c \pm \omega_{lo})$ are produced at the output of the mixer, where $\omega_c$ is the input signal frequency and $\omega_{lo}$ is the frequency of the local oscillator. The sum frequency is rejected by sharply tuned circuits and the difference frequency component, the IF, is subsequently amplified in a fixed-tuned bandpass amplifier.

The design and evaluation of tuned amplifiers and lowpass, audio output amplifiers has been covered in the earlier chapters. The operation and design of the demodulation block (function) is introduced in the next chapter.

To formalize the mixer operation, assume that both the input signal and the local oscillator output are unmodulated, single-tone sinusoids.

$$V_c = V_{cA} \cos \omega_c t \qquad (13.53)$$

$$V_{lo} = V_{loA} \cos \omega_{lo} t \qquad (13.54)$$

If the multiplier (mixer) has a gain constant, K, the output is

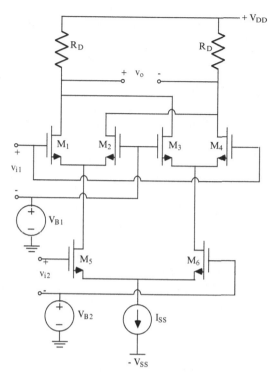

**Fig. 13.7.** A MOS analog multiplier using source-coupled pairs.

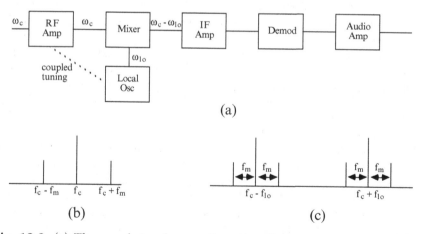

**Fig. 13.8.** (a) The superheterodyne configuration. Frequency spectrum of (b) the input, and (c) the multiplier output.

$$V_{out} = \frac{K}{2} V_{cA} V_{loA} \left[ \cos(\omega_c - \omega_{lo})t + \cos(\omega_c + \omega_{lo})t \right] \tag{13.55}$$

The difference frequency, $\omega_c - \omega_{lo}$, is labeled $\omega_{if}$.

If the input is a modulated signal, the modulation also is translated to a band about the new carrier frequency, $\omega_{if}$. For example, if the input is amplitude modulated,

$$V_s = V_{sA}[1 + m \cos \omega_m t] \cos \omega_c t \tag{13.56}$$

$$= V_{sA} \cos \omega_c t + \frac{m}{2} V_{cA} \cos(\omega_c - \omega_m)t + \frac{m}{2} V_{cA} \cos(\omega_c + \omega_m)t$$

The input can be represented as in Figure 13.8b with the carrier frequency term and an upper and a lower sideband, each sideband containing the modulation information.

For a linear multiplier, each of the input components is multiplied by the local oscillator input and the output of the multiplier contains six terms as shown in Figure 13.8c, the difference-frequency carrier with two sidebands and the sum-frequency carrier with two side bands. The latter combination is usually rejected by the passband of the IF amplifier.

## 13.7 The Fully Balanced (Quad) Mixer

A compensated, predistorted analog multiplier such as that shown in Figure 13.5b can be used directly as a mixer. One input, $v_2$, is the RF signal from the RF amplifier/tuner. The other input, $v_1$, is from the local oscillator. At the output of the multiplier, a high Q resonant circuit is included which is tuned to the difference frequency. (This tuned circuit is usually also the input circuit for the intermediate-frequency amplifier.) Good mixer performance can be provided for modest frequency applications. Charge storage and interaction effects for the usual IC realization do not permit operation for frequencies above one-tenth of the $f_T$ of the devices.

Because of the need for the high-Q tuned circuit at the output of the mixer, which rejects all frequency components other than the spectrum about the difference frequency, the linearizing circuitry of the Gilbert Cell can be omitted and only the 'small-signal' multiplier of Figure 13.2b, which is shown again in Figure 13.9a with an output tuned circuit, need be used as the mixer. The circuit is a doubly balanced, four-quadrant analog multiplier and is called the quad mixer.

The amplitudes of the input signals into a quad mixer need not be small relative to the $V_t$ of the devices. For large amplitudes of sinusoidal signals, each tanh function of (13.31) can be represented as a Fourier series in terms of the fundamentals of each input and their harmonics, cf., Section 2.5. The output of the multiplier is the infinite set of the beats of the signals and all

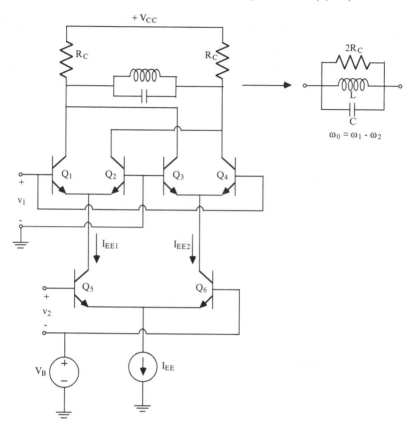

**Fig. 13.9.** (a) A quad-mixer circuit.

possible harmonics. If the output tuned circuit of the quad mixer has a high Q, the only appreciable output voltage presented to the IF amplifier is the difference-frequency, beat component of the original two input sinusoids.

Typically, the input from the local oscillator has an amplitude greater than $4V_t \approx 100$ mV. With this large an input, the transistors of the top ECPs of Figure 13.9a quickly switch from their active to the off regions and vice versa, that is, the transistors act as fast switches. The output collector currents of the top ECPs are virtually rectangular wave trains [39]. The amplitude of the lower ECP can also be large, but proper operation of the mixer is also obtained for small amplitudes.

For a numerical example, the Spice input file for the quad mixer of Figure 13.9a is given in Figure 13.9b. The difference frequency of the mixer is chosen to be 1.59 MHz. The two input frequencies are chosen to be the multiples of 8 and 9 of the difference frequency. This provides a convenient check of the output components of the mixer using Spice's Fourier analysis. The sinusoidal input to the upper ECPs, $v_1$, has an amplitude of 90 mV at a frequency of

12.732 MHz (8 times the difference frequency). The input amplitude to the lower ECP, $v_2$, is 2 mV at a frequency of 14.324 MHz (9 times the difference frequency). The output tuned circuit develops the differential output voltage. Note that the resistance of the tuned circuit is $2R_C = 10 \text{ k}\Omega$, since the tuned circuit is placed between the two collector nodes.

```
QUAD MIXER, FIG 13.9A
.TRAN 30N 13U 10U 3.5N
.PLOT TRAN V(23,24)
.FOUR 1.5916MEG V(23,24)
V1 5 0 SIN (0 90M 12.732MEG)
V2 1 4 SIN (0 2M 14.324MEG)
VB 4 0 -2
VCC 100 0 10
RL1 100 23 5K
RL2 100 24 5K
LT 23 24 100U
CT 23 24 100P
Q1 23 5 22 MODN
Q2 24 0 22 MODN
Q3 23 0 25 MODN
Q4 24 5 25 MODN
Q5 22 1 12 MODN
Q6 25 4 12 MODN
.MODEL MODN NPN IS=1E-16 BF=100
IEE 12 0 300U
.OPTIONS NOPAGE NOMOD RELTOL=1E-5
.WIDTH OUT=80
.OPTIONS ITL5=0 LIMPTS 5000
.END
```

(b)

```
FOURIER COMPONENTS OF TRANSIENT RESPONSE V(23,24)
DC COMPONENT =    5.614D-06
HARMONIC    FREQUENCY      FOURIER      NORMALIZED     PHASE      NORMALIZED
   NO         (HZ)        COMPONENT     COMPONENT      (DEG)     PHASE (DEG)

   1        1.592D+06     3.036D-02     1.000000      160.108       0.000
   2        3.183D+06     9.451D-06     0.000311      -16.705     -176.813
   3        4.775D+06     3.561D-06     0.000117       -8.217     -168.326
   4        6.366D+06     3.855D-06     0.000127       -2.037     -162.146
   5        7.958D+06     1.773D-06     0.000058       23.756     -136.352
   6        9.550D+06     2.786D-06     0.000092      -30.677     -190.785
   7        1.114D+07     2.472D-06     0.000081       42.797     -117.311
   8        1.273D+07     2.092D-06     0.000069        4.945     -155.163
   9        1.432D+07     5.696D-07     0.000019       57.621     -102.487

    TOTAL HARMONIC DISTORTION =        0.038778   PERCENT
```

(c)

**Fig. 13.9.** (b) Spice input file and output voltage waveform. (c) Fourier components of the output voltage.

Spice simulation of mixers and other frequency-translation circuits is computer-time consuming because of the wide ranges of frequencies and critical time intervals. The parameters of the .TRAN line of the input file must be carefully chosen. The .TRAN statement has the form:

$$\text{.TRAN TSTEP TSTOP TSTART TMAX} \qquad (13.57)$$

For the present example, the longest time interval permitted during the simulation, TMAX, should be no greater than approximately 1/20 of the period of the highest frequency input signal; therefore, TMAX = 3.5 ns. The simulation must take place over several cycles of the desired difference-frequency output. In this Spice input, 20 cycles are used; thus, TSTOP = 13 $\mu$s. Such a large compute time insures that startup transients have disappeared. Since only a few cycles of the difference-frequency output need be inspected, TSTART = 10 $\mu$s. Finally, the print step, TSTEP, is chosen to be 1/20 of the period of the difference frequency.

Since a large number of iterations are needed for the simulation, .options ITL5=0 must be used in Spice2 to eliminate the limit on the number of iterations. Because of the complexity of the waveform for the output of a mixer, high-order interpolation in the Fourier analysis of Spice3 should not be used. Linear interpolation, as used in Spice2, provides a filtering action.

The output voltage waveform is shown in Figure 13.9b and the Fourier components of the output are given in Figure 13.9c. The harmonic content of the output voltage is very low. In particular, the output components at the frequencies of the two inputs are negligible. Cancellation from the mixer and high rejection by the tuned circuit are present.

The amplitude of the output sinusoid is seen to be 30.4 mV. An estimate of this amplitude can be obtained from (13.31) and from Figure 2.16 (or Figure 10.16). From the results of Chapter 2, the fundamental component of the output currents of an ECP with heavy overdrive can be obtained from Figure 2.16. For $d/2 = 90$ mV/$2V_t$, $b_1 = 0.53$. This value can be used in (13.31) to obtain

$$v_o = -R_C I_{EE} \frac{v_{i2}}{2V_t} b_1 \qquad (13.58)$$

where a small-signal approximation of $\tanh (v_{i2}/2V_t)$ is used. For $R_C = 5$ k$\Omega$, $I_{EE} = 300$ $\mu$A, $v_i/2 = 2$ mV, and $V_t = 25.9$ mV, $v_o = 30.7$ mV.

To improve the linearity of the RF input stage of the quad mixer shown in Figure 13.9(a) emitter degeneration is typically used [39]. The modified quad-mixer circuit with resistive emitter degeneration is shown in Figure 13.9(d).

## 13.8 Single-Device Mixers

As mentioned earlier, the quad mixer is limited to frequencies well below the $f_T$ of the BJTs. For higher frequencies, as well as for low-cost situations, single-device mixers are used. In a single-device mixer, the input signals are introduced into the device and the square-law term of the device's characteristic

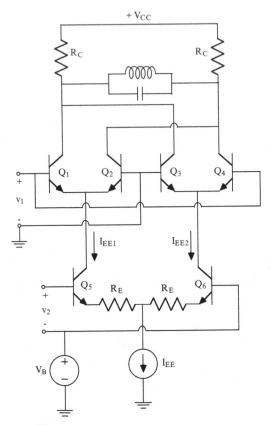

**Fig. 13.9.** (d) A quad-mixer circuit with emitter degeneration for the lower ECP.

provides the multiplication action. In Chapter 3, intermodulation distortion is introduced in this manner. There, the interaction of the two signals in the non-linearity of the device is to be minimized. In a mixer, the desired output is one of the $IM_2$ components, the difference-frequency component developed from the second-order term of the power series of the nonlinearity. The other terms of the nonlinear device characteristic also produce harmonics and beats which can be considered distortion terms which must be rejected by the fixed-tuned amplifier which follows the mixer. The filtering problem is usually severe.

### 13.8.1 BJT Mixers

Consider the BJT circuit of Figure 13.10a. The input consists of the series combination of a dc bias voltage source, $V_{BS}$, an input sinusoidal voltage source $v_s$ and a local oscillator voltage source $v_{lo}$, also sinusoidal. $V_{BS}$ is the quiescent bias value of $V_{BE}$. The transfer characteristic of the BJT is taken to be the usual ideal exponential.

$$I_C = I_S \exp\left(\frac{V_{BS} + v_s + v_{lo}}{V_t}\right) \tag{13.59}$$

$$= I_{CA} \exp\left(\frac{v_s}{V_t}\right) \exp\left(\frac{v_{lo}}{V_t}\right)$$

where

$$I_{CA} = I_S \exp\left(\frac{V_{BS}}{V_t}\right)$$

The term, $I_{CA}$, is the quiescent dc value of the collector current. Since $v_s$ and $v_{lo}$ have the form of $X \cos \omega_i t$, each exponential can be expanded as in Chapters 3 and 11 in a Fourier series with Bessel Function coefficients, cf. (11.28).

$$\exp\left(\frac{V_{sA} \cos \omega_s t}{V_t}\right) = I_0(d) + 2I_1(d) \cos \omega_s t + \cdots \tag{13.60}$$

$$\exp\left(\frac{V_{loA} \cos \omega_{lo} t}{V_t}\right) = I_0(y) + 2I_1(y) \cos \omega_{lo} t + \cdots \tag{13.61}$$

where $d$ and $y$ are normalized values of the sinusoidal amplitudes, $V_{sA}$ and $V_{loA}$, respectively.

$$d = \frac{V_{sA}}{V_t} \tag{13.62}$$

$$y = \frac{V_{loA}}{V_t} \tag{13.63}$$

The $I_n(x)$ are modified Bessel functions of order n. If the Fourier series terms are introduced into (13.59), the multiplication carried out and if trigonometric identities are introduced, the result has the form

$$I_C = I_{CA} I_0(d) I_0(y) + \cdots 2I_{CA} I_1(d) I_1(y) \cos(\omega_s - \omega_{lo})t + \cdots \tag{13.64}$$

$$= I_{DC} \left[1 + \cdots + 2\frac{I_1(d)}{I_0(d)} \frac{I_1(y)}{I_0(y)} \cos(\omega_s - \omega_{lo})t + \cdots\right]$$

where $I_{DC} = I_{CA} I_0(d) I_0(y)$ is the dynamic average value of the collector current and only the difference-frequency component is explicitly included. The desired mixing occurs and there are a multitude of other terms to reject.

We next specify that the local oscillator magnitude be larger than $V_t = 25.85$ mV at room temperature and that the input voltage be much less than $V_t$.

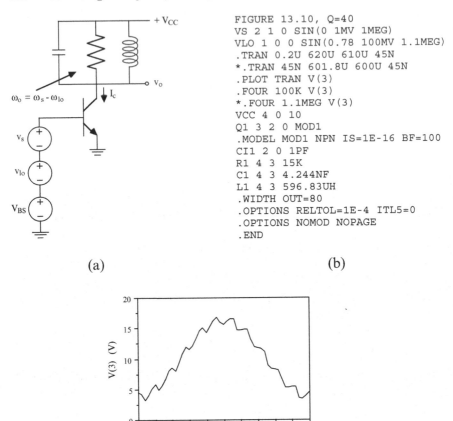

(a)

```
FIGURE 13.10, Q=40
VS 2 1 0 SIN(0 1MV 1MEG)
VLO 1 0 0 SIN(0.78 100MV 1.1MEG)
.TRAN 0.2U 620U 610U 45N
*.TRAN 45N 601.8U 600U 45N
.PLOT TRAN V(3)
.FOUR 100K V(3)
*.FOUR 1.1MEG V(3)
VCC 4 0 10
Q1 3 2 0 MOD1
.MODEL MOD1 NPN IS=1E-16 BF=100
CI1 2 0 1PF
R1 4 3 15K
C1 4 3 4.244NF
L1 4 3 596.83UH
.WIDTH OUT=80
.OPTIONS RELTOL=1E-4 ITL5=0
.OPTIONS NOMOD NOPAGE
.END
```

(b)

```
FOURIER COMPONENTS OF TRANSIENT RESPONSE V(3)
DC COMPONENT =    9.999D+00
HARMONIC   FREQUENCY     FOURIER     NORMALIZED      PHASE      NORMALIZED
   NO        (HZ)       COMPONENT    COMPONENT       (DEG)     PHASE (DEG)

    1      1.000D+05    6.228D+00    1.000000      -89.976        0.000
    2      2.000D+05    7.772D-03    0.001248        4.196       94.172
    3      3.000D+05    2.883D-03    0.000463       82.645      172.621
    4      4.000D+05    5.096D-03    0.000818      106.306      196.282
    5      5.000D+05    1.449D-03    0.000233     -158.424      -68.448
    6      6.000D+05    1.705D-03    0.000274      -19.667       70.308
    7      7.000D+05    1.990D-02    0.003196        5.753       95.728
    8      8.000D+05    1.331D-03    0.000214      -10.905       79.070
    9      9.000D+05    8.830D-04    0.000142     -139.899      -49.923

     TOTAL HARMONIC DISTORTION =       0.358472  PERCENT
```

(c)

**Fig. 13.10.** (a) A single-device BJT mixer. (b) Spice input file. (c) Output voltage waveform and its Fourier components.

$$V_{loA} \gg V_t, \; y \gg 1 \qquad (13.65)$$

$$V_{sA} \ll V_t, \; d \ll 1 \qquad (13.66)$$

These lead to the approximations:

$$\frac{I_1(y)}{I_0(y)} = 0.95 \qquad (13.67)$$

$$\frac{I_1(d)}{I_0(d)} = \frac{d}{2} \qquad (13.68)$$

With these approximations the difference component of the collector current becomes

$$I_C|_{if} \cos \omega_{if} t \approx 0.95 I_{DC} \frac{V_{sA}}{V_t} \cos \omega_{if} t \qquad (13.69)$$

where $\omega_{if} = \omega_s - \omega_{lo}$.

In some cases, it is convenient to define a new transfer coefficient, $G_{mcon}$, the conversion transconductance. This is taken as the ratio of the amplitude of the IF component of the collector current and the amplitude of the signal input voltage. From the last equation,

$$G_{mcon} = 0.95 \frac{I_{DC}}{V_t} \qquad (13.70)$$

If $y$ is not much greater than one, the plots of the $I_n(x)$ in Figure 11.5 can be used to obtain the value of $\frac{I_1(y)}{I_0(y)}$ to replace the value 0.95 in (13.69) and (13.70).

For a numerical example, let the circuit be current biased with $I_{DC} = 0.5$ mA. For $V_{sA} = 1$ mV and $V_{loA} = 100$ mV, $I_C|_{if} = 18$ $\mu$A and $G_{mcon} = 18 \times 10^{-3}$ ℧.

If the output resonant circuit of the mixer is tuned to the difference frequency and has a center-frequency resistance of $R = 10$ k$\Omega$, the output voltage of the IF component is 184 mV.

Of the other components of the output, a troublesome one is the fundamental of the local oscillator because of the assumed and usual large amplitude. From the Bessel-function expansions,

$$I_C|_{lo} = I_{DC} 2 \frac{I_1(y)}{I_0(y)} \cos \omega_{lo} t \qquad (13.71)$$

For $y \gg 1$,

$$I_C|_{lo} = 1.9 I_{DC} \cos \omega_{lo} t \qquad (13.72)$$

For the above numerical example, the magnitude of the local-oscillator output component is very large, 0.95 mA. The ratio of the magnitude of this term to the magnitude of the desired difference frequency term is

$$\frac{I_C|_{lo}}{I_C|_{if}} = 2\frac{V_t}{V_{sA}} \tag{13.73}$$

For the example above where $V_{sA} = 1$ mV, this ratio is over 50. Therefore, the tuned output circuit must provide adequate rejection at the LO frequency. One cannot depend only on the rejection of the following IF amplifier. Large-signal inputs into this amplifier may cause substantial additional IM distortion to be generated. If the output voltage of the mixer at the local oscillator frequency is to be less than, say, ten times that of the IF component, the rejection from the output tuned circuit must be approximately 500 at the local oscillator frequency relative to the IF frequency response.

For a single-tuned, parallel $RLC$, resonant circuit, the magnitude function with frequency from (8.6) has the form

$$| Z(j\omega) | = \frac{R\omega\omega_b}{\sqrt{(\omega_o^2 - \omega^2)^2 + (\omega\omega_b)^2}} \tag{13.74}$$

where $\omega$ is the frequency of interest, $\omega_o = \frac{1}{\sqrt{LC}}$ is the resonant frequency and $\omega_b = \frac{1}{RC}$ is the -3 dB bandwidth of the magnitude response. The rejection produced by the tuned circuit at $\omega$, Rej($\omega$), is the reciprocal of the impedance magnitude at $\omega$ relative to $R$, the magnitude at the center frequency, $\omega_o$. For frequencies well beyond the band edges of the passband, $| \omega - \omega_o | \gg \omega_b$,

$$\text{Rej}(\omega) = \left(\frac{| Z(j\omega) |}{R}\right)^{-1} \tag{13.75}$$

$$\approx Q \left|\frac{\omega_o}{\omega} - \frac{\omega}{\omega_o}\right|$$

For the example above and if $f_{if} = 0.5$ MHz, $f_{lo} = 1.5$ MHz and $bw = 16.7$ kHz (corresponding to a Q of the tuned circuit of 30), the rejection at the local oscillator frequency is 80. Since the collector-current LO component is 50 times greater than the IF component, the difference-frequency output voltage relative to the fundamental of the local oscillator at the output is only $50/80 = 0.63$. To reduce this to 0.1, the Q of the tuned circuit has to be increased by a factor of 6. However, this large a value of Q may be difficult to achieve unless crystal filters are used. Alternately, the local oscillator magnitude must be reduced. This in turn makes the mixer more susceptible to variations of the circuit's supply biases. For $y \gg 1$, the mixer is relatively insensitive to the amplitude of the local oscillator and subsequently the bias supply levels. As mentioned above, if $y$ is not greater than one, plotted tabular values for the Bessel functions must be used the equations above. Curves of the first few Bessel Functions are included in Figure 11.5.

The Spice input listing for the circuit of the BJT mixer of Figure 13.10a is given in Figure 13.10b. The input and local oscillator signal amplitudes are 1 mV and 100 mV, respectively. Note that the input frequency is 1 MHz and the LO frequency is 1.1 MHz. The difference frequency is therefore 100 kHz. The Q of the tuned output circuit $\frac{\omega_o}{\omega_b}$ is chosen to be 40. Frequency-response calculations estimate that the ratio of the difference-frequency output voltage to the local-oscillator fundamental in the output is 8.44.

The output voltage waveform of the mixer is shown in Figure 13.10c. In this waveform, the LO output component certainly is present. The harmonic content for the lower harmonics can be misleading considering the distorted waveform of the output voltage. Since the difference frequency and local oscillatory frequency are integrally related, the Fourier series algorithm picks points which predict a smooth sinewave. Greater detail of the effects of the output due to the local oscillator is seen more clearly for the output plot of Figure 13.10d where TSTEP is reduced to 1/20 of the LO period. (See the * lines in the Spice input file.) Some adjustments are also made to other .TRAN parameters to obtain the plot. The LO output definitely is seen as a ripple superimposed on the IF output. From the .FOUR calculation from Spice3, where 12 harmonics can be obtained, the ratio of difference frequency component of the output voltage to the LO component is 8.5.

If the Q of the output tuned circuit is increased to 100, the output waveform of Figure 13.10e is obtained. The waveform from Figure 13.10c for $Q = 40$ is superimposed. The decreased effect of the local oscillator component is clear.

### 13.8.2 MOSFET Mixers

It is clear that MOSFET circuits can be used as mixers since the transfer characteristics of the devices are not only nonlinear, but also provide the necessary square-law terms to achieve the multiplication of two inputs. The MOS devices are especially interesting because of their nearly square-law characteristics. Because the higher-order terms in the transfer characteristic are not present in single-device circuits, or are at least small, higher-order distortion components, such as $IM_3$ terms, are not produced.

A simple MOSFET mixer circuit is shown in Figure 13.11a. The input is simply the sum of two sinusoidal signals and a dc bias voltage. The ideal I-V model for the MOSFET biased in saturation is

$$I_D = \frac{k'}{2} \frac{W}{L} (V_{GS} - V_T)^2 \tag{13.76}$$

The total input voltage is

$$V_{GS} = V_{GG} + v_a + v_b \tag{13.77}$$

The I-V characteristics of (13.76) can be expanded similar to the procedure of Section 3.5 for a MOS amplifier.

```
FIGURE 13.10D, Q=40
VS 2 1 0 SIN(0 1MV 1MEG)
VLO 1 0 0 SIN(0.78 100MV 1.1MEG)
*.TRAN 0.2U 620U 610U 45N
.TRAN 45N 601.8U 600U 45N
.PLOT TRAN V(3)
*.FOURIER 100K V(3)
.FOURIER 1.1MEG V(3)
VCC 4 0 10
Q1 3 2 0 MOD1
.MODEL MOD1 NPN IS=1E-16 BF=100
CI1 2 0 1PF
R1 4 3 15K
C1 4 3 4.244NF
L1 4 3 596.83UH
.WIDTH OUT=80
.OPTIONS RELTOL=1E-6 ITL5=0
.OPTIONS NOMOD NOPAGE
.END
```

(d)

```
FIGURE 13.10E, Q=100
VS 2 1 0 SIN(0 1MV 1MEG)
VLO 1 0 0 SIN(0.78 100MV 1.1MEG)
.TRAN 0.2U 620U 610U 45N
*.TRAN 45N 601.8U 600U 45N
.PLOT TRAN V(3)
.FOURIER 100K V(3)
*.FOURIER 1.1MEG V(3)
VCC 4 0 10
Q1 3 2 0 MOD1
.MODEL MOD1 NPN IS=1E-16 BF=100
CI1 2 0 1PF
R1 4 3 15K
C1 4 3 66.667NF
L1 4 3 37.995UHUH
.WIDTH OUT=80
.OPTIONS RELTOL=1E-4 ITL5=0
.OPTIONS NOMOD NOPAGE
.END
```

(e)

**Fig. 13.10.** (d) Spice input file and output voltage waveform shown in greater detail. (e) Spice input file and output voltage waveform for $Q = 100$ (the waveform for $Q = 40$ is superimposed).

$$I_D = \frac{k'}{2}\frac{W}{L}(V_{GG} - V_T)^2 + 2\frac{k'}{2}\frac{W}{L}(V_{GG} - V_T)(v_a + v_b) \qquad (13.78)$$
$$+ \frac{k'}{2}\frac{W}{L}(v_a + v_b)^2$$

We now let each signal input be a single-tone sinusoid with amplitudes $V_{aA}$ and $V_{bA}$. $V_{bA}$ is associated with the local oscillator input. From the square-law term, we obtain the difference-frequency output current.

$$I_D|_{if} = \frac{k'}{2}\frac{W}{L}V_{aA}V_{bA} \qquad (13.79)$$

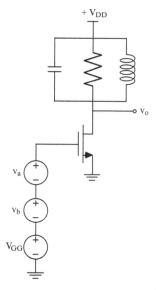

**Fig. 13.11.** A single-device MOSFET mixer circuit.

The conversion transconductance, the ratio of $I_D|_{if}$ to the amplitude $V_{aA}$ of the input signal, is

$$G_{mcon} = \frac{I_D|_{if}}{V_{aA}} = \frac{k'}{2}\frac{W}{L}V_{bA} \tag{13.80}$$

The fundamental of the LO in the drain-current component from (13.78) is

$$I_D|_{lo} = 2\frac{k'}{2}\frac{W}{L}(V_{GG} - V_T)V_{bA} \tag{13.81}$$

For a numerical example, let $k' = 100 \ \mu A/V^2$, $V_T = 0.7$ V, $\frac{W}{L} = \frac{80}{1}$, $V_{aA} = 10$ mV at 1500 kHz and $V_{bA} = 0.4$ V at a frequency of 1045 kHz. The difference frequency is 455 kHz. The dc drain current is 1 mA and the sinusoidal components are $I_D|_{if} = 8 \ \mu A$ and $I_D|_{lo} = 1.6$ mA. The ratio of the IF output current component to the LO output is 0.005. If the output circuit has a Q of 50, the rejection ratio provided by the output tuned circuit at the LO frequency is 93. Therefore, the relative IF component of the output voltage is still only 0.465 of the LO component. The LO input magnitude $V_{bA}$ must be lowered or the Q of the output circuit increased. Since a Q of 50 is already a practical maximum, unless a crystal filter is used, the LO amplitude must be reduced for adequate performance of the mixer.

Because the local oscillator is usually a strong signal, an appreciable bias shift can be obtained due to the square-law term of the drain current characteristic. Without a LO signal, the dc current is

$$I_D = \frac{k'}{2}\frac{W}{L}(V_{GG} - V_T)^2 \tag{13.82}$$

where $V_{GG}$ is the magnitude of the dc bias from gate to source of the MOS-FET. With a LO signal of $V_{bA} \cos \omega_{lo} t$, the square-law term of (13.78) produces a dc current.

$$I_D = \frac{1}{2}\frac{k'}{2}\frac{W}{L}V_{bA}^2 \tag{13.83}$$

The shift in the dc value with the local oscillator signal for the example above is 0.32 mA.

## 13.9 Modulators

Modulation is a process wherein an information-carrying signal is introduced into a carrier signal. In amplitude modulation, the amplitude of the carrier carries the information of the modulating signal.

$$V_o = V_{cA}(1 + m \cos \omega_m t) \cos \omega_c t \tag{13.84}$$
$$= V_{cA}\cos \omega_c t + \frac{V_{cA}}{2}m\cos(\omega_c - \omega_m)t + \frac{V_{cA}}{2}m\cos(\omega_c + \omega_m)t$$

where the separation into the carrier and the two sidebands is also shown. To achieve amplitude modulation, a multiplication is needed. At low frequencies, analog multipliers can be used directly. Usually the amplitudes of the two signals need not be large, and the predistortion circuits of the Gilbert cell are not needed. The modulator may have the configuration of Figure 13.5a which is repeated in Figure 13.12a. As pointed out earlier, MOS devices can also be used in this configuration. For high frequencies, single-device modulators are used and the square-law term of the transfer characteristic of the device provides the multiplication function.

The carrier frequency input may be a large differential signal $V_1(t)$ which overdrives the coupled pairs at the top of the analog multiplier. The output is then a chopped version of the modulation input as shown in Figure 13.12b. Notice that if the modulating input is absent, the output is zero. The carrier is suppressed in this balanced arrangement and this modulator is called a balanced modulator. An analysis of this process and the output waveform provides the frequency spectrum shown in Figure 13.12c. The carrier and its harmonics are not present, only the sidebands about the carrier fundamental and its odd harmonics. If the carrier is desired, a dc component is needed in the carrier input signal.

Other applications of the balanced modulator are introduced in the next chapters.

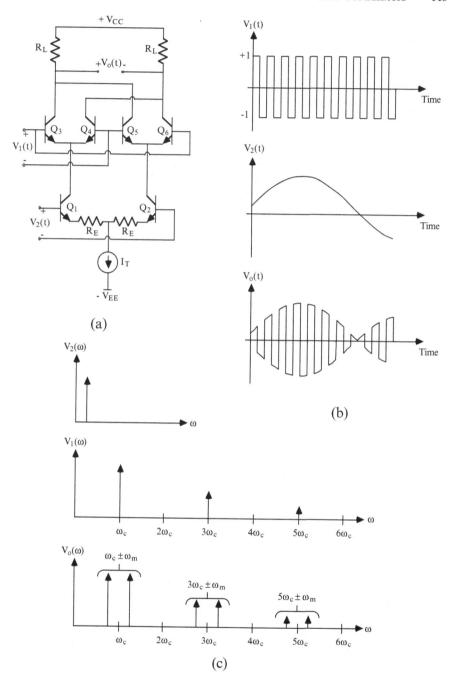

**Fig. 13.12.** (a) A modulator circuit. (b) Time-domain voltage waveforms. (c) Frequency spectrum of input and output signals.

In frequency modulation, other types of modulating circuits are needed to change the carrier frequency or its phase as a function of the modulating signal. In these modulators, voltage- and/or current-dependent devices or circuits are used which vary the resonant frequency of an oscillator, cf., Sections 12.10 and 15.2.

## Problems

**13.1.** An analog multiplier is shown in Figure 13.13. $v_1(t) = 0.01 \cos 10^6 t + 1$ V, $v_2(t) = -0.01 \cos 10^6 t + 1$ V, and $v_3(t) = 1 \cos 10^7 t + 0$ V.
(a) Determine the multiplier coefficient.
(b) What is the magnitude of the difference-frequency component of the output voltage?
(c) Verify your analysis results with Spice.

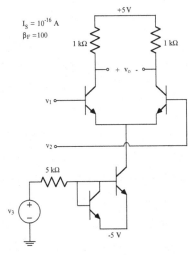

**Fig. 13.13.** Circuit for the analog multiplier of Problem 13.1.

**13.2.** A simple analog multiplier is shown in Figure 13.14.
(a) If $L$ and $C$ are removed and if $V_1$ and $V_2$ are small low frequency incremental voltages, what is the multiplier constant.
(b) The multiplier is now used as a mixer with $L$ and $C$ values given in the figure. The two inputs are $V_1(t) = 0.012 \cos(8 \times 10^7 t)$ V and $V_2(t) = 0.25 \cos(7 \times 10^7 t)$ V. What is the amplitude of the difference-frequency component of $V_{out}(t)$?
(c) What are the output amplitudes at the fundamental frequencies of the input signals?
(d) Verify your analysis results with Spice.

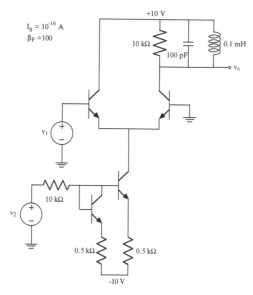

**Fig. 13.14.** Circuit for the analog multiplier of Problem 13.2.

**13.3.** A mixer circuit is shown in Figure 13.15. Assume the LO is a large signal at 2400 MHz and the RF is a small signal at 2400.6 MHz.
(a) Derive the expression for the conversion transconductance of this mixer and calculate its value.
(b) Calculate the amplitude of the IF output voltage for an input RF signal of 1 mV.
(c) Calculate the amplitude of the output voltage at the LO frequency for an LO amplitude of 1 V.

**13.4.** An analog multiplier is shown in Figure 13.16
(a) Determine the drain-source voltages for all transistors at the quiescent operating point coefficient.
(b) Assume that the bias state of the circuit is adjusted to provide $I_{SS} = 0.5$ mA. Determine the multiplier constant when the two inputs $v_x$ and $v_y$ are small signals.
(c) Verify your analysis results with Spice.

**13.5.** An analog multiplier used as a frequency translator is shown in Figure 13.17. $v_1(t) = 0.1 \cos(2\pi 10^6 t)$ V and $v_2(t) = 2 \cos(2\pi 1.1 \times 10^6 t)$ V.
(a) Determine the amplitude of the difference-frequency component of the output voltage across the tuned circuit which is tuned to the difference frequency. Clearly state any assumptions that you make.
(b) Estimate the ratios of the amplitudes of the difference-frequency component of the output voltage to the output components at the fundamental frequencies of the input signals.

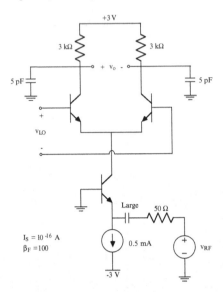

**Fig. 13.15.** Mixer circuit for Problem 13.3.

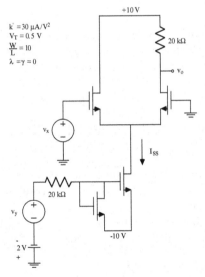

**Fig. 13.16.** Circuit for the analog multiplier of Problem 13.4.

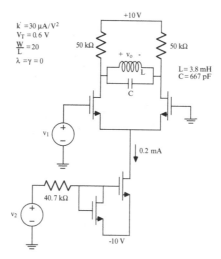

+10 V

$k' = 30\,\mu A/V^2$
$V_T = 0.6\ V$
$\frac{W}{L} = 20$
$\lambda = \gamma = 0$

50 kΩ

+ $v_o$ −

50 kΩ

L

L = 3.8 mH
C = 667 pF

C

$v_1$

0.2 mA

40.7 kΩ

$v_2$

−10 V

**Fig. 13.17.** Circuit for the analog multiplier of Problem 13.5.

**13.6.** A mixer circuit is shown in Figure 13.18. Assume the LO is a large signal and the RF is a small signal.
(a) Derive the expression for the conversion transconductance of this mixer and calculate its value.
(b) What is the IF frequency for this mixer?
(c) Calculate the amplitude of the IF output voltage for an input RF signal of 10 mV.
(d) Calculate the amplitude of the output voltage at the LO frequency for an LO amplitude of 1 V at a frequency of 100 MHz.

**13.7.** A mixer circuit is shown in Figure 13.19. Assume the LO is a large signal at 996 MHz and the RF is a small signal at 925 MHz.
(a) Calculate the conversion transconductance of this mixer.
(b) Calculate the amplitude of the IF output voltage for an input RF signal of 10 mV.
(c) Calculate the amplitude of the output voltage at the LO frequency for an LO amplitude of 1 V.

**13.8.** A single-device BJT mixer is shown in Figure 13.20. $v_1(t) = 0.001\cos(2\pi 10^6 t)$ V and $v_2(t) = 0.01\cos(2\pi 1.5 \times 10^6 t)$ V.
(a) For the output transformer what is the desired value of the magnetizing inductance $L_m$.
(b) What is the value of $Q$ for the output tuned circuit?
(c) Determine the amplitude of the difference frequency output.
(d) What is the amplitude of the output at 1.5 MHz?

**13.9.** A single-device MOS mixer is shown in Figure 13.21. $v_s(t) = 0.1\cos(2\pi 10^7 t)$ V and $v_{lo}(t) = 1\cos(2\pi 1.1 \times 10^7 t)$ V.

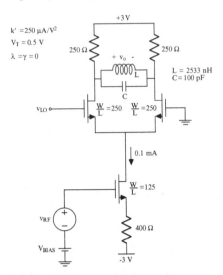

**Fig. 13.18.** Mixer circuit for Problem 13.6.

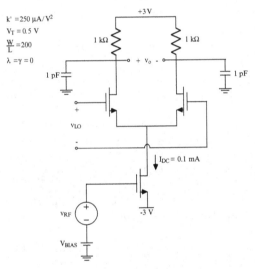

**Fig. 13.19.** Mixer circuit for Problem 13.7.

(a) Design the tuned circuit to resonate at the difference frequency.
(b) Determine the amplitude of the difference frequency output.
(c) What is the ratio of the desired output amplitude of Part (b) and the fundamental output due to the local oscillator?

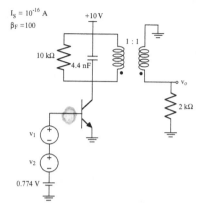

**Fig. 13.20.** Circuit for single-device mixer of Problem 13.8.

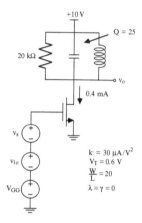

**Fig. 13.21.** Circuit for single-device mixer of Problem 13.9.

# 14

# Demodulators and Detectors

## 14.1 AM Demodulation using Analog Multipliers

Demodulation is the process of retrieving the information from a modulated carrier. It is necessary to treat AM and FM separately because of the different techniques used. Nonetheless, both AM and FM demodulation make use of analog multipliers and/or nonlinear device characteristics. In essence, the carrier is extracted and is multiplied with the original modulated signal, sometimes after its phase is shifted, to produce the modulation signal. In the process, the sidebands are frequency-translated down to a lowpass spectrum.

For AM, an input signal modulated with a single-tone sinusoid has the waveform as sketched in Figure 14.1a and can be expressed as

$$v_i = V_i(t) \cos \omega_c t = V_{iA}[1 + m \cos \omega_m t] \cos \omega_c t \qquad (14.1)$$
$$= V_{iA} \cos \omega_c t + V_{iA} \frac{m}{2} \cos (\omega_c - \omega_m) t + V_{iA} \frac{m}{2} \cos (\omega_c + \omega_m) t$$

In the second expression, the two AM sidebands are placed in evidence and appear in the frequency spectrum as shown in Figure 14.1b. Assume now that this signal is used as the input to an analog multiplier connected as a squarer, as shown in Figure 14.1a. The output has the form

$$v_o = V_{iA}^2 \left[ a_1 \cos^2 X + a_2 \cos^2 Y + a_3 \cos^2 Z \right. \qquad (14.2)$$
$$+ a_4 \cos X \cos Y + a_5 \cos Y \cos Z + a_6 \cos Z \cos X]$$
$$= V_{iA}^2 [b_o + b_1 \cos 2X + b_2 \cos 2Y + b_3 \cos 2Z$$
$$+ b_4 \cos(X \pm Y) + b_5 \cos(Y \pm Z) + b_6 \cos(Z \pm X)]$$

where the $a_i$ and $b_i$ are constants. The spectrum of the output is shown in Figures 14.1c. From (14.2), the output can be viewed as being produced by the beating of the carrier with each sideband and the two sidebands beating

D.O. Pederson and K. Mayaram, *Analog Integrated Circuits for Communication*, DOI 10.1007/978-0-387-68030-9_14,
© 2008 Springer Science+Business Media, LLC

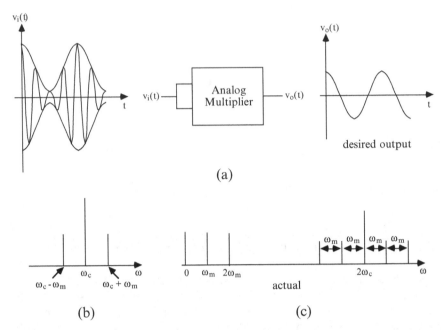

**Fig. 14.1.** (a) An AM signal input to a multiplier and the desired output. Frequency spectrum of (b) the input, and (c) actual output.

with each other to produce the modulation and a second harmonic of the modulation, plus the harmonics of the carrier and the two sidebands. In order to eliminate the terms at a higher frequency than the modulation term, a lowpass frequency filter must be added after the multiplier. The harmonics of the original signal are eliminated readily. However, the second harmonic of the modulation signal is troublesome. For a specific sinusoid tone, a filter can be designed to achieve a large rejection. But the modulation is seldom a single sinusoid. Usually the modulation lies in a lowpass spectrum. The harmonics of the upper portions of the spectra can be rejected; however, the harmonics of the lower-frequency components may pass through relatively unaffected. For these terms, the second-harmonic distortion factor is

$$HD_2 = \frac{m}{4} \tag{14.3}$$

For a large modulation index, the second-harmonic distortion is severe. An example illustrating this aspect is introduced later.

A possible solution to minimize the beating of the two sidebands which produces the second harmonic of the modulating signal is to reject partially one of the sidebands of the input signal with an off-tuned bandpass filter. This is illustrated in Figure 14.2.

A simple AM demodulator based on multiplication can be obtained by using a nonlinear device characteristic. An example is the BJT demodulator

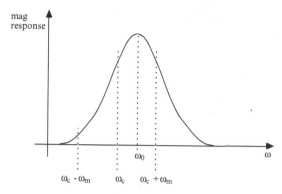

**Fig. 14.2.** An off-tuned bandpass filter response.

shown in Figure 14.3a. In the circuit on the right, the AM input signal is represented as a series of three sinusoidal voltage sources together with the dc bias source. The square-law term from the power-series expansion of the BJT's transfer characteristic provides the required multiplication. This circuit, however, also has the same second-harmonic characteristic as the example above that uses an analog multiplier as a squarer. In addition, the three input signals appear in the output as amplified components. Usually, the rejection of the carrier terms by the lowpass filter is adequate because of the large separation in frequency of the modulation signal and the carrier.

The BJT demodulator serves as a simple vehicle to introduce Spice simulation of demodulation. The input file is given in Figure 14.3b. The amplitude of the AM carrier is 10 mV, and its frequency is 1 MHz. The modulation is a single tone with a frequency of 10 kHz and a modulation index of 80%. The -3 dB band-edge frequency of the low-pass filter is $1/(2\pi R_L C_L) = 14.3$ kHz.

As mentioned above, the input to Spice must be considered to be the individual carrier and sideband signals. As with Spice simulation of mixers and other frequency translation circuits, care must be taken to simulate for a sufficient time to establish a steady-state and obtain a good waveform for the low modulation-frequency component. Finally, the minimum time step must be small enough to provide an accurate simulation of the carrier.

The output waveform from the Spice simulation is shown in Figure 14.3c. In spite of the use of a simple lowpass $RC$ filter at the output, it is clear from an inspection of the output waveform that there is a strong second harmonic. The Fourier components of the output waveform are given in Figure 14.3d, and it is seen that $HD_2 = 14.4\%$. Since the modulation index for this example is 80%, the estimate of $HD_2$ from (14.3) is 20%. If the lowpass filtering of the second harmonic is included, the estimate is 11%. The filter rejects the carrier component of the output voltage by a factor of 75.

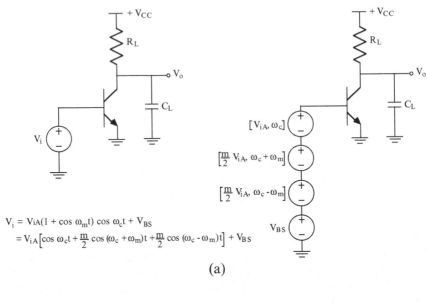

$V_i = V_{iA}(1 + \cos \omega_m t) \cos \omega_c t + V_{BS}$

$\quad = V_{iA}\left[\cos \omega_c t + \dfrac{m}{2} \cos (\omega_c + \omega_m)t + \dfrac{m}{2} \cos (\omega_c - \omega_m)t\right] + V_{BS}$

(a)

```
FIG 13.3B
V1 1 0 0.774
V2 2 1 SIN(0 10M 1MEG)
V3 3 2 SIN(0 4M .99MEG)
V4 4 3 SIN(0 4M 1.01MEG)
Q1 5 4 0 MOD1
.MODEL MOD1 NPN IS=1E-16 BF=100
RL 5 6 1K
CL 5 0 0.012U
VCC 6 0 10
.TRAN 5U 200U 0 0.1U
.PLOT TRAN V(5)
.FOUR 10K V(5)
.OPTIONS LVLTIM=1 ITL5=0
.END
```

(b)

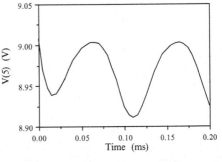

(c)

```
FOURIER COMPONENTS OF TRANSIENT RESPONSE V(5)
DC COMPONENT =    8.958D+00
```

| HARMONIC NO | FREQUENCY (HZ) | FOURIER COMPONENT | NORMALIZED COMPONENT | PHASE (DEG) | NORMALIZED PHASE (DEG) |
|---|---|---|---|---|---|
| 1 | 1.000D+04 | 4.877D-02 | 1.000000 | -127.032 | 0.000 |
| 2 | 2.000D+04 | 7.000D-03 | 0.143521 | -146.667 | -19.634 |
| 3 | 3.000D+04 | 1.533D-04 | 0.003144 | -166.430 | -39.397 |
| 4 | 4.000D+04 | 1.769D-04 | 0.003627 | -179.359 | -52.327 |

```
TOTAL HARMONIC DISTORTION =    14.374665  PERCENT
```

(d)

**Fig. 14.3.** (a) A BJT demodulator circuit. (b) Spice input file. (c) Output voltage waveform. (d) Fourier components of the output voltage.

## 14.2 Synchronous AM Detection

A better AM demodulator using an analog multiplier can be obtained if the carrier alone is extracted and multiplied with the original AM modulated signal as shown in Figure 14.4. The carrier alone is extracted using an amplitude limiter. The limiter is often an emitter- or source-coupled pair which is overdriven to eliminate the amplitude modulation for $m > 0$. A tuned filter may be used after the limiter at the carrier frequency, which will restore the sinusoidal carrier tone. The new carrier signal may have phase shift with respect to the input signal.

$$v_c = V_{cA} \cos (\omega_c t + \phi) \tag{14.4}$$

The input signal from (14.1) is

$$v_i = V_i(t) \cos \omega_c t \tag{14.5}$$

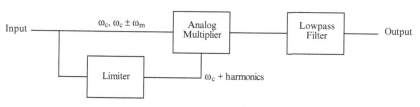

**Fig. 14.4.** AM demodulation using an analog multiplier.

The amplitude modulation is implicit in the function $V_i(t)$. After multiplication and filtering out the higher-frequency terms, the output of the multiplier is

$$v_o(t) = K' V_{cA} V_i(t) \frac{1}{2} [\cos \phi + \cos (2\omega_c t + \phi)] \tag{14.6}$$

where $K'$ is a constant. The $2\omega_c t$ term can also be rejected by filtering, and the AM information $V_i(t)$ is recovered. Note that the filtering of the carrier in the limiter should not produce a phase shift $\phi$ equal to $\pm 90°$.

To illustrate this type of demodulation, a Spice simulation is made of the circuit of Figure 14.5a. The circuit is a fully balanced, cross-coupled set of ECPs. The modulated input signal can be modeled as a set of three voltages sources driving the bottom pair. A voltage-controlled voltage source $E1$ samples the total input and drives the upper pairs. With a large drive to the upper pairs, the limiting action of Figure 14.4 is obtained. The capacitor $C_o$ and the collector load resistors constitute a lowpass filter.

The Spice input file is given in Figure 14.5b for an input voltage with a carrier amplitude of 10 mV and a modulation index of 80%. The carrier has a frequency of 1 MHz, and the modulation frequency is 10 kHz. Note that RELTOL $= 1 \times 10^{-6}$ is specified in anticipation of small distortion. The output waveform is given in Figure 14.5c. The distortion components of the latter are given in Figure 14.5d. The second-harmonic distortion is small, and THD $= 1.2\%$.

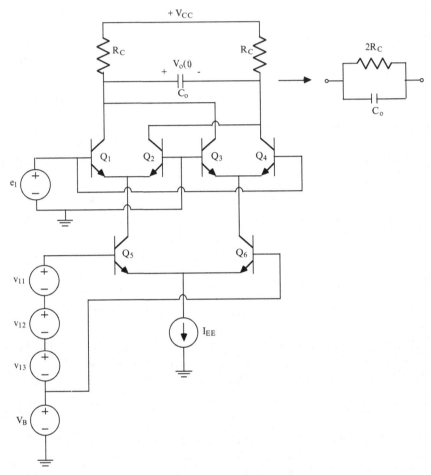

**Fig. 14.5.** (a) Circuit for AM demodulation.

## 14.3 Peak Detectors

In the AM demodulation scheme of the last section, the squaring of the modulated carrier is achieved, and a 'beating' of the modulated input with the carrier is produced that translates the modulation component of the input

```
FIG 14.5, LO DRIVES TOP ECPS
.TRAN 5U 800U 600U 0.1U
.PLOT TRAN V(23,24)
.FOUR 10K V(23,24)
V11 1 2 SIN(0 10M 1MEG)
V12 2 3 SIN(0 4M 1.01MEG)
V13 3 4 SIN(0 4M 0.99MEG)
VB   4  0 -2
VCC 100 0 10
RL1 100 23   5K
RL2 100 24   5K
CO 23 24 0.005U
Q1 23 5 22 MODN
Q2 24 0 22 MODN
Q3 23 0 25 MODN
Q4 24 5 25 MODN
Q5 22 1 12 MODN
Q6 25 4 12 MODN
.MODEL MODN NPN IS=1E-16 BF=100
E1 5  0  1  4  50
IEE 12 0 300U
.OPTIONS NOPAGE NOMOD RELTOL=1E-6
.WIDTH OUT=80
.OPTIONS ITL5=0 LIMPTS=5000
.END
```

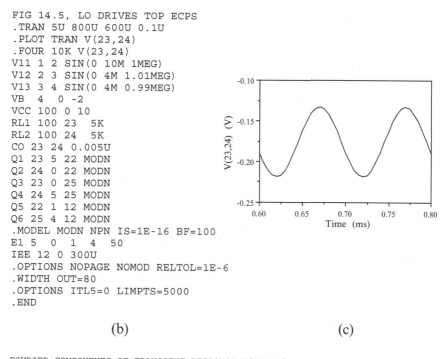

(b)                                    (c)

```
FOURIER COMPONENTS OF TRANSIENT RESPONSE V(23,24)
 DC COMPONENT =  -1.765D-01
 HARMONIC    FREQUENCY     FOURIER     NORMALIZED     PHASE      NORMALIZED
   NO          (HZ)       COMPONENT    COMPONENT      (DEG)     PHASE (DEG)

     1       1.000d+04    4.298d-02    1.000000    -162.363       0.000
     2       2.000d+04    4.837d-04    0.011255      15.836     178.200
     3       3.000d+04    5.165d-05    0.001202      93.919     256.282
     4       4.000d+04    6.717d-05    0.001563     132.260     294.623
     5       5.000d+04    1.664d-04    0.003872     171.540     333.903
     6       6.000d+04    5.399d-05    0.001256    -175.665     -13.302
     7       7.000d+04    2.201d-05    0.000512     -42.097     120.266
     8       8.000d+04    1.406d-05    0.000327     -49.225     113.138
     9       9.000d+04    9.396d-06    0.000219     128.781     291.144

     TOTAL HARMONIC DISTORTION =     1.214681   PERCENT
```

(d)

**Fig. 14.5.** (b) Spice input file. (c) Output voltage waveform. (d) Fourier components of the output voltage.

down to the lowpass region. The sketch of the frequency spectra as in Figure 14.1 aids the visualization of the translation.

In peak detection, a voltage-controlled switch is the key nonlinear element. It is necessary to remain in the time domain, observing modulated waveforms, in order to understand simply the operation of the demodulator.

The simplest peak detector including a voltage-controlled switch is shown in Figure 14.6a, with an input circuit tuned to the carrier frequency (usually the IF), and a lowpass, parallel $RC$ filter. In practice, the controlled switch is realized as a pn diode. The diode-peak detector circuit is shown in Figure 14.6b, where the input is replaced with the series of voltage sources representing an AM input. Initially, the parameters of the diode are chosen with a (relatively) large value for $I_S$ ($10^{-5}$A) and a diode resistance of $R_S = 100\ \Omega$. These values provide an approximate piece-wise-linear I-V characteristic of the diode similar to Curve A of Figure 14.6c, in contrast to the usual exponential characteristic for a pn diode as sketched as Curve B for a much smaller value of $I_S$, say $10^{-16}$A.

The waveform of the input to the circuit of Figure 14.6b is shown in Figure 14.6d. Consider that the input has been applied for some time, and that startup transients have died out. At a particular value of time, the capacitor $C_L$ of the lowpass filter has a residual voltage (charge) across it. When the input signal decreases below this value, the switch opens (the diode is off). The charge on the capacitor then discharges through $R_L$ while the input is below the capacitor voltage. When the input voltage, the modulated carrier, becomes larger than the capacitor voltage, the switch closes (the diode is on), and the capacitor voltage follows the input voltage and receives charge. The output voltage is sketched in Figure 14.6e where the positive excursions of the modulated input are included and the envelope of the waveform has been emphasized. The output voltage is a somewhat distorted approximation of the envelope.

During the *off* segment of operation, when the input voltage is less than the capacitor voltage, the $RC$ time constant of the filter must be larger than the period of the carrier. During the *on* segment of diode operation, the time constant of the total circuit is very small when the diode is forward-biased because of the low resistance path through the *on* diode and an assumed low source resistance. The capacitor then can follow closely changes of the modulated carrier.

The distortion in the output waveform consists principally of higher-order harmonics of carrier-frequency terms; however, some harmonics of the modulation frequencies also are present. It is difficult to estimate analytically these harmonic terms. Although the nonlinear circuit is quite simple, the nonlinear differential equation which describes the circuit is solvable simply only by numerical means. It is easier to use Spice simulation to establish typical performance.

In Figure 14.7a, the Spice input file for the peak-detector circuit is given. The input is a series of three sinusoids to provide the AM wave. The AM input has the same frequencies and modulation as the prior examples of this chapter; however, the carrier amplitude is increased to 5 V. The diode has the parameters introduced above to approximate a piece-wise-linear switch. The 10 $\mu$s time constant for the lowpass filter is chosen to be one-tenth the period of the modulation tone (1/10 kHz = 100 $\mu$s). The ratio of the modulation frequency to the carrier frequency is 1/100.

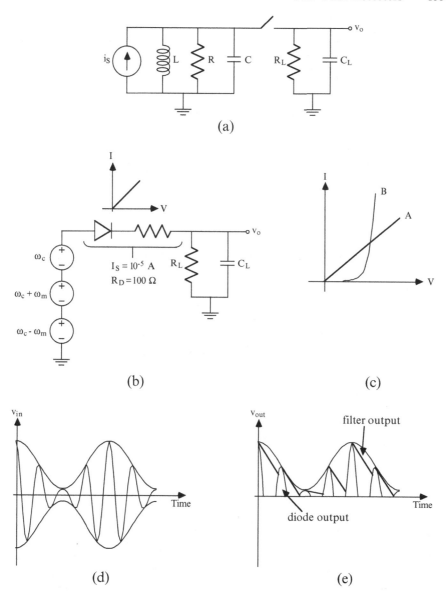

**Fig. 14.6.** (a) A simple peak-detector circuit. (b) A diode peak detector. (c) I-V characteristics of the diode. (d) Input voltage waveform. (e) Output voltage waveform.

```
FIG 14.7
VA 1 5 SIN 0 2 0.99MEG
VB 6 0 SIN 0 2 1.01MEG
VC 5 6 SIN 0 5 1.00MEG
R1 1 2 1
D1 2 3 M1
.MODEL M1 D IS=1E-5 RS=100
*.MODEL M1 D IS=1E-16
RL 3 0 1K
CL 3 0 10N
.TRAN 5U 400U 200U 0.1U
*.TRAN .05U 4U
.FOUR 10K V(3)
.PLOT TRAN V(3)
.OPTIONS ITL5=0 RELTOL=1E-6
.END
```

(a)

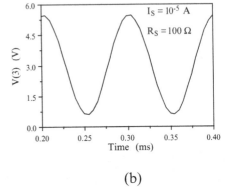

(b)

```
FOURIER COMPONENTS OF TRANSIENT RESPONSE V(3)
DC COMPONENT =    3.106D+00
```

| HARMONIC NO | FREQUENCY (HZ) | FOURIER COMPONENT | NORMALIZED COMPONENT | PHASE (DEG) | NORMALIZED PHASE (DEG) |
|---|---|---|---|---|---|
| 1 | 1.000D+04 | 2.502D+00 | 1.000000 | 80.048 | 0.000 |
| 2 | 2.000D+04 | 1.322D-02 | 0.005281 | 165.815 | 85.767 |
| 3 | 3.000D+04 | 7.873D-03 | 0.003146 | -16.895 | -96.943 |
| 4 | 4.000D+04 | 1.297D-02 | 0.005183 | -29.193 | -109.241 |
| 5 | 5.000D+04 | 3.679D-02 | 0.014703 | -4.768 | -84.816 |
| 6 | 6.000D+04 | 1.583D-02 | 0.006324 | 14.022 | -66.026 |
| 7 | 7.000D+04 | 1.539D-03 | 0.000615 | -1.099 | -81.147 |
| 8 | 8.000D+04 | 4.289D-04 | 0.000171 | -81.291 | -161.339 |
| 9 | 9.000D+04 | 2.343D-03 | 0.000936 | -179.481 | -259.529 |

```
TOTAL HARMONIC DISTORTION =          1.794729  PERCENT
```

(c)

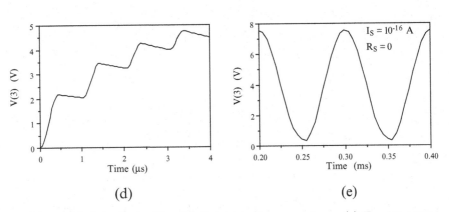

(d)                                        (e)

**Fig. 14.7.** (a) Spice input file. (b) Output voltage waveform. (c) Fourier components of the output voltage. (d) Voltage waveform for an initial time segment. (e) Waveform with a regular diode model.

The output voltage waveform is shown in Figure 14.7b, and the distortion components in the lowpass region are given in Figure 14.7c. A very clean lowpass output has been achieved. The output waveform for an initial time segment of the carrier is shown in Figure 14.7d. The buildup and 'holding' of the output voltage are clearly seen.

The output waveform of the detector and its harmonic components when a regular diode model is used in the circuit, are shown in Figures 14.7e and f. The diode parameters, as given in the commented line of Figure 14.7a, are $I_S = 10^{-16}$A and $R_S = 0$. The output voltage of the fundamental is larger, and the waveform shows more distortion than the previous case. The distortion is reduced somewhat when $C_L$ is reduced to 8 nF.

```
FOURIER COMPONENTS OF TRANSIENT RESPONSE V(3)
DC COMPONENT =    3.995D+00
HARMONIC   FREQUENCY    FOURIER     NORMALIZED     PHASE      NORMALIZED
   NO        (HZ)      COMPONENT    COMPONENT      (DEG)      PHASE (DEG)

   1      1.000D+04    3.728D+00    1.000000      88.077       0.000
   2      2.000D+04    7.755D-02    0.020805     129.018      40.940
   3      3.000D+04    3.659D-02    0.009815     -58.921    -146.999
   4      4.000D+04    2.094D-02    0.005619    -156.863    -244.941
   5      5.000D+04    8.895D-02    0.023863     -20.103    -108.180
   6      6.000D+04    2.366D-02    0.006348      12.467     -75.610
   7      7.000D+04    2.591D-02    0.006951     -37.516    -125.594
   8      8.000D+04    2.003D-02    0.005372     -37.478    -125.555
   9      9.000D+04    1.154D-02    0.003096      48.642     -39.435

   TOTAL HARMONIC DISTORTION =        3.545764  PERCENT
```

**Fig. 14.7.** (f) Fourier components of the output voltage.

If the time constant of the lowpass filter is too large relative to the period of the modulation, or alternately, if the band-edge frequency of the filter, considered independently of the input and diode, is close to or less than the frequency of the modulation, a problem develops as sketched in the waveform of Figure 14.8a. For this situation, the value of the capacitor has been increased. On the decreasing portion of the lowpass output, the output fails to follow the envelope of the input. The capacitor does not discharge fast enough. An estimate of the relation to avoid this distortion can be developed as follows. Let the input envelope voltage be

$$v_i = V_{iA} \left[1 + m \cos \omega_m t\right] \tag{14.7}$$

The output across the capacitor when the diode is *off* is of the form

$$v_o = V \exp\left(\frac{-t'}{T}\right) \tag{14.8}$$

where $V$ is the value of the filter voltage at $t' = 0$, $t'$ is the time since diode turnoff, and $T$ is the time constant of the filter, $R_L C_L$. We desire that the magnitude of the slope of the envelope of the modulated input be smaller

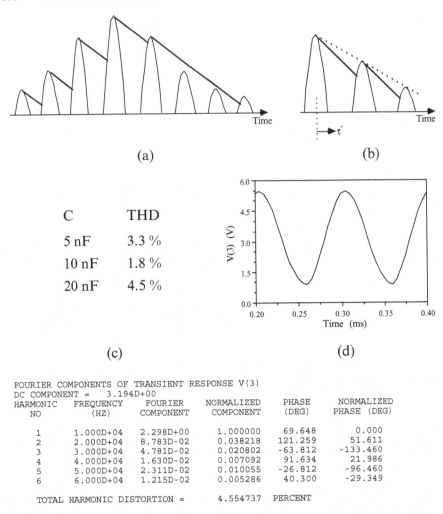

C          THD

5 nF       3.3 %

10 nF      1.8 %

20 nF      4.5 %

(a)                              (b)

(c)                              (d)

```
FOURIER COMPONENTS OF TRANSIENT RESPONSE V(3)
DC COMPONENT =    3.194D+00
HARMONIC   FREQUENCY    FOURIER     NORMALIZED     PHASE      NORMALIZED
   NO        (HZ)      COMPONENT    COMPONENT      (DEG)     PHASE  (DEG)

    1      1.000D+04    2.298D+00   1.000000       69.648       0.000
    2      2.000D+04    8.783D-02   0.038218      121.259      51.611
    3      3.000D+04    4.781D-02   0.020802      -63.812    -133.460
    4      4.000D+04    1.630D-02   0.007092       91.634      21.986
    5      5.000D+04    2.311D-02   0.010055      -26.812     -96.460
    6      6.000D+04    1.215D-02   0.005286       40.300     -29.349

     TOTAL HARMONIC DISTORTION =      4.554737  PERCENT
```

(e)

**Fig. 14.8.** (a) A problem output voltage waveform. (b) Constraint on the $RC$ time constant. (c) Total harmonic distortion as a function of $C_L$. (d) Output voltage waveform for $C_L = 20$ nF. (e) Fourier components of the output voltage for $C_L = 20$ nF.

than the magnitude of the slope of the discharge of the capacitor for $t' \geq 0$. This is illustrated in Figure 14.8b. In the following, the absolute values can be neglected since the development includes the correct sign.

$$\frac{dv_i}{dt'} \geq \frac{dv_o}{dt'} \tag{14.9}$$

which for the time segment of interest is the same as

$$\left| \frac{dv_i}{dt'} \right| \leq \left| \frac{dv_o}{dt'} \right|$$

where the inequality changes direction due to negative values. This leads to

$$V_{iA} m \omega_m \sin \omega_m t' \leq \frac{V}{T} \exp\left( \frac{-t'}{T} \right) \tag{14.10}$$

Next, we set (14.7) equal to (14.8) at $t'$ and combine with (14.10). The result is

$$\frac{1 + m \cos \omega_m t'}{m \sin \omega_m t'} \geq \omega_m T \tag{14.11}$$

From the minimum of the left-hand side of (14.11), we obtain the condition

$$\cos \omega_m t' = -m \tag{14.12}$$

which leads to

$$\sin \omega_m t' = \sqrt{1 - m^2} \tag{14.13}$$

Using these in (14.11), we obtain

$$\omega_m T \leq \frac{\sqrt{1 - m^2}}{m} \tag{14.14}$$

As a numerical example, the values from the Spice listing of Figure 14.7a are used (with the idealized diode). Since $T = R_L C_L$, the largest value of $C_L$ should be

$$C_L \leq \frac{1}{R_L 2\pi f_m} \frac{\sqrt{1 - m^2}}{m} \tag{14.15}$$
$$= 12 \text{ nF}$$

For the simulation leading to Figure 14.7b, $C_L$ is equal to 10 nF, and the output shows no 'failure-to-follow' distortion. In the table in Figure 14.8c, values of the THD of the output waveforms are given as $C_L$ is varied. With $C_L = 20$ nF, the output waveform and its harmonic components are given in Figures 14.8d and e. *'Failure-to-follow'* distortion is evident. The best performance is obtained with a value of $C_L$ near that given by the equality in (14.15).

If the AM signal source has a finite source resistance, say from the tuned output circuit of an ECP stage rather than from the output of a buffering emitter follower, the results of this section must be modified, principally in the amplitude of the average (dc) output voltage and the amplitude of the modulation output. Consider first the situation where only a carrier is present ($m = 0$). Because of the diode 'switch', only sinusoidal tips of current flow through the closed switch into the capacitor. For the high frequency, fast transient of the carrier, $C_L$ appears as a short circuit. However, these pulses of current produce and support an overall average charge in the steady state. The average value of the charged $C_L$ is labeled $V_{oave}$. With the diode switch open, the capacitor discharges through $R_L$. $C_L$ then is charged again when the switch closes at the next voltage peak of the input source. The diode current has an average value which must flow into $R_L$ since a dc current cannot flow through $C_L$.

Let $a$ be the ratio of the average current through the diode to the peak current magnitude. (For a half sinusoid, $a = 0.32$. For sinusoidal tips, $a$ is less.) The average current can be written as $I_{ave} = a(V_{cA} - V_{oave})/R'_s$ where $V_{cA}$ is the amplitude of the carrier and $R'_s$ is the value of the source resistance, $R_s$, plus the diode series resistance. In the steady state, the average output voltage is $V_{oave} = R_L I_{ave}$. Therefore, $V_{oave} = V_{cA}/(1 + R'_s/aR_L)$. For $a = 0.32$, $R'_s = 10.1$ k$\Omega$, $R_L = 10$ k$\Omega$, and $V_{cA} = 10$ V, $V_{oave} = 2.41$ V. The results from a Spice simulation as shown in Figure 14.8f are $V_{oave} = 2.1$ V with $C_L = 1$ nF; therefore, $a_{eff} = 0.27$. (The Fourier components of a sequence of sinusoidal tips can be obtained. An iterative procedure can be used to find an estimate of $a_{eff}$.)

For an amplitude modulated (AM) input, a similar argument can be used to obtain the amplitude of the output modulation superimposed on the average output. For $m = 0.8$ in the above example, the estimate of the amplitude using $a = 0.27$ is 1.69 V. Spice simulation as shown in Figure 14.8f provides 1.59 V. Again the product $R_L C_L$ must not be too large with respect to the minimum period of the modulation $1/f_m$. Equation (14.15) can be used as a guide.

A more complicated lowpass filter can be used in the peak detector. To achieve appreciably better results, a $RLC$-pi filter should be used as shown in Figure 14.9a. However, the size and cost of the inductor often are not warranted.

It is possible to replace the single diode of Figure 14.6b or Figure 14.9a with multiple diode arrangements, such as push-pull parallel and bridge configurations as shown in Figures 14.9b and c. In effect, the bottom portions of the modulated waveform are brought up above the axis as sketched in Figure 14.9d, and a more efficient performance is achieved. For AM detectors, power-

```
DIODE PEAK DETECTOR
*RS > 0, FIG13.8F
VA 1 5 SIN 0 4 0.99MEG
VB 6 0 SIN 0 4 1.01MEG
VC 5 6 SIN 0 10 1.00MEG
R1 1 2 10K
D1 2 3 M1
.MODEL M1 D IS=1E-5 RS=100
*.MODEL M1 D IS=1E-16
RL 3 0 10K
CL 3 0 1N
.TRAN 5U 400U 200U 0.1U
*.TRAN .05U 4U
.FOUR 10K V(3)
.PRINT TRAN V(3)
.OPTIONS ITL5=0 RELTOL=1E-5
.END
```

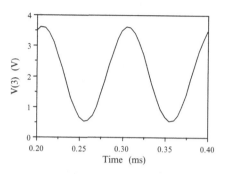

```
FOURIER COMPONENTS OF TRANSIENT RESPONSE V(3)
DC COMPONENT =    2.107D+00
```

| HARMONIC NO | FREQUENCY (HZ) | FOURIER COMPONENT | NORMALIZED COMPONENT | PHASE (DEG) | NORMALIZED PHASE (DEG) |
|---|---|---|---|---|---|
| 1 | 1.000D+04 | 1.588D+00 | 1.000000 | 66.199 | 0.000 |
| 2 | 2.000D+04 | 6.234D-03 | 0.003925 | 56.123 | -10.075 |
| 3 | 3.000D+04 | 1.863D-03 | 0.001173 | 113.154 | 46.955 |
| 4 | 4.000D+04 | 7.634D-03 | 0.004807 | 18.162 | -48.037 |

```
TOTAL HARMONIC DISTORTION =     1.181639  PERCENT
```

**Fig. 14.8.** (f) Spice input file, output voltage waveform, and Fourier components for $R_S > 0$.

conversion efficiency is not of great importance and the simpler single-diode circuit is adequate.

## 14.4 Automatic Gain Control

Usually in an AM radio, another $RC$ filter circuit is added to the peak detector as shown in Figure 14.9e. The time constant for the $R_2 - C_2$ combination is made very large relative to the period of the modulation. What is desired is an output which is proportional to the carrier magnitude. This output is used as a very low-frequency feedback signal to change the gain of the proceeding tuned amplifiers, usually the IF amplifier. Using dc coupling to the ECPs or SCPs of these amplifiers, we can change the biasing of these stages and therefore their gain. Thus, as the carrier level increases as detected by the new circuit in the peak detector, the gain of the bandpass stages in the IF amplifier can be decreased to maintain an approximately constant output of the modulation component output of the detector. This arrangement is called an 'automatic gain control' (AGC) feature. Examples of the bias control of basic stages are given in Sections 12.7 and 12.10.

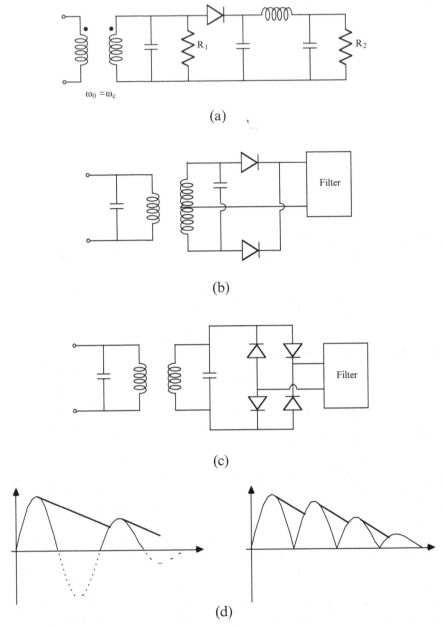

**Fig. 14.9.** (a) A peak detector with a $RLC$-pi filter. (b) A push-pull parallel diode arrangement. (c) A bridge configuration of the diodes. (d) Output voltage waveforms.

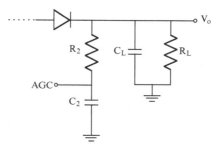

**Fig. 14.9.** (e) Circuit for automatic gain control.

## 14.5 FM Demodulation, Off-Peak Detection

To obtain the information content from a frequency-modulated (FM) signal, a simple procedure is to convert the frequency deviation into an amplitude deviation. The resulting signal is then AM demodulated with a peak detector. A simple circuit to achieve this is shown in Figure 14.10a. The input tuned circuit, say from the output stage of an IF amplifier, is tuned below or above the carrier frequency (IF) of the signal. As the frequency of the input signal to the FM detector varies with the modulation, the voltage across the tuned circuit varies as the magnitude response varies with frequency. This is illustrated in Figure 14.10b. An amplitude modulation of the signal is produced. The peak detector produces the low pass, information signal.

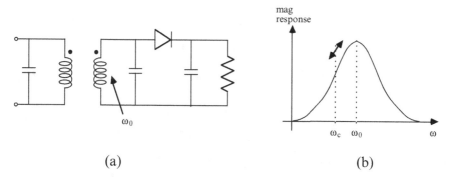

(a)                                                    (b)

**Fig. 14.10.** (a) A simple scheme for FM demodulation. (b) Conversion of frequency deviation into amplitude deviation.

The magnitude response of the tuned circuit, well away from the passband, varies approximately as $1/\omega$ or $\omega$. However, some distortion is introduced. An example shows that the performance is acceptable. For the example, an emitter-coupled pair is used as the driver stage as shown in Figure 14.11a. Device parameters and circuit values are given in the figure. The FM input signal is

$$v_i = V_{ia} \sin[\omega_c t + \text{MDI} \sin \omega_m t] \tag{14.16}$$
$$= 100 \text{ mV} \sin[2\pi \times 10^7 t + 7.5 \sin 2\pi \times 10^4 t]$$

where $MDI = \Delta f / f_m$. The carrier frequency is 10 MHz, the frequency deviation is $\pm 75$ kHz, and the modulation is a single tone of 10 kHz. (Remember that for FM, $\omega_i(t)t$ cannot be used as the argument of the principal sinusoid where $\omega_i$ is the instantaneous frequency, $\omega_i(t) = \omega_c + \Delta\omega \cos \omega_m t$. The argument must be a phase function. Only for a constant frequency, $\phi(t) = \omega_c t$, can we deal directly with the constant, $\omega_c$).

For an estimate of performance, note that the input signal amplitude is 100 mV. From earlier performance studies of the ECP, we can expect that overdrive exists and that the collector current of $Q_2$ varies over the range 0 mA to $I_{EE} = 2$ mA. The amplitude of the fundamental component, from the developments of Section 10.7, is $0.64 I_{EE} = 1.28$ mA. The carrier frequency of the input is 10 MHz, and the output tuned circuit is tuned above this at 12 MHz with a $Q$ of 20. At the resonant frequency, the magnitude of the impedance is 10 k$\Omega$, if we assume that the loading across the transformer due to the peak detector is negligible because of the diode on-off behavior. For a distortion estimate, we calculate the output voltage at the carrier frequency and at the upper and lower maximum frequency deviations. These three frequencies are

$$f_c = 10 \text{ MHz} \tag{14.17}$$
$$f_u = 10 \text{ MHz} + 75 \text{ kHz} = 10.075 \text{ MHz}$$
$$f_l = 10 \text{ MHz} - 75 \text{ kHz} = 9.925 \text{ MHz}$$

The magnitude of the tuned circuit at these frequencies can be found from

$$|Z(j2\pi f)| = \frac{R}{\sqrt{1 + Q^2 \left(\frac{f}{f_o} - \frac{f_o}{f}\right)^2}} \tag{14.18}$$

where $f_o = 12$ MHz. At the three frequencies of interest,

$$|Z(2\pi \times 10^7)| = 1.351 \text{ k}\Omega \tag{14.19}$$
$$|Z(2\pi \times 1.0075 \times 10^7)| = 1.409 \text{ k}\Omega$$
$$|Z(2\pi \times 0.9925 \times 10^7)| = 1.298 \text{ k}\Omega$$

The corresponding initial estimates of the voltages across the tuned circuit are found by multiplying by the amplitude of the fundamental collector current, 1.28 mA. The voltages presented to the detector are these same values. If we assume that the peak detector has no voltage drop and if the envelope

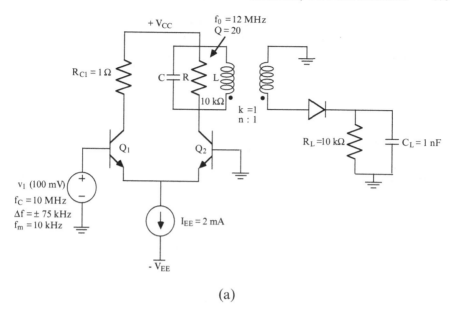

(a)

```
FM OFF PEAK DETECTION, FIG 14.11B
V1 1 0 0 SFFM 0 100M 10MEG 7.5 10K
R1 1 2 .1
Q1 3 2 4 MOD1
Q2 6 7 4 MOD1
R2 7 0 .1
RC1 5 3 1
RC2 5 6 10K
L1 5 6 6.63U
C1 5 6 26.54P
E1 8 10 6 01
D1 8 9 MOD2
.MODEL MOD1 NPN BF=100 IS=1E-16
.MODEL MOD2 D IS=1E-5 RS=100
RL 9 0 10K
CL 9 0 1N
IEE 4 10 2M
VEE 10 0 -10
VCC 5 0 10
.OPTIONS ACCT ITL5=0 RELTOL=1E-3
.OPTIONS NOPAGE NOMOD
.TRAN 5U 200U 0 5N
.PLOT TRAN V(9) (1.1,1.5)
.FOUR 10K V(9)
.END
```

(b)

**Fig. 14.11.** (a) A circuit for FM off-peak detection. (b) Spice input file.

detector has little distortion, a three-point distortion analysis can be used to estimate the second-harmonic distortion. The mid and extreme values of the output voltage are

$$V_4 = 1.804 \text{ V} \tag{14.20}$$
$$V_0 = 1.729 \text{ V}$$
$$V_2 = 1.661 \text{ V}$$

The fundamental and the second-harmonic output voltages are

$$b_1 = \frac{V_4 - V_2}{2} = 71.5 \text{ mV} \tag{14.21}$$
$$b_2 = \frac{V_4 + V_2 - 2V_0}{4} = 1.75 \text{ mV}$$

The second-harmonic distortion is

$$HD_2 = \frac{b_2}{b_1} = 2.5\% \tag{14.22}$$

Of course, many simplifying assumptions have been made and the amplitude of the AM output is low; however, the estimate illustrates that the off-peak detection circuit is capable of achieving the desired demodulation circuit function with reasonable performance, albeit not high fidelity.

In the above analysis, hard limiting by the input signal is assumed. A better estimate of the collector current can be obtained by using the curves in Figure 2.16 of Chapter 2 for the fundamental and the harmonics from the ECP with a sinusoidal input. However, such results only introduce a scale factor. For the above example, the normalized input drive is $d \approx 4$. From Figure 2.16, $I_{C2}\,|_{fund} \approx 0.56 I_{EE}$. Therefore, the estimate of the amplitude of the AM is 62.3 mV instead of 71.5 mV. Similarly, a value can be added for the detector loading, but this involves nonlinear estimates in the time domain which can be questionable with our calculations above in the frequency domain. It is best we stay with the simple evaluation above.

Spice simulations of this detector involve very large computer times to isolate the low-frequency modulation component from the multiplicity of sideband and carrier components. The input file for the circuit of Figure 14.11a is given in Figure 14.11b. Note that the input is a FM time function with a carrier frequency of 10 MHz, and MDI = 7.5 and a modulating frequency of 10 kHz. Therefore, the frequency deviation is ±75 kHz. In the .TRAN specification, the print interval is taken to be 1/20 of the period of the modulating frequency while the maximum time step is 1/20 of the period of the carrier. The output transformer is replaced with a voltage-controlled voltage source to eliminate loading. The output voltage waveform and its harmonic components are given in Figure 14.11c. From the latter, $b_1 = 62$ mV and $HD_2 = 2.3\%$. These values are very close to the estimates. However, $HD_3 = 4.2\%$ and $THD = 7.9\%$. Hopefully, the higher-order harmonics are attenuated by lowpass filtering in the following audio amplifiers.

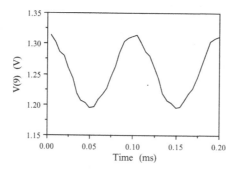

| DC COMPONENT = | 1.254d+00 | | | | |
|---|---|---|---|---|---|
| HARMONIC NO | FREQUENCY (HZ) | FOURIER COMPONENT | NORMALIZED COMPONENT | PHASE (DEG) | NORMALIZED PHASE (DEG) |
| 1 | 1.000d+04 | 5.881d-02 | 1.000000 | 86.302 | 0.000 |
| 2 | 2.000d+04 | 1.309d-03 | 0.022262 | 70.522 | -15.780 |
| 3 | 3.000d+04 | 5.248d-04 | 0.008923 | 142.553 | 56.251 |
| 4 | 4.000d+04 | 2.025d-04 | 0.003443 | -49.351 | -135.653 |
| 5 | 5.000d+04 | 7.603d-04 | 0.012928 | -45.964 | -132.266 |
| 6 | 6.000d+04 | 7.253d-04 | 0.012332 | 6.345 | -79.957 |
| 7 | 7.000d+04 | 4.594d-04 | 0.007811 | -146.506 | -232.808 |
| 8 | 8.000d+04 | 7.320d-04 | 0.012446 | -32.570 | -118.872 |
| 9 | 9.000d+04 | 7.634d-04 | 0.012980 | 159.212 | 72.910 |

TOTAL HARMONIC DISTORTION =    3.592607  PERCENT

**Fig. 14.11.** (c) Output voltage waveform and its Fourier components.

## 14.6 Discriminators

In the next section, FM demodulation is studied using analog multipliers and bandpass filters. In the next chapter, FM demodulation is used as an application example for phase-locked loops. In this section, we look briefly at other historically important FM demodulation configurations.

An improvement in the linearity of the magnitude response of the tuned circuit of the off-peak detector can be obtained if two resonant circuits are used, one tuned above the carrier frequency and the other tuned below. An idealized situation is used in Figure 14.12a where two peak-detectors are used. One output (secondary) tuned circuit has a center frequency above the carrier frequency, and the other is tuned below. A critical assumption is that there is no coupling between the two tuned circuits. The dashed lines in the figure signify this restriction. The magnitude response of each output tuned circuit is sketched in Figure 14.12b. A negative-going plot is used for the lower detector

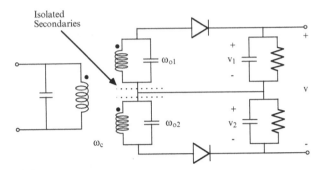

**Fig. 14.12.** (a) Improvement in the linearity of the magnitude response by use of two peak detectors.

since the output from the detectors involves a subtraction. The modulation output is taken as the difference of the two detector outputs.

$$V_o = \mid v_1 \mid - \mid v_2 \mid \tag{14.23}$$

$$\approx \frac{1}{\sqrt{1 + Q_1^2 \left(\frac{\omega}{\omega_{o1}} - \frac{\omega_{o1}}{\omega}\right)^2}} - \frac{1}{\sqrt{1 + Q_2^2 \left(\frac{\omega}{\omega_{o2}} - \frac{\omega_{o2}}{\omega}\right)^2}}$$

where it is assumed that the outputs of the detectors are proportional to the magnitude responses of the tuned circuits. As shown in Figure 14.12c, the difference of the magnitude responses of the two tuned frequencies with input frequency can be quite linear near the carrier frequency. An analysis of this configuration leads to the result that for best linearity and with equal bandwidths for the two tuned circuits, the resonant frequencies should be separated by 2.45 times the -3 dB bandwidth of the tuned circuits.

Because of the 'out-band' frequency response, FM demodulation can occur at three tuning frequencies: the desired mid-position of Figure 14.12c and at both upper and lower outband positions where 'off-peak detection' occurs. This three-point tuning possibility is observed with many FM receivers.

Demodulator circuits with two (double) tuned circuits are often called frequency discriminator circuits. A practical version utilizes push-pull coupling between the secondary windings as shown in Figure 14.12d. This is a form of the Foster-Seeley discriminator. In operation, the voltage across the peak detectors is the vector sum or difference of the two secondary responses plus the voltage from the primary. The primary has an ac ground as shown in Figure 14.12d, and the lowpass filter brings the high-frequency ground to the diodes, as shown in Figure 14.12e. The small coupling capacitor completes the circuit. Because of the 90° midband phase shift for double-tuned circuits relative to the primary, the components are in phase quadrature as shown in Figure 14.12f. The vector sum and difference vary in amplitude with the frequency

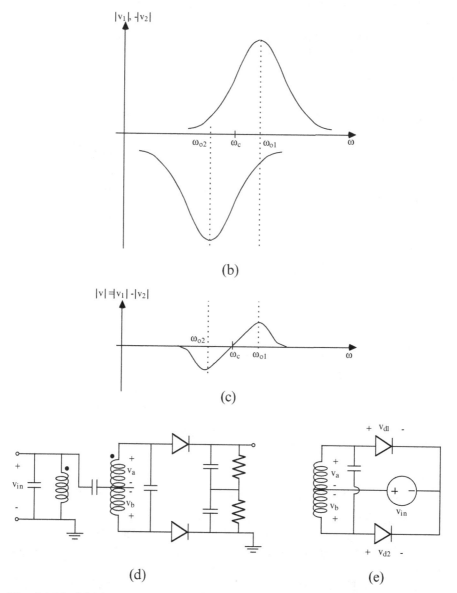

**Fig. 14.12.** (b) Magnitude response of each output tuned circuit. (c) The difference of the magnitude responses. (d) A practicl demodulator circuit. (e) Circuit at high frequencies.

deviation of the input, because of the phase shift of the tuned circuits. Again the difference of the detector outputs is taken as the demodulation output. In use, a very good amplitude limiter must be used ahead of the demodulator to eliminate any input AM.

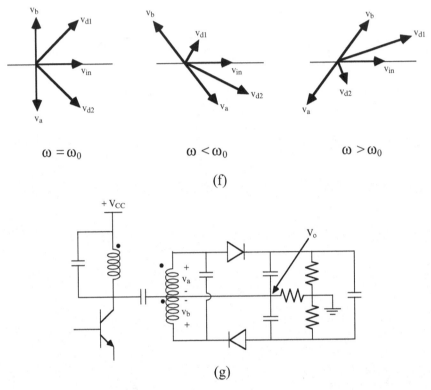

**Fig. 14.12.** (f) Phasor diagram. (g) The ratio-detector circuit.

A modified form of the discriminator is the ratio detector and is illustrated in Figure 14.12g. Note that one diode is reversed. This circuit is not sensitive to input amplitude changes and eliminates the need for the proceeding limiter. The overall linearity performance of the demodulator is not as good, however, as that of the Foster-Seeley circuit. The ratio detector has been used over the years in relatively inexpensive receivers.

## 14.7 FM Demodulators using Multipliers

A. Bilotti in 1968 first proposed that excellent FM demodulation can be achieved with an analog multiplier together with a bandpass phase-shift network as shown in Figure 14.13 [40]. The modulated input is first amplitude limited to remove amplitude modulation from noise or other causes. As established in Chapter 8, several tuned circuits can produce a bandpass response with $-90°$ phase shift at the center frequency. The simple lowpass filter is one example. The input to the analog multiplier is the limited signal and the phase-shifted signal. The output from the multiplier is the desired signal (frequency) modulation.

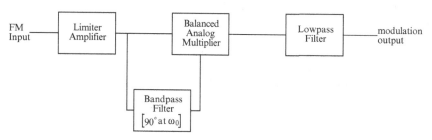

**Fig. 14.13.** FM demodulation using an analog multiplier.

Assume that the output of the limiter is

$$v_1(t) = V_{1A} \sin \phi_c(t) \tag{14.24}$$

where $\phi_c(t)$ contains the FM modulation as well as the carrier, cf., (14.16). The output from the bandpass filter is this signal plus added phase. For a filter realized from a lowpass function, the phase function is

$$\angle \frac{v_o}{v_i}(j\omega) = -\frac{\pi}{2} + \tan^{-1}\left(\frac{\omega_o^2 - \omega^2}{\omega_b \omega}\right) \tag{14.25}$$

$$\approx -\frac{\pi}{2} + 2Q\frac{\Delta\omega}{\omega}$$

where $\omega_o$ is the center frequency and $\omega_b$ is the width of the pass band. The second term of (14.25) results from the assumption that $\omega \approx \omega_o$ and that $\Delta\omega = \omega - \omega_o$ is small relative to $\omega_o$ and uses $\tan^{-1} x \approx x$. The output from the analog multiplier has the form

$$v_o(t) = KV_{1A}^2 \sin \phi_c(t) \sin\left(-\frac{\pi}{2} + 2Q\frac{\Delta\omega}{\omega} + \phi_c(t)\right) \tag{14.26}$$

where $K$ is a constant.

The second term can be replaced by the cosine function of the same argument without the term $\frac{-\pi}{2}$. Using a trigonometric identity, we obtain for one component of the output,

$$v_o(t) = K V_{1A}^2 \sin\left(2Q\frac{\Delta\omega}{\omega}\right) \tag{14.27}$$

$$\approx K'\frac{\Delta f}{f}$$

where $K'$ is a new constant. A great number of other terms also are produced. However, all occur at high frequencies and can be removed by the (true) lowpass filter which follows the multiplier. The final output is thus a signal proportional to the frequency deviation of the input signal, i.e., the FM modulation.

The analog multiplier above can consist of cross-coupled, emitter-coupled pairs together with appropriate biasing circuits. The complete circuit is readily implemented as an integrated circuit and is used extensively as the FM demodulator in commercial FM receivers and in TV audio sections.

## Problems

**14.1.** An AM peak detector is shown in Figure 14.14. The input AM signal has a single-tone modulation frequency of 1 kHz and a modulation factor of 0.8. For $n_i = 2$ (overall) and $R_L = 8$ k$\Omega$, determine the maximum value of $C$ to avoid failure-to-follow distortion.

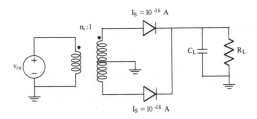

**Fig. 14.14.** AM peak detector circuit for Problem 14.1.

**14.2.** A simple AM demodulator is shown in Figure 14.15. The input voltage has a carrier amplitude of 4 V with a modulation factor of 0.8. The carrier frequency is 1 MHz and the single-tone modulation frequency is 10 kHz.
(a) For $R_S = 1$ $\Omega$, estimate the amplitude of the demodulated output signal.
(b) Repeat Part (a) for $R_S = 10$ k$\Omega$.

**14.3.** For the circuit shown in Figure 14.16,
$v_i(t) = 5[1 + 0.6\cos(2\pi 10^2 t)]\cos(2\pi 10^6 t)$.
(a) For $R_S = 0$ $\Omega$, choose $C$ for proper operation for AM demodulation.
(b) Sketch the output waveform in the time domain, $v_{out}(t)$.
(c) Sketch the frequency spectrum of the output waveform, i.e, $|v_{out}(j\omega)|$.
(d) Estimate the effect of using $R_S = 1$ k$\Omega$.

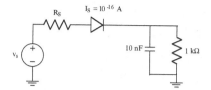

**Fig. 14.15.** AM demodulator circuit for Problem 14.2.

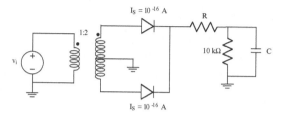

**Fig. 14.16.** Circuit for Problem 14.3.

**14.4.** An AM peak detector is shown in Figure 14.17. The input voltage has a carrier amplitude of 4 V with a modulation factor of 0.8. The carrier frequency is 0.5 MHz and the single-tone modulation frequency is 10 kHz.
(a) Use the series 1 $\Omega$ resistor as shown in the figure to connect the input and the diode and determine a suitable value for the filter capacitor $C_L$.
(b) Determine the harmonic distortion of the output voltage at the modulation frequency.
(c) Replace the 1 $\Omega$ series resistance with the tuned transformer circuit shown in the figure. The center frequency is 0.5 MHz with a Q of 20. Determine the THD of the modulation output. Compare the results with those of (b).

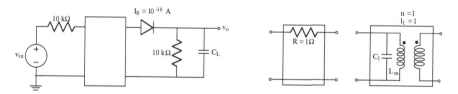

**Fig. 14.17.** AM peak detector circuit for Problem 14.4.

**14.5.** A MOS circuit is shown in Figure 14.18 with
$v_{in}(t) = 1[1 + 0.4\cos(2\pi10^2 t)]\cos(2\pi10^5 t)$.
(a) Show why this circuit can be used as an AM demodulator.
(b) Choose the value of $C$ for proper operation as an AM demodulator.
(c) What is the output voltage amplitude at the modulation frequency of $10^2$ Hz?

(d) Estimate HD2 in the output waveform.
(e) Verify your analysis with Spice.

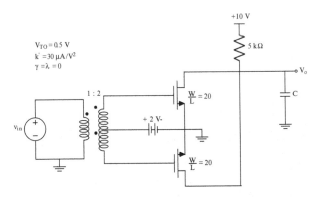

**Fig. 14.18.** Circuit for Problem 14.5.

**14.6.** A circuit for FM off-peak detection is shown in Figure 14.19.
(a) What is the output voltage amplitude at the modulation frequency of 10 kHz?
(b) Estimate HD2 in the output waveform.
(c) Verify your analysis with Spice.

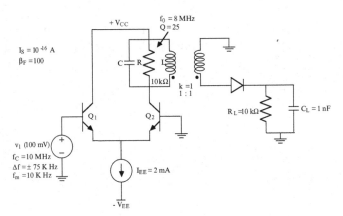

**Fig. 14.19.** FM off-peak detection circuit for Problem 14.6.

# 15

# Phase-Locked Loops

## 15.1 Basic Configurations and Applications

A phase-locked loop (PLL) is an electronic system which synchronizes an internal oscillator, in frequency and phase, with an external signal. As brought out below, the PLL is extremely useful for signal processing and signal synthesis [41]. PLLs have been used extensively in the electronic tuning of radios and in the signal processing within TVs. Today, PLLs are key building blocks of frequency synthesizers and clock generators and are used in all communication circuits [42], [43].

A phase-locked loop (PLL) is a negative feedback configuration which can be represented by the block diagram in Figure 15.1a. To appreciate the operation of a PLL, it is helpful first to compare the PLL with the block diagram and electrical behavior of a negative feedback amplifier, as shown in Figure 15.1b, cf., Section 5.1. For the feedback amplifier, an analog subtraction is achieved at the input and, to use feedback control-system notation, the output of the subtractor is an 'error signal', $v_e(t) = v_i(t) - v_f(t)$, the difference between the input signal and the feedback signal. If the loop gain, $a_L$, of the feedback amplifier is large, the feedback signal is almost equal to the input signal, and the output of the subtractor, the error signal, is almost zero. Therefore, with a large loop gain the feedback signal can be considered to follow or track the input signal. A similar tracking is achieved in a PLL involving other circuit variables, such as the phases of the input and the feedback signals.

The basic PLL of Figure 15.1a contains an analog multiplier at its input instead of an analog subtractor (adder), a voltage-controlled oscillator (VCO) and a filter amplifier. For the simple case of a single-tone, sinusoidal input signal and with the loop of the PLL opened at its output, the results of Chapter 13 for mixer circuits can be used to establish the basic operation of the PLL. For a sinusoidal input of frequency $f_i$ and an assumed sinusoid from the VCO of frequency $f_{osc}$, the output of the analog multiplier is sinusoids of the difference frequency, $f_i - f_{osc}$, and of the sum frequency, $f_i + f_{osc}$.

D.O. Pederson and K. Mayaram, *Analog Integrated Circuits for Communication*, DOI 10.1007/978-0-387-68030-9_15,
© 2008 Springer Science+Business Media, LLC

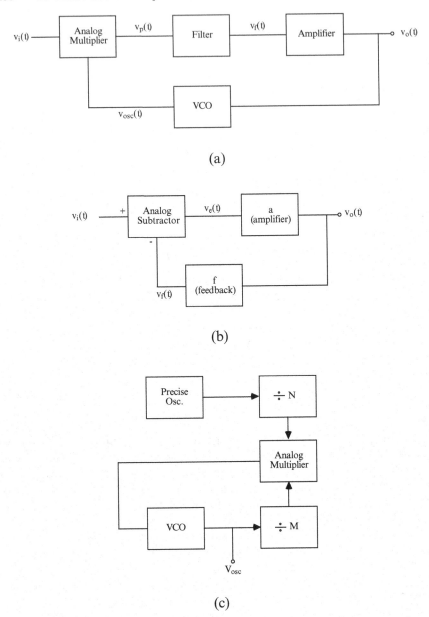

**Fig. 15.1.** (a) Block diagram of a phase-locked loop. (b) Block diagram of a negative feedback amplifier. (c) The PLL as a frequency synthesizer.

$$V_{am} = Kv_i \times v_{osc} \tag{15.1}$$
$$= KV_{iA} \cos \omega_i t \times V_{oscA} \cos(\omega_{osc}t + \Delta\phi)$$
$$= \frac{K}{2} V_{iA} V_{oscA} \cos[(\omega_i - \omega_{osc})t - \Delta\phi]$$
$$+ \frac{K}{2} V_{iA} V_{oscA} \cos[(\omega_i + \omega_{osc})t + \Delta\phi]$$

Note that a phase angle, $\Delta\phi$, is introduced. $\Delta\phi$ is the relative phase of $v_{osc}(t)$ with respect to $v_i(t)$. The frequency of the VCO, $\omega_{osc}$, is controlled by the output voltage, $V_o$, of the PLL as indicated in Figure 15.1a. If the output voltage is zero, the VCO is said to oscillate at its free-running frequency, $f_{free}$. Now the loop is closed and for the moment, assume that the input signal has a frequency equal to $f_{free}$. For this case, the output voltage of the PLL must be zero to produce the free-running frequency of the VCO. Therefore, the loop gain must force the difference component of (15.1) to be zero. The sum frequency term of (15.1) is the second harmonic of $f_i = f_{osc}$ and must be rejected by the filter of the PLL. Note also that the loop gain of the feedback must produce a 90° phase shift in the VCO frequency to force the cosine difference-frequency term in (15.1) to zero. These aspects are expanded upon in the example of the next section.

If the frequency of the input signal is increased with respect to $f_{free}$ and after the steady state is reached, the output from the analog multiplier is a nonzero constant voltage plus a sum-frequency term. The greater-than-zero output from the analog multiplier is produced by an additional phase shift in the oscillator output relative to the input. After filtering, the dc output voltage must have the necessary magnitude to shift the frequency of the VCO to equal that of the input signal. Therefore, a matching or tracking of the input frequency has occurred. This again is the result of the large loop gain of the PLL and is comparable to the matching (tracking) of the feedback signal with the input signal in the feedback amplifier. If the input signal to the PLL has a (relatively) slowly varying frequency, as in frequency modulation where the modulation frequency is usually much less than the carrier frequency, the loop can track the input and an output voltage can be obtained which is proportional to the input modulation. Thus, the PLL with proper design choices may function as a FM demodulator.

The function of the analog multiplier in the PLL is to achieve a phase comparison of its two inputs. Thus, it is often referred to as a phase comparator (PC) or phase detector.

Another important application of the PLL is as a frequency synthesizer as illustrated in Figure 15.1c. This function is widely used as a digitally controlled local oscillator for superheterodyne receivers. In lock the VCO of the PLL oscillates at a rational fraction of the input frequency. To see this, note in the figure that the frequency of the input signal, $f_i$, is divided down by an integer multiple, $N$, usually by a digital divider (controlled shift register). This is the frequency of one input to the PC. In lock, the frequency of the other input to the PC, originating from the VCO, must also equal $\frac{f_i}{N}$. The frequency from the VCO is divided by an integer M by another digital divider. From the two inputs to the phase comparator

$$\frac{f_{osc}}{M} = \frac{f_i}{N} \tag{15.2}$$

$$f_{osc} = \frac{M}{N} f_i \tag{15.3}$$

If the input signal to the PLL is a precise, crystal-controlled oscillator, an output from the VCO of the PLL can be a precise rational fraction of the input frequency. The rational fraction can be electronically controlled by simple digital count-up, count-down register circuits. This frequency-synthesizer scheme is used as the local oscillator in many electronically tuned, superheterodyne receivers, e.g., in almost all TV and FM receivers, as well as AM/FM automobile radios.

The digital dividers of this synthesizer scheme usually introduce waveforms which are rectangular pulse trains. From a Fourier-series viewpoint, each signal from the dividers contains the fundamental and harmonics. Many additional intermodulation (beat) terms are produced by the analog multiplier; however, these are rejected by the lowpass filter of the PLL. This aspect is brought out in a later section.

The concepts and principles of PLLs were established in the early 1930s. However, actual physical realizations of PLLs were initially very expensive and the applications were restricted to use in precision radio receivers. With the advent of analog ICs, including emitter-coupled pairs, analog multipliers, and voltage-controlled oscillators, entire PLLs can be implemented simply on one chip and became readily available in the late 1960s. These IC PLLs now are used extensively as demodulators, frequency synthesizers, etc.

## 15.2 A Simple Circuit Model of a PLL

The basic properties and operation of a PLL can be illustrated using a simple circuit model for a PLL that can be constructed using the elements available in the Spice simulators. Circuit simulation then provides waveforms and numerical data to document the operation and properties of the PLL. One circuit possibility is shown in Figure 15.2a. The Spice input file is given in Figure 15.2b. The subcircuits of the PLL model (often referred to as a macromodel) include the VCO implemented with a Wien-type oscillator and a lowpass, RC filter. The amplifier of the PLL is achieved by setting the desired value of the gain as the constant of a voltage-controlled voltage source, which also models the PC. In the following subsections, each of these subcircuits is described.

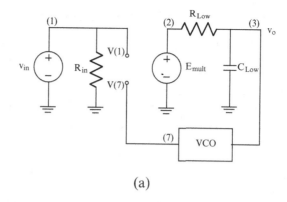

(a)

```
PLL MACROMODEL, FIG 15.2B
.TRAN 0.5U 20U 0 0.1U
.PLOT TRAN V(3)
* INPUT *
VIN 1 0 SIN(0 1 1.0MEG)
*VIN 1 0 SIN(0 1 1.1MEG)
*VIN 1 0 SIN(0 1 0.9MEG)
*VIN 1 0 SFFM(0 1 10MEG 7.5 10K)
RIN 1 0 1K
* PHASE DETECTOR MULTIPLIER*
EMULT 2 0 POLY(2) 1 0 7 0     0 0 0 0
* LOW PASS FILTER   BW=20KHZ*
RLOW 2 3 1K
CLOW 3 0 7.958NF
* WIEN-TYPE VOLTAGE CONTROLLED OSCILLATOR
*   KO=100KHZ/VOLT    FO=1MHZ  *
* WEIN FEEDBACK CIRCUIT*
RB 4 10 1K
GVAR2 4 10 POLY(2) 4 10 3 0    0 0 0 0 1E-4
VTRIG 10 0 PULSE(0 5 0 0 0 1NS)
C2 4 0 159.15PF
RA 4 6 1K
GVAR1 4 6 POLY(2) 4 6 3 0     0 0 0 0 1E-4
C1 6 7 159.15PF
* POSITIVE-GAIN BLOCK*
EGAIN 5 0 4 0 3.05
RO 5 7 10
D1 7 8 MD
V1 8 0 10V
D2 0 9 MD
V2 9 7 10V
.MODEL MD D IS=1E-16
* CONTROL CARDS *
.OPTIONS NOPAGE ITL5=0 LIMPTS=2000
.OPTIONS RELTOL=1E-4
```

(b)

**Fig. 15.2.** (a) Circuit model for a PLL. (b) Spice input file.

### 15.2.1 Voltage-Controlled Oscillator (VCO)

The VCO of the PLL macromodel is an idealized Wien-type oscillator as shown in Figure 15.2c. A prototype of this configuration is given in Figure 10.11 and operational results are given in Section 10.6. The two voltage-controlled conductors, described below, provide a linear change of oscillation frequency with the output voltage of the filter. The gain element of the oscillator is produced with a simple voltage-controlled voltage source providing a gain of 3.05. Two diode, voltage-source clamp combinations are used to limit the output voltage of the idealized amplifier to $\pm 10$ V and provide the necessary nonlinearity for the oscillator. A pulse voltage source is included in one conductor combination of the VCO to initialize the circuit for transient operation and to reduce the time needed to let the startup transients die off.

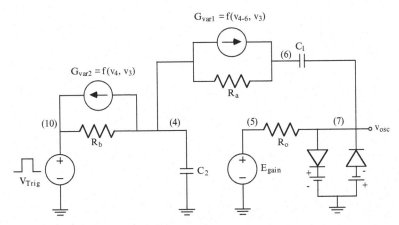

**Fig. 15.2.** (c) The VCO of the PLL macromodel.

In the Wien oscillator, it is convenient to let the resistors and capacitors of the feedback network circuit be equal. $R_1 = R_2 = R$ and $C_1 = C_2 = C$. The steady-state frequency of the oscillator is then approximately

$$\omega_{osc} = \frac{1}{RC} = \frac{G}{C} \tag{15.4}$$

where $G = \frac{1}{R}$. It is now convenient to use the conductance notation rather than the resistance notation for the oscillator. The two conductances, $G_1 = \frac{1}{R_1}$ and $G_2 = \frac{1}{R_2}$, which have identical values, are each implemented, as shown in Figure 15.2c, as the parallel combination of a fixed resistance and variable conductance,

$$G_1 = \frac{1}{R_a} + G_{var1} \tag{15.5}$$

$$G_2 = \frac{1}{R_b} + G_{var2}$$

The voltage control of the frequency of the Wien oscillator is produced by realizing $G_{var1}$ and $G_{var2}$ with voltage-controlled, two-dimensional current sources. These two current sources can be chosen to be proportional both to the voltage across them and the output voltage of the PLL, $V_o = V(3)$, producing the voltage-controlled conductor function.

$$I_x = (K_G V_o) V_x \qquad (15.6)$$
$$= G_{equiv} V_x$$

where $G_{equiv} = K_G V_o$.

In terms of the Spice element, for the element $G_{var1}$:

$$I_x = G_{var1} \; 4 \; 10 \; \text{poly}(2) \; 4 \; 10 \; 3 \; 0 \; 0 \; 0 \; 0 \; 0 \; K_G \qquad (15.7)$$

The nodes of the element are 4 and 10. The control voltage is $V(3,0)$. The value of the constant $K_G$ in (15.6) and (15.7), for the example listed in Figure 15.2b, is $1 \times 10^{-4}$ Ʊ/V.

Using the above results in (15.4), we obtain

$$\omega_{osc} = \frac{1/R_a + K_G V_o}{C} \qquad (15.8)$$
$$= \frac{1}{R_a C} + \frac{K_G V_o}{C}$$
$$= \omega_{free} + K_o V_o$$

where $\omega_{free} = \frac{1}{R_a C}$ and $K_o = \frac{K_G}{C}$.

For the values of the Spice example of Figure 15.2, $f_{free} = \frac{\omega_{free}}{2\pi} = 1$ MHz and $k_o = 100$ kHz/V, where the lower-case constant denotes the cyclic value: $k_o = \frac{1}{2\pi} K_o$.

$$f_{osc} = \frac{\omega_{osc}}{2\pi} = 1 \text{ MHz} + 0.1 \text{ (MHz/V)} V_o \qquad (15.9)$$

## 15.2.2 Filter

In the PLL shown in Figure 15.2a, a one-pole, lowpass filter is included, but a more complicated filter is easily substituted.

## 15.2.3 Phase Comparator (PC)

The PC of the PLL of Figure 15.2a is implemented with a two-dimensional voltage-controlled voltage source, $E_{mult}$. The output voltage for this element should be the function:

$$V(2) = KV_iV_{osc} \tag{15.10}$$

The Spice element to provide this multiplier in Figures 15.2a and b is $E_{mult}$ and has the description:

$$V(2) = E_{mult} \; 2 \; 0 \; \text{poly(2)} \; 1 \; 0 \; 7 \; 0 \; 0 \; 0 \; 0 \; 0 \; p4 \tag{15.11}$$

where $V_i = V(1,0), V_{osc} = V(7,0)$. The value of the coefficient $p4$ is the 'gain constant' of the PLL, A, and is chosen for this example to be 1.

As done in the last section, let

$$v_i(t) = V_{iA} \cos \omega_i t \tag{15.12}$$
$$v_{osc}(t) = V_{oscA} \cos(\omega_{osc}t + \Delta\phi)$$

Now set $\omega_i = \omega_{osc}$. The output from $E_{mult}$ is

$$V_p = V(2) = \frac{V_{iA}V_{oscA}}{2} \cos \Delta\phi \tag{15.13}$$
$$+ \frac{V_{iA}V_{osc}}{2} \cos[(\omega_i + \omega_{osc})t + \Delta\phi] \tag{15.14}$$

As mentioned in the last section, these sum and difference frequency terms are the same as those in the output of a mixer. After filtering out the high-frequency component, the output signal (voltage) is proportional to the cosine of the phase difference between the two sinusoids. A plot of the output voltage with the phase difference, $V_p$ versus $\Delta\phi$, for this simple case is shown in Figure 15.3 and is the negative of a segment of the cosine function.

$$V_p = -K_p \cos \Delta\phi, \; 0 < \Delta\phi < \pi \tag{15.15}$$

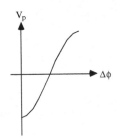

**Fig. 15.3.** Plot of the output voltage with the phase difference.

The sign of the function above is necessary to produce a positive output voltage with a positive change of the frequency difference. The constant $K_p$ is the slope of this function for $V_p = 0 \left(\Delta\phi = \frac{\pi}{2}\right)$.

$$K_p = -K_p \left. \frac{d\cos\Delta\phi}{d\Delta\phi} \right|_{\Delta\phi=\frac{\pi}{2}} \tag{15.16}$$

For the values of the Spice example, $V_{iA} = 1$ V and $V_{oscA} = 10$ V.

$$K_p = \frac{1}{2}V_{iA}V_{oscA} = 5 \text{ V} \tag{15.17}$$

### 15.2.4 PLL Operation

As noted, for equal input and VCO free-running frequencies, the phase difference of the two signals must be 90° to provide a zero dc output voltage for the PC. This can be readily verified from Spice simulation. In the Spice input file of Figure 15.2b, several input signals can be used. For the operative input for the present analysis, without the leading *, the input frequency is 1 MHz. The transient response of the PLL is shown in Figure 15.4a. After a transient startup condition, the dynamics of which are studied in the next section, the output voltage settles to zero volts indicating that the PLL is locked to the input frequency. The input frequency is equal to the free-running frequency of the VCO.

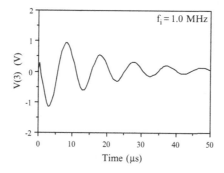

**Fig. 15.4.** (a) The transient response of a PLL.

A more detailed plot of the input signal and the output of the VCO is given in Figure 15.4b. In lock, the expected 90° phase difference between the two sinusoids can be seen with $v_{osc}$ leading $v_i$ by 90°. This aspect is also established analytically in the next section.

The output waveform of Figure 15.4c is produced when the input frequency to the PLL is changed to 1.1 MHz. After the transients have decayed, the output voltage settles toward +1 V and controls the VCO to provide an output frequency of 1 MHz + $k_o(1$ V$) = 1.1$ MHz. It is to be noted that the 'startup transients' in this case are quite different than that of the first

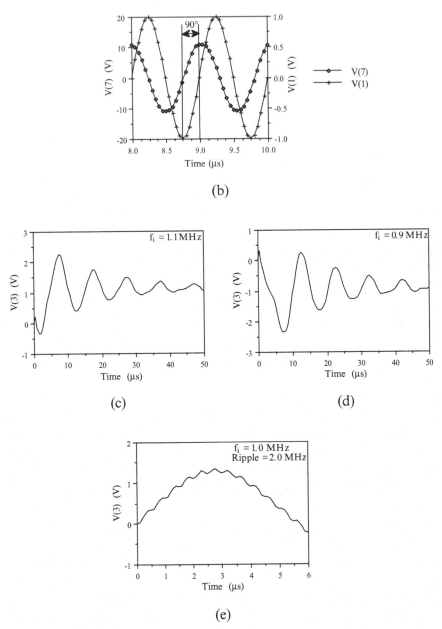

(b)

(c)

(d)

(e)

**Fig. 15.4.** (b) Plot of the input signal and the VCO output. Output voltage waveform for an input frequency of (c) 1.1 MHz, and (d) 0.9 MHz. (e) Output of the PLL showing the ripple content.

example where $f_i = f_{free}$. This is due to the capture phenomena which is analyzed in a later section.

Similarly, for an input frequency of 0.9 MHz, the output voltage settles to $-1$ V to produce a VCO frequency of 0.9 MHz. The output waveform of the PLL is shown in Figure 15.4d.

The rejection of the filter is not perfect and a portion of the sum frequency from the analog multiplier (PC) output appears at the PLL output. This is shown, for an input frequency of 1 MHz, in the more detailed output of Figure 15.4e.

## 15.3 The Small-Signal Analysis of the PLL

The block diagram of a PLL is repeated in Figure 15.5a. Note that in this diagram, all input and output variables are denoted as voltage variables in the time domain. As mentioned in the last sections, the analog multiplier serves as the phase comparator. The low-frequency output of the PC is equal to the voltage $V_p$ which is proportional to the phase difference of its two input signals, $v_i$ and $v_{osc}$. As brought out earlier, although the two input signals to the phase comparator may be voltages and currents, the desired 'error' signal is the phase difference between the signals.

Instantaneous frequencies are of prime concern, but the error signal of the PC must be the differences of the phases of the two inputs, not the frequency difference. Comparable to the situation for frequency modulation, as mentioned in Chapter 14, frequency differences cannot be used directly since this introduces an averaging, and information about the instantaneous frequency is lost.

In Figure 15.5b, it is assumed that the PLL is in a locked condition and that a definite dc bias state is present. An incremental evaluation of the PLL about this dc state is now made. In the figure, variables of the block diagram are chosen to be in the (complex) frequency domain. In Figure 15.5c, the individual blocks of the PLL are shown with their input and output variables in the frequency domain, assuming linear operation about the dc operating point. The output of the phase comparator for this small-signal situation is the slope of the transfer characteristic, (15.16), about the operating point, $\Delta\phi = \frac{\pi}{2}$.

$$v_p(s) = K_p(\phi_i - \phi_{osc}) \qquad (15.18)$$

where $\Delta\phi = \phi_i - \phi_{osc}$ and $\sin x \approx x$ for small $x$. $K_p$ is often called the conversion (gain or transfer) constant of the comparator.

The filter is usually lowpass and its transfer function is denoted

$$F(s) = \frac{v_f(s)}{v_p(s)} \qquad (15.19)$$

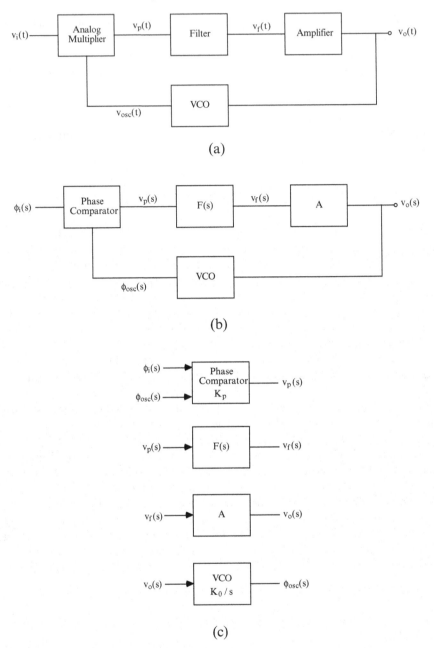

**Fig. 15.5.** Block diagram of the PLL in (a) the time domain, and (b) the frequency domain. (c) The individual blocks of the PLL.

The loop amplifier can be chosen to have a constant gain A; any frequency effects are assumed to be included in the filter.

$$v_o(s) = Av_f(s) \tag{15.20}$$

The voltage-controlled oscillator (VCO) provides a sinusoidal output voltage with a controlled frequency. In the time domain,

$$\omega_{osc}(t) = \omega_{free} + K_o v_o(t) \tag{15.21}$$

where $\omega_{free}$ is the free-running frequency of the VCO and $K_o$ is the control constant. In the frequency domain, the incremental frequency can be written:

$$\Delta\omega_{osc}(s) = K_o v_o(s) \tag{15.22}$$

Again, since we are concentrating on the small-signal behavior about the operating state, the constant free-running term can be neglected. The phase function of the oscillator output is of primary interest. This is obtained in the time domain by using the relation that the phase function is the integral of the frequency function.

$$\phi_{osc}(t) = \int \Delta\omega_{osc}(t)dt \tag{15.23}$$

In the frequency domain,

$$\phi_{osc}(s) = \frac{1}{s}\Delta\omega_{osc}(s) \tag{15.24}$$

Therefore, the phase output from the VCO can be expressed

$$\phi_{osc}(s) = \frac{1}{s}K_o v_o(s) \tag{15.25}$$

Choosing the phase as the output variable of the VCO introduces an inherent integration, $1/s$.

To repeat, the PLL is assumed to be in a locked state, i.e., the VCO is locked to the input frequency. The closed-loop transfer function of the PLL is established as follows: The amplifier output is

$$v_o(s) = AF(s)K_p(\phi_i - \phi_{osc}) \tag{15.26}$$

The oscillator output in terms of its phase response is given in (15.25). The closed-loop transfer function in terms of the phase, $\phi_i$, of the input signal is

$$\frac{v_o(s)}{\phi_i(s)} = \frac{K_p F A}{1 + \frac{K_p F A K_o}{s}} \tag{15.27}$$

The frequency of the input signal can be introduced using a relation for $\Delta\omega_i(s)$ comparable to (15.24). Not that $\Delta\omega_i$ must be used, i.e., the change of input frequency about the reference operating state.

$$\frac{v_o(s)}{\Delta\omega_i(s)} = \frac{v_o(s)}{s\phi_i(s)} \tag{15.28}$$

$$= \frac{K_p F A}{s + K_p F A K_o}$$

In the next section, this transfer function is used to explore for the locked condition the dynamics of the PLL and its response characteristics.

Note that (15.27) has the form of a closed-loop feedback function,

$$A_f = \frac{a}{1 - af} \tag{15.29}$$

$$= \frac{a}{1 + a_L}$$

where the 'open-loop gain', $a$, and 'loop-gain' function, $a_L$, are defined as:

$$a = K_p F A \tag{15.30}$$

$$a_L = -af = \frac{K_L}{s} \tag{15.31}$$

$$K_L = K_p F A K_o \tag{15.32}$$

For the situation where no filter is included in the PLL, $F = 1$, and $a_L$ has the form of a hyperbolic function and can be considered as a degenerate or limiting form of a lowpass function. The product $K_L = K_p F A K_o$ can be identified as the magnitude of transfer function around the loop, i.e., the magnitude of the loop-gain constant. The loop-gain function is dimensionless. However, note that the loop-gain constant, $K_L$, has the dimension of radial frequency.

An alternate output of the closed-loop PLL is $\Delta\phi$, the phase difference between the input sinusoid and that of the VCO output. Using (15.26) and (15.27), we obtain

$$\Delta\phi = \phi_i - \phi_{osc} \tag{15.33}$$

$$= \frac{\phi_i s}{s + K_L}$$

For $K_L$ very large,

$$\Delta\phi = \frac{\phi_i s}{K_L} \tag{15.34}$$

Therefore, for a large magnitude of the loop-gain constant, the magnitude of the phase error reduces to zero with a $+90°$ phase shift with respect to the input sinusoid. This result is consistent with the physical reasoning and observations of the last sections, cf., Figures 15.3 and 15.4b.

## 15.4 Dynamics of the PLL in the Locked Condition

In this section, the dynamical response characteristics of the PLL are examined for typical filter components for the PLL. In the simplest case, no lowpass filter is included, and $F = 1$. Of course, there is no filtering of high-order intermodulation terms produced by the PC. All of these terms appear unattenuated in the output voltage. Nonetheless, for reference purposes, the results from the closed-loop response for this case are included.

For $F = 1$, the inherent integration in the VCO variables introduces the only frequency effect. The closed-loop transfer function of the PLL from (15.28) is

$$\frac{v_o}{\Delta\omega_i} = \frac{\frac{1}{K_o}K_L}{s + K_L}$$
$$= \frac{H}{s + \omega_a} \tag{15.35}$$

where $H$ and $\omega_a$ are auxiliary constants. $K_L$ is the loop-gain constant introduced in the last section.

$$K_L = K_p A K_o \tag{15.36}$$

For $F = 1$, the closed-loop transfer function of the PLL, $\frac{V_o}{\Delta\omega_i}$, has a single real pole at $-\omega_a = -K_L$, as shown in Figure 15.6a. Note that for $K_L = 0$, which is the open-loop case, the pole lies at the origin which moves out into the LHP as $K_L$ increases. The negative-real axis is the locus of the closed-loop pole with $K_L$. (Remember for negative feedback that the locii on the real axis lie to the left of an odd number of poles and zeros.) The closed-loop PLL transfer function, (15.35), for any value of $K_L$, has a first-order, lowpass transfer characteristic. The -3 dB frequency of the magnitude of the transfer function for $s = j\omega$ is $K_L$, as illustrated in Figure 15.6b. (With this identification, $K_L$ is often referred to as the closed-loop bandwidth for $F = 1$, or more simply as the 'loop bandwidth'.)

For the above example, $K_L$ is the effective output bandwidth for frequency deviations of the input signal, $\Delta\omega_i = 2\pi\Delta f_i$. For example, if the input is a FM signal, the PLL in lock follows the frequency variation of the input at a cyclic

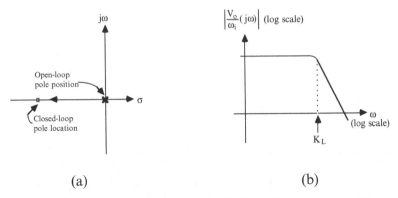

(a)                                        (b)

**Fig. 15.6.** (a) Pole locations with no filter. (b) Magnitude response of the closed-loop transfer function.

rate equivalent to the modulation frequency $f_m$. The low-frequency component of the control voltage to the VCO is a lowpass signal with a frequency variation equal to the modulation frequency of the FM signal. Therefore, FM demodulation is achieved if this voltage is taken as the output of the PLL. To the extent that the PC and VCO have linear transfer characteristics, the demodulation is linear. If the passband width of the modulation is less than $K_L$, the -3 dB bandwidth of the closed-loop PLL, little (lowpass) frequency distortion is introduced although the output contains unrejected components of the higher-order intermodulation terms.

For a numerical example, we choose the parameters of the PLL macro-model of Figures 15.2a and b. $f_{free} = 1$ MHz and $k_o = 0.1$ MHz/V. In radial measure,

$$\omega_{osc} = 2\pi(1 \times 10^6 + 0.1 \times 10^6 V_o) \tag{15.37}$$
$$= 6.28 \times 10^6 + 6.28 \times 10^5 V_o$$

As defined in (15.16), the value of $K_p$ is the slope of the transfer characteristic of the PC.

$$K_p = -\frac{1}{2}V_{iA}V_{oscA}\left.\frac{d\cos\Delta\phi}{d\Delta\phi}\right|_{\Delta\phi=\frac{\pi}{2}} \tag{15.38}$$
$$= +5\text{ V}$$

For these choices and with $A = 1$, the closed-loop bandwidth is

$$K_L = K_p A K_o = 3.14 \times 10^6 \text{ r/s} \tag{15.39}$$
$$k_L = \frac{K_L}{2\pi} = 500 \text{ kHz}$$

### 15.4.1 One-Pole Filter

Usually, the presence of a lowpass filter in the loop is inevitable because of the charge-storage effects in the amplifier. Further, it is often desirable to introduce filtering with other RC elements to reject high-frequency intermodulation components generated in the PC. For the overall response of the PLL, the filtering is desired for applications where additional, interfering signals are present at the input to the PLL. In a common situation for an IC PLL, the filter has a single-pole response and takes the form shown in Figure 15.7a. For this situation,

$$F(s) = \frac{1}{1 + \frac{s}{\omega_1}} \tag{15.40}$$

where $\omega_1 = \frac{1}{R_f C_f}$. The closed-loop transfer function of the PLL becomes

$$\frac{v_o}{\Delta \omega_i} = \frac{\frac{1}{K_o}}{1 + \frac{s}{K_L} + \frac{s^2}{\omega_1 K_L}} \tag{15.41}$$

$$= \frac{K_L \omega_1}{K_o} \left( \frac{1}{s^2 + \omega_1 s + \omega_1 K_L} \right)$$

The transfer function of the closed-loop PLL has a two-pole response and the PLL is called a second-order loop. The poles of the closed-loop response are shown in Figure 15.7b where the locii of the poles are also shown as the parameter $K_L$ is varied. The closed-loop poles, if complex, are

$$s_1, s_2 = \frac{-\omega_1}{2} \pm j \sqrt{\omega_1 K_L - \left( \frac{\omega_1}{2} \right)^2} \tag{15.42}$$

$$= -\zeta \omega_n \pm j \omega_n \sqrt{1 - \zeta^2}$$

where the auxiliary constants are

$$\omega_n^2 = \omega_1 K_L \tag{15.43}$$

$$\zeta = \frac{1}{2} \sqrt{\frac{\omega_1}{K_L}}$$

A set of steady-state magnitude responses are plotted for a sequence of the closed-loop pole locations in Figure 15.7c. Peaking occurs for $\zeta < 0.707$. For $\zeta = 0.707$, $\omega_1 = 2K_L$ and a maximally flat magnitude response is obtained with a -3 dB bandwidth of $\omega_{-3dB} = \omega_n = 1.414 K_L$. Here again, the -3 dB bandwidth of the loop is dictated by the value of the loop gain, $K_L$.

For a numerical example, we start with the values of the Spice input of Figure 15.2b.

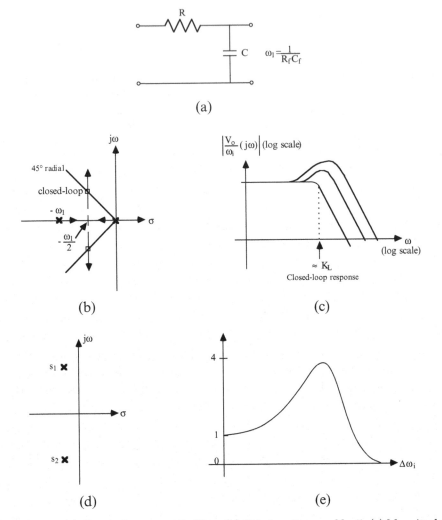

**Fig. 15.7.** (a) Circuit for a one-pole filter. (b) Pole locations and locii. (c) Magnitude response of the closed-loop transfer function. (d) A sketch of the closed-loop pole locations. (e) The magnitude response.

$$\omega_1 = \frac{1}{R_f C_f} = \frac{1}{(1 \text{ k}\Omega)(8 \text{ nF})} \quad (15.44)$$
$$= 2\pi(20 \text{ kHz})$$
$$K_L = 2\pi(500 \text{ kHz})$$
$$\omega_n = 2\pi(100 \text{ kHz})$$

$$\zeta = \frac{1}{10}$$
$$s_1, s_2 = 2\pi(0.1 \pm j1)10^5$$
$$= 2\pi(10 \pm j100)10^3$$

A sketch of the closed-loop pole locations is given in Figure 15.7d. The magnitude response for this example is given in Figure 15.7e. A 400% peaking is observed, and the -3 dB bandwidth is approximately to $\Delta\omega_i = 2\pi(155 \text{ kHz})$. This type of magnitude response can permit 'outband' signals to enter the PLL and appear at the output.

Clearly, the closed-loop poles of the example above are not positioned for a maximally flat, closed-loop magnitude function. For this type of response, the pole of the filter should be located at $-\omega_1 = -2K_L = -2\pi 10^6$. The -3 dB frequency for this condition would be 1.414 MHz and the filtering of the higher-order intermodulation components of the PC output would not be large.

We can check on the validity of the above evaluation by determining the closed-loop pole positions from the data of Figure 15.4a. First recall the form of the closed loop response for the small-signal situation.

$$v_o(s) = A(s)\Delta\omega_1(s) \tag{15.45}$$

$\Delta\omega_1$ must be considered to be a constant and for the small-signal case is the deviation from $\omega_{free}$.

$$\Delta\omega_1(s) = \frac{\Delta\omega_1}{s} \tag{15.46}$$

For the case at hand, where a one-pole filter is used,

$$A(s) = \frac{H_1}{s^2 + as + b} = \frac{H_1}{(s - s_1)(s - s_2)} \tag{15.47}$$

where $H_1$ is an auxiliary constant. For the example in question, the closed-loop poles, $s_1, s_2$, are a complex pair. The output voltage in the time domain has the form:

$$v_o(t) = H_2 \exp(-\frac{at}{2})\cos(Bt) + H_3 \tag{15.48}$$

where $B = \sqrt{b - \left(\frac{a}{2}\right)^2}$. There is an oscillatory decay to the steady-state, the constant $H_3$, since the system is underdamped. From the period of the response, we obtain an estimate for the value of the imaginary part of the poles. The period is approximately 10 $\mu$s, leading to $B = \text{Imag}(s_1) = 2\pi 10^5$, which is equal to the estimated value of (15.44). The decay of the envelope of the waveform is assumed to have a simple exponential form of (15.48).

$$v(t) = A \exp(\frac{-at}{2}) \tag{15.49}$$

From the ratio of the successive positive peaks or negative peaks, $\frac{v_1}{v_2}$, we obtain

$$t_2 - t_1 = \text{period} = \frac{2}{a} \ln \left( \frac{v_1}{v_2} \right) \tag{15.50}$$

$$\frac{a}{2} = \frac{\ln \left( \frac{v_1}{v_2} \right)}{t_2 - t_1}$$

where $\frac{a}{2}$ is equal to $|\text{Real}(s_1)|$. The estimate of the magnitude of the real part of the closed-loop poles is $68 \times 10^3$ from the negative peaks and $56 \times 10^3$ from the positive peaks, for an average of $62 \times 10^3$. The value from the analysis above is $62.8 \times 10^3$. Thus, our estimate of the dynamical response is good.

### 15.4.2 One-Zero, One-Pole Filter

If a zero is added into the transfer function of the filter, an additional degree of design freedom is obtained with respect to the loop response. This technique permits one to set the loop bandwidth relatively independently of $K_L$. A common filter to provide a zero as well as a pole for $F(s)$ is shown in Figure 15.8a. The transfer function of the filter is

$$F(s) = \frac{v_o}{v_p} = \frac{1 + \frac{s}{\omega_2}}{1 + \frac{s}{\omega_1}} \tag{15.51}$$

The pole at $-\omega_1$ has the magnitude

$$\omega_1 = \frac{1}{C(R_1 + R_2)} \tag{15.52}$$

The zero at $-\omega_2$ has the magnitude

$$\omega_2 = \frac{1}{CR_2} \tag{15.53}$$

The locii of the closed-loop poles as the loop gain constant $K_L$ is increased from zero are sketched in Figure 15.8b. Note that the closed-loop poles are initially negative real and then become complex. For large values of $K_L$, the poles again become negative real. One of the closed-loop poles, say $s_2$, asymptotically approaches the zero at $-\omega_2$. The zero, however, also appears in the closed-loop response, and an approximate pole-zero cancellation occurs in the overall closed-loop response function.

$$\frac{v_o}{\Delta \omega_i} = \frac{1}{K_o} \frac{K_L \left( 1 + \frac{s}{\omega_2} \right)}{s \left( 1 + \frac{s}{\omega_1} \right) + K_L \left( 1 + \frac{s}{\omega_2} \right)} \tag{15.54}$$

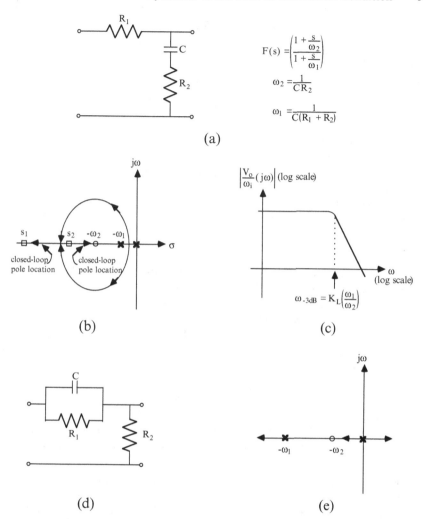

**Fig. 15.8.** (a) A one-pole, one-zero filter. (b) Pole locations and locii. (c) Magnitude response of the closed-loop transfer function. (d) Another filter providing one pole and one zero. (e) Pole locations and locii.

$$= \frac{\frac{K_L \omega_1}{K_o \omega_2}(s + \omega_2)}{s^2 + \left(1 + \frac{K_L}{\omega_2}\right)\omega_1 s + \omega_1 K_L}$$

$$= \frac{H(s + \omega_2)}{(s - s_1)(s - s_2)}$$

When the pole, $s_2$, and the zero, $-\omega_2$, approximately cancel for large values of $K_L$, the closed-loop response is approximately a one-pole response.

$$\frac{v_o}{\Delta\omega_i} = \frac{\frac{K_L\omega_1}{K_o\omega_2}}{s - s_1} \tag{15.55}$$

The conditions to achieve this response are

$$\omega_1 < \omega_2 < \frac{\omega_1}{\omega_2}K_L \tag{15.56}$$

where $\omega_1 < \omega_2$ follows from the properties of the circuit. The approximate value of the remaining pole is:

$$s_1 \approx -K_L\frac{\omega_1}{\omega_2} \tag{15.57}$$

The magnitude response is sketched in Figure 15.8c. The closed-loop bandwidth is the magnitude of the remaining pole, $|s_1|$.

$$\text{BW}_{-3dB} = |s_1| \tag{15.58}$$

To the extent that (15.56) is satisfied, the closed-loop bandwidth can be set independently of the value of the loop gain.

To continue with the numerical example of this section where $K_L = 2\pi(5 \times 10^5)$, we choose the zero of the filter function to lie at $-2\pi(20 \times 10^3)$ r/s and the pole at $-2\pi(10 \times 10^3)$ r/s. The closed-loop poles from (15.54) are $-2\pi(21 \times 10^3)$ and $-2\pi(239 \times 10^3)$. The -3 dB bandwidth of the closed-loop response is approximately 239 kHz. The estimate from (15.58) is 250 kHz.

To reduce the $-3$ dB frequency, new choices for $\omega_1$ and $\omega_2$ can be made using (15.58) as a guide. For an approximate 50 kHz bandwidth, we choose $\omega_1 = 2\pi(0.5 \times 10^3)$ r/s and $\omega_2 = 2\pi(5 \times 10^3)$ r/s. The closed-loop pole locations are $-2\pi(5.56 \times 10^3)$ r/s and $-2\pi(44.9 \times 10^3)$ r/s. $s_2$ is close to the zero at $-\omega_2$ and the magnitude response has a $-3$ dB bandwidth of approximately 45 kHz.

Another simple filter configuration providing one pole and one zero is shown in Figure 15.8d. For this circuit $F(0) = \frac{R_2}{R_1+R_2} < 1$, and the value of $K_L$ is reduced accordingly. The zero is now inside of the pole and the locii of the closed-loop poles are those shown in Figure 15.8e. For any reasonable value of the loop-gain constant, the bandwidth of the response is determined by the outside pole, $s_2$.

## 15.5 The Lock and Capture Ranges

When no control voltage is applied to the VCO, the frequency of oscillation is the free-running frequency, $\omega_{free} = 2\pi f_{free}$. At the input to the PLL there is a range of input frequencies with respect to $f_{free}$ for which the PLL can 'capture' the input and produce a locked state. There is a second range of input-signal frequencies with respect to $f_{free}$ for which the loop can maintain lock, once lock is attained. The capture range is usually less than the range to maintain lock. These limits are the result of the limited phase-voltage range provided by the phase comparator as well as the limits due to the magnitude and frequency characteristics of the loop-gain function.

### 15.5.1 Lock Range

For the simple PC of the Section 15.2, an idealized linear analog multiplier, the transfer characteristic is repeated in Figure 15.9a and has the functional form:

$$V_p = -K_p \cos(\Delta\phi) \tag{15.59}$$

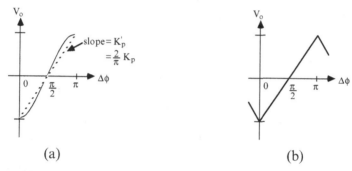

(a)                                    (b)

**Fig. 15.9.** (a) The transfer characteristics of an idealized linear analog multiplier. (b) The transfer characteristics of a different phase comparator.

The value of the slope of this characteristic for $\Delta\phi = \frac{\pi}{2}$ is $K_p$. The maximum value of the characteristic is also $K_p$ and occurs at $\Delta\phi = \pi$. In order to maintain lock, the phase difference between the VCO output and the signal input must be kept between 0 and $\pi$. At the extremes, and for $F = 1$ as well as $A = 1$, the maximum output voltage is equal to the maximum output of the PC.

$$V_{o\,max} = \pm K_p \tag{15.60}$$

This leads to the maximum change of VCO frequency to maintain lock. The maximum change of signal input frequency from $f_{free}$ for $\Delta\phi = \frac{\pi}{2}$, to either the positive or negative extreme is the lock range, $\omega_L$.

$$\omega_L = V_{o\,max}K_o = K_p K_o = K_L \tag{15.61}$$

for $F = A = 1$. Remember that in lock the phase difference is a constant, i.e., a dc quantity, and the value of $F$ for dc is 1.

For the PC transfer characteristic of (15.59) and Figure 15.9a, the values of the slopes of the characteristic become zero at the extremes. Therefore, the small-signal value of $K_L$ reduces to zero and the value of the lock range, (15.61), based on the value for $\Delta\phi = 0$ is very optimistic. As an approximate correction, a 'large excursion' value, $K'_L$, can be used in place of $K_L$. The approximation is sketched in Figure 15.9a. In the expressions above, an approximate slope of $K'_p = \frac{2}{\pi}K_p$ can be used. leading to a corrected value, $K'_L = \frac{2}{\pi}K_L$.

In a later section, a different phase comparator is introduced for which the constants in the development above must be changed. The new PC is the result of overdriving a practical analog multiplier (mixer) or presenting it with large inputs having rectangular waveforms. For this situation, the transfer characteristic has the shape shown in Figure 15.9b. The slope of the characteristic is labeled $K_p$ and because of the linear curve, the extremes occur at approximately $\frac{\pi}{2}K_L$. Therefore, $K_L$ in the above expression for $\omega_L$ is replaced with $\frac{\pi}{2}K_L$.

### 15.5.2 Capture Range

The range of input frequencies for which an initially unlocked loop will achieve lock is always less then the lock range. In the simplest terms, if lock is not present, the difference frequency output from the PC is a time-varying function, not a constant; therefore, the filter response reduces the magnitude of $V_o$ and the maximum range of $\Delta\omega_{osc}$. Once in lock, the PC output after filtering is again a dc quantity.

The capture range is difficult to estimate accurately. Even with a simple analog multiplier, its nonlinear transfer function, as given in (15.59), leads to a nonlinear differential equation describing the PLL. A closed-form solution for this equation does not exist.

For a physical description of the capture process, let the loop be initially open with the frequency of the input signal close but not equal to the free-running frequency of the VCO. Let the input frequency be higher than $f_{free}$. At the output of the PC, the difference-frequency and sum-frequency sinusoids appear. As usual we assume that the sum-frequency term is rejected by the lowpass filter. The output voltage then contains only the difference-frequency sinusoid.

The loop is now closed. The frequency of the VCO becomes a function of the time-varying output voltage, but because of the lowpass filter, the output voltage cannot follow closely the changes of the output of the PC. Nonetheless, as the output voltage increases with time during a half cycle of the input from the PC, the frequency of the VCO also tends to increase and the amplitude of the difference-frequency sinusoid from the PC becomes smaller. However, for the negative half-cycle, the opposite is true. The waveform of the PC output and the PLL output become quite distorted with respect to a sinusoid, as shown in the waveform of Figure 15.4c. The slow-moving portions of the waveform, when $f_i \approx f_{osc}$, hug the steady-state dc value. The fast portions of the waveform become smaller and smaller 'blips' until they disappear when capture occurs and the PLL is in lock. From another point of view, the distorted waveform of the PC output produces an average dc value which brings the average frequency of the VCO to equal $f_i$.

For a crude estimate of the capture range, the following analysis is proposed. First, assume that the PLL is in lock with an input frequency of $\omega_{i1}$. In lock, the frequency of the signal input is equal to that of the VCO.

$$\omega_{i1} = \omega_{osc} = \omega_{free} + K_o V_o \qquad (15.62)$$

Therefore,

$$\Delta\omega_1 = \omega_i - \omega_{free} = K_o V_o \qquad (15.63)$$

In lock, $V_o$ is a constant or changing very slowly relative to the carrier frequency due to a frequency modulation of the input. As brought out above, $\Delta\omega_1$ has a maximum value equal to the lock range.

$$\omega_L = K_o V_o \mid_{max} \qquad (15.64)$$
$$= K_L$$

Now consider that the PLL is not locked and let the input signal frequency be chosen to be different than the free-running frequency of the VCO. At the output of the PC, the response in time is

$$v_p(t) = K_p \cos[(\omega_i - \omega_{osc})t - \Delta\phi] + \text{sum term} \qquad (15.65)$$

This voltage is the input to the loop filter. At the output of the filter, the difference-frequency signal is

$$v_o(t) = K_p |F(j\Delta\omega_2)| \cos(\Delta\omega_2 t - \Delta\phi + \theta_f) \qquad (15.66)$$

where

$$\Delta\omega_2 = \omega_i - \omega_{osc} \qquad (15.67)$$

and the phase function is

$$\theta_f = \angle F(j\Delta\omega_2) \qquad (15.68)$$

If capture is achieved at an input frequency, $\omega_{i1}$, the difference frequency $\Delta\omega_2$ must approach the lock range, $K_L$. However, because of the reduced magnitude of $v_o$ due to the filter, the maximum change of the VCO frequency is not as large as that of the lock range. The maximum magnitude of the voltage which appears at the VCO input is

$$| v_o | \leq K_p \, | \, F(j\Delta\omega_2) \, | \tag{15.69}$$

$$= \frac{\Delta\omega_2}{K_o}$$

In the limiting case, just after capture, $\Delta\omega_2$ is approximately equal to the lock range reduced by the filter magnitude response, leading to

$$\omega_c = \Delta\omega_2 = K_o K_p \, | \, F(j\Delta\omega_2) \, | \tag{15.70}$$

$$= K_L \, | \, F(j\omega_c) \, |$$

where $\omega_c$ is the capture frequency, with respect to $\omega_{free}$.

### 15.5.3 Capture Examples

For a first-order loop where no explicit filter is used, the lock range and the capture ranges are (approximately) equal since $F = 1$ and $\omega_L = K_L$.

For a second-order loop containing a one-pole filter, the capture range is less than the lock range. To illustrate this with a numerical example, let the filter have a single-pole response with a pole at $-\omega_1$. The magnitude response of this type of filter has the form:

$$|F(j\Delta\omega)| = \frac{1}{\sqrt{1 + \left(\frac{\Delta\omega}{\omega_1}\right)^2}} \tag{15.71}$$

where $\Delta\omega$ is the difference frequency of the PC. Using this in the magnitude criterion for the capture range of (15.69) and setting $\Delta\omega = \omega_c$, we obtain

$$\left(\frac{\omega_c}{K_L}\right)^2 = \frac{1}{1 + \left(\frac{\omega_c}{\omega_1}\right)^2} \tag{15.72}$$

If a variable substitution is made, the results can be put into a simple form. Let

$$\omega_c^2 = \omega_1^2 u \tag{15.73}$$

The criterion becomes

$$u^2 + u - \left(\frac{K_L}{\omega_1}\right)^2 = 0 \tag{15.74}$$

For the values of the continuing Spice example, $\omega_1 = 2\pi(20 \times 10^3)$ r/s and $K_L = 2\pi(500 \times 10^3)$ r/s, $u = 25$ and $\omega_c = 2\pi(100 \text{ kHz})$. This is significantly smaller than the lock range for this example, $\omega_L = 2\pi(500 \text{ kHz})$. Spice simulation can be used to check this estimate. For the input file in Figure 15.2b, the signal input frequency is 1.1 MHz $= f_{free} + f_c$, the limiting case on the basis of the estimate above. As noted in Figure 15.4c, capture is obtained. At an input frequency of 1.2 MHz, no capture is achieved. From a series of Spice runs, the capture range for this example is found to be approximately 160 kHz.

For $\omega_1 \ll K_L$, the approximate solution of (15.74) is $u \approx \frac{K_L}{\omega_1}$ and $\omega_c \approx \sqrt{\omega_1 K_L}$.

For a PLL with a one-pole filter, a closed-loop maximally flat response can be achieved. From the earlier results, this requires that $\omega_1 = 2K_L$. Using this requirement in (15.74), we obtain

$$u = 0.207 \tag{15.75}$$

$$\omega_c = 0.45\omega_1 = 0.9K_L = 2\pi(450 \text{ kHz}) \tag{15.76}$$

Since the lock range is equal to $K_L$, the capture range is 90% of the lock range.

Next we consider the example from the last section where the filter provides a real zero and a real pole. The zero is chosen to have a magnitude of $2\pi(20$ kHz) and the pole has a magnitude of $2\pi(10$ kHz). The free-running frequency of the VCO is $1MHz$, $K_o = 2\pi(100$ kHz/V) and $K_p = 5$ V/r with $A = 1$. The loop gain is $K_L = K_p A K_o = 2\pi(500$ kHz). From (15.69), for capture to occur,

$$\omega_c = K_L |F(j\omega_c)| \tag{15.77}$$

where

$$|F(j\omega_c)| = \sqrt{\frac{1 + (\frac{\omega_c}{\omega_2})^2}{1 + (\frac{\omega_c}{\omega_1})^2}} \tag{15.78}$$

Again it is convenient to introduce a variable change. We let

$$\omega_c^2 = \omega_1^2 u \tag{15.79}$$

The constraint equation becomes

$$u^2 - u \left[ \left( \frac{K_L}{\omega_2} \right)^2 - 1 \right] - \left( \frac{K_L}{\omega_1} \right)^2 = 0 \tag{15.80}$$

For the numerical values above, where $\omega_1 = 2\pi(10$ kHz) and $\omega_2 = 2\pi(20$ kHz), the solution of the quadratic is

$$u = 628 \tag{15.81}$$

The value of the capture range is

$$\omega_c = \pm\omega_1\sqrt{u} = 2\pi(250 \text{ kHz}) \tag{15.82}$$

This estimate of the capture range is less by a factor of two than the lock range, $2\pi(500 \text{ kHz})$.

For a second example, the magnitudes of the pole and zero are changed. From the results of the last section, an approximate single-pole, closed-loop response is produced if

$$\omega_1 < \omega_2 < \frac{\omega_1}{\omega_2} K_L \qquad (15.83)$$

These relations are satisfied if we decrease $\omega_1$ to $2\pi(0.5 \times 10^3)$ r/s and $\omega_2$ to $2\pi(5 \times 10^3)$ r/s. The approximate $-3$ dB bandwidth of the response and the approximate capture range are the same

$$\omega_{-3dB} = \omega_c = \frac{\omega_1}{\omega_2} K_L \qquad (15.84)$$

For the values above: $\omega_{-3dB} = \omega_c \approx 2\pi(50 \text{ kHz})$. The exact solution provides $\omega_{-3dB} = 45$ kHz. The solution of the quadratic, (15.80), yields $u = 1 \times 10^4$, and $\omega_c = 2\pi(50 \text{ kHz})$.

## 15.6 PLLs with Overdriven PCs and Relaxation Oscillators

A simple implementation of a phase comparator is an emitter-coupled pair, such as introduced in Section 13.1, Figure 13.1b. One input is introduced differentially across the two upper-base nodes and the other input is via the common-emitter current. With this scheme, if amplitudes of the two inputs are small, linear operation is produced and the output voltage is sensitive to amplitude changes of the inputs.

As brought out in Sections 13.2 and 13.7, better operation is achieved with the doubly balanced (quad) mixer, the configuration of which is repeated in Figure 15.10a. Of particular interest is the situation with large signal inputs. If the amplitude of the differential input voltage is large, the transistors are alternately on or off. A switching behavior of the ECP is obtained. The output of this type of PC consists of the sum and difference frequencies of the inputs plus corresponding components from the harmonics of the overdriven input. As with the quad mixer, the output filter rejects these high-order terms and the output of the filter (and the PLL) is approximately the same as that of an ideal analog multiplier.

It is simple to assume first that one overdriven input to the ECP is that from the VCO since we have control of the amplitude of its output, again comparable to the case for the ECP mixer. A further step may be taken, in that a near-harmonic oscillator for the VCO is not necessary. Rather a relaxation configuration can be chosen, as studied in Section 12.10, Figure 12.22. The output is a rectangular pulse train with a frequency controlled by

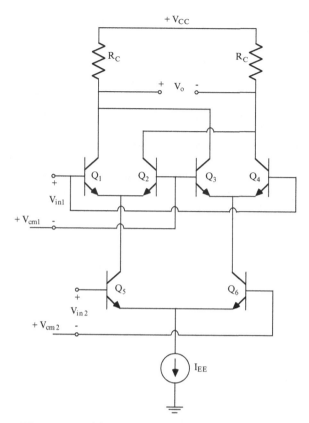

**Fig. 15.10.** (a) A doubly balanced (quad) mixer.

a single capacitor and by the values of the current sources. The output voltage amplitude extremes are well defined.

For the circuit of Figure 15.10a, the signal input to the lower ECP input is first amplified to also overdrive this ECP of the quad. All of the ECPs are thus overdriven. In effect, the quad is driven by rectangular pulse trains.

For the PLL, the two inputs to the PC can now be considered to be rectangular pulse trains as shown in Figure 15.10b. In the locked PLL, both inputs have the same frequency. The output waveform for the PC is shown in the figure and consists of a pulse train with positive and negative portions. When the two inputs are both positive or negative, the output is positive with a fixed height. When the two inputs have different signs, the output is negative. For one cycle the negative area of the waveform is denoted $A_1$ and the positive area $A_2$. The dc component of the output waveform is clearly the difference of two areas.

$$V_{oave} = -\frac{1}{\pi}(A_1 - A_2) \tag{15.85}$$

$$= I_{EE}R_C\left[2\left(\frac{\Delta\phi}{\pi}\right)-1\right]$$

where $\frac{1}{\pi}$ gives the average value of this waveform in terms of the peak value which is $I_{EE}R_C$ for a bipolar doubly balanced multiplier. The phase variable, $\Delta\phi$, is introduced according to the defined phasing of the two inputs as shown in the figure. This phase-voltage relation is plotted in Figure 15.10c and can be considered the transfer characteristic of this PC. The slope of this characteristic is the constant $K_p$ for this circuit. In comparison with the ideal multiplier, the shape of the transfer characteristic is linear over the entire range, $0 < \Delta\phi < \pi$. Outside of this range, the transfer characteristic becomes a triangular waveform as sketched in the figure. Of course, proper operation of the PLL is restricted to the principal region, $0 < \Delta\phi < \pi$.

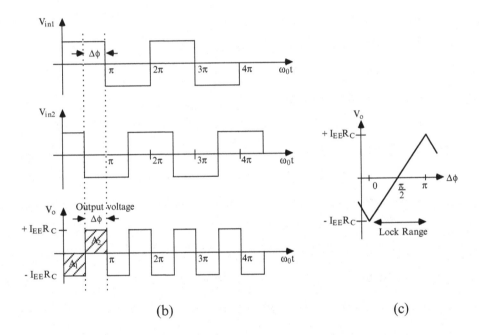

(b)                                                    (c)

**Fig. 15.10.** (b) The input and output voltages. (c) Phase-voltage relation of the comparator.

The slope of the linear transfer characteristic is labeled $K_p$. Therefore, the maximum output at $\Delta\phi = \pi$ is $\frac{\pi}{2}K_p$, as shown in Figures 15.9b and 15.10c. (In terms of the maximum value shown in Figure 15.10c, $K_p = \frac{2}{\pi}I_{EE}R_C$ V/r). In the various developments to this point in the chapter, the basic operation of the PLL and the small-signal dynamics are unchanged with the use of this new PC. In the formulas for the lock and capture ranges, the earlier values for $K_L$

and $K_p$ must be replaced by $\frac{\pi}{2}K_L$ and $\frac{\pi}{2}K_p$ to obtain the appropriate values for the response and characteristics of the PLL. For example, the appropriate output function for the PC in place of (15.65) is

$$v_p(t) = \frac{\pi}{2}K_p \cos[(\omega_i - \omega_{osc})t - \Delta\phi] \qquad (15.86)$$

Of course with a linear characteristic over the entire operating range of the PC, the lock and capture ranges are more definite since there is not the reduced slope values at the extremes of the characteristic as is the case for the cosine function of the earlier PC, cf., Figure 15.9a.

## 15.7 PLL Design Example

For another example, consider the PLL configuration shown in Figure 15.11. The parameters and constants of the loop components are given in the figure. Note that a frequency divider is included with a programmable scale of 5 to 15. With an input frequency of $f_i = 100$ kHz, a controlled output frequency can be obtained from the output of the VCO from 0.5 MHz to 1.5 MHz. The VCO must have a tunable range from 0.5 MHz to 1.5 MHz and is assumed to have a free-running value of 1.0 MHz. The value of the VCO transfer coefficient is given as $1 \times 10^7$ r/V-s or 1.6 MHz/V-s. The VCO output is

$$f_{osc} = (1.0 + 1.6V_o) \text{ MHz} \qquad (15.87)$$

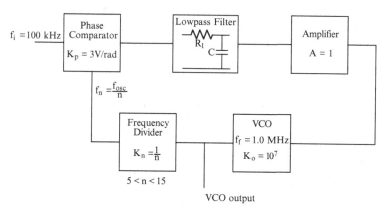

**Fig. 15.11.** A phased-lock loop configuration.

We check out the design with the VCO assumed oscillating at 1.5 MHz. The appropriate divide ratio to produce an input to the PC equal to the signal input frequency of 100 kHz is 15. The phase shift of the oscillator is also divided by 15. Therefore, the loop gain is

$$K_L = K_p A K_o K_n \tag{15.88}$$

$$= (3)(1)(1 \times 10^7)\frac{1}{15} = 2 \times 10^6 \text{ r/s}$$

where from the arguments above and from Figure 15.11, $K_n = \frac{1}{n} = \frac{1}{15}$. The input voltage to the VCO must be

$$V_o = \frac{f_{osc} - f_{free}}{K_o} \tag{15.89}$$

$$= \frac{1.5 - 1.0}{1.6} = +0.3125 \text{ V}$$

If the filter is to have a single pole, the magnitude of the pole should be $2K_L = 4 \times 10^6$ r/s to achieve a flat-magnitude response for the PLL. If the filter resistor has a value of 1 k$\Omega$, the required capacitor must be 250 pF. The passband width of the closed-loop PLL is $1.414K_L = 2.83 \times 10^6$ r/s$= 2\pi(450$ kHz). A filter with a one-zero, one-pole response, can be used to reduce the $-3$ dB bandwidth of the closed-loop PLL.

## 15.8 PLL Parameters for a Typical IC Realization

The 560B IC is an early, commercially available realization produced by several manufacturers. The circuit schematic is shown in Figure 15.12. The reader is referred to Section 10.4.4 of [6]for the bias analysis and component-transfer-constant evaluation of the 560B. From these evaluations, the following values are obtained for the PLL components.

$$f_{free} = 0.26\frac{10^3}{C} \tag{15.90}$$

where $C$ is an external capacitor. The transfer constant for the VCO is

$$K_o = 0.93\omega_o \text{ r/V} - \text{s} \tag{15.91}$$

For a free-running frequency of 10 MHz, $\omega_o = 2\pi(1 \times 10^7)$, $K_o = 5.85 \times 10^7$ r/V-s and $C = 26$ pF. The transfer constant for the PC is

$$K_p = 2.55 \text{ V/r} \tag{15.92}$$

The value of the gain constant for the 560B is $A = 1$. These values provide a loop bandwidth of

$$K_L = 1.5 \times 10^8 \text{ r/s} \tag{15.93}$$

$$k_L = 24 \text{ MHz}$$

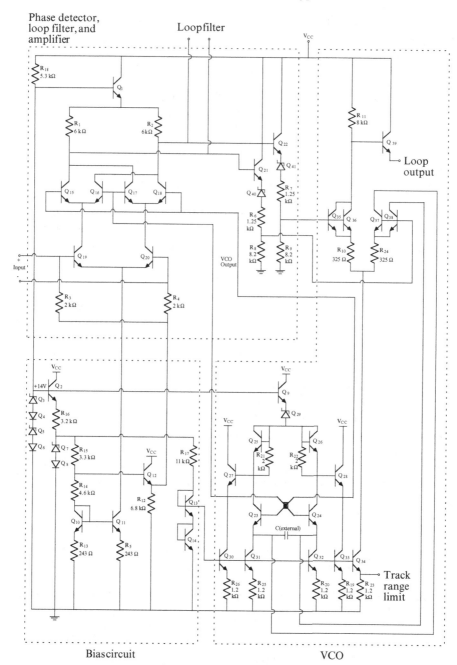

**Fig. 15.12.** Circuit schematic of the 560B PLL.

Since the PC has the triangular transfer characteristic of Figure 15.10c, the appropriate constant to use for the lock and capture ranges is $\frac{\pi}{2}K_L$. The lock range for this PLL will be

$$\omega_L = \frac{\pi}{2}K_L = 2.36 \times 10^8 \text{ r/s} \qquad (15.94)$$

$$f_L = 37.5 \text{ MHz}$$

Assume that the application of interest is that the input be a FM signal with a carrier frequency of $f_o = 10$ MHz, a maximum frequency deviation of $\Delta f = \pm 75$ kHz, and a modulation frequency of $f_m = 10$ kHz. We choose the capture range to be approximately equal to the maximum deviation. Therefore, a one-pole, one-zero filter is proposed to achieve the capture range and to reduce the closed-loop bandwidth of the PLL. As a guide, we assume that the approximate capture range formula holds, comparable to (15.84),

$$\omega_c \approx \frac{\omega_1}{\omega_2}\frac{\pi}{2}K_L \qquad (15.95)$$

where the factor $\frac{\pi}{2}$ is appropriate for the PC characteristic of Figure 15.10c. From (15.95), $\omega_2 \approx 500\omega_1$. From (15.84),

$$f_{-3dB} = \frac{2}{\pi}75 \text{ kHz} \approx 48 \text{ kHz} \qquad (15.96)$$

This value of bandwidth is well above the modulation frequency. Choices of $f_1 = 4$ Hz and $f_2 = 2$ kHz satisfy the constraints of (15.56).

The Spice simulation of a PLL such as the 560B can be accomplished. However, the size of the complete circuit, as shown in Figure 15.12, leads to time-consuming and expensive computer runs. In order to study more economically the properties and performance of PLLs, we can use the simplified, 'macromodel' of the PLL which is developed in Sections 15.1 and 2. In particular the basic behavior of the PLL under different inputs and with different transfer constants of the PLL components can be studied.

## 15.9  A PLL Example with a FM input

In this section the PLL macromodel of Figure 15.2a is used to investigate the use of the PLL as a demodulator of frequency modulation. It is to be noted that a linear, simple analog multiplier is used together with a Wien-type, near-harmonic oscillator, not the quad multiplier and relaxation oscillator of the last section. Nonetheless, typical behavior can be illustrated.

As brought out Chapter 14, Spice includes a source which is frequency modulated. The defining equation for this source is

$$v_{in} = V_{inA}\sin[2\pi f_o t + \text{MDI}\sin(2\pi f_m t)] \qquad (15.97)$$

where $f_o$ is the frequency of the carrier, $f_m$ is the modulation frequency and MDI is the modulation index $= \frac{\Delta f}{f_m}$ . This source permits us to evaluate a PLL configuration used as an FM demodulator.

The Spice input file for the PLL macromodel of this chapter is repeated in Figure 15.13a. For the macromodel, the PC is a simple two-dimensional voltage-controlled voltage source, and the VCO is a macro-model of a Wien-type oscillator. The PLL constants are: $K_p = 5$ r/s, $A = 1$, the filter bandwidth can be easily set by a choice of $R_f$ and $C_f$. For the VCO, $K_o = 2\pi(100$ kHz)/V. The value of the loop-gain constant is $K_L = 2\pi(500$ kHz). Since a linear multiplier is used, the transfer characteristic of the phase comparator is sinusoidal as shown in Figure 15.9a. In a large-signal situation, as in capture, a modified value of $K_p$ can be used: $K'_p = \frac{2}{\pi}K_p$ leading to $K'_L = 2\pi(318$ kHz).

In Figure 15.13a, the Spice input file contains the frequency-modulated input. The carrier frequency is $f_o = 10$ MHz, the signal frequency is $f_m = 10$ kHz. The frequency deviation is $\Delta f = (\pm)75$ kHz; therefore, the modulation index is MDI $= 7.5$. The filter of the PLL is chosen to have a $-3$ dB bandwidth of 20 kHz with $R_f = 1$ k$\Omega$ and $C_f = 8$ nF.

The output waveform of the PLL for a time sequence appropriate for the modulating signal is shown in Figure 15.13b. The harmonic-distortion components are given in Figure 15.13c: $HD2 = 0.47\%, HD3 = 0.17\%$, with $THD = 1.34\%$ due to the upper harmonics and intermodulation components. Additional filtering in the lowpass amplifier which follows the demodulator in a superheterodyne receiver will aid in rejecting these high-frequency components. It is to be noted that to obtain these results a long simulation time is needed, even for this simple macromodel, to ensure that steady-state performance has been attained.

Note from the values in Figure 15.13c that the small harmonic components of the output voltage waveform depend strongly on the values of the TSTOP and TSTEP parameters of the .TRAN specification. The set mentioned above should be the more accurate.

When the value of $C_f$, the filter capacitor, is reduced to achieve a filter design to produce a closed-loop, maximally flat magnitude response, the lack of adequate filtering of the sum-frequency component of the PC is observed in Figure 15.13d.

## Problems

**15.1.** A phase-locked loop has the parameters, $K_p = 1$ V/rad, $K_0 = 250$ kHz/V, and the free-running frequency of the VCO is 1 MHz.
(a) For $AF(s) = 10$, estimate the lock range and the close-loop bandwidth.
(b) For $AF(s) = \frac{10}{1+j\frac{\omega}{2\pi 4 \times 10^6}}$, estimate the lock range and the closed-loop bandwidth.

```
PLL MACROMODEL, FM DEMOD, FIG 15.13A
.TRAN 5U 600U 400U 0.01U
.PLOT TRAN V(3)
.FOUR 10K V(3)

* INPUT *                                          fc = 10 MHz
VIN 1 0 SFFM(0 1 10MEG 7.5 10K)
RIN 1 0 1K                                         Δf = ±75 KHz
*KL = 500 KHZ                                      fm = 10 KHz

* PHASE DETECTOR MULTIPLIER *
* Kp = 5 V/R
EMULT 2 0 POLY(2) 1 0 7 0     0 0 0 0 1

* LOW PASS FILTER  BW=20KHZ *
RLOW 2 3 1K
CLOW 3 0 8NF

* WIEN-TYPE VOLTAGE CONTROLLED OSCILLATOR*
*  KO=100KHZ/VOLT   FO=1MHZ  *

* WEIN FEEDBACK CIRCUIT *
RB 4 10 1K
GVAR2 4 10 POLY(2) 4 10 3 0    0 0 0 0 1E-5
VTRIG 10 0 PULSE(0 5 0 0 0 1NS)
C2 4 0 15.915PF
RA 4 6 1K
GVAR1 4 6 POLY(2) 4 6 3 0     0 0 0 0 1E-5
C1 6 7 15.915PF

* POSITIVE-GAIN BLOCK *
EGAIN 5 0 4 0 3.05
RO 5 7 10
D1 7 8 MD
V1 8 0 10V
D2 0 9 MD
V2 9 7 10V
.MODEL MD D IS=1E-16

* CONTROL CARDS *
.OPTIONS NOPAGE NOMOD ITL5=0 LIMPTS=2000
.OPTIONS RELTOL=1E-4
.WIDTH OUT=80
.END
```

**Fig. 15.13.** (a) Spice input file for the PLL macromodel.

**15.2.** A phase-locked loop block diagram is shown in Figure 15.14.
(a) Determine the loop gain $K_L$.
(b) What is the closed-loop -3 dB bandwidth assuming that the closed-loop poles are located on the $\pm 45^o$ radials from the negative real axis.
(c) Estimate the voltage input to the VCO in the locked condition.

**15.3.** In a phase-locked loop, the free-running frequency of the VCO is 2 MHz. The VCO can provide a change of 25 kHz with each volt of the output voltage of the PLL. The phase detector output is 20 mV/degree, and the voltage gain

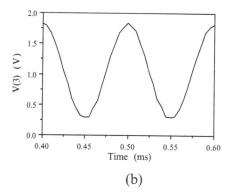

(b)

FOURIER COMPONENTS OF TRANSIENT RESPONSE V(3)    $T_{step} = 1\mu s$, $T_{max} = 0.01\mu s$, $T_{stop} = 400\ \mu s$
DC COMPONENT =   1.041d+00

| HARMONIC NO | FREQUENCY (HZ) | FOURIER COMPONENT | NORMALIZED COMPONENT | PHASE (DEG) | NORMALIZED PHASE (DEG) |
|---|---|---|---|---|---|
| 1 | 1.000d+04 | 7.693d-01 | 1.000000 | 88.899 | 0.000 |
| 2 | 2.000d+04 | 4.405d-03 | 0.005726 | -13.530 | -102.429 |
| 3 | 3.000d+04 | 5.497d-03 | 0.007146 | -102.384 | -191.283 |
| 4 | 4.000d+04 | 5.643d-03 | 0.007335 | -98.698 | -187.596 |
| 5 | 5.000d+04 | 3.477d-03 | 0.004520 | -23.612 | -112.510 |
| 6 | 6.000d+04 | 2.076d-03 | 0.002699 | 94.207 | 5.308 |
| 7 | 7.000d+04 | 4.725d-03 | 0.006141 | -167.228 | -256.126 |
| 8 | 8.000d+04 | 4.582d-03 | 0.005956 | -138.912 | -227.811 |
| 9 | 9.000d+04 | 3.679d-03 | 0.004782 | 58.327 | -30.572 |

TOTAL HARMONIC DISTORTION =       1.616844   PERCENT

FOURIER COMPONENTS OF TRANSIENT RESPONSE V(3)    $T_{step} = 5\mu s$, $T_{max} = 0.01\mu s$, $T_{stop} = 600\ \mu s$
DC COMPONENT =   1.043d+00

| HARMONIC NO | FREQUENCY (HZ) | FOURIER COMPONENT | NORMALIZED COMPONENT | PHASE (DEG) | NORMALIZED PHASE (DEG) |
|---|---|---|---|---|---|
| 1 | 1.000d+04 | 7.698d-01 | 1.000000 | 89.023 | 0.000 |
| 2 | 2.000d+04 | 3.336d-03 | 0.004333 | 143.615 | 54.592 |
| 3 | 3.000d+04 | 1.899d-03 | 0.002467 | -128.026 | -217.049 |
| 4 | 4.000d+04 | 5.147d-03 | 0.006686 | -74.577 | -163.599 |
| 5 | 5.000d+04 | 5.124d-03 | 0.006657 | -115.642 | -204.665 |
| 6 | 6.000d+04 | 1.877d-03 | 0.002438 | 77.565 | -11.457 |
| 7 | 7.000d+04 | 3.076d-03 | 0.003995 | -146.069 | -235.092 |
| 8 | 8.000d+04 | 3.759d-03 | 0.004883 | 157.698 | 68.675 |
| 9 | 9.000d+04 | 8.678d-03 | 0.011273 | -130.472 | -219.494 |

TOTAL HARMONIC DISTORTION =       1.693274   PERCENT

(c)

**Fig. 15.13.** (b) Output voltage waveform. (c) Fourier components of the output voltage.

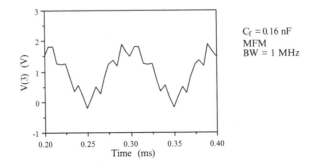

$C_f = 0.16$ nF
MFM
BW = 1 MHz

```
FOURIER COMPONENTS OF TRANSIENT RESPONSE V(3)
DC COMPONENT =    1.026d+00
HARMONIC   FREQUENCY   FOURIER    NORMALIZED    PHASE    NORMALIZED
   NO        (HZ)     COMPONENT   COMPONENT    (DEG)    PHASE (DEG)

    1      1.000d+04  8.876d-01   1.000000     88.657      0.000
    2      2.000d+04  9.256d-02   0.104284   -100.117   -188.774
    3      3.000d+04  1.027d-01   0.115722    -90.341   -178.998
    4      4.000d+04  1.159d-01   0.130623    -97.317   -185.974
    5      5.000d+04  3.981d-02   0.044847   -121.913   -210.570
    6      6.000d+04  2.660d-02   0.029971    144.997     56.340
    7      7.000d+04  6.006d-02   0.067663   -105.293   -193.950
    8      8.000d+04  7.225d-02   0.081403    -85.331   -173.987
    9      9.000d+04  2.002d-02   0.022555     61.256    -27.401

    TOTAL HARMONIC DISTORTION =     23.654138  PERCENT
```

**Fig. 15.13.** (d) Output voltage waveform and its Fourier components for a reduced value of the filter capacitor.

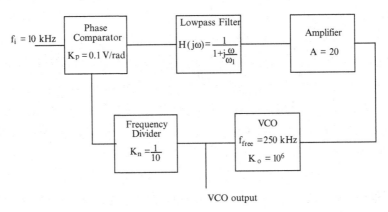

VCO output

**Fig. 15.14.** PLL block diagram for Problem 15.2.

of the amplifier at low frequencies is 12.

(a) If the amplifier provides no filtering, what is the -3 dB bandwidth of the closed-loop PLL.

(b) If the output load of the amplifier has a resistance of 12 k$\Omega$, what value of shunt $C$ is needed to achieve a maximally-flat-magnitude response of the closed-loop PLL.

(c) What is the -3 dB bandwidth for Part (b)?

**15.4.** In an IC phase-locked loop, the free-running frequency of the VCO is 10 MHz. The maximum deviation of the input frequency is 50 kHz about a carrier frequency equal to the free-running frequency of the VCO. For the phase detector of the PLL the output is 6 V/rad.

(a) What value of $K_0$ for the VCO is needed to achieve proper operation?

(b) Sketch the closed-loop magnitude response of $V_o/\omega_i$ if no lowpass amplifier is included and if a one-pole filter is used in the PLL with a -3 dB frequency of 50 kHz.

**15.5.** A phase-locked loop has the parameters, $K_p = 3$ V/rad, $K_0 = 15$ kHz/V. For a filter with a one-pole lowpass response, establish the corner frequency of the filter and the gain constant $A$ of the PLL to achieve a closed-loop, double-pole response with a closed-loop, -3 dB bandwidth of 50 kHz. (*Hint:* A two-pole bandwidth shrinkage factor can be used.)

**15.6.** A phase-locked loop has the parameters, $K_p = 1$ V/rad, $K_0 = 250$ kHz/V, and the VCO has a free-running frequency of 1 MHz.

(a) Design the $AF(s)$ block to achieve a lock range of 3.14 MHz and a capture range of 2.64 MHz producing a maximally-flat-magnitude response for the closed-loop PLL. (*Hint:* First establish the value of A. Then determine the filter characteristic. A first-order filter should suffice.)

(b) Design the filter for Part (a).

(c) Determine the closed-loop magnitude response of the PLL, $V_o/\omega_i$, if a zero at 6.26 MHz is added to the filter.

(d) What is the lock range and the approximate capture range when a zero at 6.26 MHz is added to the filter?

# Erratum

# Analog Integrated Circuits for Communication: Principles, Simulation and Design

### Second Edition

### Donald O. Pederson and Kartikeya Mayaram

© 2008 Springer Science+Business Media, LLC

**Erratum to: DOI 10.1007/978-0-387-68030-9**

Note to Instructors:

The website address is missing a ˜. The correct link is http://eecs.oregonstate.edu/~karti/book/spicefiles

Chapter 3:

In Figure 3.10, the Fourier components of the typical circuit are to be ignored.

Chapter 6:

For Problem 6.3, the transistor parameters are missing. Please use kp = $30\mu A/V^2$, Vt = 0.7V, W/L = 50.

Chapter 11:

In Problem 11.3, the $10\,k\Omega$ resistance in series with the inductor should be $10\,\Omega$.

In Problem 11.7, part (e) should be labeled (d) and (f) should be labeled (e).

In Problem 11.8, the 2nd part (a) should be labeled (b), (b) -> (c), (c) -> (d).

In Problem 11.8, part (f) should read: Using the result of (e) calculate the amplitude of the gate voltage.

Chapter 14:

p. 459, last line of $2^{nd}$ paragraph: 14.3 kHz should be 13.3 kHz.

p. 476, last line on page: $b_1$ = 59 mV and $HD_2$ = 2.3%.

p. 477, first line on page: $HD_3$ = 0.9% and THD = 3.6%.

In Problem 14.6, $+V_{CC}/-V_{EE}$ are $+10/-10$ V.

D.O. Pederson and K. Mayaram, *Analog Integrated Circuits for Communication*, DOI 10.1007/978-0-387-68030-9_16, © 2008 Springer Science+Business Media, LLC

# References

1. S. Haykin, *Communication Systems*, 4th edition, John Wiley & Sons, Inc., 2001.
2. B. P. Lathi, *Modern Digital and Analog Communication Systems*, 3rd edition, Oxford University Press, 1998.
3. A. Vladimirescu, *The SPICE Book*, John Wiley & Sons, Inc., 1994.
4. B. Razavi, "Challenges in portable RF transceiver design," *IEEE Circuits and Devices Magazine*, pp. 12-25, Sept. 1996.
5. B. Razavi, *RF Microelectronics*, Prentice Hall PTR, 1998.
6. P. R. Gray, P. J. Hurst, S. H. Lewis, and R. G. Meyer, *Analysis and Design of Analog Integrated Circuits*, 4th edition John Wiley & Sons, Inc., 2001.
7. R. S. Muller and T. I. Kamins, *Device Electronics for Integrated Circuits*, 2nd edition, John Wiley & Sons, Inc., 1986.
8. I. Getreu, *Modeling the Bipolar Transistor*, Tektronix, Inc., 1976.
9. Y. Tsividis, *Operation and Modeling of The MOS Transistor*, 2nd edition, McGraw-Hill, 1999.
10. D. Foty, *MOSFET Modeling with SPICE: Principles and Practice*, Prentice Hall, 1997.
11. W. Liu, *MOSFET Models for SPICE Simulation, Including BSIM3v3 and BSIM4*, John Wiley & Sons, Inc., 2001.
12. K. S. Kundert, *The Designer's Guide to SPICE & SPECTRE*, Kluwer Academic Publishers, 1995.
13. K. S. Kundert, J. K. White, and A. Sangiovanni-Vincentelli, *Steady-State Methods for Simulating Analog and Microwave Circuits*, Kluwer Academic Publishers, 1990.
14. K. S. Kundert, "Introduction to RF simulation and its application," *IEEE J. Solid-State Circuits*, pp. 1298-1319, Sept. 1999.
15. K. Mayaram, D. C. Lee, S. Moinian, D. A. Rich, J. Roychowdhury, "Computer-aided circuit analysis tools for RFIC simulation: algorithms, features, and limitations," *IEEE Trans. Circuits and Systems - II*, pp. 274-286, April 2000.
16. C. D. Hull and R. G. Meyer, "A systematic approach to the analysis of noise in mixers," *IEEE Trans. Circuits and Systems - I*, pp. 909-919, Dec. 1993.
17. A. Hajimiri and T. H. Lee, *The Design of Low Noise Oscillators*, Kluwer Academic Publishers, 1999.
18. A. Hajimiri and T. H. Lee, "A general theory of phase noise in electrical oscillators," *IEEE J. Solid-State Circuits*, pp. 179-194, Feb. 1998.

D.O. Pederson and K. Mayaram, *Analog Integrated Circuits for Communication*, DOI 10.1007/978-0-387-68030-9,
© 2008 Springer Science+Business Media, LLC

19. T. H. Lee, *The Design of CMOS Radio-Frequency Integrated Circuits*, 2nd edition, Cambridge University Press, 2004.

20. K. K. Clarke and D. T. Hess, *Communication Circuits: Analysis and Design*, Addison-Wesley Publishing Company, 1971.

21. N. M. Nguyen and R. G. Meyer, "Si IC-compatible inductors and LC passive filters," *IEEE J. Solid-State Circuits*, pp. 1028-1031, Aug. 1990.

22. J. R. Long, "Monolithic transformers for silicon RF design," *IEEE J. Solid-State Circuits*, pp. 1368-1382, Sept. 2000.

23. W. H. Hayt and J. E. Kemmerly, *Engineering Circuit Analysis*, McGraw-Hill, 5th edition, 1993.

24. M. H. Rashid, *Introduction to PSpice Using OrCAD for Circuits and Electronics*, Pearson Prentice Hall, 3rd edition, 2004.

25. B. van der Pol, "A theory of the amplitude of free and forced triode vibrations," it Radio Review, vol. 1, pp. 701-710 and 754-762, Nov. 1920.

26. B. van der Pol, "The nonlinear theory of electric oscillations," *Proc. Institute of Radio Engineers*, vol. 22, pp. 1051-1086, Sept. 1934.

27. G. R. Boyle, D. O. Pederson, B. M. Cohn, and J. E. Solomon, "Macromodeling of integrated circuit operational amplifiers," *IEEE J. Solid-State Circuits*, pp. 353-364, Dec. 1974.

28. T. Okanobu, T. Tsuchiya, K. Abe, and Y. Ueki, "A complete single chip AM/FM radio integrated circuit," *IEEE Trans. Consumer Electronics*, pp. 393-408, Aug. 1982.

29. K. Mayaram and D. O. Pederson, "Analysis of MOS transformer-coupled oscillators," *IEEE J. Solid-State Circuits*, pp. 1155-1162, Dec. 1987.

30. K. Mayaram, "Output voltage Analysis for the MOS Colpitts oscillators," *IEEE Trans. Circuits and Systems - I*, pp. 260-263, Feb. 2000.

31. M. A. Unkrich and R. G. Meyer, "Conditions for start-up in crystal oscillators," *IEEE J. Solid-State Circuits*, pp. 87-90, Feb. 1982.

32. R. G. Meyer and D. C.-F. Soo, "MOS crystal oscillator design," *IEEE J. Solid-State Circuits*, pp. 222-228, April 1984.

33. E. A. Vittoz, M. G. R. Degrauwe, and S. Bitz, "High-performance crystal oscillator circuits: theory and application," *IEEE J. Solid-State Circuits*, pp. 774-783, June 1988.

34. A. Lienard, "Etude des oscillations entretenues," *Rev. Gen. Elect.*, vol. 23, pp. 901-946, 1928.

35. H. Schmitt, "A thermionic trigger," *J. Sci. Instrum.*, vol. 15, pp. 24-26, Jan. 1938.

36. W. H. Eccles and F. W. Jordan, "A trigger relay utilizing three-electrode thermionic vacuum tubes," *Radio Rev.*, vol. 1, pp. 143-146, Dec. 1919.

37. B. Gilbert, "A precise four-quadrant multiplier with subnanosecond response," *IEEE J. Solid-State Circuits*, pp. 365-373, Dec. 1968.

38. J. N. Babanezhad and G. C. Temes, "A 20-V four-quadrant CMOS analog multiplier," *IEEE J. Solid-State Circuits*, pp. 1158-1168, Dec. 1985.

39. K. L. Fong and R. G. Meyer, "Monolithic RF active mixer design," *IEEE Trans. Circuits and Systems - II*, pp. 231-239, March 1999.

40. A. Bilotti, "FM detection using a product detector," *Proceedings of the IEEE*, pp. 755-757, April 1968.

41. F. M. Gardner, *Phaselock Techniques*, 3rd edition, John Wiley & Sons, Inc., 2005.

42. B. Razavi, *Monolithic Phase-Locked Loops and Clock Recovery Circuits: Theory and Design*, IEEE Press, 1996.
43. R. E. Best, *Phase-Locked Loops: Design, Simulation, and Applications*, McGraw-Hill, 2003.

# Index

D.O. Pederson and K. Mayaram, *Analog Integrated Circuits for Communication*, DOI 10.1007/978-0-387-68030-9,

© 2008 Springer Science+Business Media, LLC